DAT

WESTFIELD
Road,

ENGINEERING DESIGN
A Materials and Processing Approach

McGraw-Hill Series in Mechanical Engineering

Consulting Editors

Jack P. Holman, *Southern Methodist University*
John R. Lloyd, *Michigan State University*

Anderson: *Modern Compressible Flow: With Historical Perspective*
Arora: *Introduction to Optimum Design*
Bray and Stanley: *Nondestructive Evaluation: A Tool for Design, Manufacturing and Service*
Dally: *Packaging of Electronic Systems: A Mechanical Engineering Approach*
Dieter: *Engineering Design: A Materials and Processing Approach*
Eckert and Drake: *Analysis of Heat and Mass Transfer*
Edwards and McKee: *Fundamentals of Mechanical Component Design*
Heywood: *Internal Combustion Engine Fundamentals*
Hinze: *Turbulence*
Hutton: *Applied Mechanical Vibrations*
Juvinall: *Engineering Considerations of Stress, Strain, and Strength*
Kays and Crawford: *Convective Heat and Mass Transfer*
Kane & Levinson: *Dynamics: Theory and Applications*
Martin: *Kinematics and Dynamics of Machines*
Phelan: *Fundamentals of Mechanical Design*
Raven: *Automatic Control Engineering*
Rosenberg & Karnopp: *Introduction to Physics*
Schlichting: *Boundary-Layer Theory*
Shames: *Mechanics of Fluids*
Sherman: *Viscous Flow*
Shigley: *Kinematic Analysis of Mechanisms*
Shigley and Uicker: *Theory of Machines and Mechanisms*
Shigley and Mischke: *Mechanical Engineering Design*
Stoecker and Jones: *Refrigeration and Air Conditioning*
Vanderplaats: *Numerical Optimization: Techniques for Engineering Design, with Applications*
White: *Viscous Fluid flow*

Also Available from McGraw-Hill

Schaum's Outline Series in Mechanical Engineering

Most outlines include basic theory, definitions, and hundreds of solved problems and supplementary problems with answers.

Titles on the Current List Include:

Acoustics
Basic Equations of Engineering
Continuum Mechanics
Engineering Economics
Engineering Mechanics, 4th edition
Fluid Dynamics, 2d edition
Fluid Mechanics & Hydraulics, 2d edition
Heat Transfer
Introduction to Engineering Calculations
Lagrangian Dynamics
Machine Design
Mathematical Handbook of Formulas & Tables
Mechanical Vibrations
Operations Research
Statics & Mechanics of Materials
Strength of Materials, 2d edition
Theoretical Mechanics
Thermodynamics, 2d edtion

Schaum's Solved Problems Books

Each title in this series is a complete and expert source of solved problems containing thousands of problems with worked out solutions.

Related Titles on the Current List Include:

3000 Solved Problems in Calculus
2500 Solved Problems in Differential Equations
2500 Solved Problems in Fluid Mechanics and Hydraulics
1000 Solved Problems in Heat Transfer
3000 Solved Problems in Linear Algebra
2000 Solved Problems in Mechanical Engineering Thermodynamics
2000 Solved Problems in Numerical Analysis
700 Solved Problems in Vector Mechanics for Engineers: Dynamics
800 Solved Problems in Vector Mechanics for Engineers: Statics

Available at your College Bookstore. A complete list of Schaum titles may be obtained by writing to:

Schaum Division
McGraw-Hill, Inc.
Princeton Road, S-1
Hightstown, NJ 08520

ENGINEERING DESIGN

A Materials and Processing Approach

Second Edition

George E. Dieter

University of Maryland

McGraw-Hill, Inc.

New York   St. Louis   San Francisco   Auckland   Bogotá   Caracas
Hamburg   Lisbon   London   Madrid   Mexico   Milan   Montreal   New Delhi
Paris   San Juan   São Paulo   Singapore   Sydney   Tokyo   Toronto

This book was set in Times Roman.
The editors were Lyn Beamesderfer and John M. Morriss;
the production supervisor was Janelle S. Travers.
The cover was designed by Rafael Hernandez.
Project supervision was done by The Universities Press.
Arcata Graphics/Halliday was printer and binder.

ENGINEERING DESIGN
A Materials and Processing Approach

1 2 3 4 5 6 7 8 9 0 HAL HAL 9 5 4 3 2 1 0

ISBN 0-07-016906-3

Library of Congress Cataloging-in-Publication Data
Dieter, George Ellwood.
  Engineering design: a materials and processing approach / George
E. Dieter.—2nd ed.
    p.   cm.—(McGraw-Hill series in materials science and
engineering) (McGraw-Hill series in mechanical engineering)
  Includes indexes.
  ISBN 0-07-016906-3
  1. Engineering design.   I. Title.   II. Series.   III. Series:
McGraw-Hill series in mechanical engineering.
TA174.D495   1991
620'.00425—dc20   90-38003

ABOUT THE AUTHOR

George E. Dieter is currently Dean of Engineering and Professor of Mechanical Engineering at the University of Maryland. The author received his B.S. Met.E. degree from Drexel University and his D.Sc. degree from Carnegie-Mellon University. After a career in industry with the DuPont Engineering Research Laboratory, he became Head of the Metallurgical Engineering Department at Drexel University, where he later became Dean of Engineering. Professor Dieter later joined the faculty of Carnegie-Mellon University as Professor of Engineering and Director of the Processing Research Institute. He moved to the University of Maryland four years later.

A former member of the National Materials Advisory Board, Professor Dieter is a fellow of the American Society for Metals, an AAAS, ASEE, and member of AIME (TMS) NSPE, and SME. He also is the author of Mechanical Metallurgy, published by McGraw-Hill, now in its third edition.

CONTENTS

PREFACE TO THE SECOND EDITION

The theme of the first edition of *Engineering Design* was the importance of the connection between design and manufacturing. In the eight years since the first edition was published, this connection has become one of the driving forces of engineering in the World. Also in the intervening years, the role of computers in design has accelerated and become all-pervasive. These two aspects of design have been reinforced and strengthened in this new edition.

As before, *Engineering Design* is intended to provide the senior engineering student with a broad realistic understanding of the design process. It draws on a diverse set of topics: decision making, optimization, engineering economy, planning, applied statistics, reliability, and quality engineering, and focuses them on the design process.

A number of topics have been added or greatly expanded. These include: ethics in engineering, societal considerations in engineering, technological innovation, market identification, competitive benchmarking, protection of intellectual property, human factors in design, industrial design, expert systems, and Taguchi methods. A major revision has been given to Chapter 3, which has been retitled "Design Methods." Emphasis is given to conceptual design to introduce the ideas of French and Pugh's ideas of the product design specification and the concept selection technique. Also, the student is introduced to Nam Suh's principles of design. The section in this chapter on assessing alternatives has been completely rewritten.

Chapter 13 has been greatly expanded and focused on the quality issues of engineering design. It opens with Deming's philosophy of total quality and includes an in-depth discussion of Taguchi's methods, including an example on robust design. The coverage of control charts and statistical process control is increased.

Special thanks goes to Shapour Azarm, Herbert Foerstel, Richard McCuen, Ioannis Pandelidis, and Marvin Roush, my colleagues at the University of Maryland, for advice and consultation. I must also thank the following reviewers for their many helpful comments and suggestions: Louis Bucciarelli, Massachusetts Institute of Technology; Kevin Craig, Rochester Polytechnic Institute; Darrell Gibson, Rose-Hulman Institute of Technology; Gaza Kardos, Carleton University–Canada; and George Schade, University of Nebraska.

George E. Dieter

PREFACE TO THE FIRST EDITION

Engineering Design is intended to provide the senior engineering student with a realistic understanding of the engineering design process. It is written from the viewpoint that design is the central activity of the engineering profession, and it is more concerned with developing attitudes and approaches than in presenting design techniques and tools. Like other texts on this subject, it develops design as an interdisciplinary activity that draws on such diverse subjects as decision making, optimization, engineering economy, planning, and applied statistics. Chapters are presented on each of those subjects, all of which are needed to some degree by the designer.

However, this text goes beyond other design books in giving special emphasis to materials selection and materials processing and manufacturing. These are critical aspects of the design process that have been given little attention in other design texts. Moreover, because they have been deemphasized in most engineering curricula, it is felt that special attention to them is warranted and that they can be learned successfully within a design context. Although the text should be applicable to students in all fields of engineering, special emphasis has been given to the materials, mechanical, and metallurgical fields in the selection of examples and illustrations.

The only real way to learn design is to do design. The best way to use this book is as part of a project design course in which the students are engaged in a major design problem. When possible, the actual experience in doing design should be supplemented by lectures on such subjects as engineering economy and reliability, depending on the particular mix of other courses the students have taken. When the class schedule will not permit both lectures and design experience, selected chapters should be assigned as outside reading as the design progresses. Clearly, in a subject as broad as engineering design, everything that is needed cannot be included in a book of modest size. This is not a handbook or cookbook. Rather , it aims more at developing good design attitudes and habits. One of those habits is self-reliance—the ability of the student to learn independently. Therefore, special emphasis has been given to selected further readings and far more references have been included than is usual in a basic textbook so as to provide a convenient launching point for the student's independent study.

The contents of this text and their use in a project design approach have been shaped by over fifteen years of experience in teaching design to metallurgical and mechanical engineering students at three institutions. Special recognition is due to Howard A. Kuhn, John C. Purcupile, Dwight A. Baughman, Clifford L. Sayre, Jr., and Richard W. McCuen for their valuable ideas, suggestions, and interactions. The writer also acknowledges the special association of nearly twenty years with the many fine engineers at the Materials Technology Laboratory of TRW, Inc., Cleveland, who have done much to add realism to this book. Finally, special thanks goes to Jean Beckmann for her painstaking efforts to create a perfect manuscript.

George E. Dieter

ENGINEERING DESIGN
A Materials and Processing Approach

CHAPTER
1

THE
DESIGN
PROCESS

1-1 INTRODUCTION

What is design? If you search the literature for an answer to that question, you will find about as many definitions as there are designs. Perhaps the reason is that the process of design is such a common human experience. Webster's dictionary says that to design is "to fashion after a plan," but that leaves out the essential fact that to design is to create something that has never been. Certainly an engineering designer practices design by that definition, but so does an artist, a sculptor, a composer, a playwright, or many another creative member of our society.

Thus, although engineers are not the only people who design things, it is true that the professional practice of engineering is largely concerned with design; it is frequently said that design is the essence of engineering. To design is to pull together something new or arrange existing things in a new way to satisfy a recognized need of society. An elegant word for "pulling together" is *synthesis*. We shall adopt the following formal definition of design:[1] "Design establishes and defines solutions to and pertinent structures for problems not solved before, or new solutions to problems which have previously been solved

---

[1] J. F. Blumrich, *Science,* vol. 168, pp. 1551–1554, 1970.

1

in a different way." The ability to design is both a science and an art. The science can be learned through techniques and procedures to be covered in this course, but the art can be learned only by doing design. It is for this reason that your design education must involve some realistic experience.

The emphasis that we have given to the creation of new things in our discussion of design should not unduly alarm you. To become proficient in design is a perfectly attainable goal for an engineering student, but its attainment requires the guided experience that we intend this course to provide. Design should not be confused with discovery. *Discovery* is getting the first sight of, or the first knowledge of something, as when Columbus discovered America. We can discover what has already existed but has not been known before. But a *design* is the product of planning and work. A design is produced to satisfy a need that someone has. It is something that has not always existed; instead, it is created expressly to satisfy a need.

We should note that a design may or may not involve *invention*. To obtain a legal patent on an invention requires that the design be a step beyond the limits of the existing knowledge (beyond the state of the art). Some designs are truly inventive, but most are not.

Good design requires both analysis and synthesis. In order to design something we must be able to calculate as much about the thing's behavior as possible by using the appropriate disciplines of science or engineering science and the necessary computational tools. *Analysis* usually involves the simplification of the real world through models. It is concerned with the separation of the problem into manageable parts, whereas *synthesis* is concerned with assembling the elements into a workable whole.

At your current stage in your engineering education you are much more familiar and comfortable with analysis. You have dealt with courses that were essentially disciplinary. For example, you were not expected to use thermodynamics and fluid mechanics in a course in mechanics of materials. The problems you worked in the course were selected to illustrate and reinforce the principles. If you could construct the appropriate model, you usually could solve the problem. Most of the input data and properties were given, and there usually was a correct answer to the problem. However, real-world problems rarely are that neat and circumscribed. The real problem that your design is expected to solve may not be readily apparent. You may need to draw on many technical disciplines (solid mechanics, fluid mechanics, electromagnetic theory, etc.) for the solution and usually on nonengineering disciplines as well (economics, finance, law, etc.). The input data may be fragmentary at best, and the scope of the project may be so huge that no individual can follow it all. If that is not difficult enough, usually the design must proceed under severe constraints of time and/or money. There may be major societal constraints imposed by environmental or energy regulations. Finally, in the typical design you rarely have a way of knowing the correct answer. Hopefully, your design works, but is it the best, most efficient design that could have been achieved under the conditions? Only time will tell.

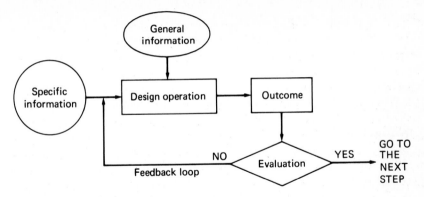

FIGURE 1-1
Basic module in the design process. (*After Asimow.*)

We hope by now you have some idea of the design environment and the design process. One thing that should be clear is how engineering extends well beyond the boundaries of science. If perhaps you have become a bit despondent over the complexity described above, cheer up. The increased boundaries and responsibilities of engineering create almost limitless opportunities for you. In your professional career you will have the opportunity to create dozens of original designs and have the satisfaction of seeing them become working realities. "A scientist will be lucky if he makes one creative addition to human knowledge in his whole life, and many never do so. A scientist can discover a new star but he cannot make one. He would have to ask an engineer to do it for him."[1]

1-2 THE DESIGN PROCESS—A SIMPLIFIED APPROACH

We frequently talk about "designing a system." By a system we mean the entire combination of hardware, information, and people necessary to accomplish some specified mission. A system may be an electric power distribution network for a region of the nation, a procedure for detecting flaws in welded pressure vessels, or a combination of production steps to produce automobile parts. A large system usually is divided into *subsystems*, which in turn are made up of *components*.

There is no single universally acclaimed sequence of steps that leads to a workable design. Different writers or designers have outlined the design process in as few as 5 steps or as many as 25. One of the first to write introspectively about design was Morris Asimow.[2] He viewed the heart of the design process as consisting of the elements shown in Fig. 1-1. As portrayed

---

[1] G. L. Glegg, "The Design of Design," Cambridge University Press, New York, 1969.
[2] M. Asimow, "Introduction to Design," Prentice–Hall, Inc., Englewood Cliffs, N.J., 1962.

there, design is a sequential process consisting of many design operations. Examples of the operations might be 1) exploring the alternative systems that could satisfy the specified need, 2) formulating a mathematical model of the best system concept, 3) specifying specific parts to construct a component of a subsystem, and 4) selecting a material from which to manufacture a part. Each operation requires information, some of it general technical and business information that is expected of the trained professional and some of it very specific information that is needed to produce a successful outcome. An example of the last kind of information might be 1) a manufacturer's catalog on miniature bearings, 2) handbook data on high-temperature alloys, or 3) personal experience gained from a trip to observe a new manufacturing process. Acquisition of information is a vital and often very difficult step in the design process, but fortunately it is a step that usually becomes easier with time. (We call this process experience.)[1] The importance of developing sources of information is considered more fully in Chap. 14.

Once armed with the necessary information, the design engineer (or design team) carries out the design operation by using the appropriate technical knowledge and computational and/or experimental tools. At this stage it may be necessary to construct a mathematical model and conduct a simulation of the component's performance on a digital computer. Or it may be necessary to construct a full-size prototype model and test it to destruction at a proving ground. Whatever it is, the operation produces a design outcome that, again, may take many forms. It can be a ream of computer printout, a rough sketch with critical dimensions established, or a complete set of engineering drawings ready to go to the manufacturing department. At this stage the design outcome must be evaluated, often by a team of impartial experts, to decide whether it is adequate to meet the need. If so, the designer may go on to the next step. If the evaluation uncovers deficiencies, then the design operation must be repeated. The information from the first design is fed back as input, together with new information that has been developed as a result of questions raised at the evaluation step.

The final result of the chain of design modules, each like Fig. 1-1, is a new working object (often referred to as *hardware*) or a collection of objects that is a new system. However, many design projects do not have as an objective the creation of new hardware or systems. Instead, the objective may be the development of new information that can be used elsewhere in the organization. It should be realized that few system designs are carried through to completion; they are stopped because it has become clear that the objectives of the project are not technically and/or economically feasible. However, they create new information, which, if stored in retrievable form, has future value.

The simple model shown in Fig. 1-1 illustrates a number of important

---

[1] Experience has been defined, perhaps a bit lightheartedly, as just a sequence of non-fatal events.

aspects of the design process. First, even the most complex system can be broken down into a sequence of design objectives. Each objective requires an evaluation, and it is common for the decision-making phase to involve repeated trials or iterations. The need to go back and try again should not be considered a personal failure or weakness. Design is a creative process, and all new creations of the mind are the result of trial and error. In fact, if it were possible to work a design straight through without iteration, the design would indeed be very routine. This iterative aspect of design may take some getting used to. You will have to acquire a high tolerance for failure and the tenacity and determination to persevere and work the problem out one way or the other.

The iterative nature of design provides an opportunity to improve the design on the basis of a preceding outcome. That, in turn, leads to the search for the best possible technical condition, e.g., maximum performance at minimum weight (or cost). Many techniques for *optimizing* a design have been developed, and some of them are covered in Chap. 5. And although optimization methods are intellectually pleasing and technically interesting, they often have limited application in a complex design situation. In the usual situation the actual design parameters chosen by the engineer are a compromise among several alternatives. There may be too many variables to include all of them in the optimization, or nontechnical considerations like available time or legal constraints may have to be considered, so that trade-offs must be made. The parameters chosen for the design are then close to but not at optimum values. We usually refer to them as *optimal values,* the best that can be achieved within the total constraints of the system.

In your scientific and engineering education you may have heard reference to the scientific method, a logical progression of events that leads to the solution of scientific problems. Percy Hill[1] has diagramed the comparison between the scientific method and the design method (Fig. 1-2).

The scientific method starts with a body of existing knowledge. The scientist has curiosity that causes him to question these laws of science; and as a result of his questioning, he eventually formulates a hypothesis. The hypothesis is subjected to logical analysis that either confirms or denies it. Often the analysis reveals flaws or inconsistencies, so that the hypothesis must be changed in an iterative process. Finally, when the new idea is confirmed to the satisfaction of its originator, it must be accepted as proof by fellow scientists. Once accepted, it is communicated to the community of scientists and it enlarges the body of existing knowledge. The knowledge loop is completed.

The design method is very similar to the scientific method if we allow for differences in viewpoint and philosophy. The design method starts with

---

[1] P. H. Hill, "The Science of Engineering Design," Holt, Rinehart and Winston, New York, 1970.

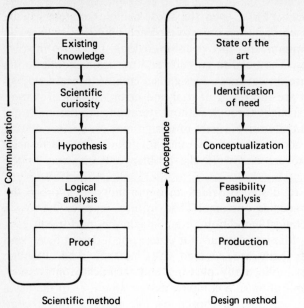

FIGURE 1-2
Comparison between the scientific method and the design method. (*After Percy Hill.*)

knowledge of the state of the art. That includes scientific knowledge, but it also includes devices, components, materials, manufacturing methods, and market and economic conditions. Rather than scientific curiosity, it is really the needs of society (usually expressed through economic factors) that provide the impetus. When a need is identified, it must be conceptualized as some kind of model. The design concept must be subjected to a feasibility analysis, almost always with iteration, until an acceptable product is produced or the project is abandoned. When the design enters the production phase, it begins to compete in the world of technology. The design loop is closed when the product is accepted as part of the current technology and thereby advances the state of the art of the particular field.

1-3 THE DESIGN PROCESS STEPS

To further illustrate the design process, we consider the process to consist of the following steps:

- Recognition of a need
- Definition of a problem
- Gathering of information
- Conceptualization
- Evaluation
- Communication of the design

As mentioned earlier, others writers on design may choose to expand upon these steps. The design process generally proceeds from top to bottom in the list just given, but it must be understood that in practice some of the steps will be carried out in parallel and that feedback leading to iteration is a common fact of design.

Recognition of a Need

Needs are identified at many points in a business or agency. Most organizations have research or development components whose job it is to create ideas that are relevant to the needs of the organization. Needs may come from inputs of operating or service personnel or from customers through sales or marketing representatives. Other needs are generated by outside consultants, purchasing agents, government agencies, or trade associations or by the attitudes or decisions of the general public.

Needs usually arise from dissatisfaction with the existing situation. They may be to reduce cost, increase reliability or performance, or just change because the public has become bored with the product.

Definition of a Problem

Probably the most critical step in the design process is the definition of the problem. The true problem is not always what it seems to be at first glance. Because this step requires such a small part of the total time to create the final design, its importance is often overlooked. Figure 1-3 illustrates how the final design can differ greatly depending upon how the problem is defined.

It is advantageous to define the problem as broadly as possible. If the definition is broad, you will be less likely to overlook unusual or unconventional solutions. Broad treatment of problems that previously were attacked in piecemeal fashion can have a big payoff. However, you should realize that the degree to which you can pursue a broad problem formulation toward a final design will depend on factors often outside your control. Pursuit of a broad formulation may bring you into direct conflict with decisions already made by your employer or client, or it may lead you into areas of responsibility of other persons in the organization. In most cases the extent to which you are able to follow a broad problem formulation will depend on the importance of the problem, the limits on time and money that have been placed on the problem, and your own position in the organization.

One approach that you should not take is to consider the existing solution to the problem to be the problem itself. That approach immediately submerges you in the trees of the forest, and you will find yourself generating solutions to a problem that you have failed to define.

The definition of a problem should include writing down a formal *problem statement,* which should express as specifically as possible what the design is intended to accomplish. It should include objectives and goals,

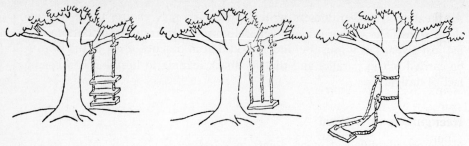

As proposed by the project sponsor   As specified in the project request   As designed by the senior designer

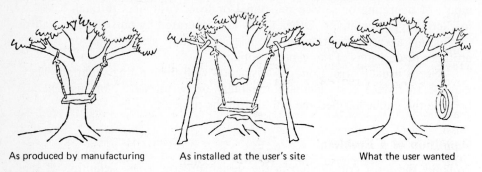

As produced by manufacturing   As installed at the user's site   What the user wanted

FIGURE 1-3
Note how the design depends on the viewpoint of the individual who defines the problem.

definitions of any special technical terms, the constraints placed on the design, and the criteria that will be used to evaluate the design.

Perhaps the best way to proceed is to develop a problem statement at the initial problem definition step and then, in the second iteration after much information has been gathered, develop a much more detailed problem statement that is usually called the *problem analysis*.

Setting down the objectives and goals often raises questions of what to include and what to exclude. One way to approach those questions was suggested by Ira and Marthann Wilson.[1] They suggest four categories of objectives and goals:

1. *Musts*. The set of requirements that must be met
2. *Must nots*. A set of constraints stating what the system must not be or do
3. *Wants*. The requirements that are worth stating but are not hard and fast
4. *Don't wants*.

---

[1] I.G. Wilson and M.E. Wilson, "From Idea to Working Model," Wiley-Interscience, New York, 1970.

Gathering of Information

Perhaps the greatest frustration you will encounter when you embark on your first design problem will be due to the dearth or plethora of information. No longer will your responsibility stop with the knowledge contained in a few chapters of a text. Your assigned problem may be in a technical area in which you have no previous background, and you may not have even a single basic reference on the subject. At the other extreme you may be presented with a mountain of reports of previous work and your task will be keep from drowning in paper. Whatever the situation, the immediate task is to identify the needed pieces of information and find or develop that information.

An important point to realize is that the information needed in design is different from that usually associated with an academic course. Textbooks and articles published in the scholarly technical journals usually are of lesser importance. The need often is for more specific and current information than is provided by those sources. Technical reports published as a result of government-sponsored R&D, company reports, trade journals, patents, catalogs, and handbooks and literature published by vendors and suppliers of material and equipment are important sources of information. Often the missing piece of information can be supplied by a telephone call to a key supplier. Discussions with in-house experts (often in the corporate R&D center) and outside consultants may prove helpful.

The following are some of the problems connected with obtaining information:

Where can I find it?
How can I get it?
How credible and accurate is the information?
How should the information be interpreted for my specific need?
When do I have enough information?
What decisions result from the information?

Conceptualization

The conceptualization step is to determine the elements, mechanisms, processes, or configurations that in some combination or other result in a design that satisfies the need. It is the key step for employing inventiveness and creativity. Some ideas for stimulating creativity are given in Chap. 3.

Very often the conceptualization step involves the formulation of a model which may be either of the two general types: analytical and experimental. In most of your engineering courses the emphasis has been on the development of analytical models based on physical principles, but experimental models are no less important. Modeling is considered in greater detail in Chap. 4.

A vital aspect of the conceptualization step is synthesis. *Synthesis* is the process of taking the elements of the concept and arranging them in the proper order, sized and dimensioned in the proper way. Synthesis is a creative process and is present in every design.

Design is very individualized. There are no ironclad rules for teaching successful design, and unfortunately very little has been written about the conceptualization step that is at the heart of the design process. Gordon L. Glegg, in his delightful little book on design, does list some guidelines that have served him well through a productive engineering design career.[1] They are worth passing on to the novice designer.

1. Don't follow traditional design procedures unless you have examined other approaches and found them to be wanting.
2. Often one must complicate the design of some component to simplify the design of the overall system.
3. Make allies of the materials from which you construct the design.
4. When faced with an overwhelmingly complex design task, subdivide the problem into a number of smaller problems.
5. Keep abreast of developments in the physical sciences and feed them back into your practical design solutions.
6. Remember that invention is the most fickle of a designer's muses, so do not look down at inventions which are unsupported by science or analysis.

Evaluation

The evaluation step involves a thorough analysis of the design. The term *evaluation* is used more in the sense of weighing and judging than in the sense of grading. Typically the evaluation step may involve detailed calculation, often computer calculation, of the performance of the design by using an analytical model. In other cases the evaluation may involve extensive simulated service testing of an experimental model or perhaps a full-sized prototype.

An important consideration at every step in the design, but especially as the design nears completion, is checking. In general, there are two types of checks that can be made: mathematical checks and engineering-sense checks. Mathematical checks are concerned with checking the arithmetic and the equations used in the analytical model. Incidentally, the frequency of careless math errors is a good reason why you should adopt the practice of making all your design calculations in a bound notebook. In that way you won't be missing a vital calculation when you are forced by an error to go back and check things out. Just draw a line through the part in error and continue. It is of special importance to ensure that every equation is dimensionally consistent.

---

[1] G. L. Glegg, op. cit.

Engineering-sense checks have to do with whether the answers "feel right." Even though the reliability of your feeling of rightness increases with experience, you can now develop the habit of staring at your answer for a full minute, rather than rushing on to do the next calculation. If the calculated stress is 10^6 psi, you know something went wrong! (Incidentally, fear of losing their feel for the magnitude of the results is one reason why American engineers are resisting the change to SI units.) Limit checks are a good form of engineering-sense check. Let a critical parameter in your design approach some limit (zero, infinity, etc.), and observe whether the equation behaves properly.

We have stressed the iterative nature of design. Optimization techniques (see Chap. 5) most likely will be employed during the evaluation step to select the best values of key design parameters. The management decision as to when to stop the optimization process and "freeze the design" will be dictated chiefly by considerations of time and money. An important question during the evaluation is whether results of the design can be generalized to a class of design rather than simply be a solution to a specific problem.

Communication of the Design

It must always be kept in mind that the purpose of the design is to satisfy the needs of a customer or client. Therefore, the finalized design must be properly communicated, or it may lose much of its impact or significance. The communication is usually by oral presentation to the sponsor as well as by a written design report. A recent survey showed that design engineers spend 60% of their time in discussing designs and preparing written documentation of designs, while only 40% of the time is spent in analyzing designs and doing the designing. Detailed engineering drawings, computer programs, and working models are frequently part of the "deliverables" to the customer. It hardly needs to be emphasized that communication is not a one-time thing to be carried out at the end of the project. In a well-run design project there is continual oral and written dialog between the project manager and the customer. This extremely important subject is considered in greater depth in Chap. 15.

1-4 A DETAILED MORPHOLOGY OF DESIGN

The typical design project will break itself down into a number of time phases. In accordance with Asimow's morphology of design,[1] we can consider the following phases.

---

[1] M. Asimow, op. cit.

Phase I. Feasibility Study

The purpose of the feasibility study is to initiate the design and establish the line of thinking. The goal in this phase is to validate the need, produce a number of possible solutions, and evaluate the solutions on the basis of physical realizability, economic worthwhileness, and financial feasibility. This stage sometimes is called *conceputal design*.

Phase II. Preliminary Design

This phase lays the basis for good detail design by means of a structured development of the design concept. The preliminary embodiment of all the main functions that must be performed by the product must be undertaken. This involves the clear determination of the physical processes which govern the main flows and conversions of material, energy, and information. This design stage is often called *embodiment design*. It may also be called the experimental stage, since it involves building and testing experimental models.

The importance of this stage of the design process is not always appreciated. Once a concept has been formulated there is a tendency to rush into detailed design before the concept has been carefully developed and its full implications understood. An important task in preliminary design is to quantify the parameters so as to establish the optimal solution. Thus, it is often necessary to produce several layouts to scale to obtain information about the advantages and disadvantages of different design variants. It is also at this stage that a final check is made on function, strength, spatial compatibility, design aesthetics, and on the financial viability of the project. Once beyond this design phase major changes become very expensive.

Phase III. Detail Design

In this phase the design is brought to the stage of a complete engineering description of a tested and producible product. The arrangement, form, dimensions, tolerances, and surface properties of all individual parts are determined and the materials and manufacturing processes are specified.

This is a challenging and time-consuming task because of the many complex interrelationships involved. The quality and cost advantage of a product are determined by the level of excellence of detail design. Also, it is at this stage that manufacturing specialists should work closely with engineering designers to assure that the product can be produced at acceptable cost and quality.

Phases I, II, and III carry the design from the realm of possibility to probability to the real world of practicability; they constitute the *primary design*. However, the design process is not finished with the delivery of a set of detailed engineering working drawings. Many other technical and business decisions must be made and are really part of the design process. A great deal of thought and planning must go into how the design will be manufactured, how it will be marketed, and finally, how it will be retired from service and replaced by a new, improved design. Generally, these phases of the design process are carried out elsewhere in the organization than in the engineering department, which created the primary design. As the project proceeds into the new phases, the expenditure of money and personnel time increases very greatly.

Phase IV. Planning for Manufacture

A great deal of detailed planning must be done to provide for the production of the design. A method of manufacture must be established for each component in the system. As a usual first step, *a process sheet* is established; it contains a sequential list of manufacturing operations that must be performed on the component. Also, it specifies the form and condition of the material and the tooling and production machines that will be used. The information on the process sheet makes possible the estimation of the production cost of the component. High costs may indicate the need for a change in material or a basic change in the design. Close interaction with manufacturing, industrial, materials, and mechanical engineers is important at this step. This topic is discussed more fully in Chap. 7.

The other important tasks performed in phase IV are the following:

1. Designing specialized tools and fixtures
2. Specifying the production plant that will be used (or designing a new plant) and laying out the production lines
3 Planning the work schedules and inventory controls (production control)
4. Planning the quality-control system
5. Establishing the standard time and labor costs for each operation
6. Establishing the system of information flow necessary to control the manufacturing operation

All of these tasks are generally considered to fall within industrial or manufacturing engineering.

Phase V. Planning for Distribution

Important technical and business decisions must be made to provide for the effective distribution to the consumer of the systems that have been produced. In the strict realm of design, the shipping package may be critical. Concepts such as shelf life may also be critical and may need to be addressed in the earlier stages of the design process. A system of warehouses for distributing the product may have to be designed if none exists.

The economic success of the design often depends on the skill exercised in marketing the product. If the product is of the consumer type, the marketing effort is concentrated on advertising and news media techniques, but highly technical products may require that the marketing step be a technical activity supported by specialized sales brochures and performance test data.

Phase VI. Planning for Use

The use of the design by the consumer is all-important, and considerations of how the consumer will react to the product pervade all steps of the design process. The following specific topics can be identified as being important user-oriented concerns in the design process: ease of maintenance, reliability, product safety, convenience in use (human factors engineering), aesthetic appeal, economy of operation, and duration of service.

Obviously, these consumer-oriented issues must be introduced into the design process at the very beginning. Phase VI of design is less well defined than the others, but it is becoming increasingly important with the advent of consumer protection and product safety legislation. More strict interpretation of product liability laws is having a major impact on design.

An important phase VI activity is the acquisition of reliable data on failures, service lives, and consumer complaints and attitudes to provide a basis for product improvement in the next design.

Phase VII. Planning for Retirement of the Product

The final step in the design process is the disposal of the product when it has reached the end of its useful life. *Useful life* may be determined by actual deterioration and wear to the point at which the design can no longer function, or it may be determined by technological obsolescence, in which a competing design performs the function either better or cheaper. In consumer products, it may come about through changes in fashion or taste.

In the past, little attention has been given in the design process to product retirement. In some areas, such as abandoned junk automobiles and roadside litter of cans and bottles, legislation has forced design changes aimed at changing the product retirement phase. More important, we are facing an era of decreasing raw materials, just as we are now in an era of decreasing energy supplies. This will necessitate the major changes in design philosophy with respect to phase VII. Products will be designed with ease of scrap recovery as one of the major considerations. More emphasis will be given to designing for several levels of use, so that when the service life at a higher level is terminated, the product will be adaptable to further use at a less-demanding level.

To conclude this section, look at Fig. 1-4 showing the typical steps in the evolution of a new product. The reader should be able to recognize the steps in the design process in this diagram. Note how information flow from design

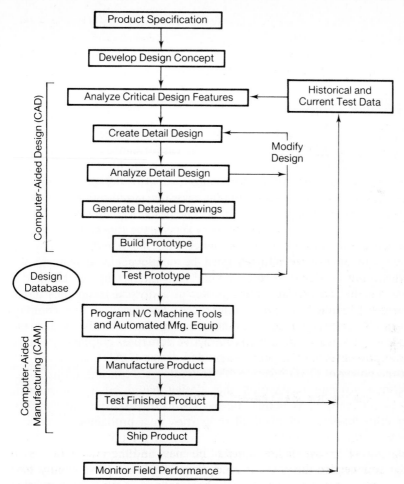

FIGURE 1-4
Typical steps in the design of a new product.

analysis, prototype testing, and product performance in the field can be used to improve the product. In modern design practice many of the design steps are facilitated by the use of computers (computer-aided design—CAD) and there is a growing use of computer-controlled machine tools (computer-aided manufacturing—CAM). The link between these two functions is the design database in digital form.

1-5 FURTHER CONSIDERATIONS IN DESIGN

The preceding description of the design process has followed the time sequence of events in the usual project. However, design is so multifaceted and concerned with trade-offs among so many factors that it is not possible to gain a true perspective of the process from a single-dimensional traverse through design space. Another way to look at design[1] is to consider the various design factors grouped into three categories: 1) functional requirements, 2) the total life cycle, and 3) further major factors.

Functional Requirements

It is obvious that, to be satisfactory, the design must meet the required performance specifications. Development of the performance specifications is the major task in problem definition. This is a critical stage in design; for if it is done skillfully, it will aid greatly in producing a superior design. A common failing in problem definition is to overemphasize the specification of certain property values and performance parameters. The design is then forced to take a predetermined path. A better approach is to specify the *functions* the design should have so that the creativity of the designer is not hampered at the outset.

A variety of analysis techniques must be employed in arriving at the actual configuration or shape of the design. The digital computer has had a major impact in this area by providing powerful analytical techniques based on finite-element analysis and finite differences. The result is that complex geometry and loading conditions can now be handled readily. When the analytical techniques are coupled with interactive computer graphics, we have the exciting capability known as computer-aided design (CAD); see Sec. 1-10. Because of the power of CAD it is possible to treat stress concentrations fully.[2] There is also a growing recognition that design based on a deterministic philosophy is not realistic. A better approach is to use statistical analysis for considering both loading and material properties in a probabilistic manner[3] (see Chap. 11).

The increasingly competitive nature of business and the greatly increased emphasis on product liability have placed more emphasis on reliability (see Chap. 12). In this technical context, *reliability* is defined as the probability that a device will perform adequately in service for the period of time intended under a specific set of operating conditions. Closely related are the important probabilistic factors of maintainability, availability, and repairability. *Maintainability* is concerned with the ease, economy, and safety of performing maintenance functions to minimize maintenance time and maximize the time that the system is in productive service. *Availability* is the probability that a system or piece of equipment will operate satisfactorily when called upon.

---

[1] C. O. Smith, "Engineering Design," NBS Special Publication 487, pp. 1–15, August 1977.

[2] R. E. Peterson, "Stress-Concentration Design Factors," John Wiley & Sons, Inc., New York, 1974.

[3] E. B. Haugen, "Probabilistic Approaches to Design," John Wiley & Sons, Inc., New York, 1968.

Repairability refers to the ease which a failed part may be repaired and put back into service. These important topics are grouped into a larger discipline called logistics engineering.[1]

An important part of the design process is to anticipate the modes of failure for the product and incorporate a defense against failure in the design (see Chap. 13). A variety of techniques, such as fault-tree analysis, failure modes and effects analysis, and hazards analysis, have been developed for this purpose (see Chap. 12).

A related, but distinctly separate methodology, is *value analysis,*[2] a formalized technique for reviewing a design from the standpoints of function and cost to assure maximum value. Maximum value is attained when the required function (without unnecessary functional capacity) is achieved at the lowest cost. A formal design review is an important factor in enhancing product performance. It should take place at critical stages in the initial design process or after a design has been in service for a reasonable period. A redesign, based on feedback from the field and customer engineers, can lead to substantial improvement in cost or function. Standardization and simplification of design are important steps that can lead to cost reduction in design. Cutting back on the grades of steel stocked in the warehouse or the number of sizes of bearings can result in savings with no loss in function. The process of redesign offers the opportunity for *functional substitution.* Here the objective is not simply to replace but to find a new and better way to achieve the same function. For example, an adhesive joint could have the same function as a bolted or riveted joint.

Total Life Cycle

The total life cycle of a part follows the sequence described in Sec. 1-4 as the detailed morphology of design. It starts with the conception of a need and ends with the retirement and disposal of the product.

Material selection is a key element in the total life cycle (see Chap. 6). In selecting materials for a given application, the first step is evaluation of the service conditions. Next, the properties of materials that relate most directly to the service requirements must be determined. Except in almost trivial conditions, there is never a simple relation between service performance and material properties. The design may start with the consideration of static yield strength, but properties that are more difficult to evaluate, such as fatigue,

---

[1] B. Blanchard, "Logistics Engineering and Management," Prentice-Hall, Inc., Englewood Cliffs, N.J., 1974.

[2] C. Fallon, "Value Analysis to Improve Productivity," John Wiley & Sons, Inc., New York 1971; L. D. Miles, "Techniques of Value Analysis and Engineering," 2d ed., McGraw-Hill Book Company, New York, 1972.

creep, toughness, ductility, and corrosion resistance, may have to be considered. We need to know whether the material is stable under the environmental conditions. Does the microstructure change with temperature? Does the material corrode slowly or wear at an unacceptable rate?

Material selection cannot be separated from *producibility* (see Chap. 7). There is an intimate connection between design and material selection and the production processes. The objective in this area is a trade-off between the opposing factors of minimum cost and maximum durability. *Durability* is concerned with the number of cycles of possible operation, i.e., the useful life of the product.

Current societal issues of energy conservation, material conservation, and protection of the environment result in new pressures in selection of materials and manufacturing processes. Energy costs, once nearly totally ignored in design, are now among the most prominent design considerations. Design for materials recycling also is becoming an important consideration.

Increasing attention is being given to design aspects that extend beyond the production stage. Growing knowledge about fracture of materials and the development of fracture mechanics[1] has permitted fracture calculations to become part of the design process. For example, it is possible to determine whether a particular crack is of a critical size, i.e., whether it will propagate to failure under given stress conditions. That introduces the need for improved nondestructive inspection (NDI) for cracks. Consideration of crack inspection at the early stages of design can greatly improve the NDI capability. Packaging and storage of the product are factors that often are neglected in design. Consumer interest activities have resulted in greater design attention in these areas of the product life cycle.

Further Major Factors

Specifications and standards have an important influence on design practice (see Chap. 14). The standards produced by such societies as ASTM and ASME represent voluntary agreement among many elements (users and producers) of industry. As such, they often represent minimum or least-common-denominator standards. When good design requires more than that, it may be necessary to develop your own company or agency standards. On the other hand, because of the general nature of most standards, a standard sometimes requires a producer to meet a requirement that is not essential to the particular function of the design.

Matters of safety and health have recently become regulated to an increased degree by federal agencies. The requirements of the Occupational Safety and Health Administration (OSHA), the Consumer Product Safety

---

[1] S. T. Rolfe and J. M. Barsom. "Fracture and Fatigue Control," Prentice-Hall, Inc., Englewood Cliffs, N.J., 1977.

Commission (CPSC), and the Environmental Protection Administration (EPA) place direct constraints on the designer. Several aspects of the CPSC regulations have far-reaching influence on product design. Although the intended purpose of a particular product normally is quite clear, the unintended uses of that product are not always obvious. Under the CPSC regulations, the designer has the obligation to foresee as many unintended uses as possible, then develop the design in such a way as to prevent hazardous use of the product in an unintended but foreseeable manner. When unintended use cannot be prevented by functional design, clear, complete, unambiguous warnings must be permanently attached to the product. In addition, the designer must be cognizant of all advertising material, owner's manuals, and operating instructions that relate to the product to ensure that the contents of the material are consistent with safe operating procedures and do not promise performance characteristics that are beyond the capability of the design.

An important design consideration is adequate attention to human factors engineering, which uses the sciences of biomechanics, ergonomics, and engineering psychology to assure that the design can be operated efficiently by humans. It applies physiological and anthropometric data to such design features as visual and auditory display of instruments and control systems. It is also concerned with human muscle power and response times. Reference should be made to specialized texts[1] for further information on human factors engineering.

Another important design area involves *aesthetics*, which deals with the shape, texture, and color of the product and also such factors, as balance, unity, and interest. This aspect of design usually is the reponsibility of the industrial designer, as contrasted with the engineering designer. The industrial designer is an applied artist. However, he should work closely with the engineering design team. Decisions concerning the appearance of the final product should be an integral part of the initial design concept.

The final major factor in design is cost. It is possibly the most important factor; for if preliminary estimates of cost look unfavorable, the design project may never be initiated. Cost enters into every aspect of the design process. Therefore, we have considered the subject of economic decision making (engineering economics) in some detail in Chap. 8. Procedures for estimating costs are given in Chap. 9.

Design Strategies

Within the realm of mechanical and electrical engineering design there are four conditions which represent the extremes of the situations in which engineering

---

[1] E. J. McCormick, "Human Factors in Engineering Design," 4th ed., McGraw-Hill Book Company, New York, 1976; W. E. Woodson, "Human Factors Design Handbook," McGraw-Hill Book Company, New York, 1981; A. Garg and D. Kohli, *Trans. ASME, J. Mech. Design,* vol 101, pp. 587–593, 1979.

design is required.[1] These situations are:

1. One-of-a-kind design
2. Design for mass production
3. Large expensive systems
4. Design to code

The first situation is typified by the design of a welding fixture for a new product. This design would likely be characterized by a minimum of analysis or optimization. Rather the designer would depend on experience and intuition, figuring on the ability to quickly fix any problem which turned up in service.

In design for mass production, emphasis is on cost and quality. Design would proceed through the various phases with extensive analysis, prototype testing, and optimization. Because of heavy emphasis on patent position there would be strong attention to documenting the solution to the design.

With a large expensive system like a 100 MW steam turbine the ability to learn from testing of prototypes simply is not affordable. Here we must learn as much as possible from analysis and from field experience. Design in situations like this is likely to be incremental.

In some situations, public health or safety issues are so paramount that available designs have been circumscribed by codes or standards. The design of a steam boiler is such an example, where the code specifies the limiting stress and the methods for calculating it.

1-6 A DESIGN EXAMPLE

The design process will be illustrated here with an example. The example is conceptually simple, yet it introduces sufficient analytical detail and engineering decision making to be realistic. Moreover, students with a mechanical, civil, or materials engineering background should have the training necessary to understand the details of the design solution.

Problem. Design the ladle hooks to be used with the transfer ladles for a steel-melting furnace.

Problem discussion. The problem statement is too fragmentary. Many questions must be answered before we have sufficient information to complete the problem statement.

What is a ladle hook and how does it function?

---

[1] H.O. Fuchs, *Trans. ASME, J. Mech. Design,* vol. 102, p. 1, 1980.

How much molten steel does an average ladle hold, and what does the steel weigh?

What is the empty weight of a ladle?

What are the dimensions of the trunnions where the ladle hook meets with the ladle?

What are the service conditions? What are the maximum and minimum temperatures? Is impact loading a factor? Should fatigue be considered? What about corrosion and wear?

What is the expected life of a ladle hook?

What are the limitations on cost?

Are there any special precautions with regard to safety?

Should the hook be so designed as to be adaptable to use in handling scrap boxes or charging the furnace with ferroalloy additions?

Other questions might be suggested, but the above will do for starters. Certainly, the list does point up the need for information. The experienced steel mill engineer could answer the questions from his experience base or, if not from direct experience, from engineering judgment plus a few quick phone calls to friends or friendly equipment suppliers. The student must turn to more impersonal sources and so must learn to use the published information base (see Chap. 14).

Figure 1-5 shows a teeming ladle that transfers molten steel from the melting furnace to the ingot mold. Note that the ladle is supported from the crane by two ladle hooks. The hooks make contact with the ladle at bearing surfaces called trunnions.

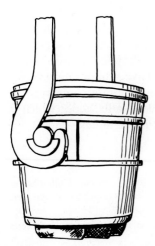

FIGURE 1-5
Teeming ladle and ladle hooks.

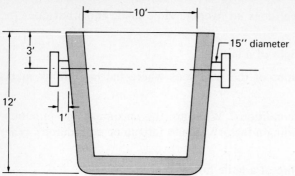

FIGURE 1-6
Details of ladle used for design.

Steel mill cranes typically have a capacity of 100 to 200 tons.[1] We will assume a total ladle weight of 150 tons. Since the ladle is made of heavy steel plate lined with refractory brick, its capacity for holding molten metal is only about 60 percent of the total weight. The specific dimensions of the ladle are given in Fig. 1-6; they come from specific information on the design of hot-metal ladles.[2]

Next we need to establish the conditions of service. The ladle hook will operate in an unheated mill building. Winter temperature may go as low as −10°F, yet the radiation from the molten steel in the uncovered ladle may result in temperatures as high as 600 to 1000°C in parts of the hook surface. Impact loading and fatigue loading (very low frequency and low number of cycles) will be present, but they are not likely to be controlling factors. Corrosion is likely to be a minor problem, but wear could be a factor where the hook pins to the crane and where it supports the ladle at the trunnions.

Steel mill equipment is big and rugged and built to last. An average life of 10 to 15 years would not be unreasonable for a ladle hook. On the other hand, it is the type of equipment that would probably be replaced before other more major items are.

We can offer no special guidelines on cost other than that a ladle hook is a fairly standard item. Any "gold-plating" would have to be well justified. In opposition is the fact that failure of a ladle hook that dropped a ladle full of molten steel would be catastrophic. The design must have a high safety factor yet be reliable and economical. One solution is to incorporate some type of redundancy into the design. We will settle for a standard ladle hook, which is used only for the single purpose of lifting 150-ton hot-metal ladles.

---

[1] H. M. McGannon (ed.): "The Making, Shaping and Treating of Steel," 9th ed., United States Steel Corp., Pittsburgh, 1971. Alternatively, scanning back issues of the periodical *Iron and Steel Engineer* should provide similar information.

[2] American Iron and Steel Engineers Standard No. 9, Specification for Design of Hot Metal Ladles, December 1959.

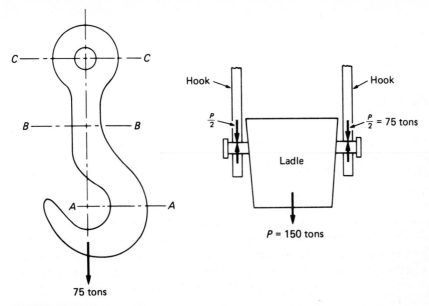

FIGURE 1-7
Details of ladle hook and ladle.

With the above elaboration we can develop the complete problem statement as follows:

Problem statement. Design a hook for lifting hot-metal ladles with a maximum weight of 150 tons. The hook should be compatible with the ladle details given in Fig. 1-6. The hook eye should receive an 8-in-diameter pin for attaching to the crane.

Problem analysis. We first need to determine the types of load and their critical locations in the ladle hook. The hook is a load-connecting member between the crane and the ladle. From this information, and Fig. 1-5, we can deduce that the hook will have the general configuration shown in Fig. 1-7.

There are three critical stress regions in the hook.

Section A (at bight of hook). Hook is subjected to 1) direct tensile loading, 2) bending stress as a curved beam, and 3) lateral bending stress due to deflection of trunnion and concentration of load toward outer surface of hook thickness.

Section B (at shank section of hook). Hook is a subjected to 1) direct axial stress and 2) lateral bending stress.

Section C (at eye of hook). Hook is subjected to 1) direct stress with stress concentration factor and 2) lateral bending.

Using some intuition, we can hypothesize that section *A* will be the critical section and that stresses there will establish the hook thickness. Section *B* will establish the width of the shank. Section *C* will determine the dimensions of the eye.

Material Selection and Working Stress

It is almost axiomatic that a load-bearing member 8 to 10 ft tall, in a situation where dead weight is not critical, will be made from steel. Moreover, the need for high reliability in large section, coupled with modest cost, suggests that a structure built up from steel plates is preferable to a monolithic cast- or forged-steel body. Therefore, we need to make a selection from the standard grades of carbon- and low-alloy-steel plates that are commercially available.

Structural-quality steel plate is designated by various ASTM specifications (Table 1-1). Steels of this type fall into three categories: 1) plain hot-rolled carbon steels, 2) high-strength low-alloy steel (HSLA) in which grain refiners have been added to control the hot-rolled grain size, and 3) alloy steels which are quenched and tempered for high strength. Since welding may be used in fabrication, only weldable grades with less than 0.30 percent carbon will be considered. The A36 has impact and brittle fracture characteristics that make it unsatisfactory for this application. The A514 is ruled out by its cost and the fact that its higher yield strength really cannot be justified. The two HSLA steels differ chiefly in their resistance to atmospheric corrosion. The A 441 is about twice as good as A 36, and A 242 is about twice as good again. Since corrosion resistance is a useful property, but not a crucial one in this application, we decide to use the cheaper A 441 steel.

Thus, we have decided to use a commercially available HSLA steel. We have considered brittle fracture and corrosion but not fatigue in our selection. The possibility of high-temperature effects, or of unrecognized stress concentration owing to fabrication or such metallurgical causes as decarburization and large inclusions must be allowed for. We know that safety is paramount. Therefore, we decide to reduce the material yield strength by a *factor of safety*.

TABLE 1-1
Characteristics of candidate structural steels*

ASTM spec.	Description	C	Mn	Other	YS, psi	UTS, psi	Relative cost
A 36	Carbon steel	0.29	1.0	—	36,000	60,000	1.0
A 441	HSLA steel	0.22	1.25	0.02 V	50,000	70,000	1.15
A 242	HSLA steel	0.15	1.10	0.05 V, 0.3 Cu	50,000	70,000	1.25
A 514	Alloy steel	0.15	0.80	Ni–Cr–Mo	100,000	120,000	2.0

* "Metals Handbook," 9th ed., pp. 181–190, American Society for Metals, Metals Park, Ohio, 1978.

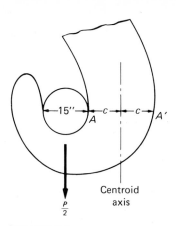

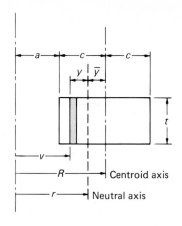

FIGURE 1-8
The ladle hook as a curved beam.

Because weight is not an important consideration, we have the luxury of using a conservatively high factor of safety of 4.

$$\text{Working stress} = \frac{\text{yield strength}}{\text{factor of safety}} = \frac{50,000}{4} = 12,500 \text{ psi}$$

Although the method of manufacturing the designed hook will be considered after we have a better understanding of the hook's size and detailed dimensions, for present purposes we can assume that the ladle hook will be manufactured from built-up thicknesses of steel plate held together by rivets and/or welds.

Detailed Stress Analysis

Stresses at section A. Three different stresses can exist at the bight section of the hook. Of the three, the bending stress in the curved section of the hook that is due to the weight of the ladle is the most significant. Because of geometry, the hook at Section A is classified as a curved beam. A characteristic of a curved beam is that the neutral axis does not coincide with the centroid axis (Fig. 1-8). Even if your mechanics of materials course did not cover curved beams, you should have received enough background to read intelligently about the subject and apply the new information to this problem.

The bending stress in section $A - A'$ is given by[1]

$$\sigma = \frac{My}{A\bar{y}(r - y)}$$

[1] S. Timoshenko and D. Young, "Elements of Strength of Materials," 5th ed., pp. 167–171, Van Nostrand Reinhold, 1968

where $r = v + y$ (see Fig. 1-8)

$$A = 2ct$$

The maximum tensile stress occurs at point A. The bending moment is

$$M = \frac{P}{2} R = \frac{300,000}{2} (7.5 + c)$$

$$\sigma_{max} = \left[\frac{150,000(7.5 + c)}{A\bar{y}a}\right](c - \bar{y})$$

since the minimum value of $(r - y)$ is a and the maximum value of y is $(c - \bar{y})$. Remember that y is measured from the neutral axis.

By definition:

$$\bar{y} = R - \frac{A}{\int \frac{dA}{v}} = (7.5 + c) - \frac{2ct}{\int \frac{dA}{v}}$$

$$\int \frac{dA}{v} = \int_a^{a=2c} \frac{t\,dv}{v} = t \ln v \Big|_{7.5}^{7.5+2c} = t \ln \frac{7.5 + 2c}{7.5}$$

$$\bar{y} = (7.5 + c) - \frac{2ct}{t \ln \dfrac{7.5 + 2c}{7.5}}$$

If we let $2c = 24$ in, then $\bar{y} = 2.8$ and $\sigma_{max} = 53,392/t$.

The direct (axial) stress in section $A - A'$ is:

$$\sigma_d = \frac{150,000}{24t} = \frac{6250}{t} \quad \text{(neglects any area removed for rivets)}$$

The lateral (sidewise) bending stress can be determined by assumming that the force on the hook is applied at 1 in from the outer surface of the plate.

$$\sigma_{lb} = \frac{My}{I} = \frac{(P \times l)t/2}{I}$$

$$l = \frac{t}{2} - 1$$

$$I = \tfrac{1}{12}bh^3 = \tfrac{1}{12}2ct^3 = 2t^3$$

$$\sigma_{lb} = \frac{150,000(t/2 - 1)}{2t^3} = \frac{37,500}{t^2}\left(\frac{t}{2} - 1\right)$$

These three stresses act simultaneously and at the critically stressed location of A:

$$\sigma_{max} = \frac{53,392}{t} + \frac{6250}{t} + \frac{37,500}{t^2}\left(\frac{t}{2} - 1\right) = 12,500$$

since we already have decided to limit the working stress to 12,500 psi.

Hook thickness, in	Sum of stresses at point A, psi
5	14,178
6	12,023
7	10,433

Thus, if the dimensions of the hook at section $A - A'$ are $2c = 24$ in and $t = 6$ in, the stress level is kept to a conservative value of 12,500 psi.

Stresses in the shank, section B

$$\text{Direct stress } \sigma_d = \frac{150,000}{tw} = \frac{150,000}{6w} = \frac{25,000}{w}$$

where w is the width of the hook shank.

Lateral bending

$$\sigma_{lb} = \frac{150,000(\frac{6}{2} - 1)6}{\frac{1}{12}(w)(6)^3} = \frac{100,000}{w}$$

$$\therefore 12,500 = \frac{25,000}{w} + \frac{100,000}{w}$$

$$w = 10 \text{ in}$$

Stresses in section C, through the eye of the hook. The direct stress is subjected to a stress concentration of 3 because of the hole for the connecting pin.

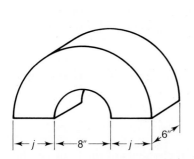

$$\sigma_d = \frac{3P}{A} = \frac{3(150,000)}{2j(6)} = \frac{37,500}{j}$$

$$\sigma_{lb} = \frac{P(t/2 - 1)t/2}{\frac{1}{12}jt^3} = \frac{150,000 \times 2 \times 3 \times 12}{216j}$$

$$= \frac{50,000}{j}$$

$$\sigma_{total} = 12,500 = \frac{37,500}{j} + \frac{50,000}{j} = \frac{87,500}{j}$$

$$j = 7 \text{ in}$$

Manufacturing Methods

Cost and reliability of the product are the chief considerations in selecting the method for manufacturing the ladle hook. The prime decision is whether to use a monolithic hook with the 6 in thickness produced by forging or casting or to build up the thickness from several layers of hot-rolled steel plate. The latter method was chosen because it is cheaper than forging or casting when production lots are relatively small. Also, in thickness of less than $1\frac{1}{2}$ in, hot-rolled steel plate has relatively uniform and predictable mechanical properties. In addition, the laminated hook provides a redundant structure. Should one of the plates develop a fatigue crack or fail from brittle fracture, the fracture will not propagate immediately to all of the plates. Thus, there should be time to locate the crack during a routine inspection and repair the hook.

The hook will be made from five plates of 1.2-in-thick A 441 steel.[1] Hot-rolled plate is rolled to slightly more than the nominal thickness. Plates that are from 1 to 2 in thick have a thickness tolerance of +0.070 in, −0.000. Thus, five 1.2-in-thick plates will exceed the required thickness of 6 in.

The processing route sheet for fabricating the ladle hook will read as follows:

1. Acid-pickle, shot-blast, or vapor-blast plate surfaces to remove mill scale (if necessary).
2. Flame-cut ladle hook to size.
3. Grind flame-cut edges smooth.
4. Repair any gouges or cracking from flame cutting by weld-depositing metal from a low-hydrogen-metal electrode. Grind weld deposit flat and check for cracks with magnaflux or ultrasonics.
5. Stack plates, clamp, and drill and ream holes for 1-in-diameter rivets.
6. Rivets will be driven hot and will have a button head design.
7. No rivets will be permitted near the bight section of the hook or in the eye of the hook. In those portions of the hook the plates are to be welded together.
8. A hard-facing alloy will be weld-deposited at the inner surface of the hook, which contacts the trunnion. The hard surface deposit will be ground to provide a smooth wear surface.
9. The flame-cut eye will be machined to close tolerance to receive a press-fitted high-carbon-steel bushing. The bushing will be welded around its circumference to lock it in place.

Final Dimensions

We have determined the dimensions of the critical elements of the hook on the basis of the stresses that are expected to be present. These dimensions have been based on keeping the nominal stresses at a level below 12,500 psi. At this point the ladle hook should be drawn to scale to check the validity of the calculated dimensions and establish the remaining dimensions. Some of the

---

[1] 1978 Annual Book of ASTM Standards, part 3, Steel Plates, Sheet, Strip, Wire.

latter will be set by functional requirements; others will be set by engineering common sense. For example, the distance between the crane pin and the trunnion must be great enough to accommodate one-half the width of the ladle if it is tipped to its maximum extent.

The last design detail we need to consider is the rivets. To establish the number of rivets required, we use the criterion that the total area of rivets should be sufficient to take the full load in shear. If the rivet steel is assumed to have a 30,000-psi yield strength in tension, it will be about 15,000 psi in shear. Assuming a factor of safety of 2, the working stress will be 7500 psi.

$$A = \frac{P}{\sigma_y} = \frac{150,000}{7500} = 20 \text{ in}^2 \text{ of rivet}$$

Using 1-in-diameter rivets

$$A = 0.785 \text{ in}^2$$

$$\text{Number of rivets required} = \frac{20}{0.785} = 26$$

1-7 AN ADVANCED-DESIGN EXAMPLE

The efficiency of the gas turbine engine is determined by the turbine inlet temperature. Over 30 years of vigorous metallurgical research on superalloys, coupled with ingenious design to introduce cooling mechanisms, has succeeded in increasing the turbine inlet temperature by 700°F. Turbines currently operate at 2100°F, but the technology appears to have plateaued at the top of a S-shaped growth curve. However, turbine inlet temperatures of 3000°F (which would increase the efficiency from 81 to 90 percent) appear possible if the main turbine components could be built from high-strength ceramic materials such as silicon nitride (Si_3N_4) and silicon carbide (SiC). This is a good example of how quantum advances in engineering technology often depend upon major advances in materials technology.

The covalent bonding of ceramic materials, coupled with high melting point and relative resistance to oxidation, make ceramics good candidates for high-temperature materials. In addition, they are relatively cheap and abundant and are not dependent on foreign countries for supply, as are such strategic materials as cobalt and chromium.

However, ceramic materials are brittle, and in the past that quality has severely limited their use in structural applications. Their behavior, as contrasted with that of ductile metals, is illustrated in Fig. 1-9. Ceramics are brittle elastic materials; the yield strength and fracture strength of a ceramic coincide. In a ductile metal, on the other hand, yielding precedes fracture. Localized stresses that exceed the yield strength in a metal are removed by local plastic deformation that redistributes the stress over a wider area. Thus, localized fracture usually is prevented. Ceramics, in contrast, have no yield strength and fail when the local stress exceeds the material fracture strength; they are unforgiving and fracture upon the slightest overload. For that reason,

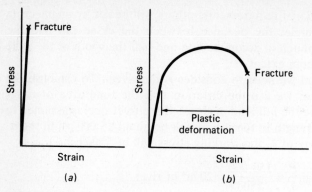

FIGURE 1-9
Schematic stress-strain curves for (*a*) a brittle ceramic and (*b*) a ductile metal.

they are sensitive to the distribution of flaws, and so their measured strength properties exhibit wide variability (scatter).

A number of converging factors have made it possible to seriously consider building high-temperature aircraft structural components from ceramic materials:

1. The development of higher-purity ceramic materials with improved high-temperature strength. The candidate materials are Si_3N_4 and SiC.
2. The development of processing methods to produce the new materials with low porosity and more consistent properties.
3. The development of the finite-element method, which permits the calculation of stresses to whatever level of detail and refinement that is desired. Because ceramics have no ductility, there is no margin for error in design calculations. Moreover, the design must be carried out as to minimize loading due to impact, shock, and vibration.
4. The science of fracture mechanics has provided a theoretical and analytical framework for developing brittle material design procedures.
5. The Weibull analysis has provided a statistical theory for allowing for the extreme scatter in mechanical properties of ceramic materials.

Fracture Mechanics

Fracture mechanics[1] is a principal connection between the theory of materials and engineering design. The theoretical fracture strength of a solid derived

---

[1] G. E. Dieter, "Mechanical Metallurgy," 3rd ed., chap. 11, McGraw-Hill Book Company, New York, 1986; D. Broek, "Elementary Engineering Fracture Mechanics," Martinus Nijhoff Publishers, The Hague, 1982; J. J. Burke and V. Weiss (eds.), "Applications of Fracture Mechanics to Design," Plenum Press, New York, 1975.

from a simple model of the force between atoms is

$$\sigma_f \approx \left(\frac{\gamma_s E}{a_0}\right)^{1/2} \tag{1-1}$$

where g_s = surface energy of the fracture surface

E = Young's modulus

a_0 = interatomic spacing.

Since actual materials have strength of one-fifth to one-hundredth of the theoretical value, we are led to the conclusion that real materials contain flaws or cracks that lower their fracture strength to less than the ideal. In 1920, A. A. Griffiths used an energy approach to show that the fracture strength in a material with crack of length $2c$ was given by

$$\sigma_f \sim \left(\frac{2\gamma_s E}{\pi c}\right)^{1/2} \tag{1-2}$$

In the 1950s and 1960s, G.R. Irwin extended Griffiths' theory into a useful method of engineering analysis. He established that the material behavior could be denoted by an experimentally measureable parameter, the stress-intensity factor K_c. Now the fracture strength of a material is described by

$$\sigma_f \approx \frac{K_c}{(Y/Z)\sqrt{c}} \tag{1-3}$$

where Y and Z are geometric factors that account for the crack shape, orientation, and location with respect to the loading. Normally, Y will depend on whether the crack extends from the surface ($Y \approx 1.99$) or is beneath the surface ($Y \approx 1.77$). The parameter Z depends upon crack shape. $Z = 1$ for a long shallow crack and increases with increasing ratio of crack depth to width. For a circular crack $Z = \pi/2$.

Equation (1-3) can be utilized in several important ways. Once the properties of the material are measured through K_c and the size and shape of flaws are established by nondestructive evaluation (NDE) techniques, the level of stress that will cause fracture is established. Alternatively, when design is based on a specific stress level, Eq. (1-3) establishes the largest flaw size that can be tolerated.

Statistical Analysis

The larger scatter in mechanical properties that is characteristic of brittle materials like ceramics requires statistical analysis. The function most widely used with brittle materials is the Weibull distribution.[1] This statistical

[1] C. Lipson and N. J. Sheth, "Statistical Design and Analysis of Engineering Experiments," pp. 36–44, McGraw-Hill Book Company, New York, 1973.

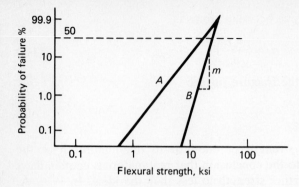

FIGURE 1-10
Typical Weibull plots for brittle materials. Material B has a higher Weibull modulus, and thus less scatter.

technique will be considered in more detail in Chap. 11, but it is discussed here briefly in its design context.

If the probability of failure is plotted against strength level on a Weibull probability paper, the data will plot as a straight line (Fig. 1-10). The mathematical equation for the lines is

$$\log \log \frac{1}{1 - P} = m \log \sigma + \log \frac{V \log e}{\sigma_0^m} \tag{1-4}$$

where P = probability of failure at a stress σ
m = Weibull modulus, a measure of the scatter
V = volume of material
σ_0 = a constant that represents the value of strength below which $P = 0$

The Weibull modulus m is an important parameter of a brittle material. A high value of m indicates less scatter in strength values. Note from Fig. 1-10 that the mean failure stress (the stress at $P = 0.50$) has little significance in the design of brittle materials. Rather, brittle materials must be designed on the basis of some low probability of failure, for example, $P = 0.01$. The volume of material enters into the Weibull plot because brittle materials exhibit a strong size effect. Since failure is initiated at flaws and the likelihood of a critical-size flaw increases with the volume of the part, the strength is reduced in large section sizes. The inter-relation between these design factors is illustrated by the data[1] given in Table 1-2.

This statistical approach is used two ways in design. The first, as shown above, is to characterize the mechanical properties of a brittle material accurately. The second is to replace the factor-of-safety concept by a more precise strength-reliability relation. Figure 1-11, which is derived directly from Weibull's theory,[2] shows the probability of failure vs. relative fracture strength

[1] W. Duckworth and G. Bansal, *ASME Jnl. Eng. for Power*, vol. 100, pp. 260–266, 1978.

[2] G. G. Trantina and H. G. de Lorenzi, *Trans. ASME*, ser. A, vol. 99, pp. 559–566, 1977.

TABLE 1-2
Allowable stress, ksi

	$m = 8$		$m = 32$	
Failure probability	V_1	$V_2 = 100 V_1$	V_1	$V_2 = 100 V_1$
0.50	50	28.1	50	43.3
0.05	36.4	20.3	46.1	39.9
0.01	29.4	16.5	43.8	37.9

(mean fracture strength/maximum stress) for a material with a Weibull modulus of 7 and for large and small stressed volumes. This plot shows that $\sigma_{max} = \sigma_f/1.5$ at $P = 0.5$, but it changes to $\sigma_{max} = \sigma_f/5$ for $P = 0.0001$. Even greater reductions in working stress are required for the large stressed volume. Although this type of statistical design is relatively new, it is developing rapidly in sophistication. For example, the statistical analysis has been incorporated into finite-element computer programs for calculating stresses in brittle materials.

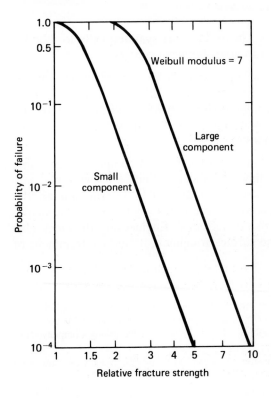

FIGURE 1-11
Probability of failure vs. relative fracture strength (fracture strength/maximum stress in structure). (From *Trantina and de Lorenzi*, 1977.)

1-8 DESIGN DRAWINGS

Engineering drawings are vital for communicating the design. Although at one time it could be assumed that most engineering graduates were proficient in engineering drawing, today the education of an engineer provides scant exposure to the subject. Clearly, the time devoted to engineering drawing in the present education of an engineer is not commensurate with the practical importance of the subject. Every engineer who deals more than casually with design should be well grounded in the elements of orthographic projection, be able to read the language of engineering drawings fluently, and be able to produce an acceptable sketch that can be converted by a draftsman into an engineering drawing. Although formal instruction helps, it should be possible to obtain this ability from self-study.[1]

An engineering drawing can be considered as a coding technique to transmit information. For example, the following word specification defines the shape and dimensions of a simple angle bracket.[2]

Title: Angle Bracket

Material: Hot-rolled steel, AISI 1010.

Given a rectangular parallelepiped of the subject strip steel, having a width of 1 in ($\pm \frac{1}{16}$ in), length of 4 in ($\pm \frac{1}{64}$ in), and a thickness of $\frac{1}{8}$ in ($\pm \frac{1}{64}$ in), perform the following operations:

1. Drill in the flat plane two holes $\frac{1}{4}$ in in diameter symmetrically located on the longitudinal centerline and 3 in apart between centers.
2. Enlarge each of these holes by an 82° included angle countersunk to the full depth of the hole.
3. Bend the bar at its midpoint to a 90° angle (with a $\frac{1}{8}$-in vertex fillet) around an axis parallel to the flat plane, said axis also perpendicular to the longitudinal centerline. Other dimensions are $\pm \frac{1}{32}$ in.

Note in Fig. 1-12 how much more succinctly and completely the same information is conveyed by an engineering drawing. Such a drawing not only conveys an instant impression of shape but also gives information on the material to be used and the dimensions that are necessary to specify the shape, together with the tolerances on the dimensions. Engineering drawings also often contain instructions on 1) the surface roughness or surface treatment of

[1] T. E. French, C. L. Svensen, J. D. Helsel, and B. Urbanick, "Mechanical Drawing," 8th ed., McGraw-Hill Book Company, New York, 1974; F. E. Giesecke, A. Mitchell, H. C. Spencer, I. L. Hill, and R. O. Loving, "Engineering Graphics," 2d ed., Macmillan Publishing Company, New York, 1975.

[2] T. T. Woodson, "Introduction to Engineering Design," McGraw-Hill Book Company, New York, p. 167, 1966.

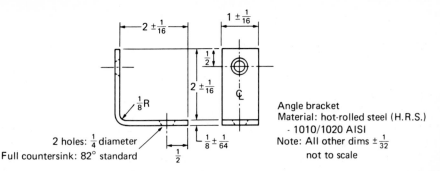

FIGURE 1-12
Engineering drawing of angle bracket. (*From T. T. Woodson, "Introduction to Engineering Design," McGraw-Hill Book Co., 1968; used with permission of McGraw-Hill Book Co.*)

the part, 2) the required heat treatment, and 3) the inspection or testing of the part. When the information is too detailed or voluminous, the drawing has a reference to a specification or standard that will supply the needed information.

Different kinds of engineering drawings are used for various purposes. A *detail drawing* gives a complete description of the shape of a part using up to three orthographic views and possibly one or more section views. It provides all of the information for producing the part. The detail drawing specifies the material, finished dimensions, surface finish, and any special processing (such as heat treatment). Usually a separate drawing is made for each component. An *assembly drawing* shows how the components are assembled into a system. Such a drawing normally will include a parts list that identifies component part numbers, part names, and the required number of pieces. *Schematic drawings* show the manner in which components are connected together, as in a piping system or electronic control system. The components are shown in symbolic form in this type of drawing.

Three aspects of engineering drawing that are often slighted in an introduction to the subject but are vital in design practice are dimensions, tolerances, and specification of surface finish. Careful attention to those aspects of engineering drawing can greatly improve the cost and quality of a design.

Tolerances

Tolerances must be placed on the dimensions of a part to limit the permissible variations in size because it is impossible to manufacture a part exactly to given dimensions. A small tolerance results in greater ease of interchangeability of parts, but it also greatly adds to the cost of manufacture (Fig. 1-13).

Tolerances can be expressed in either of two ways. A *bilateral tolerance* is specified as a plus or minus deviation from a basic dimension, e.g., 2.000 ± 0.004 in. This system is being replaced by the *unilateral tolerance*, in

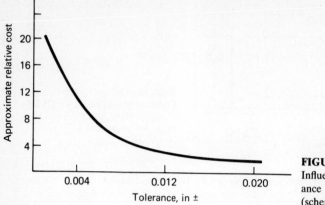

FIGURE 1-13
Influence of dimensional tolerance on cost of manufacture (schematic).

which the deviation in one direction from the basic dimension is given. For example,

$$2.000 + 0.008 \qquad \text{or} \qquad 5.005 + 0.000$$
$$\qquad\quad - 0.000 \qquad\qquad\qquad\quad - 0.005$$

In the case of bilateral tolerance, the dimension of the part would be permitted to vary between 1.996 and 2.004 in for a total tolerance of 0.008 in. If unilateral tolerance is specified, the dimension could vary between 2.000 and 2.008, and again the total tolerance is 0.008 in. Unilateral tolerances have the advantages that they are easier to check on drawings and that a change in the tolerance can be made with the least disturbance to other dimensions.

The American National Standards Institute (ANSI) has established eight classes of fit that specify the amount of allowance and the tolerance on the hole and a mating shaft, see Table 1-3. The ANSI system considers that the hole size d is the basic dimension, because most holes are produced by using standard-size drills and reamers. Therefore, the shaft can be more easily

TABLE 1-3
ANSI recommended allowances and tolerances

Class of fit	Description	Clearance	Interference	Hole tolerance	Shaft tolerance
1	Loose fit	$0.0025\sqrt[3]{d^2}$		$+0.0025\sqrt[3]{d}$	$-0.0025\sqrt[3]{d}$
2	Free fit	$0.0014\sqrt[3]{d^2}$		$+0.0013\sqrt[3]{d}$	$-0.0013\sqrt[3]{d}$
3	Medium fit	$0.0009\sqrt[3]{d^2}$		$+0.0008\sqrt[3]{d}$	$-0.0008\sqrt[3]{d}$
4	Snug fit	0		$+0.0006\sqrt[3]{d}$	$-0.0004\sqrt[3]{d}$
5	Wringing fit		0	$+0.0006\sqrt[3]{d}$	$+0.0004\sqrt[3]{d}$
6	Tight fit		$0.00025d$	$+0.0006\sqrt[3]{d}$	$+0.0006\sqrt[3]{d}$
7	Medium force fit		$0.0005d$	$+0.0006\sqrt[3]{d}$	$+0.0006\sqrt[3]{d}$
8	Shrink fit		$0.001d$	$+0.0006\sqrt[3]{d}$	$+0.0006\sqrt[3]{d}$

produced to a nonstandard dimension. Consider a basic hole size of 2.000 in and a class 3 (medium) fit.

Allowance $\quad 0.0009\sqrt[3]{2^2} = 0.0014$ in

Tolerance $\quad \pm 0.0008\sqrt[3]{2} = 0.0010$ in

Hole

   Maximum dimension 2.001 in

   Minimum dimension 2.000 in

Shaft

   Maximum dimension $2.000 - 0.0014 = 1.9986$ in

   Minimum dimension $1.9986 - 0.001 = 1.9976$ in

Therefore, the maximum clearance between shaft and hole is

$$2.0010 - 1.9976 = 0.0034 \text{ in}$$

and the minimum clearance between shaft hole is

$$2.00 - 1.9986 = 0.0014 \text{ in}$$

For additional details on tolerances see machine design books[1] and specialized treatises.[2]

Dimensions

The engineering drawing provides the manufacturing department with the information necessary for producing the part. Therefore, it is important that the dimensions of the part be clear and complete. The dimensions given should be sufficient in number to make it unnecessary for shop personnel to perform involved calculations for setting up the production equipment. On the other hand, too mnay dimensions can cause problems by resulting in ambiguity and leaving the manufacturing department with a choice. Figure 1-14 shows a part with redundant dimensioning. Figure 1-14a gives the detail drawing for a simple part. According to the information given, the manufacturing department could make the part to the dimensions given in Fig. 1-14b and not be outside tolerances for dimensions AB and AC. However, dimension BC $(2.750 - 0.990 = 1.760)$ is automatically outside tolerance. Obviously, all three dimensions cannot be specified with close tolerances without one of them running a high probability of falling out of specification. The solution is to specify only two out of the three dimensions on the drawing. For example, if

---

[1] M. F. Spotts, "Design of Machine Elements, " 5th ed., chap. 13, Prentice-Hall, Inc., Englewood Cliffs, N.J., 1978; A. D. Deutschman, W. J. Michels, and C. E. Wilson, "Machine Design," Macmillan Publishing Company, New York, 1975, pp. 189–207; M. F. Spotts," Dimensioning and Tolerancing for Quantity Production," Prentice-Hall Inc; Englewood Cliffs, N.J., 1983.

[2] "Dimensions and Tolerancing for Engineering Drawings," ANSI Y14.5M–1982, ASME.

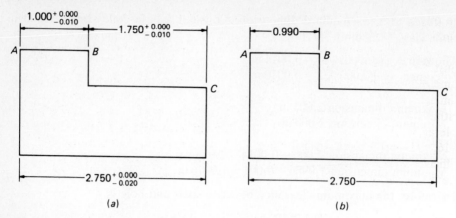

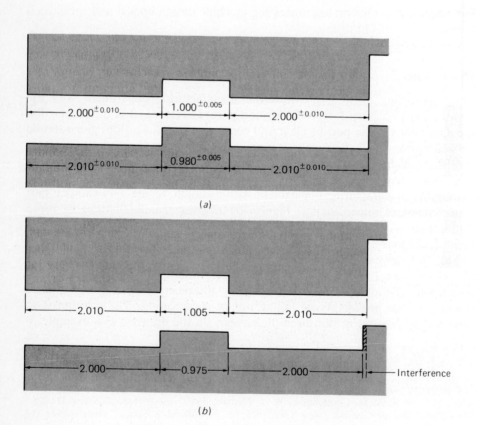

FIGURE 1-14
Example of redundant dimensioning.

the overall length and the length of the shoulder are most important for the functioning of the part, then AC and AB should be given on the drawing.

Another difficult situation arises when successive points on a drawing must be dimensioned. Figure 1-15 shows two mating parts. In Fig. 1-15a the dimensions with tolerances run from point to point in a chain or cumulative dimensioning way. Note that it is possible for production parts manufactured within tolerance to not fit because of interference due to the accumulation of tolerances (Fig. 1-15b). However, if all dimensions start at a datum line, Fig. 1-16, all parts made to within tolerance will assemble properly and the difficulty with interference is eliminated. Thus, in Fig. 1-16b, all tolerances for the upper surface are at their largest value and those for the lower surface are

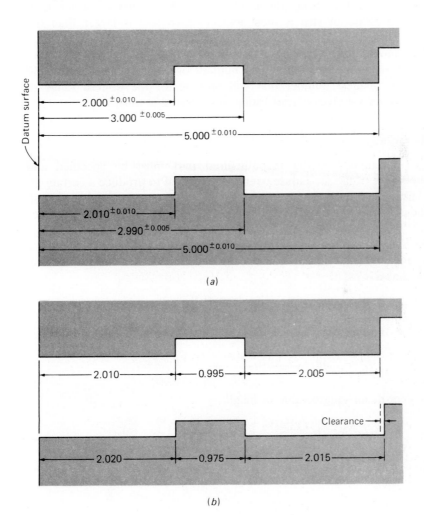

(a)

(b)

at their largest value, as was the case in Fig. 1-16*b*, but now the two surfaces mate. Another reason for dimensioning from datum surfaces is that it usually is more convenient for setting up machine tools. Note that datum surfaces must be actual physical surfaces from which gaging can be done. Centerlines, axes of rotation, and other imaginary lines that may appear convenient on the drawing cannot be used as datums.

Preferred Sizes

Standard components such as bolts, bearings, and electric motors need to be made according to a rational scheme by which the size (or weight, speed, power, etc.) covers the usual range of needs. A geometric rather than arithmetic progression of size is most logical. Each size is larger than the preceding size by a fixed percentage. At the small-size end of the range there will be more items than at the large-size end.

From an economic standpoint, the number of standardized sizes should be kept to the smallest number that will provide for the desired range of applications. Table 1-4 gives a brief listing of preferred numbers.

Surface Roughness

The surface roughness of the manufactured part must be specified and controlled because of fatigue failure, wear, or the need to produce a certain fit. However, like the situation with tolerances, overrefinement of surface finish costs money. Therefore, we need a way to measure and specify surface roughness.[1]

No surface is absolutely smooth and flat; on a highly magnified scale, it looks like Fig. 1-17. Several parameters are used to describe the state of surface roughness.

R_t is the height from maximum peak to deepest trough.
R_a is the centerline average (CLA), the airthmetic average based on the deviation from the mean surface

$$R_a = \frac{y_1 + y_2 + y_3 + \cdots + y_n}{n}$$

R_{rms} is the root-mean-square value of height.

$$R_{rms} = \left(\frac{y_1^2 + y_2^2 + y_3^2 + \cdots + y^2}{n} \right)^{1/2}$$

Surface roughness measurements typically are expressed in microinches

---

[1] See Surface Texture, ANSI Std. B46.1, ASME, 1985.

TABLE 1-4
Table of preferred sizes

5 Series 60 percent steps	10 Series 25 percent steps	20 Series 12 percent steps
	10	10
		11.2
10		
		12.5
	12.5	14
		16
	16	18
16		
		20
	20	22.4
		25
	25	28
25		
		31.5
	31.5	35.5
		40
	40	46
40		
		50
	50	56
		63
	63	71
63		
		80
	80	90

Preferred numbers below 10 are formed by dividing the numbers between 10 and 100 by 10. Preferred numbers above 100 are obtained by multiplying by 10.

($1\,\mu\text{in} = 0.025\,\mu\text{m} = 0.000001\,\text{in}$). Until recently, surface roughness was characterized by the rms value, but currently the CLA value is preferred. The rms value is about 11 percent greater than the value based on the arithmetic average.

There are other important characteristics of a surface besides the height

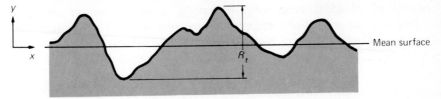

FIGURE 1-17
Cross-sectional profile of surface roughness with vertical direction magnified.

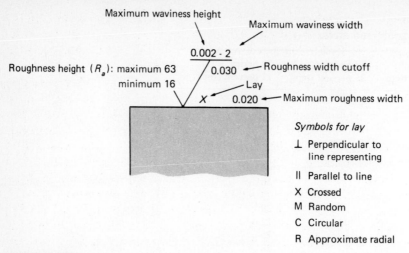

FIGURE 1-18
Symbols used to specify surface finish characteristics. Roughness given in microinches.

of the roughness. Surfaces may exhibit a directionality characteristic called *lay*. Surfaces may have a strong directional lay (e.g., from machining grooves), a random lay, or a circular pattern of marks. Another characteristic of the surface is its *waviness*, which occurs over a longer distance than the peaks and valleys of roughness.

All these surface characteristics are specified on the engineering drawing by the scheme shown in Fig. 1-18.

It is important to realize that specifying a surface by average roughness height is not an ideal approach. Two surfaces can have the same value of R_a (or R_{rms}) and vary considerably in the details of surface profile. There is much yet to be learned about the control and specification of surfaces.[1]

There is a close relation between surface roughness and the tolerances that can be achieved. Generally speaking, tolerances must exceed R_t unless the surface roughness is to be smoothed out in a force fit. Since $R_t \approx 10R_a$, a surface roughness of $R_a = 125\ \mu$in would exceed a tolerance of 0.001 in.

1-9   COMPUTER-AIDED ENGINEERING

The advent of plentiful computing is having a major impact on how engineering is practiced. While engineers were one of the first professional groups to adapt the computer to their needs, the early applications chiefly were

[1] Tool and Manufacturing Engineers Handbook, "vol. 4, chap. 5, Soc. of Mfg. Engrs, Dearborn, MI., 1987.

computational intensive ones, using a high-level language like FORTRAN. The first computer applications were conducted in batch mode, with the code prepared on punch cards. Overnight turnaround was the norm. Later, remote access to computer mainframes through terminals became common, and the engineer could engage in interactive (if still slow) computation. The development of the microprocessor and the proliferation of personal computers and engineering workstations with computational power equivalent to that of a mainframe ten years ago has created a revolution in the way an engineer approaches and carries out problem solving and design. This great change has not progressed uniformly in all dimensions, but the trend is clear and sure.

The greatest impact of computer-aided engineering to date has been in engineering graphics. The automation of drafting in two dimensions (CAD) has become commonplace (see Sec. 1-10). Current technology can readily transform 2D line drawings into 3D solid models with shaded images in true perspective. Such geometric modeling capabilities tie in nicely with analysis capabilities introduced through extensive use of finite element modeling (FEM). This makes possible interactive simulations in such problems as stress analysis, kinematics of mechanical linkages, and numerically controlled tool path generation for machining operations. The computer extends the designer's capability in several ways. First, by organizing and handling time-consuming and repetitive operations, it frees the designer to concentrate on more complex design situations. Second, it allows the designer to analyze complex problems faster and more completely. Both of these factors make it possible to carry out more iterations of design. Finally, through a computer-based information system the designer can share more information sooner with people in the company, like purchasing agents, tool and die designers, manufacturing engineers, and process planners, who need the design information. The link between computer-aided design (CAD) and computer-aided manufacturing (CAM) is particularly important, and often difficult to achieve.

A successful implementation of CAD–CAM is the Boeing 647 airliner. The upper and lower wing panel on this aircraft has stringers attached to it with 18,000 rivets each. The location of these rivets is extremely critical. More than 1000 pieces of input information are needed to specify the exact panel location on the wing. Wing panel assembly had been a problem because it was impossible to accurately visualize the results until the stringer was essentially attached. Using CAD–CAM allowed visualization of the entire panel prior to engineering release. It also greatly reduced error and improved fitup so well that on the first panel assembly the actual labor hours required were 72 percent less than originally estimated.

While commercially available database management systems (DBMS) were developed for business users they can be helpful in many engineering design situations. For example, a design engineer could ask the DBMS to list all steel beams in a structure where the static load is greater than an allowable maximum load. Commercially available DBMS can be used for such functions as keeping lists of vendors, managing literature references, keeping track of inventory or equipment, or developing a database of material properties. However, engineering database management has several important differences from a business database. Interaction with a business database usually deals with a complete database, one that contains essentially all of the data and does not undergo significant change. Engineering design does not deal with a complete database. A design database starts out nearly empty and is filled up as application programs, specifications and other constraints generate more data. Only when the design is finished is a complete database achieved.[1]

Spreadsheet programs are useful because of their ability to quickly make multiple calculations without requiring the user to reenter all of the data. Each combination of row and column in the spreadsheet matrix is called a cell. The quantity in each cell can represent either a number entered as input or a number that the spreadsheet program calculates according to a prescribed equation. The power of the spreadsheet is due to its ability to automatically recalculate results when new inputs have been entered in some cells. This can serve as a simple optimization tool as the values of one or two variables are changed and the impact on the output is readily observed. The usefulness of a spreadsheet in cost evaluations is self-evident. Most spreadsheets contain built-in mathematical functions, and it is possible to use them as an alternative to programming to solve problems in numerical analysis.[2]

The solution of equations with a spreadsheet requires that the equation have all input terms on one side of the equal sign. Therefore a class of equation-solving programs has been developed for the personal computer for small-scale computations. They are useful for varying parameters in a mathematical model or for solving equations numerically.[3] The best known examples are TK Solver, MathCAD, and Eureka.

All common higher-level languages, FORTRAN, BASIC, Pascal, and C are available on microcomputers. Rather than programming with one of these languages to use numerical methods to solve a mathematical model a current approach is to use a symbolic language that manipulates the symbols representing the equation. Formerly restricted to use on a mainframe, several symbolic computation programs are now available for computer workstations.[4] The spread of symbolic computation will have major impact on computer modeling for engineering design.

1-10 COMPUTER-AIDED DESIGN

The widespread use and decreasing cost of the digital computer have brought about a revolution in the practice of engineering design. There are two

---

[1] W. J. Rasdorf, *Computers in Mech. Engr.*, March 1987, pp. 62–69; D. N. Chorafas and S. J. Legg, "The Engineering Database," Butterworths, Boston, 1988.

[2] M. E. Palmer, M. G. Pecht, and J. V. Horan, *Computers in Mech. Engr.*, Sept. 1985, pp. 49–56.

[3] K. R. Foster, *Science*, June 3, 1988, pp. 1353–1358.

[4] K. R. Foster and H. H. Bau, *Science*, Feb. 3, 1989, pp. 679–684.

different aspects to this fantastic change in design practice.

1. Through on-line interaction with the computer in real time, the designer is able to utilize the computer and its graphics input-output devices to perform many of the routine aspects of design at far greater speed and lower cost. For example, the designer is able to draw objects on a graphics display terminal and, by utilizing computer software, portray the object in a three-dimensional view, an oblique view, or in any cross section.

2. By employing computer software codes based on the finite-element method,[1] the designer is able to perform powerful analytical procedures. The actual structural members under analysis can be displayed graphically. Graphical simulation, such as how a structure deforms under load, can be observed. The interactive mode of communication with the computer through the graphics terminal permits easy iteration procedures and design optimization.

A typical computer-aided design system consists of a central processing unit (CPU), magnetic tape or disk memory for storage of design data, a plotter for producing computer output on paper, and an interactive graphics terminal. A storage graphics terminal (CRT), which can display information on the screen for substantial periods of time, is a very versatile device. The designer may "draw" his part on the CRT by moving a cursor over the screen with a light pen or joystick and using a keyboard to call up specific geometrical shapes stored in the computer memory. Alternatively, graphic data may be input to the computer with a *digitizer tablet* that converts line drawings into digital data for analysis, storage, and computation.

The input is displayed instantaneously on the CRT. It may be modified as desired. By using software subroutines, orthographic views, or three-dimensional views, enlarged, rotated, or mirrored views can be generated. Any design change made on one view is automatically added to the others. By using a storage CRT, the designer can zoom in on a specific detail and enlarge it. Symbols for complex electrical, piping, or instrumentation diagrams are available. Other subroutines calculate and display such properties as curvature, slope, principal axis of stress, centroids, weight per linear foot, surface area, and volume. Even more complex software generates the tapes for operating a numerically controlled (N/C) machine tool that produces the part designed on the computer. The linking of computer-aided design and computer-aided manufacturing is called CAD–CAM. The output of computer-aided design

---

[1] O. Zienkiewicz, "The Finite Element Method," 3d ed., McGraw-Hill, New York, 1977; R. H. Gallagher, "Finite Element Analysis," Prentice-Hall, Inc., Englewood Cliffs, N.J., 1974: K. H. Huebner, "The Finite Element Method for Engineers," John Wiley & Sons, Inc., New York, 1975.

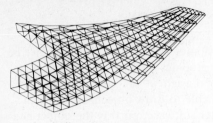

First vibration mode   Second mode

FIGURE 1-19
Vibration modes of aircraft wing using FEM.

may be stored for subsequent processing, or it may be plotted out on paper to produce a working engineering drawing.[1]

Until the middle 1960s, stress analysis in relation to design was fairly limited in scope. Although civil engineers could routinely analyze trusses, columns, and beams because those geometries could be handled with simple theory, mechanical engineers, who dealt in more complex three-dimensional bodies, designed by simplified analysis supplemented by experimental testing using strain gage and brittle coating techniques.[2] The development of the finite-element method (FEM) coupled with computer analysis created a new and powerful tool for the analysis of engineering problems. Now the analysis can be performed on a complex shape with the actual loads rather than use a simplified geometry and/or loads for which a solution is available.

In FEM modeling, the structure is represented by a mesh of elements interconnected at node points. Since these elements can be put together in virtually any fashion, they can be arranged to simulate complex shapes. The coordinates of nodes are combined with the elastic properties of the material to produce a stiffness matrix. This matrix is combined with the applied loads to determine the deflections at the nodes, and thus the stresses are determined. The analysis results in thousands of linear simultaneous equations. Therefore, this approach required the computational capability of the digial computer before it could become a reality.

The finite-element method originally was developed for the calculation of stress, but the matrix approach is also successful for problems in vibration, acoustics, fluid dynamics, and heat flow. Figure 1-19 shows the finite-element

[1] A. J. Medland," The Computer-Based Design Process," Kogan Page Ltd, London, 1986; Y. C. Pao, "Elements of Computer-Aided Design and Manufacturing," John Wiley & Sons, New York, 1984; S. A. Meguid, "Integrated Computer-aided Design of Mechanical Systems," Elsevier Applied Science Publ., New York, 1987.

[2] J. W. Dally and W. F. Riley, "Experimental Stress Analysis," 2d ed., McGraw-Hill Book Company, New York, 1978.

model of an aircraft wing and the distortion of this wing under modes of vibration.

The finite-element method was first applied in the aircraft and nuclear industries, where the complexity and importance of the problems could justify the large expenditure in computer time and engineering effort. Continued development has increased FEM capability and decreased its cost. The FEM is widely used for design in the automotive, construction and farm equipment, and machine tool industries. Some of the major developments[1] which have resulted in the widespread diffusion of the FEM have been:

1. Use of digitizing tablets and interactive computer displays to make it easier to input the location of grid points.

2. Automatic mesh generation techniques to assist in model building.

3. Development of isoparametric elements with curved sides and nodes at midwall that appreciably reduce the number of elements needed for a solution. Such a reduction not only makes the model simpler and easier to understand but also reduces modeling time and computer costs.

4. The ability to combine empirical test data in FEM models.

5. The growth of FEM consultants and computer programs to assist beginners in applying FEM to their problems.

The application of FEM to the complex problem of a truck frame is illustrated in Fig. 1-20. The first step in applying FEM is to construct a stick-figure or beam model. Considerable experience is required to construct a model that represents the structure accurately. The purpose of the model is to determine the overall structure deflections and identify the areas where large stresses are likely to occur. Once the critical locations are identified, a detailed FEM model is constructed for those parts of the structure. A "plate," or two-dimensional, model is simpler, but it may not be as realistic as a "solid," or three-dimensional, model. The output can consist of a computer-generated drawing of the part with the stress contours plotted on it .

1-11 DESIGNING TO STANDARD

The use of standards provides a definitive solution to a repetitive problem with the best technical means available at the time. There is a definite economy in time spent searching for solutions and a substantial cost saving in inventory and manufacturing. Published standards take many forms. They may describe standard dimensions and sizes of small parts like screws and bearings, the

---

[1] J. K. Krouse, *Machine Design*, pp. 98–103, Jan. 12, 1978.

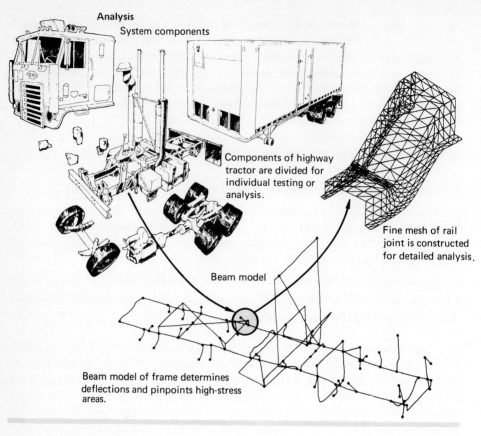

Analysis

System components

Components of highway tractor are divided for individual testing or analysis.

Fine mesh of rail joint is constructed for detailed analysis.

Beam model

Beam model of frame determines deflections and pinpoints high-stress areas.

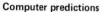

Computer predictions

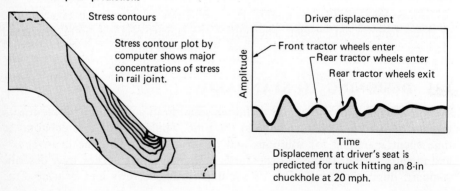

Stress contours

Stress contour plot by computer shows major concentrations of stress in rail joint.

Driver displacement

Amplitude

Front tractor wheels enter
Rear tractor wheels enter
Rear tractor wheels exit

Time

Displacement at driver's seat is predicted for truck hitting an 8-in chuckhole at 20 mph.

FIGURE 1-20
Example of application of FEM in design. (*From* Machine Design.)

minimum properties of materials, or the allowable discharge of pollutants. Other standards, usually called *design codes,* prescribe methods of analysis and calculation for certain routine design problems. The reader is referred to Sec. 14-8 for a fuller discussion of the procedures for establishing standards and sources of standards.

The engineering design process is concerned with balancing four goals: proper function, optimum performance, adequate reliability, and low cost. The greatest cost saving comes from reusing existing parts in design. The main savings come from eliminating the need for new tooling in production and from a significant reduction in the parts that must be stocked to provide service over the lifetime of the product. In much of new product design only 20 percent of the parts are new, about 40 percent are existing parts used with minor modification, while the other 40 percent are existing parts reused without modification.

Computer-aided design has much to offer in design standardization. A 3D model represents a complete mathematical representation of a part which can be readily modified with little design labor. It is a simple task to make drawings of families of parts which are closely related.

A formal way of recognizing and exploiting similarities in design is through the use of *group technology* (GT). GT is based on similarities in geometrical shape and/or similarities in their manufacturing process. Coding and classification systems[1] are used to identify and understand part similarities. A computerized GT database makes it possible to easily and quickly retrieve designs of existing parts that are similar to the part being designed. This helps combat the tendency toward part proliferation which is encouraged by the ease of use of a CAD system. The installation of a GT system aids in uncovering duplicative designs; it is a strong driver for part standardization. GT may also be used to create standardization in part features. For example, the GT database may reveal that certain hole diameters are used frequently in a certain range of parts while others are infrequently used. By standardizing on the more frequently used design features simplifications and cost savings in tooling can be achieved. Finally the information on manufacturing costs should be fed back to the designer so that high cost design features are avoided.

An important aspect of standardization in CAD–CAM is in interfacing and commmunicating information between various computer devices and manufacturing machines. The National Bureau of Standards has been instrumental in promulgating the Initial Graphics Exchange Specification (IGES). This is being replaced by the Product Data Exchange Specification (PDES). Both of these represent a neutral data format for transferring geometric data between equipment from different vendors of CAD systems. In

---

[1] W. F. Hyde, "Improving Productivity by Classification, Coding, and Data Base Standardization," Marcel Dekker Inc, New York, 1981.

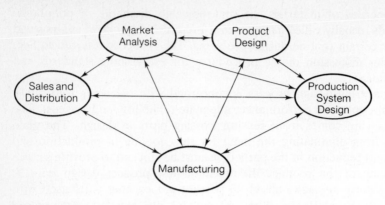

FIGURE 1-21
Model of concurrent engineering.

a similar vein, the Manufacturing Automation Protocol (MAP) proposed by General Motors is a common standard for communicating data between various computer systems in factory floor automation.

1-12 CONCURRENT ENGINEERING

Concurrent engineering,[1] also called simultaneous engineering, is a design approach in which product design and product manufacturing are merged in an intimate way. It is a response in the United States to greatly increased overseas competitiveness in design and manufacturing. It is a recognition that many of the high costs of manufacturing are designed in at the product development stage.

Actually the concurrent engineering concept is broader than simply integrating design and manufacturing at the start of the design process. It really involves the model shown in Fig. 1-21 in which the people responsible for market analysis (see Sec. 2-9) are intimately involved as are those concerned with distribution and sales. An important aspects is to incorporate the idea of designing for a lifetime of use, i.e. life cycle engineering (see Sec. 2-13).

The reason that manufacturing goals and requirements must be considered at the very beginning of a project is to minimize the costs of making changes in design late in the product development cycle. There is approximately a ten-fold increase in the cost of making an engineering change on going from R&D to design to production to use-after-sales. Thus, a major goal

---

[1] J. L. Nevins and D. E. Whitney (eds.), "Concurrent Design of Products and Processes," McGraw-Hill Publ. Co., New York, 1989.

of concurrent engineering is to move engineering changes back into the early stages of design. The importance of this goal is reinforced by the fact that it is widely held that 80 percent or more of manufacturing decisions are directly determined by the product design. Also, early design decisions affect life cycle cost in very significant ways. All the process optimization in the world cannot make up in cost savings for careless design decisions in areas like materials selection, selection of fasteners, ease of assembly, etc. In an era of increasing automation with related high capital costs it is not unreasonable to find that products must be designed to fit the factory as much or more than the factory is designed to fit the product.

The benefits of concurrent engineering extend well beyond achieving reduced manufacturing cost. In a product that has been designed for ease of manufacture there is less risk of deviating from design intent and causing quality defects. By investing more engineering effort in the early stage of design to develop more complete design information, fewer engineering changes are required downstream toward production and sales. Not only does this save money but it most likely will reduce the time to introduce the product in the market. Generally, the first quality products to enter a new market obtain the largest share of the market and an even larger share of the total profits generated by that class of product. In recent years much of the success of overseas competitors has been attributed to a rapid product development cycle through the use of concurrent engineering.[1]

The difficulty of integrating product and process design in large corporations is due to the fact that historically product design has been done serially. Product concept, product design, and product testing have been done prior to manufacturing system design, process planning, and production. Commonly these serial functions are carried out in distinct and separate organizations with little interaction. Thus, it is easy to see how the design team will make decisions, many of which are irreversible, without having adequate knowledge of the capabilities of the manufacturing process. The result is suboptimal design of each part of the system. To correct this situation requires the adoption of the concurrent engineering concept. The development of a team with representation of all the needed viewpoints and knowledge base may run counter to existing management structures, but a number of corporations have been able to adopt concurrent engineering with great success. Since many products have a large fraction of the components made by outside vendors, it is important that these suppliers be made a part of the product development team. Because it utilizes a multidisciplinary design team concurrent engineering places a higher requirement for communication skills for all participants (see Chap. 15).

---

[1] A study by the consulting firm of McKinsey & Co. determined that going 50 percent over budget during development to get a product out on time reduced total profit by only 4 percent, while staying on budget and getting the product to the market six months late reduced profits by a third.

1-13 DESIGN REVIEW

The design review is a vital aspect of the design process. It provides an opportunity for specialists from different disciplines to interact with generalists to ask critical questions and exchange vital information. A design review is a retrospective study of the design up to that point in time. It provides a systematic method for identifying problems with the design, aids in determining possible courses of action, and initiates action to correct the problem areas.

To accomplish these objectives the review team should consist of representatives from design, manufacturing, marketing, purchasing, quality control, reliability engineering, and field service. The chairman of the review team is normally a chief engineer or project manager with broad technical background and broad knowledge of the company's products. In order to ensure freedom from bias the chairman of the design review team should not have direct responsibility for the design under review.

Depending on the size and criticality of the project, full scale design reviews should be held at three or four times in the life of the project. The first review should be held when concept feasibility has been established. The problem definition and initial specifications should be critically examined with respect to the needs of the marketplace. If the feasibility study has been done by the research laboratory, then this review is especially critical to pass on information to the engineering group. Sometimes an intermediate design review is conducted before the detail drawings have been completed. This review would look critically at the interfaces between the specialty design teams, e.g. mechanical, electronics, materials, and begin to discuss tooling and packaging. A design review after the detail drawings are complete will ensure that the design is ready for prototype testing. A review after the completion of prototype testing is critical to ensure that there are no loose ends to the project. The purpose of this review is to fine-tune the design prior to authorizing full scale production. This review focuses on achieving the performance, producibility, cost, and reliability goals. There may also need to be a final acceptance review prior to handing the project over to the customer.

It is helpful to prepare a checklist for the design review. The major headings should consist of:

1. Design requirements
2. Functional requirements
3. Environmental requirements
4. Manufacturing requirements
5. Operational requirements
6. Reliability-related requirements

For each item under these headings answer yes or no as to whether the condition has been fulfilled by the design.

Redesign

A common situation is redesign. As a result of decisions made at design reviews the details of the design are changed many times as prototypes are developed and tested. There are two categories of redesigns: *fixes* and *updates*. A fix is a design modification that is required due to less than perfect performance once the product has been introduced into the marketplace. On the other hand, updates are usually planned as part of the product's life cycle before the product is introduced to the market. An update may add capacity and improve performance to the product, or improve its appearance to keep it competitive.

BIBLIOGRAPHY

Burgess, J. A.: "Design Assurance for Engineers and Managers," Marcel Dekker Inc., New York, 1984.
Glegg, G. L.: "The Design of Design," Cambridge University Press, New York, 1969.
Hill, P. H.: "The Science of Engineering Design," Holt, Rinehart and Winston Inc., New York, 1970.
Hubka, V.: "Principles of Engineering Design," Butterworth & Co, London, 1982.
Leech, D. J., and B. T. Turner: "Engineering Design for Profit," John Wiley & Sons, New York, 1985.
Lewis, W. P., and A. E. Samuels: "Fundamentals of Engineering Design," Prentice-Hall Inc., Englewood Cliffs N.J., 1990.
Middendorf, W. H.: "Design of Devices and Systems," 2d ed., Marckel Dekker Inc., New York, 1989.
Pahl, G., and W. Beitz: "Engineering Design," Springer-Verlag, New York, 1984.
Ray, M. S.: "Elements of Engineering Design," Prentice-Hall Inc., Englewood Cliffs N.J., 1985.
Woodson, T. T.: "Introduction to Engineering Design," McGraw-Hill Book Co., New York, 1966.

CHAPTER
2

DESIGN
IN A
BROADER
CONTEXT

2-1 SPECTRUM OF ENGINEERING ACTIVITIES

Engineering is sometimes defined as the application of science to the solution of a problem of society at a profit. The accrediting agency for U.S. engineering curricula, the Accreditation Board for Engineering and Technology[1] (ABET), prepared the following more formal definition:

> Engineering is the profession in which a knowledge of the mathematical and natural sciences gained by study, experience, and practice is applied with judgment to develop ways to utilize, economically, the materials and forces of nature for the benefit of mankind.

Both of these definitions explain that engineering is based on mathematics and science but is focused on the solutions of specific problems. Also, the solutions, because they are needed or desired by society, must meet certain legal, environmental, and economic constraints. Thus, engineering practice is

---

[1] Formerly known as the Engineers' Council for Professional Development.

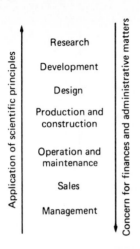

FIGURE 2-1
Spectrum of engineering functions.

highly interactive and constrained, which also means that it is very challenging. Because engineering is so interwoven in the fabric of modern society, we find that engineers engage in a very broad spectrum of activities. At the core of engineering practice is the process of design.

A simple grouping of engineering practice into its various functions is given in Fig. 2-1. We start with research, which is closest to the academic experience; and as we progress downward in the hierarchy, we find that more emphasis on the job is given to financial and administrative matters and less emphasis is given to strictly technical matters.

Research and development (R&D) often is considered the glamor end of the engineering spectrum. We tend to think of research as limited to a scientist isolated in a laboratory and motivated by curiosity to explore nature. But that is basic research of a very pure type. Actually, R&D itself covers quite a spectrum of effort. The Department of Defense (DOD), which sponsors a very large amount of R&D, classifies its activities as follows:

6.1 Research
6.2 Exploratory development
6.3 Advanced development
6.4 Engineering development
6.5 Management and support

With downward progression in this classification, the work becomes more directed (applied) to a specific objective.

The terminology *basic and applied research* is not very precise. Figure 2-2 is an attempt to distinguish clearly between basic research, applied research, and development on the basis of five criteria: the constituency that will authorize the research, the relation of the research to the disciplines and to specific problems, and the input and output of the research.

	Basic Research	Applied Research	Development
Constituency (program direction, proposal review, advisory committees, etc.	Researchers	Mostly researchers; some users	Some researchers; mostly users
Relation to disciplines	Discipline oriented, especially basic sciences, e.g. physics, psychology	Discipline oriented, especially applied disciplines, e.g. engineering	Often multidisciplinary
Relation to specific problems	None; extend knowledge in discipline	Some, especially to broad classes of problems	Primary; problem-focused
Inputs (What determines the activity?)	Perceive knowledge gaps; join other researchers in the discipline in current pursuits	Perceive new applications; fill in missing science; perceive new techniques of analysis and design	Perceive societal need or problem
Outputs (measures of success)	Publications and citations in scientific literature	Publications and citations in professional literature	Measurable impact on problem or need; people doing something differently

FIGURE 2-2
Classification of research and development. (*Adapted from National Science Foundation report.*)

Design is the function of creating a technical plan to solve a problem. It involves both analysis and synthesis. As you have seen from Chap. 1, it involves much more than calculations on a computer and lines on a drawing. It also involves careful planning of how the thing designed will be manufactured and how it will be maintained.

The production and construction function translates the design into a manufacturing process or a process plant or civil engineering structure. Engineers working in this functional area usually are responsible for taking the design through initial production to prove that the product can be manufactured within planned cost. They must be generalists who can work with many experts and who can organize many diverse details and work to schedule and cost. It is important to realize that the design department is not the only place where designing is done. Important preliminary designing may be done in the development stage and in the production stage of a project.

The operations function produces the product for which the plant was designed. An engineer is often a factory manager responsible for supervising the plant's work force, materials flow, and machines. An important aspect of the job is meeting production schedules while maintaining product quality standards. There is also an important plant engineering function of providing plant utility services, maintenance, and modification.

Sales engineering is technical marketing. In a technology-based industry, sales engineers are the liaison between the production facilities and the user. They work closely with customers to recognize customer needs and then recommend the products and systems that best fill those needs. A special group of sales engineers, often called application engineers, are more technically based; they attempt to feed back customer needs to improve the product or expand its level of acceptance. Field engineers supervise the installation and service of large or highly complex equipment systems such as rolling mills and microwave communications systems.

Management is a logical career progression for an engineer. The graduate engineer is thrust into a management situation almost immediately through an engineering project assignment that requires the supervision of drafting personnel, technicians, and other engineers. Well over 50 percent of engineers have some management responsibility, either as managers of production workers or other engineers or in general management. Because of the increasing technical nature of business, engineers are finding their way to the top corporate executive levels in increasing numbers.

2-2 ORGANIZATION OF THE ENGINEERING FUNCTION

Having considered the spectrum of engineering functions, we turn to a brief look at how engineers are organized within an industrial corporation. There are three general types of organizational structure:[1]

1. Functional organization
2. Project organization
3. Project-functional staff organization

In *functional organization* the functional specialities are grouped into separately identifiable organizational units. The theory behind the functional organization is that similar work should be performed within one organizational component. Figure 2-3 shows the organizational chart for a manufacturing company organized along functional lines. All research, development, and design work reports to a single vice president; all manufacturing activity is the

---

[1] B. S. Blanchard, "Engineering Organization and Management," pp. 228–237; Prentice-Hall, Inc., Englewood Cliffs, N.J., 1976; D. I. Cleland and D. F. Kocaoglu, "Engineering Management," McGraw-Hill Book Company, New York, 1981.

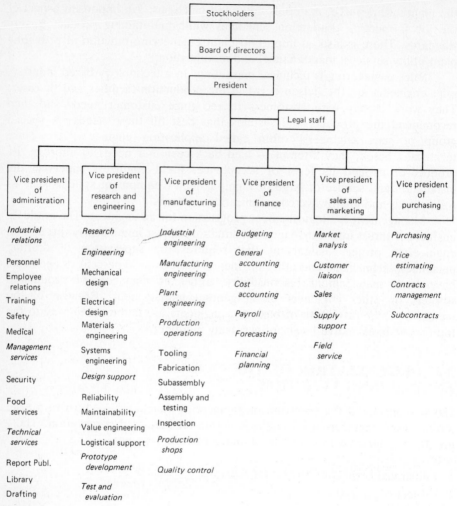

FIGURE 2-3
Example of a functional organization.

responsibility of another vice president; etc. It is important to take the time to read the many functions under each vice president that are needed even in a manufacturing enterprise that is modest in size.

In a *project organization* the functions are organized around a single product or major system. Thus, the various functions listed in Fig. 2-3 are redistributed, according to need, among a number of project managers. An example of project organization is shown in Fig. 2-4. An important feature is that the project organization is dynamic and time-limited. The life of a given project organization, e.g., project *Y*, is only long enough to accomplish the project's objectives; then the people are reassigned.

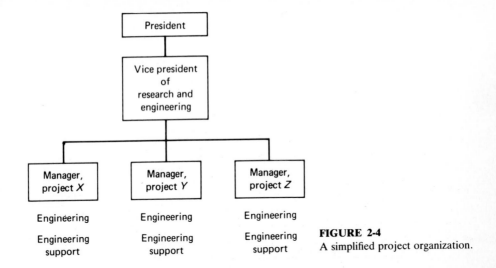

FIGURE 2-4
A simplified project organization.

In an organizational structure a distinction should be made between line and staff functions. A line organization represents a chain-of-command relationship wherein supervisors report to managers and so on up the organizational hierarchy. *Line functions* are those that are directly concerned with achieving the prime objectives of the organization. *Staff functions* support one or more different line organizations. They provide special expertise or information that is often vital to achieving the objectives of the line organization. However, since staff personnel are not in the chain of command, they do not have prime responsibility for achieving the major objectives of the organization.

The *project-functional staff organization* is the result of merging the traditional functional organization and project team (Fig. 2-5). Because such an organization consists of a project team superimposed on the existing vertical functional structure of the organization, it usually is called a *matrix organization*.

In a matrix organization the engineering personnel assigned to the project design teams come from the functional disciplines. A project manager is responsible for planning, organizing, motivating, directing, and controlling the resources that will be applied to the project. To do so, the manager works with the managers of functional groups such as electrical design. The matrix organization produces an interplay of checks and balances (or controlled conflict) between the managers who have specific program responsibility and those who have specific functional responsibility. The project manager is responsible for the completion of the project on time and within budget and for the attainment of the goals set forth for the project. The functional manager is responsible for providing specialized resources to accomplish the project. Generally, the functional manager retains the administration of the people who are assigned to him but who work on the project.

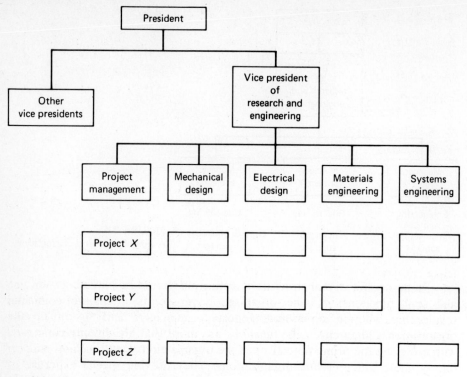

FIGURE 2-5
Project-functional staff (matrix) organization.

A functional form of organization generally is used where well-established, standard products are being produced. Coordination between functional departments occurs at the presidential level or between the vice presidents. Innovation and marked shifts in product line may be difficult to achieve. The project management organization provides a means of focusing resources around a specific task to be accomplished. It is frequently used on large, complex construction projects. Sometimes a functional organization will set up one or more project teams to develop a new product line or accomplish a specific task that is ordinarily difficult to fit into a functional organization. The matrix organizational structure originated in the aerospace industry, where there is a need for highly interdisciplinary problem solving but also a need for sophisticated scheduling and project management. Many corporations have adopted various forms of the matrix concept to accomplish their specific objectives.[1]

---

[1] D. I. Cleland and D. F. Kocaoglu, op. cit., pp. 36–40.

2-3 THE ENGINEERING PROFESSION

Engineering generally is recognized as a profession, along with medicine, dentistry, law, ministry, architecture, and education. Three attributes make an occupation a profession:

1. It involves intellectual effort.
2. It calls for creative thinking.
3. It is motivated by a desire for service.

Collectively, a group of people form a true profession only as long as they command the respect of the public and inspire confidence in their integrity and a belief that they are serving the general welfare.

The nature of professional service varies widely. The physician, lawyer, and clergyman have direct, individual relations with their clients, but an engineer usually is salaried in someone else's employ. About 98 percent of engineers work for either industry or government, and only a small, but important, percentage is in direct contact with the public as consulting engineers. Thus, the service aspect of engineering may be less obvious to the general public than in other professions. Countering that aspect is the growing influence of high technology on the everyday life of people. Thus, when there is an incident at a nuclear power plant or a fleet of commercial aircraft is grounded because of an unexplained metallurgical failure, the entire population is made aware of the professional contribution of engineers. And although there has been some vocal opposition to high technology in recent years, the general public is overwhelmingly in favor of the higher living standards and the creature comforts made possible by technology. Engineers generally get high marks in public opinion polls aimed at determining the esteem in which the various professions are held.

The number of engineers in the United States is variously reported to be between 1,200,000 and 2,500,000. The lack of precision points up one important characteristic of the profession. Engineering is a very open profession; it has no clearly defined educational requirements, and there is no generally accepted definition of engineer. In contrast, medicine requires four years of study beyond the B.S. degree, plus an internship for professional practice. Engineering has no specified educational requirement, and admission to the profession is determined by what a person knows and can do. Practically speaking, however, most people entering engineering since World War II have had a B.S. degree, and there is a strong trend toward education to the M.S. level.

In the eyes of the lay public, if not some employers, the distinction between the role played by the engineer and the roles of others who are occupied in technology (Fig. 2-6) is not always clear. The engineer makes greater use of mathematics and science in the solution of problems than does

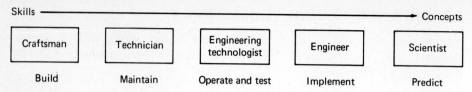

FIGURE 2-6
Spectrum of technological occupations.

the engineering technology graduate. Each utilizes the skilled technician and craftsman to translate engineering ideas into something concrete.

A necessary preliminary to the practice of most professions is licensing by or registration with the state. The purpose of licensing is to assure that those who practice a profession that affects life, health, rights, and property are competent. All 50 states have enacted laws for the registration of professional engineers (PE), and only a person registered by a state has the legal right to use the PE title. Professional registration is required for private practice and the design of public works such as bridges, highways, and buildings. But because of a variety of exemptions in the laws, usually known as the industry exemptions, a PE license is not required for the design of the vast majority of products manufactured by industry. Thus, fewer than 25 percent of practicing engineers are registered.

The growing trend toward public and private accountability and the increased attention to consumer protection and product liability are making professional registration more important. Periodically, attempts are made to eliminate the industry exemption in the state registration laws. Also, midcareer job changes may result in the need for professional registration, which may then be difficult to accomplish after many years away from study and examination. These are some of the reasons why the engineering senior should seriously consider obtaining professional registration.

Although the registration laws vary among the states, there is substantial commonality in the basic steps for obtaining professional registration. The first part of the examination is the fundamentals of engineering examination to become an engineer-in-training (EIT). The graduates of an ABET approved program may take the EIT examination prior to graduation, whereas up to twelve years of professional experience is required for individuals who do not have a degree. Following attainment of EIT status, graduates of an ABET approved program must obtain four years of engineering experience, preferably under the supervision of a professional engineer, to qualify for the second part, principles of practice of engineering. Satisfactory completion of part two leads to professional registration.

Since a true professional is concerned with competence, some professions have introduced the idea of *certification* as a way to protect the public and police the profession. Certification generally is administered by a professional or technical society rather than the state. It provides a mechanism for peer recognition of competency in a special field, usually by passing an examination or demonstrating competence through education and experience. Most certification programs require recertification at specified intervals to demonstrate continuing competence.

The organization of the engineering profession is diverse and highly decentralized. Engineering has broken down into over 100 specialized professional groups. There is no single, powerful organization, such as the American Medical Association, that can speak for the entire engineering profession with an authoritative voice. As a result, engineering has suffered both in its public image and in its dealing with Congress and the federal bureaucracy. A major step toward correcting this situation was the formation in December 1979 of the American Association of Engineering Societies (AAES). AAES is an umbrella over 22 engineering organizations that represent nearly a million engineers. In its membership are all of the major engineering professional and technical societies. AAES has as its objectives to:

Advance the practice of engineering in the public interest

Provide for regular and orderly communications among its member organizations

Organize and conduct forums for the consideration of issues of interest to member organizations

Identify needs and opportunities for service in the engineering community

Recommend programs of studies and research to engineering organizations

Undertake activities that a member organization acting individually could not accomplish as well

Foster interactions of the engineering community with other segments of our culture

To accomplish those objectives, the AAES is organized into four councils:

Public Affairs Council. Responsible for broad policy coordination and technical inputs for use by the public and member societies in interactions with all branches of government.

Engineering Affairs Council. Responsible for matters concerned with the practice of engineering in the public interest. There are committees dealing with employment guidelines, pensions, ethics, public awareness, and engineering manpower statistics.

Educational Affairs Council. Concerned with promotion and advancement of engineering education.

International Affairs Council. Responsible for international activities such as cooperation with engineering organizations of other countries.

The technical side of engineering is divided into professional-technical societies devoted to particular engineering disciplines and broader, interdisciplinary technical societies.

Society	Year organized
American Society of Civil Engineers (ASCE)	1852
American Institute of Mining, Metallurgical & Petroleum Engineers (AIME)	1871
American Society of Mechanical Engineers (ASME)	1880
Institute of Electrical and Electronics Engineers (IEEE)	1884
American Institute of Chemical Engineers (AIChE)	1908

The oldest professional societies are the five founder societies listed in the accompanying table. As technology advanced, other specialized societies were established for agricultural engineers (ASAE, 1907), metals engineers (ASM, 1913), aeronautical engineers (AIAA, 1932), manufacturing engineers (SME, 1932), ceramic engineers (NICE, 1938), plastics engineers (SPE, 1941), industrial engineers (IIE, 1948), and nuclear engineers (ANS, 1954).

Some of the more important interdisciplinary societies are those listed in the accompanying table. They are primarily concerned with advancing the profession by holding technical meetings for the interchange of information and by publishing technical journals. They also honor professional accomplishments. Since the societies generally are tax exempt under Section 501C of the Internal Revenue Act, they are prohibited from lobbying and other political acts.

Society	Year organized
American Association of Cost Engineers (AACE)	1956
American Institute of Plant Engineers (AIPE)	1954
American Society for Engineering Education (ASEE)	1893
American Society for Nondestructive Testing (ASNT)	1941
American Society for Quality Control (ASQC)	1946
American Society of Heating, Refrigerating and Air Conditioning (ASHRAE)	1894
American Society of Lubrication Engineers (ASLE)	1944
American Society for Testing and Materials (ASTM)	1898
Instrument Society of America (ISA)	1946
National Association of Corrosion Engineers (NACE)	1945
Operations Research Society of America (ORSA)	1952
Society for Experimental Mechanics (SEM)	1943
Society of Automotive Engineers (SAE)	1905

Important guidance for the engineering profession and coordination with the federal government for the solution of societal problems is provided by the National Academy of Engineering (NAE). The NAE is a private honorary organization of engineers who are selected for membership for reason of their contributions to engineering. Election to the NAE is the highest honor that can be bestowed on a U.S. engineer. Established in 1964, the NAE is autonomous in organization and selection of members, but it coordinates closely with the older and larger National Academy of Sciences. The National Research Council is the operating arm of the combined academies. Its full-time staff is supplemented by thousands of volunteer engineers and scientists who serve on boards, committees, and panels directed at advising the government on engineering and scientific issues of concern to the nation.

The National Research Council is organized into eight assemblies and commissions.

- Assembly of Behavioral and Social Sciences
- Assembly of Engineering
- Assembly of Life Sciences
- Assembly of Mathematical and Physical Sciences
- Commission on Human Resources
- Commission on International Relations
- Commission on Natural Resources
- Commission on Sociotechnical Systems

Some of the components of these assemblies and commissions that are of special interest are the National Materials Advisory Board (NMAB), the Transportation Research Board, the Building Research Advisory Board, and the Manufacturing Studies Board.

2-4 ETHICS IN ENGINEERING

Ethics are the principles of conduct that govern an individual or a profession. They provide the framework of the rules of behavior that are moral, fair, and proper for a true professional. A code of ethics serves to remind individuals how important integrity is in a self-regulating profession. It lays out the issues that are deemed most important in the collective wisdom of the profession. By publishing and enforcing a code of ethics, the profession serves notice that its ethical precepts are to be taken seriously. Obviously, engineers should respect the same fundamental ethical principles in the context of the special expertise and public trust of the engineering profession.

The peculiarities of the engineering profession as compared with the professions of law and medicine that were discussed in a preceding section carry over into the area of ethics. Because engineering lacks the homogeneous character of such professions as law and medicine, it is not surprising to find that there is no widely accepted code of engineering ethics. Most professional societies have adopted their own codes, and ABET and NSPE have adopted broader-based but not universally accepted ethical codes. Again, because engineers who are employees of either business or government are in the great majority, they face ethical problems that self-employed professionals avoid. These arise from the conflict between the engineer's desire to gain a maximum profit for the employer (and thus achieve recognition and promotion) and the desire to adhere to a standard of ethics that places the public welfare ahead of corporate profit. For example, what can an employed engineer[1] do to expose and correct the corrupt practices of an employer? What should an engineer do if employed in a business atmosphere in which kick-backs and bribes are an accepted practice?

The code of ethics adopted by the Institute of Electrical and Electronics Engineers (IEEE) is a good example of a modern engineering code of ethics; see Fig. 2-7.

The published codes of ethics frequently prove to be of little value in the gray areas of ethics. However, ethics can be learned like any other subject. Ethics is based on moral philosophy. While it is not possible to develop a single comprehensive moral principle that can provide guidance in solving all ethical dilemmas, it is possible to lay down a set of moral values for personal behavior:

Respect the rights of others

Be fair

Do not lie or cheat

Keep promises and contracts

Avoid harming others

Prevent harm coming to others

Help others in need

Obey all laws

Ethical theory considers two extreme types of behavior. *Altruism* is a form of moral behavior in which individuals act for the sake of other people's interests. Ethical altruism is the view that individuals ought to act with each others' interests in mind. This is the viewpoint best summarized by the Golden Rule: Do unto others as you would have others do unto you. *Egoism* is a form of moral behavior in which individuals act for their own advantage. Ethical egoism is the view that individuals ought always to act to satisfy their own interests.[2] Most day-to-day practice of engineering is done in the individual's

---

[1] T. S. Perry, *IEEE Spectrum*, pp. 56–61, September 1981.

[2] R. H. McCuen, *Issues in Engineering—Jnl. of Prof. Activities*, ASCE, vol. 107, no. E12, pp. 111–120, April 1981.

self-interest and is not in conflict with the codes of ethics. However, the codes of ethics are meant to alert the practicing professional that he or she has altruistic obligations that must be properly balanced with self-interest.

An engineer in a design situation potentially can be faced with a miriad of ethical situations. Consider the following short list.

Specification of components or materials can lead to conflict of interest for certain suppliers

Authorizing the release of production parts that are only marginally out of specification

Condoning the use of pirated design software

Firing a hardworking veteran employee to make room for a star who suddenly becomes available

Ethical situations in engineering generally fall within the following general situations.

Loyalties between two groups

Duty to general public versus duty to employer

Confidentiality

Rights of personal conscience versus rights of employer

Issues of recognition for performance

Falsification of experimental data

Issues of differing standards of morals in different parts of the world

Consider the following example. A newly graduated engineer is working for a consulting firm that is engaged in research for the Department of Defense. The company is preparing a proposal to DOD and will use his résumé as part of the proposal. The engineer is in the room where report copying is done and happens to look at his résumé as the proposal is being assembled. To his surprise he observes that someone has grossly modified his résumé, listing professional experience that he does not have. What should he do?

He immediately tells his superior about the "mistakes" in his résumé, believing at the time that it was an error. The superior states that he will take care of it, but a week later the young engineer finds out that the falsified résumé had been included in the proposal. He talks again with his superior, but receives an evasive answer. What should he do? Should he resign immediately, or stick around and hope this was an isolated incident? Should he find other employment and then report the situation to the DOD agency that received the proposal, or should he just forget about it and get on with his career with a new employer?

An important ethical situation which periodically attracts wide attention

IEEE Code of Ethics

Preamble

Engineers, scientists and technologists affect the quality of life for all people in our complex technological society. In the pursuit of their profession, therefore, it is vital that IEEE members conduct their work in an ethical manner so that they merit the confidence of colleagues, employers, clients and the public. This IEEE Code of Ethics represents such a standard of professional conduct for IEEE members in the discharge of their responsibilities to employers, to clients, to the community and to their colleagues in this Institute and other professional societies.

Article I

Members shall maintain high standards of diligence, creativity and productivity, and shall:

1. Accept responsibility for their actions;
2. Be honest and realistic in stating claims or estimates from available data;
3. Undertake technological tasks and accept responsibility only if qualified by training or experience, or after full disclosure to their employers or clients of pertinent qualifications;
4. Maintain their professional skills at the level of the state of the art, and recognize the importance of current events in their work;
5. Advance the integrity and prestige of the profession by practicing in a dignified manner and for adequate compensation.

Article II

Members shall, in their work:

1. Treat fairly all colleagues and co-workers, regardless of race, religion, sex, age or national origin;
2. Report, publish and disseminate freely information to others, subject to legal and proprietary restraints;

3. Encourage colleagues and co-workers to act in accord with this Code and support them when they do so;
4. Seek, accept and offer honest criticism of work, and properly credit the contributions of others;
5. Support and participate in the activities of their professional societies;
6. Assist colleagues and co-workers in their professional development.

Article III

Members shall, in their relations with employers and clients:

1. Act as faithful agents or trustees for their employers or clients in professional and business matters, provided such actions conform with other parts of this Code;
2. Keep information on the business affairs or technical processes of an employer or client in confidence while employed, and later, until such information is properly released, provided such actions conform with other parts of this Code;
3. Inform their employers, clients, professional societies or public agencies or private agencies of which they are members or to which they may make presentations, of any circumstance that could lead to a conflict of interest;
4. Neither give nor accept, directly or indirectly, any gift, payment or service of more than nominal value to or from those having business relationships with their employers or clients;
5. Assist and advise their employers or clients in anticipating the possible consequences, direct and indirect, immediate or remote, of the projects, work or plans of which they have knowledge.

Article IV

Members shall, in fulfilling their responsibilities to the community:

1. Protect the safety, health and welfare of the public and speak out against abuses in these areas affecting the public interest;
2. Contribute professional advice, as appropriate, to civic, charitable or other nonprofit organizations;
3. Seek to extend public knowledge and appreciation of the profession and its achievements.

Approved February 18, 1979, by the Board of Directors of the Institute of Electrical and Electronics Engineers, Inc.

FIGURE 2-7
The IEEE Code of Ethics. (*Reprinted by permission.*)

is *whistleblowing*. Whistleblowing refers to making a public accusation about misconduct within an organization. In the usual case the charges are made by an employee or former employee who has been unable to obtain the attention of the organization's management to the problem. Sometimes whistleblowing is confined to within the organization where the whistleblower's supervision is bypassed in an appeal to higher management. An important issue is to determine the conditions under which engineers are justified in blowing the whistle. DeGeorge[1] suggests that it is morally permissible for engineers to engage in whistleblowing when the following conditions are met.

1. The harm that will be done by the product to the public is considerable and serious.
2. Concerns have been made known to their superiors, and getting no satisfaction from their immediate superiors, all channels have been exhausted within the corporation, including the board of directors.
3. The whistleblower must have documented evidence that would convince a reasonable impartial observer that his or her view of the situation is correct and the company position is wrong.
4. There must be strong evidence that releasing the information to the public would prevent the projected serious harm.

Clearly a person engaging in whistleblowing runs considerable risk of being labeled a nut or of being charged with disloyalty, and possibly being dismissed. The decision to blow the whistle requires great moral courage. Federal government employees have won protection under the Civil Service Reform Act of 1978, but protection under state laws or active support from the engineering professional societies is still spotty. Some far-sighted companies have established the office of ombudsman or an ethics review committe to head off and solve these problems internally before they reach the whistleblowing stage.

Solving Ethical Conflicts

In solving ethical conflicts it is important to distinguish between internal and external appeal.[2] Except under very unusual circumstances it is important that all internal options should be explored before seeking an appeal external to the organization. The various steps for solving ethical conflicts are listed as follows.

---

[1] R. T. De George, *Business and Prof. Ethics Jnl.*, vol. 1, no. 1, pp. 1–14, 1981.
[2] R. H. McCuen, Dept. of Civil Engineering, University of Maryland.

Procedure for Solving Ethical Conflicts

I. Internal Appeal Options
  A. Individual Preparation
    1. Maintain a record of events and details
    2. Examine the company's internal appeals process
    3. Know federal and state laws
    4. Identify alternative courses of action
    5. Specify the outcome that you expect the appeal to accomplish
  B. Communicate with Immediate Supervisor
    1. Initiate informal discussion
    2. Make a formal appeal
    3. Indicate that you intend to begin the company's internal process of appeal
  C. Initiate Appeal through the Internal Chain of Command
    1. Maintain formal communication appeal
    2. Formally inform the company that you intend to pursue an external solution
II. External Appeal Options
  A. Personal Options
    1. Engage legal counsel
    2. Contact your professional society
  B. Communicate with client
  C. Contact the media

Once the individual has studied and documented the facts and formulated a plan for appeal, the matter should be discussed with his or her immediate supervisor. Remember that in developing your position it is important to recognize the moral rights of the company, the profession, and society, and not just express concern for the impact of the issue on yourself. Failure to fully communicate your concerns to your immediate supervisor or going over his/her head to higher levels will be viewed negatively by all involved and will decrease the likelihood of a favorable resolution. Finally, be sure to inform your supervisor in writing of your intention to appeal beyond that level.

The process of appealing an ethical conflict within the company is usually similar to the process of interacting with the immediate supervisor. Formal steps should follow informal discussions, and steps within the appeal chain should not be bypassed. If the internal appeals process does not resolve the ethical conflict then the individual should formally notify the company that they intend to continue with an external review of the problem.

Before expressing any public concern legal advice should be obtained. A lawyer can identify courses of action and legal pitfalls in your external appeal. While lawyers understand the legal issues they may not have the technical background to evaluate the technical adequacy of your arguments. For this reason it would be helpful to involve an engineering professional society as an impartial judge of your arguments. Engineering societies vary widely in their willingness to become involved in these kind of activities.

If your company worked for a client in the issue that you are concerned with then the client should be approached before going public. The client may pressure the company to resolve the issue internally, or the client may provide the resources to obtain an unbiased review of the issue.

The last resort to solving ethical conflicts is public disclosure. The only case where whistleblowing is acceptable before following the appeals process described above is if there is an immediate danger to the public.

Several excellent texts are available for a deeper study of engineering ethics.[1] The aim of studying engineering ethics is to foster moral autonomy, i.e. the ability to arrive at reasoned moral views based on the responses to humane values most of us were taught as children.

2-5 SOCIETAL CONSIDERATIONS IN ENGINEERING

The first fundamental canon of the ABET Code of Ethics states that "engineers shall hold paramount the safety, health, and welfare of the public in the performance of their profession." A similar statement has been in engineering codes of ethics since the early 1920s, yet there is no question that what society perceives to be proper treatment by the profession has changed greatly in the intervening time. Today's mass communications make the general public, in a matter of hours, aware of events taking place anywhere in the world. That, coupled with a generally much higher standard of education and standard of living, has led to the development of a society that has high expectations, reacts to achieve change, and organizes to protest perceived wrongs. At the same time, technology has had major effects on the everyday life of the average citizen. Whether we like it or not, all of us are intertwined in complex technological systems: an electric power grid, a national network of air traffic controllers, and a gasoline distribution network. Much of what we use to provide the creature comforts in everyday life has become too technologically complex or too physically large for the average citizen to comprehend.

Thus, in response to real or imagined ills, society has developed mechanisms for countering some of the ills and/or slowing down the rate of social change. The major social forces that have had an important impact on the practice of engineering are occupational safety and health, consumer rights, environmental protection, the antinuclear movement, and the freedom of information and public disclosure movement. The result of those social

---

[1] M. W. Martin and R. S. Schinzinger, "Ethics in Engineering," 2d Ed., McGraw-Hill Book Co, New York, 1989; S. H. Unger, "Controlling Technology: Ethics and the Responsible Engineer," Holt, Rinehart, and Winston, New York, 1982.

forces has been a great increase in federal regulations (in the interest of protecting the public) over many aspects of commerce and business and/or a drastic change in the economic payoff for new technologically oriented ventures. Those new factors have had a profound effect on the practice of engineering and the rate of innovation.

The following are some general ways in which increased societal awareness of technology, and subsequent regulation, has influenced the practice of engineering:

- Greater influence of lawyers on engineering decisions
- More time spent in planning and predicting the future effects of engineering projects
- Increased emphasis on "defensive research and development," which is designed to protect the corporation against possible litigation
- Increased effort expended in research, development, and engineering in environmental control and safety—areas that generally do not directly enhance corporate profit but can affect profits in a negative way because of government regulation

On a more global scale, potential developments around the world can have an impact on engineering practice. It should be clear to everyone that our world supply of raw materials (fuel and minerals) is insufficient to support our growing level of technology. The major dependence of the United States on imported oil has given strong impetus to research, development, and engineering (RD&E) in synthetic fuels, solar energy, and the technology of energy conservation. Because so many strategic minerals are found concentrated in only a few places on earth, we often find the supply of critical metals is controlled by the politics of some African or Asian country. The periodic shortage of cobalt is a good example.

It seems clear that the future is likely to involve more technology, not less, so that engineers will face demands for innovation and design of technical systems of unprecedented complexity. While many of these challenges will arise from the requirement to translate new scientific knowledge into hardware, many of these challenges will stem from the need to solve problms in "socialware". By the term socialware is meant the patterns of organization and management instructions necessary to effective functioning of hardware.[1] Such designs will have to deal not only with the limits of hardware, but also with the vulnerability of any system to human ignorance, human error, avarice, and hubris. A good example of this point is the delivery system for civilian air transportation. While the engineer might think of the modern jet transport, with all of its complexity and high technology, as the main focus of concern,

---

[1] E. Wenk Jr., *Engineering Education*, Nov. 1988, pp. 99–102.

such a marvellous piece of hardware only satisfies the needs of society when embedded in an intricate system that includes airports, maintenance facilities, traffic controllers, navigation aids, baggage handling, fuel supply, meal service, bomb detection, air crew training, and weather monitoring. It is important to realize that almost all of these socialware functions are driven by federal or local rules and regulations. Thus, it should be clear that the engineering profession is required to deal with much more than technology. Techniques for dealing with the complexity of large systems have been developed in the discipline of systems engineering (Sec. 2-12).

Another area where the interaction between technical and human networks is becoming stronger is in consideration of risk, reliability, and safety (see Chap. 12). No longer can safety factors simply be looked up in codes or standards. Engineers must recognize that design requirements depend on public policy as much as industry performance requirements. This is an area of design where government influence has become much stronger.

There are five key roles of government in interacting with technology:[1]

- As a stimulus to free enterprise through manipulation of the tax system
- By influencing interest rates and supply of venture capital through changes in fiscal policy to control growth of the economy
- As a major customer for high technology
- As a funding source (patron) for research and development
- As a regulator of technology

Wenk[2] has expanded on the future interactions between engineering and society. The major conclusions of this study are summarized in Table 2-1.

Because of the growing importance of technology to society, a methodology for systematically determining the impact of technology on the social, political, economic, and physical environment is being evolved. It is called *technology assessment* (TA),[3] and Congress has established an Office of Technology Assessment (OTA) to help it in its work. Technology assessment is an attempt to determine the benefits and risk inherent in the range of technological alternatives. Its practitioners try to provide an early-warning system for environmental mishaps, define the necessary monitoring and surveillance mechanisms, and provide the decision-making tools for setting technological priorities and allocating resources. Technology assessment, although still an evolving science, already has a number of characteristics that

---

[1] E. Wenk, Jr., op. cit.

[2] E. Wenk, Jr., "Tradeoffs: Imperatives of Choice in a High-Tech World," The Johns Hopkins University Press, Baltimore, 1986.

[3] M. A. Borough, K. Chen, and A. N. Christakis, "Technology Assessment: Creative Futures," North Holland Publishing Co., New York, 1980.

TABLE 2-1

Future trends in interaction of engineering with society. From E. Wenk, Jr: "Tradeoffs," Johns Hopkins Univ. Press, 1986.

- The future will entail more technology, not less.
- Because all technologies generate side effects, designers of technological delivery systems will be challenged to prevent, or at least mitigate, adverse consequences.
- The capacity to innovate, manage information, and nourish knowledge as a resource will dominate the economic domain as natural resources, capital, and labor once did. This places a high premium on the talent to design not simply hardware, but entire technological delivery systems.
- Cultural preferences and shifts will have more to do with technological choice than elegance, novelty, or virtuosity of the hardware.
- Acting as an organizing force, technology will promote concentration of power and wealth, and tendencies to large, monopolistic enterprises.
- The modern state will increasingly define the political space for technological choice, with trends becoming more pronounced toward the "corporate state." The political-military-industrial complex represents a small-scale model of such evolution.
- Distribution of benefits in society will not be uniform, so disparity will grow between the "haves" and the "have nots."
- Conflicts between winners and losers will become more strenuous as we enter an age of scarcity, global economic competition, higher energy costs, increasing populations, associated political instabilities, and larger-scale threats to human health and the environment.
- Because of technology, we may be moving to "one world," with people, capital, commodities, information, culture, and pollution freely crossing borders. But as economic, social, cultural, and environmental boundaries dissolve, political boundaries will be stubbornly defended. The United States will sense major economic and geopolitical challenges to its position of world leadership in technology.
- Complexity of technological delivery systems will increase, as will interdependencies, requiring management with a capacity for holistic and lateral conceptual thinking for both systems planning and trouble-free, safe operations.
- Decision-making will become more difficult because of increases in the number and diversity of interconnected organizations and their separate motivations, disruptions in historical behavior, and the unpredictability of human institutions.
- Mass media will play an ever more significant role in illuminating controversy and publicizing technological dilemmas, especially where loss of life may be involved. Since only the mass media can keep everyone in the system informed, a special responsibility falls on the "fourth estate" for both objective and courageous inquiry and reporting.
- Amidst this complexity and the apparent domination of decision-making by experts and the commercial or political elite, the general public is likely to feel more vulnerable and impotent. Public interest lobbies will demand to know what is being planned that may affect people's lives and environment, to have estimates of a wide range of impacts, to weigh alternatives, and to have the opportunity to intervene through legitimate processes.
- Given the critical choices ahead, greater emphasis will be placed on moral vision and the exercise of ethical standards in delivering technology to produce socially satisfactory results. Accountability will be demanded more zealously.

Reprinted with permission from *Engineering Education,* Nov. 1988, p. 101.

differentiate it from the more traditional methods of engineering analysis used in planning.

1. Technology assessment is mostly concerned with the second-, third', and higher-order effects or impacts that are rarely considered in engineering analysis. Remote impacts often can be more important than the intended primary variable in social issues.
2. TA considers the needs of a wide range of constituencies.
3. TA is interdisciplinary. There is a need to be able to integrate widely different intellectual traditions and diverse methods of treating data.
4. TA probably is more closely related to policymaking than to technical problem solving.

We stated in the opening paragraph of this chapter that engineering is concerned with problems whose solution is needed and/or desired by society. The purpose of this section was to reinforce that point, and hopefully to show the engineering student how important a broad knowledge of economics and social science is to modern engineering practice.

2-6 THE PRODUCT LIFE CYCLE

Every product goes through a cycle from birth, into an initial growth stage, into a relatively stable period, and finally into a declining state that eventually ends in the death of the product (Fig. 2-8).

In the introductory stage the product is new and consumer acceptance is low, so sales are low. In this early stage of the product life cycle the rate of product change is rapid as management tries to maximize performance or product uniqueness in an attempt to enhance customer acceptance. When the product has entered the *growth stage*, knowledge of the product and its capabilities has reached a growing number of customers. There may be an

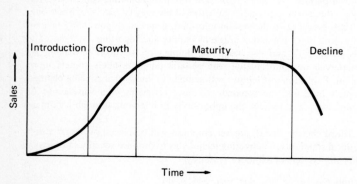

FIGURE 2-8
Product life cycle.

emphasis on custom tailoring the product for slightly different customer needs. At the *maturity stage* the product is widely accepted and sales are stable and are growing at the same rate as the economy as a whole. When the product reaches this stage, attempts should be made to rejuvenate it by incremental innovation or the development of still new applications. Products in the maturity stage usually experience considerable competition. Thus, there is great emphasis on reducing the cost of a mature product. At some point the product enters the *decline stage*. Sales decrease because a new and better product has entered the market to fulfill the same societal need.

In the product introduction phase, market uncertainty and the high cost of advanced productivity processes act as barriers to product innovation. Because product volume is low, expensive but flexible manufacturing processes are used and product cost is high. As we move into the period of product market growth, higher volume manufacturing processes reduce the unit cost. In the product maturity stage emphasis is on prolonging the life of the product by product improvement and significant reduction in unit cost. The high investment cost of advanced productivity processes becomes the barrier to further product innovation.

If we look more closely at the product life cycle, we will see that the cycle is made up of many individual processes (Fig. 2-9). In this case the cycle has been divided into the premarket and market *phases*. The former extends back to the idea concept and includes the research and development and marketing studies needed to bring the product to the market phase. The investment (negative profits) needed to create the product is shown along with the profit. The numbers along the profit vs. time curve correspond to the following processes in the product life cycle. This brief introduction should serve to

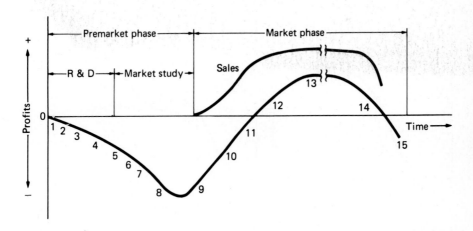

FIGURE 2-9
Expanded product life cycle.

emphasize that innovation leading to a new product is a complex, costly, and time-consuming process.

Premarket phase	Market phase
1. Idea generation	9. Product introduction
2. Idea evaluation	10. Market development
3. Feasibility analysis	11. Rapid growth
4. Technical R&D	12. Competitive market
5. Product (market) R&D	13. Maturity
6. Preliminary production	14. Decline
7. Market testing	15. Abandonment
8. Commercial production	

Technology Development Cycle

The development of a new technology follows an S-shaped curve (Fig. 2-10a). In its early stage, progress is limited by the lack of ideas. A single good idea can make several other good ideas possible, and the rate of progress becomes exponential. During this period a single individual or a small group of individuals can have a pronounced effect on the direction of the technology. Gradually the growth becomes linear when the fundamental ideas are in place and progress is concerned with filling in the gaps between the key ideas. This is the period when commercial exploitation flourishes. Specific designs, market applications, and manufacturing develop rapidly in a field that has not yet settled down. Smaller entrepreneurial firms can have a large impact and capture a dominant share of the market. However, with time the technology

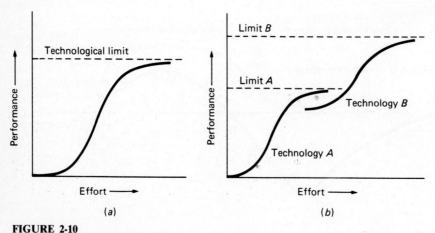

FIGURE 2-10
(a) Simple technology development cycle. (b) Transfer from one technology growth curve (A) to another developing technology (B).

begins to run dry and increased improvements come with greater difficulty. Now the market tends to become stabilized, manufacturing methods become fixed in place, and more capital is expended to reduce the cost of manufacturing. The business becomes capital-intensive; the emphasis is on production know-how and financial expertise rather than scientific and technological expertise. The maturing technology grows slowly, and it approaches a limit asymptotically. The limit may be set by a social consideration, such as the fact that the legal speed of automobiles is set by safety and fuel economy considerations, or it may be a true technological limit, such as the fact the speed of sound defines an upper limit for the speed of a propeller-driven aircraft.

The success of a technology-based company lies in recognizing when the technology on which the company's products are based is beginning to mature and, through an active R&D program, transferring to another technology growth curve that offers greater possibilities (Fig. 2-10b). To do so, the company must manage across a *technological discontinuity*. Past examples of technological discontinuity are the change from vacuum tubes to transistors and from the three- to the two-piece metal can. Changing from one technology to another may be difficult because it requires different kinds of technical skill. Technology usually begins to mature before profits, so that there is a management reluctance to switch to a new technology when business is going so well.

Occasionally a major scientific discovery will open up new opportunities for great advances in performance and reduction in cost. The microcomputer is such an example. However, more frequently small, almost imperceptible improvements will add up to equally great progress. These improvements occur through changes in operating procedures, materials, small variations in manufacturing processes, redesign of products for easier production, or substitution of less expensive components for those used in earlier design. Thus, a great deal of technological innovation is made by incremental advances in cost, performance, and quality improvements.

One can generalize that design and development can be broadly divided into idea dominated design and incremental dominated design. In recent years when the United States has not been competitive in an area of technology we have lost, usually not to radical new technology, but to better incremental improvements. Incremental development is also cyclical. When the current version of the product is in production a development team is working on the next product generation. And, when that next generation goes into production the following generation is started through the development and manufacturing cycle to build a significant product lead and achieve technological leadership.[1]

---

[1] R. E. Gomory and R. W. Schmitt, *Science,* vol. 240, pp. 1131–1132, 1203–1204, May 27, 1988.

It is now widely recognized, based on the success of the Japanese, that the key factor in the speed of this cycle is the closeness of the link between development and manufacturing. In Sec. 1-12 this was discussed through the concept of concurrent engineering. Close ties between design and manufacturing result in early knowledge of technical problems, which when overcome lead to speedy market introduction and higher quality, because the product is easier to manufacture. Because of this cycle, there is a right and a wrong time to introduce new ideas. An idea must be produced at the beginning of the cycle; halfway through is too late because it would cause many changes and delay. Thus, in a technological development there is usually a *window of opportunity* that needs to be recognized. The window of opportunity can have an important pacing function on the development of technology. In areas like consumer electronics, where the development cycle is short, new ideas can be implemented at frequent intervals and the technology develops rapidly. However, in military aircraft, where the cycle time is of the order of 10 years the technology can only advance by large increments. Many of the problems associated with these systems are due to the difficulty of applying incremental developments.

Process Development Cycle

Three stages can be identified[1] in the development of a manufacturing process.

1. *Uncoordinated development.* The process is composed of general-purpose equipment with a high degree of flexibility. Since the product is new and is developing, the process must be kept fluid.
2. *Segmental.* The manufacturing system is designed to achieve higher levels of efficiency in order to take advantage of increasing product standardization. This results in a high level of automation and process control. Some elements of the process are highly integrated; others are still loose and flexible.
3. *Systemic.* The product has reached such a high level of standardization that every process step can be described precisely. Now that there is a high degree of predictability in the product, a very specialized and integrated process can be developed.

Process innovation is emphasized during the maturity stage of the product life cycle. In the earlier stages the major emphasis is on product development and generally only enough process development is done to support the product. However, when the process development reaches the systemic stage, change is disruptive and costly. Thus, process innovations will be justified only if they offer large economic advantage.

---

[1] E. C. Etienne, *Research Mgt.*, vol. 24, no. 1, pp. 22–27, 1981.

Production and Consumption Cycle

Another way to look at life cycle is to consider production and consumption of the product. The *materials cycle* shown in Fig. 2-11 includes all of the stages starting with the mining of a mineral or the drilling for oil or the harvesting of an agricultural fiber such as cotton. These raw materials must be processed to refine or extract bulk material (e.g., a steel ingot) that is further processed into a finished engineering material (e.g., a steel sheet). At this stage an engineer designs a product that is manufactured from the material, and the part is put into useful service. Eventually the part wears out or becomes obsolete because a better product comes on the market. At this stage, our tendency has been to junk the part and dispose of it in some easy way that eventually returns the material to the earth. However, society is becoming increasingly aware of the dangers of haphazard disposal practices, so that this step in the materials cycle is becoming increasingly complicated and expensive. At the same time there is growing recognition that we are dangerously depleting our natural resources, so that there is increasing interest in the recycling of waste materials.

2-7 TECHNOLOGICAL FORECASTING

Technological forecasting[1] is an organized methodology for visualizing likely future technological trends. The field grew rapidly in the 1970s because of increased importance placed upon planning and optimum allocation of resources. Advocates of technological forecasting point out that a forecast is less useful as a prediction of the future than as a guide for making proper choices in the present. The act of forecasting should stimulate the asking of important questions and provide clear, creative thought about important issues.

Forecasting methods fall into two broad groups: intuitive and extrapolative. The former method ranges from simple personal judgment or opinion, however expert, to obtaining the opinion of a committee of experts. A highly regarded technique is the Delphi Method (named after the oracle of Delphi), which attempts to arrive at a consensus about the future by soliciting opinions from a group of individual experts through a succession of rounds of questioning. The technique is similar to a committee-of-experts approach with the important difference that the experts have no face-to-face contact. Since all communication is through a clearinghouse, psychological influences are minimized. Also, because the experts remain anonymous, concern for reputation does not bias the final results.

Extrapolative methods attempt to predict the future by extending

---

[1] J. P. Martino, "Technological Forecasting for Decision-making," 2d ed., American Elsevier Publishing Co., Inc., New York, 1983; J. S. Armstrong, "Long-Range Forecasting," John Wiley & Sons, Inc., New York, 1978.

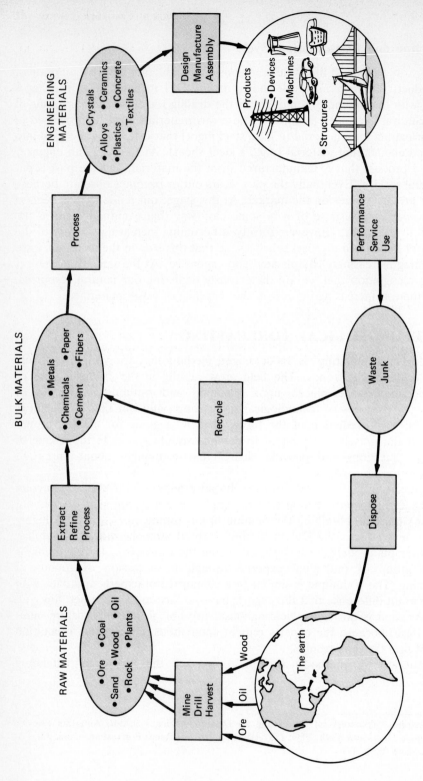

FIGURE 2-11

The total materials cycle. (*Reproduced from "Materials and Man's Needs," National Academy of Sciences, Washington, D.C., 1974.*)

historical data. Simple *trend extrapolation* is based on the assumption that the forces that created the pattern of progress are more likely to continue than change. A difficulty with extrapolation is that the longer the period of forecasting the greater the probability that the assumption will not hold. Extrapolation is not likely to predict the occurrence of step functions in technology, such as the development of the transistor in an era of vacuum-tube technology. An improvement on simple extrapolation is *growth analogy,* which recognizes that technology grows according to an S curve (Fig. 2-10*a*) and that there will be an upper limit to the growth of the technology. *Trend correlation analysis* is based on established or assumed interrelations among measures of performance within various technologies. For example, plotting the speed of military and commercial aircraft shows that commercial planes lag in speed by a predictable amount. Therefore, by looking at current military aricraft, one can gain good insight into the future of commercial airliners.

An example of technological forecasting is given in Fig. 2-12, which predicts the trend for the highest use temperature available in load-bearing plastics. Note that a semilog plot is used to linearize the data, since the beginning of the growth curve is exponential. If maximum use temperature is plotted against the year in which the plastic was developed, the trend line predicts that use will not exceed 672 K (750°F) before the year 2000; yet a Delphi exercise conducted in 1966 predicted that 672 K plastics would be available in 1975. Note, however, the significant error (uncertainty) trend for

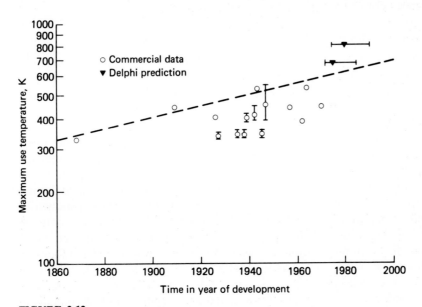

FIGURE 2-12
Technological forecast of maximum use temperature for load-bearing plastics. (*Reproduced from "Technological Forecasting and its Application to Engineering Materials," National Academy Press, Washington, D.C,. 1970.*)

the Delphi prediction. One plausible explanation for the discrepancy between the extrapolation prediction and the Delphi prediction (other than that predicting the future always is chancy) is that the experts in the Delphi exercise were highly research oriented people who already knew of promising plastics in the laboratory stage of development. However, like most research people, they were optimistic about the outcome and did not properly anticipate the long time needed to bring a new product into the marketplace.

The emphasis in this section has been on acquainting the reader with the concept of technological forecasting, but there are other types of forecasting, more business- and economics-oriented, that can be very important to the engineer. Forecasts of *demographic trends* (changes in the size and age distribution of the population) can be very important in predicting the demand for certain types of consumer goods. Predictions of the cost of energy will have great influence on the engineering effort spent to develop new energy sources or engines that are more fuel efficient. Estimates of the future cost of borrowing money will have a strong influence on the level of innovation and investment in new plants. Forecasts of the likelihood of peace in foreign affairs could have strong influence on future developments in the aerospace and other defense technology industries.

2-8 TECHNOLOGICAL INNOVATION

The advancement of technology has three phases:

Invention. The creative act whereby an idea is conceived

Innovation. The process by which an invention or idea is brought into successful practice and is utilized by the economy

Diffusion. The successive and widespread initiation of successful innovation

Without question, innovation is the most critical and most difficult of the three phases. Many studies have shown that the ability to introduce and manage technological change is a major factor in a country's leadership in world markets and also a major factor in raising the standard of living at home. Science-based innovation in the United States has spawned such key industries as aircraft, computers, plastics, and television. Relative to other nations, however, the importance of the United States role in innovation appears to be decreasing. If the trend continues, it will affect our own well-being. Likewise, the nature of innovation has changed with time. Opportunities for the lone inventor and entrepreneur have become relatively more limited. As one indication, independent investigators obtained 82 percent of all U.S. patents in 1901, whereas the corresponding number in 1967 was 24 percent. Nevertheless, small companies do make a major contribution to innovation in this country.

The purpose of this section is to acquaint you with the innovation process and the steps in new product development. Traditionally, engineers play the major role in technological innovation, yet they often do not view themselves in that role. Engineering design and technological innovation are inseparable. It is hoped that you will view your study of design in that spirit so you can make your own strong contribution as an innovator.

The steps in a technological innovation activity can be considered to be:

This model differs from one that would have been drawn in the 1960s, which would have started with basic research in the innovation chain. The research results would have led to research ideas that in turn would have led to commercial development. Although strong basic research obviously is needed to maintain the storehouse of new knowledge and ideas, it has been well established that innovation in response to a market need has greater probability of success than innovation in response to a technological research opportunity. Market pull is far stronger than technology push when it comes to innovation. Even so, the failure rate of innovation is high. Something like 1 in 10 of the ideas submitted for initial screening enter the development phase. Only from 1 in 50 to 1 in 100 of these survivors is successful in the marketplace. Innovation truly is survival of the fittest that involves technical and marketing risks.

A study of 200 innovation failures showed that 12 percent were due to technology-related problems, over 50 percent to market obstacles, and 28 percent to management problems. The predominant causes of technical failure were the existence of a superior rival technical approach, design problems, and failure to control quality. Of the market obstacles, the chief problem areas were limited sales potential, unacceptable cost/price ratio, insufficient customer benefit/cost ratio, and inability to aggregate a fragmented market. Management failure centered about poor market analysis, failure to provide adequate staff, and lack of capital resources to see the innovation through to success.

One important study of technological innovation[1] has shown that success can be determined by looking at certain aspects of the technology and business contexts of the innovation. With regard to technology, it is important to know whether the innovation removes fundamental technical constraints that limited the performance of the prior art and whether new technical constraints are inherent in the new art. Obviously, the technical balance should be favorable. As for the business context, we need to know what previously established

---

[1] G. R. White, *Technology Rev.*, pp. 15–23, February 1978.

business operations are displaced or weakened by the innovation and what new business operations are needed to support it. Again, the balance between what business operations must be stopped and initiated has to be favorable. Finally, we need to ask some important questions concerning market dynamics. Does the product that incorporates the new technology provide enhanced performance (effectiveness) to the user? Does the innovation reduce the cost of delivering the product or service? Will the innovation really substitute for an existing market, or will it result in a greatly expanded market because of reduced cost or enhanced performances?

An approach to business strategy dealing with innovation and investment uses the colorful terminology advanced by the Boston Consulting Group in their portfolio management technique. Business projects are placed in one of four categories:

Star businesses. High growth potential, high market share
Wildcat businesses. High growth potential, low market share
Cash-cow businesses. Low growth potential, high market share
Dog businesses. Low growth potential, low market share

In this context, the break between high and low market share is the point at which a company's share is equal to that of its largest competitor. For a cash-cow business, cash flow should be maximized but investment in R&D and new plant should be kept to a minimum. The cash these businesses generate should be used in star and wildcat businesses. Heavy investment is required in star businesses so they can increase their market share. By pursuing this strategy, a star becomes a cash-cow business and eventually a dog business. Wildcat businesses require generous funding to move into the star category. That only a limited number of wildcats can be funded will bring about the survival of the fittest. Dog businesses receive no investment and are sold or abandoned as soon as possible. This whole approach is artificial and highly stylized, but it is a good characterization of corporate action concerning business investment. Obviously, the innovative engineer should avoid become associated with the dogs and cash cows; for there will be little incentive for creative work.

There are other business strategies that can have a major influence on the engineering design. A company that follows a *first in the field* strategy is usually a high-tech innovator. Some may prefer to let others pioneer and develop the market, with the strategy of being a *fast follow on* that is content to have a lower market share at the avoidance of the heavy R&D expense of the pioneer. Other companies may emphasize process development with the goal of becoming the *high-volume, low-cost producer.* Yet other companies adopt the strategy of being the supplier to a few major customers that market the product to the public.

A company with an active research program usually has more potential products than the resources to develop them into marketable products. To be considered for development a product should fill a need that is presently not adequately served, or serve a current market for which the demand exceeds the supply, or has a differential advantage over an existing product (such as better performance, improved features, or lower price). A screening matrix that can be used[1] to select the best prospects for product development is shown in Fig. 2-13. Examine particularly the kind of business criteria that are used to make this decision. The range of expectations for each criterion is given in the five columns, from excellent to poor. A weighting factor is applied to certain criteria. In this rating scheme a total of 16 criteria are considered, such that a perfect product would receive a rating of 100 and a poor product would score 20. Most potential products would range from 40 to 80 on this scale, with a rating of 70 typically being required for further consideration. A screening matrix like this should be completed by managers from marketing, product design, R&D, and manufacturing, each working independently.

Studies of the innovation process by Roberts[2] have identified five kinds of people who are needed for technological innovation.

Idea generator—the creative individual

Entrepreneur—the person who "carries the ball" and takes the risks

Gatekeepers—people who provide technical communication from outside to inside the organization

Program manager—the person who manages without inhibiting

Sponsor—the person who provides financial and moral support, often senior management

Roughly 70 to 80 percent of the people in a technical organization are routine problem solvers and are not involved in innovation. Therefore, it is important to be able to identify and nurture the small number who give promise of becoming technical innovators.

Innovators tend to be the people in a technical organization who are most current with technology and who have well-developed contacts with technical people outside the organization.[3] Thus, the innovators receive information directly and then diffuse it to other technical employees. Innovators tend to be predisposed to "do things differently" as contrasted with "doing things better." They are able to deal with unclear or ambiguous situations without feeling uncomfortable. That is because they tend to have a high degree of self-reliance and self-esteem. Age is not an important factor in innovation, nor is

---

[1] R. J. Bronikowski, "Managing the Engineering Design Function," Van Nostrand Reinhold Co., New York, 1986.

[2] E. B. Roberts and H. A. Wainer, *IEEE Trans. Eng. Mgt.*, vol. EM-18, no. 3, pp. 100–109, 1971.

[3] R. T. Keller, *Chem. Eng.*, pp. 155–158, Mar. 10, 1980.

Considerations	wt. factor	Excellent 5
Est. gross profit	2	50% and over/yr
Est. IROI	1	90% and over
Current annual available business	2	$10 million +
Market potential 5 yr	1	In growth stage. Increasing sales & demand at an increasing rate
Est. market share 1 yr	1	25% and over
Est. market share 5 yr	2	50% and over
Stability	1	Product resistant to economic change
Degree of competition	1	No competitive products
Product leadership	1	Fills a need not currently satisfied. Is original
Customer acceptance	1	Readily accepted
Influence on other products	1	Complements and reinforces an otherwise incomplete line
Manufacturing content	1	Completely manufactured in-house
Patent position	1	Impregnable. Exclusive license or rights
Sales force qualification	1	Qualified sales force available
Time to introduction	2	Less than 6 mo.
Technical ability to develop and produce	1	Present technical know-how available and qualified

FIGURE 2-13
Screening matrix for selection of new products (*From R. J. Bronikowski, "Managing the Engineering Design Function," Van Nostrand Reinhold Co., New York, 1986; reprinted by permission of the publisher. All rights reserved.*)

Above average 4	Average 3	Below average 2	Poor 1
49–30%/yr.	29–20%/yr	19–10%/yr.	9% and less/yr
89–75%	74–60%	59–49%	39% and less
$10–7.5 million	$7.5–$5 million	$5–$2.5 million	Less than $2.5 million
Reaching maturity. Increasing sales but at a decreasing rate	Turning from maturity to saturation. Leveling of sales	Declining sales & profits	Demand, sales & profits declining at an increasing rate
24–15%	14–10%	9–5%	4% and less
49–30%	29–20%	19–10%	9% and less
Some resistance to economic change and out of phase	Sensitive to economic change—but out of phase	Sensitive to economic change and in phase	Highly sensitive to economic change and in phase
Only slight competition from alternative	Several competitors to different extents	Many competitors	Firmly entrenched competition
Improvement over existing competition	Some individual appeal, but basically a copy	Barely distinguished from competitors	Copy with no advantages, possibly some disadvantages
Slight resistance	Moderate resistance	Appreciable customer education needed	Extensive customer education need
Easily fits current line, but not necessary	Fits current line, but may compete with it	Competes with, and may decrease sales of current line	Endangers or replaces an otherwise successful line
Partially mfg'd, assembled & packaged in-house	Assembled and packaged in-house	Packaged in-house	No manufacturing content
Some resistance to infringement, few firms with similar patents	Probably not patentable; however, product difficult to duplicate	Not patentable, can be copied	Product may infringe on other patents
Sales force has basic know-how. Minor product orientation required	Sales force has basic know-how, requires product and application education	Sales force requires extensive product and application education	Existing sales force inadequate to handle market and/or product
6–12 mo.	12–24 mo.	24–36 mo.	over 36 mo.
Most technical know-how available	Some know-how available	Extensive technical support required	Ability to develop and produce with present technology questionable

experience in an organization so long as it has been sufficient to establish credibility and social relationships. It is important for an organization to identify the true innovators and provide a management structure that helps them develop. Innovators respond well to the challenge of diverse projects and the opportunity to communicate with people of different backgrounds.

A successful innovator is a person who has a coherent picture of what needs to be done, not necessarily a detailed picture. The innovator emphasizes goals not methods of achieving the goal. He or she can move forward in the face of uncertainty because he does not fear failure. Many times the innovator is a person who has failed in a previous venture—and knows why. The innovator is a person who identifies what he or she needs in the way of information and resources—and gets them. The innovator aggressively overcomes obstacles—by breaking them down, or hurdling over them, or running around them. Frequently the innovator works the elements of the problem in parallel—not serially.

A successful technological innovation requires a good idea or concept that satisfies a societal need. It requires a business champion in the form of supportive top management willing to take the financial risk. The necessary human and technical resources must be assembled. Clear goals and objectives which include milestones for technical accomplishment, and financial and time expenditures, must be established. The final ingredient in a successful innovation is a little bit of luck.

Chief among the factors which lead to failure of a technological innovation are loss of the business champion and/or resources, or oscillation of the commitment to the project. Other factors could be the collapse of the window of opportunity and inadequate attention to technical weaknesses in the concept. Often failure to properly gage the market is a major contributing factor in failure. Much of the success of the Japanese in world markets is attributed to their overwhelming emphasis in producing the product that is demanded by the market at the price the market is willing to pay, with constant emphasis on raising quality and lowering costs.

2-9 MARKET IDENTIFICATION

The number one cause of new-product failure is inadequate market analysis.[1] Therefore, much greater attention must be given to marketing than is usually considered necessary by the young engineer. Before initiating the design effort to produce a new product it is imperative to obtain answers to the following questions.

Who is the customer?

---

[1] R. G. Cooper, "Winning at New Products," Addison-Wesley Publishing Co., Reading MA, 1986.

What does the customer really want?

What will the customer expect in terms of product support?

When will the product be required by the customer?

What is the competition?

What is the customer willing to pay per item?

What is the potential for related products or services?

The answers to these questions will depend on the source of the new product idea. If the idea came from the field through one of the company's technical reps or salespersons, or from an existing customer, then the audience to query is fairly well defined. If the idea came internally, from corporate R&D or engineering, then the audience may be more diffuse. Sophisticated sampling and testing techniques have been developed by marketing experts to gage true customer response. Although these methods are beyond the scope of this book, the engineering designer should understand that the time is long past when it is sufficient to come up with a good product and hope that the world will beat a path to your door. It is interesting that the Japanese approach to engineering appears to be totally market-oriented. Engineering is perceived as a key element of a commercial enterprise where the aim is to produce the product demanded by the market at the price the market is willing to pay, with constant emphasis on improving quality and reducing costs.

Perhaps a single example will serve to illustrate this point. The story is told how Japanese engineers stalked U.S. shopping mall parking lots asking female owners of hatchback cars whether they preferred a raised lip or a smooth floor in their hatchbacks. The answer provided a marketing advantage for the Toyota.

2-10 COMPETITIVE BENCHMARKING

Benchmarking[1] is a technique for identifying the strengths and weaknesses of the internal operation and comparing them with competitors and industry leaders worldwide. It takes its name from the surveyor's benchmark or reference point from which elevations are measured. Benchmarking, as used in modern business context, is the search for industry best practices that lead to superior performance. Benchmarking can be used in the design and manufacturing of products, or it can be applied to any other business function such as order entry, billing, repair services, or financing. The Xerox Corporation has been a pioneer in applying the technique to design of consumer products.

The ten steps in the benchmarking process are shown in Fig. 2-14. The identification of what is to be benchmarked is not a trivial point. While it is

---

[1] R. C. Camp, "Benchmarking," Quality Press, American Society for Quality Control, Milwaukee, 1989.

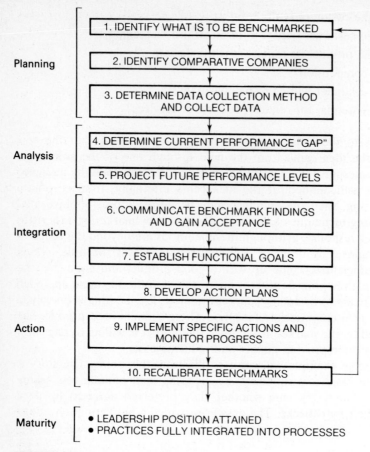

FIGURE 2-14
Steps in competitive benchmarking. (*From R. C. Camp, "Benchmarking", Quality Press, ASQC, 1989.*)

commonplace for manufacturers of a product to tear apart their competitors' product and examine it in excruciating detail, it is less common but nevertheless vital, to examine the competitors' design practices or quality control techniques. The benchmarking process is general enough to permit this type of analysis. The direct competitors to your product will be obvious companies to compare, but it is also important to make comparisons against leadership companies outside of the specific industry or product area. In deciding what to benchmark it is important to establish measurable quantities or metrics. A typical metric might be $/piece or life to failure.

Collection of data is a critical point. Much business data is available from published sources, government publications, or from trade associations. A good place to start is by benchmarking internal operations within the company. Internal information usually is more readily available than from external sources. Also, developing an internal network of experts within the company may lead to greater knowledge about activities at competing companies. One source of information about competitors is consultants. They should be generally knowledgeable about practices in the industry. Gathering more detailed information may require mailing questionnaires and plant visits. These often can be accomplished on a professional to professional basis if there is an agreement as to information sharing.

The analysis of the data will determine the current gap in performance or cost. An important point is to project the progress being made to close the gap. The results of the benchmark study must then be communicated to the employees who have the ability to close the gap and to management who have the ability to establish goals and formulate action plans. An important element is to continually monitor progress. It is important to realize that benchmarking is not a one-time effort. To prevent being blindsided by the competition it is important that it be an ongoing effort.

Benchmarking is a formalized process by which an organization looks outside to understand and adopt the best practices and technology of its competition. In this proactive search for change it combats the human instinct of "not invented here". By generating many options, as opposed to a few internally generated solutions, it is more likely to result in corporate excellence.

2-11 PROTECTION OF INTELLECTUAL PROPERTY

In a highly competitive world market where nations vie for leadership in the commercialization of the next developing technology the protection of the ideas or intellectual property on which it is all based has become very important. The term *intellectual property* generally refers to patents, copyrights, trademarks, and trade secrets. These entities fall within the broad area of property law and, as such, they can be sold or leased like other forms of property. Similarly, just as property can be stolen or trespassed upon, so intellectual property can be infringed.

Patents

Article 1, Section 8 of the Constitution of the United States says that Congress shall have the power to promote progress in science and the useful arts by securing for limited times to inventors the exclusive right to their discoveries. A patent granted by the United States government gives the patentee the right to prevent others from making, using, or selling the patented invention. Patents are awarded for a period of 17 years. A utility patent may be issued for a new and useful machine, process, article of manufacture, or composition of matter. In addition design patents are issued for new ornamental designs and plant patents are granted on new varieties of plants. Computer programs cannot be patented but are protected by copyright. However, where software is embedded in computer hardware it may be patented.

In patent law a process is defined as an operation performed by rule to produce a certain result. In addition, patent law defines a patentable process to include a *new use* of a known process, machine, manufacture, or composition of matter. Thus, a new use for a known compound which is not analogous to a known use may be a patentable process. However, not all processes are patentable. Methods of doing business or natural laws or phenomena, as well as mathematical equations and methods of solving them are not patentable subject matter.

There are three general criteria for awarding a patent.

1. The invention must be *new* or *novel*.
2. The invention must be *useful*.
3. It must be nonobvious to a person skilled in the art covered by the patent.

A key requirement is novelty. Thus, if you are not the first person to propose the idea your cannot expect to obtain a patent. If the invention was made in another country but it was known or used in the United States before the date of the invention in the United States it would not meet the test of novelty. Finally, if the invention was published anywhere in the world before the date of invention but was not known to the inventor it would violate the requirement of novelty. The requirement for usefulness is rather straightforward. For example, the discovery of a new chemical compound (composition of matter) which has no useful application is not eligible for a patent. The final requirement, that the invention be unobvious can be subject to considerable debate. A determination must be made as to whether the invention would have been the next logical step based on the state of the art at the time the discovery was made. If it was, then there is no patentable discovery.

The requirement for novelty places a major restriction on publication prior to filing a patent application. In the United States the printed publication of the description of the invention anywhere in the world more than one year before the filing of a patent application results in automatic rejection by the Patent Office. It should be noted that to be grounds for rejection the publication must give a description detailed enough so that a person with ordinary skill in the subject area could understand and make the invention. Also, public use of the invention or its sale in the United States one year or more before patent application results in automatic rejection. The patent law also requires diligence in *reduction to practice*. If development work is suspended for a significant period of time, even though the invention may have been complete at that time, the invention may be considered to be abandoned. Therefore, a patent application should be filled as soon as it is practical to do so.

In the case of competition for awarding a patent for a particular invention, the patent is awarded to the inventor who can prove the earliest date of conception of the idea and can demonstrate reasonable diligence in reducing the idea to practice. The date of invention can best be proved in a court of law if the invention has been recorded in a bound laboratory notebook with prenumbered pages and if the invention has been witnessed by a person competent to understand the idea. For legal purposes, corroboration of an invention must be proven by people who can testify to what the inventor did and the date when it occurred. Therefore, having the invention disclosure notarized is of little value since a notary public usually is not in a position to understand a highly technical disclosure. Similarly, sending a registered letter to oneself is of little value. For details about how to apply, draw up, and pursue a patent application the reader is referred to the literature on this subject.[1]

Copyrights

A copyright is the exclusive legal right to publish a tangible expression of literary or artistic work. Copyrights may be obtained for such things as books, works of art, musical compositions, motion pictures, recordings, computer software, advertisements, engineering and architectural drawings and models. For a copyright awarded to an individual, the term of the copyright is the life of the writer plus 50 years. For a copyright awarded to a corporation the term of the copyright is 100 years from creation or 75 years from publication of the work. A basic tenet of copyright law is that an idea cannot be copyrighted, only its tangible expression, i.e. the actual arrangement of words or the design.

Trademark

A trademark is any name, word, symbol, or device that is used by a manufacturer to identify his goods and distinguish them from those made or sold by others. The right to use trademarks is obtained by registration and extends indefinitely so long as the trademark continues to be used. Suits can be brought for infringement of trademark in the same way as for patents.

Trade Secret

A trade secret is any formula, pattern, device, or compilation of information which is used in a business to create an opportunity to obtain an advantage

---

[1] D. A. Burge, "Patent and Trademark Tactics and Practice," 2d ed., John Wiley & Sons, New York, 1984; W. G. Konold et al., "What Every Engineer Should Know About Patents," Marcel Dekker, New York, 1989; D. Pressman, "Patent It Yourself? How to Protect, Patent, and Market Your Inventions," McGraw-Hill Book Co., New York, 1979; "Patents and Inventions: An Information Aid for Inventors," U.S. Govt. Printing Office, Washington, D.C.

over competitors who do not have or use this information. Sometimes trade secrets are information which might be patentable but for which the corporation chooses not to seek a patent. For example, process patents are notoriously difficult to police against infringement so that process parameters often are treated as trade secrets rather than seeking patent protection. Since a trade secret has no legal protection it is essential to maintain the information in secret. For example, the formula for Coca-Cola is a trade secret which has endured for many years. Frequently trade secrets are lost when employees take employment with competing companies. To combat this loss companies frequently obtain court orders to restrain former employees from divulging trade secrets to a competitor.

Technology Licensing

The right to exclusive use of technology that is granted by a patent may be transferred to another party through a licensing agreement. A license may be either an exclusive license, in which it is agreed not to grant any further licenses, or a nonexclusive license. The licensing agreement may also contain details as to geographic scope, e.g. one party gets rights in Europe, another gets rights in South America. Sometimes the license will involve less than the full scope of the technology. Frequently consulting services are provided by the licensor for an agreed upon period.

Several forms of financial payment are common. One form is a paid up license which involves a lump sum payment. Frequently the licensee will agree to pay the licensor a percentage of the sales of the products that utilize the new technology, or a fee based on the extent of use of the licensed process. Before entering into an agreement to license technology it is important to make sure that the arrangement is consistent with U.S. antitrust laws or that permission has been obtained from appropriate government agencies in the foreign country.

2-12 SYSTEMS ENGINEERING

Most engineering products are designed to function as part of a system. For example, an automobile must fit into the road network and traffic control system of the nation, as well as conforming to the legal system in terms of speed, safety, and fuel economy. The automobile itself is a system comprising many components. These components were designed by many people, working mostly independently, yet the parts must operate interactively to produce a functioning system.

As engineering systems grow larger and more complex it becomes necessary to develop a methodology to deal with this design complexity. Examples of such systems are air traffic control or military radar systems. The field of *systems engineering*[1] has evolved to deal with these type of problems. The methodology is very similar to that defined in Chapter 1 for the design process but it becomes more formalistic and systematized because the great complexity requires this rigor. Another characteristic of systems engineering is that it is holistic. It involves a top-down synthesis, development, and operation of a real-world system that satisfies, in near optimal fashion, the full range of requirements for the system. The notion of top-down synthesis suggests the decomposition of the system into subsystems and further into components. The systems engineering manager needs to be concerned with the effectiveness of the system as a whole than with any particular part of it.

2-13 LIFE CYCLE ENGINEERING

The philosophy of life cycle engineering is that the entire life of the product should be considered in its original design. Traditionally, engineers have focused chiefly on the acquisition phase of a product's life cycle. However, in the worldwide market of today a competitive product cannot be achieved unless performance, reliability, and maintainability are addressed at the design concept stage. These issues cannot be considered only after the design is completed.

The stages in the life of a product are design, production, distribution, consumption, and retirement. The need for close integration of all aspects of manufacturing into the design stage has already been discussed—and is a pervasive theme of this book. Distribution involves the transfer of the product to the customer. The product must be designed so that it is not damaged or lost in transit, and so it will be efficiently warehoused. For highly technical products distribution may involve the use of field engineers who assist in installing the product and advise the user regarding its use. Consumption involves service in use. Everything is consumed or wears out as it is used. Good design anticipates these events and provides recovery mechanisms or for easy replacement of parts. Finally, retirement involves withdrawal of the product from use. The designer must provide for easy recycling of the materials, or for disposal of the product in a way that is not harmful to the environment or to public safety.

The Department of Defense has been very concerned with the life cycle of its weapons systems and has required its contractors to design these products with life cycle engineering having high priority. This has given rise to

---

[1] A. D. Hall, "A Methodology for Systems Engineering," Van Nostrand Reinhold, New York, 1962; R. DeNeufville and J. H. Stafford, "Systems Analysis for Engineers and Managers," McGraw-Hill Book Co., New York, 1971; B. S. Blanchard and W. J. Fabrycky, "Systems Engineering and Analysis," Prentice-Hall Inc., Englewood Cliffs, N.J., 1981; H. Eisner, "Computer-Aided Systems Engineering," Prentice-Hall Inc., Englewood Cliffs, N.J., 1988.

the development of formal methodologies to analyze reliability and maintainability (see Chapter 12). Greater attention is being given to these issues in the design of products for the civilian economy as competitive pressure intensify. Manufacturers find that features such as higher energy efficiency or lifetime warranty provide a significant competitive edge.

The steam turbines in an electric generating plant is an interesting example of design for life cycle engineering. They operate in an environment where a high degree of reliability and long life (30 to 40 years) are expected, yet the electric utility industry is highly regulated as to the costs that can be charged the user. The technology is not new, for steam turbines were introduced in the early part of the 20th century, but steam turbines have continuously evolved in capacity, operating temperature, and efficiency. A generic problem of long standing is the failure of turbine blades. The unavailability of units to make electricity due to blade failure costs utilities in the United States more than $235 million annually.[1] The problem is made more serious because of changes in power company operating strategy which have relegated many older units that were designed for steady operation at full load to cycling or peaking duty.

Three-fourths of the steam turbine blade failures occur in low-pressure turbines and about half of all blade failures occur in the last two rows of blades. The first step in life cycle redesign is to determine the causes of the failures (see Secs. 13-8 and 13-9). It has been established that when the three factors of corrosive environment, a susceptible blade material, and high fatigue stresses act either individually or in combination that failure will occur. Although corrosive substances such as chlorides occur in steam in concentrations of only a few parts per billion, when dry steam begins to condense, as it does in the last rows of blades, the contaminants become concentrated by a factor of over 10,000 times, forming a corrosive solution. An approach is to install improved condensate polishing units and to be more vigilant in monitoring and controlling water chemistry. The preferred blade material for this application is the titanium alloy Ti–6Al–4V, which has superior resistance to corrosion fatigue in chloride environments and better resistance to stress-corrosion cracking compared to the standard blade alloy 17-4 PH stainless steel. However, the cost differential for materials is about 15:1. A third approach is to reduce the level of fatigue stress by improved design. A computer code based on finite element modeling will evaluate the probability of fatigue failure under specific conditions of steady and dynamic loading at various locations on the blade. The decision as to whether one or all approaches will be adopted to minimize blade failure will be dependent on an engineering economic analysis of whether the increased initial cost, i.e. capital cost for corrosion control, improved blade material, or improved blade design,

---

[1] K. Yeager, *EPRI Jnl.*, Dec. 1985. pp. 55–57.

will be more than compensated in savings due to reduced turbine outage (see Chap. 8).

2-14 HUMAN FACTORS IN DESIGN

Good engineering designs work well with people. The best way to achieve this goal is to consider the product and the user as an operating system. The study of people as a component of an engineering system is called *human factors engineering*.[1] This area of study is also known as *ergonomics,* from the Greek *ergon* (work) and *nomos* (law, rule). Human factors is applied in engineering design from several viewpoints:

1. To design a product that is pleasing and easy to use.
2. To improve the safety, health, and comfort of operators of machinery and production equipment.
3. To increase the productivity of workers.

Human factors engineering is very interdisciplinary. It includes knowledge from anatomy, anthropometry (the science of dimensions of humans), applied psychology, biomechanics, bioengineering, and physiology.

The common personal computer represents an engineering system which is not particularly well designed from the standpoint of human factors. If placed on a standard desk the keyboard is too high for comfortable typing, which requires the hands be either level with or below the elbow. Generally the monitor sits on the computer, which places the screen too high. This forces the operator to look straight ahead or slightly upward. Looking up while your hands work below places stress on the neck and shoulders. If you place the computer vertically on the floor under the desk then it becomes difficult to turn on and off, since the power switch on many units is on the side in the rear. Finally, there is the problem of glare on the screen from overhead lights or through the window. And, oh yes, some units emanate an annoying continual hum from the cooling fan. The personal computer is a product that is young in the product cycle. It is perhaps understandable that technical features have taken precedence over human engineering. It is left as an exercise to the reader to suggest solutions to these ergonomic problems.

Clearly the first step in "integrating man with the machine" is to have data on human dimensions.[2] From this anthropometric data you can design a

---

[1] E. J. McCormick, "Human Factors in Engineering and Design," 4th ed., McGraw-Hill Book Co., New York, 1976; R. W. Bailey, "Human Performance Engineering," Prentice-Hall, Englewood Cliffs, N.J., 1982; G. Salvendy (ed.)," Handbook of Human Factors," John Wiley & Sons, New York. 1987.

[2] W. E. Woodson, "Human Factors Design Handbook," McGraw-Hill Book Co., New York, 1981; H. Schmidtke (ed.), "Ergonomic Data for Equipment Design," Plenum, New York, 1984; "Humanscale," a library of design charts, MIT Press, Cambridge, MA.

lever that all but the shortest five percent of adult males can reach from a chair without leaning forward. Other tables give the probability distribution of heights, weight, hearing abilities, sight characteristics, reaction times, etc. For example, the normal working area for a person in a seated position extends only about 16 inches from the body. The greatest work rate that a well-developed healthy male can provide on a sustained basis is about 500W. The reaction time to a random stimulus varies from about 100 to 500 microsecond. Most of the delay in responding is due to information processing by the brain, which can vary from 70 to 300 μs. The reaction time also varies with the nature of the stimulus. An auditory stimulus results in a faster reaction time than a visual stimulus. A manual response can be accomplished in about three-fourths of the time for a verbal response.

The ergonomics of human accuracy in movement often is important in design. This has to do with the biomechanics of the joints that are used. The ball joint in the shoulder requires much more control guidance for accurate movement than does the two-member pivot joint in the elbow. The least control is required by the simple hinge joints which bend the fingers.

Products often are designed to provide feedback to humans. Tactile information is transmitted by shape, size, and texture. For example, the cockpit controls on military planes have distinctive shapes for the knobs which control the throttle, the flaps, the fuel mixture, etc. Sound signals have the advantage of being pervasive, and they engender the fastest response time. For example, wailing sirens or loud buzzers are universally identified as warning signals, and because they are annoying they demand response. Sometimes natural sounds in the operation of a product provide the required feedback. The sound of a pop-up toaster is such an example. Obviously the use of sound to transmit information can be mitigated by excessive background noise or if the intended recipient has a loss of hearing.

Feedback information can be provided in more detail and at less cost by using light signals. These also have the advantage of displaying information by position, shape, and color. A light display can be passive, such as a sign, or dynamic, like a blinking colored light. New technology like solid-state diodes, fiber optics, and liquid crystal displays, is opening up many opportunities for more effective feedback with light.

Human factors engineering plays a major role in minimizing operator error. It must be clearly demonstrated that the physical and psychological responses required for operation are within the normal expectations of humans. When design changes are made to an existing product it must be established that the normal human will be able to operate it, or else clearly visible instructions must be provided. The difficulty in extracting the ignition key from certain makes of automobiles after antitheft design changes were made is but one example. Particularly important is the location of control devices.

Operator fatigue is an important consideration in human engineering. Fatigue is associated with prolonged periods of heavy work which drains the muscles of their energy reserves, but it also can result from physical or mental stress. Monotony coupled with lack of activity can also cause fatigue. Fatigue increases the probability of operator error and results in a decline in physical and mental performance. To combat fatigue it is necessary to provide work environments which are challenging but not physically and mentally exhausting or monotonous. Physical causes for fatigue can be removed through automation or the use of robots. Mental fatigue and monotony can be reduced by introducing more variety into the job and making the worker feel more important. The team approach to automotive assembly that has been adopted by a few companies is a major move in that direction.

Human factors engineering, like manufacturing, reliability, and many other issues, must be considered in the early stages of design. It is just not efficient to have completed the technical design and then go back and engineer for human performance. One important way of maximizing the human interaction is by controlling the environment in which the human must perform. If the operator is subject to cramped quarters, high or low temperature, noise, vibration, dirt, or poor lighting then performance will suffer. To address some of these concerns it may be necessary to introduce a certain degree of automation. These issues must be understood when developing the problem statement and needs analysis.

2-15 DESIGN FOR OCCUPATIONAL SAFETY AND HEALTH

The great societal interest in consumer safety and increased safety and health in the workplace has resulted in considerable federal and state legislation in these areas. The Occupational Safety and Health Act of 1970 is enforced by the Occupational Safety and Health Administration (OSHA), a part of the Department of Labor. OSHA publishes standards that specify how industry should operate to ensure that the workplace is free of hazards to workers. Thus, concern for safety and health must be an issue in design as never before. Certainly the concepts of human engineering discussed in the previous section are directly applicable to this subject. A considerable literature on this subject has developed.[1]

Industry has become more aware of the need for ergonomic solutions to reduce job-related injuries. Lower back pain is the most common worker injury. Ergonomic solutions are directly applicable. Some plants have formed ergonomics committees, similar to quality committees, consisting of a doctor,

---

[1] F. E. McElroy (ed.), "Accident Prevention Manual for Industrial Operations," Nat. Safety Council, Washington, 1981; R. De Reamer, "Modern Safety and Health Technology," John Wiley & Sons, New York, 1980; L. Slote (ed.), "Handbook of Occupational Safety and Health," John Wiley & Sons, New York, 1987; R. A. Wadden and P. A. Scheff, "Engineering Design for Control of Workplace Hazards," McGraw-Hill Book Co., New York, 1987.

engineer, senior management, floor supervisor, union representative, a production worker, and a skilled worker. One worker in an auto assembly plant was required to work in an extremely bent-over sideways position. The ergonomics committee arranged for a eight-inch pit which allowed the worker to do the same job at a slight angle. This simple improvement completely took the trauma out of the job. A type of problem called cumulative trauma disorder is increasing rapidly in production situations. These are injuries to the hand, wrist, elbow, forearm, shoulder, or back caused by repetitive motions. These start as aches and pains, but with time they develop into permanent injuries. Once again, ergonomics leads to the solution.

A major area is the prevention of injury from contact with machinery. It is sometimes a major design challenge to provide guards and safety controls that will prevent hands, arms, and other parts of the body from coming in contact with machinery in ways that do not interfere with production or which can be disarmed by overzealous workers.

Decisions of the courts have made it abundantly clear that a manufacturer is required to design, build, and sell a product that is not unreasonably dangerous to the user. Product liability under the law has become an increasingly serious problem for manufacturers. This places greater responsibility for design safety into the product. The topic of product liability is discussed in Sec. 13-11.

The emphasis on occupational safety is this section should not be interpreted that environmental issues are no longer important in design. The public continues to make it known in many ways that it is interested in a cleaner environment and this is reflected in the many regulations of the Environmental Protection Agency (EPA). A proper strategy when designing to remove pollutants or hazardous wastes is first to attempt to minimize their production in the manufacturing process, then to design for their recycle and recovery, and, finally, to design for their safe disposal.

2-16 INDUSTRIAL DESIGN

Industrial design, also often called product design, is concerned with the visual appearance of the product. The terminology is not very precise in this area. Up until now, what we have called product design has dealt chiefly with the function of the design. Even the term industrial design is usually applied to consumer products, although more producers of products aimed at the industrial market are giving greater attention to their appearance.

The reason for increased attention to industrial design is that performance alone may no longer be sufficient to sell a product in today's highly competitive marketplace. Companies are learning that a pleasing design appearance is a way of adding perceived value to their product. In a true sense, design has become a strategic weapon for business.

It is really too restricting to state that industrial design deals only with appearance.[1] That part of design which is concerned with appearance is properly called *styling*. Industrial design is most properly concerned with arranging the elements of a product to accomplish a particular use. It is the communication of the quality and function of an object, as well as the character and integrity of its makers, by visual nonverbal means. Some few, and successful, companies consistently have made a high level commitment to design over the full spectrum of their activities. Industrial designers pay especial attention to the human engineering aspects of design. A goal is to design products that are easy to use, safe, and comfortable. Their purpose should be self-evident from the moment you look at or touch them. High technology has given a special freedom to the industrial designer, because the trend to incorporate microprocessor control means that the inner mechanical mechanisms no longer dictate the outer form. Microelectronics has freed the designer to give the product any shape he/she wants.

The industrial designer is usually educated as an applied artist or architect. This is a decidedly different culture than in the education of the engineer. While engineers may see color, form, comfort, and convenience as necessary evils in the product design, the industrial designer is more likely to see these features as intrinsic in satisfying the needs of the user. The two groups have roughly opposite styles. Engineers work from the inside out. They are trained to think in terms of technical details. Industrial designers, on the other hand, work from the outside in. They start with a concept of a complete product as it would be used by a customer and work back into the details needed to make the concept work. Industrial designers often work in independent consulting firms, although large companies may have their own in-house staff. Regardless, it is important to have the industrial designers involved at the beginning of a project, for if they are called in after the details are worked out there may not be room to develop a proper concept. There currently is a trend for larger design firms that are staffed to handle a project from industrial design through to a functioning manufactured design.

BIBLIOGRAPHY

Blanchard, B. S.: "Engineering Organization and Management," Prentice-Hall, Inc., Englewood Cliffs, N.J., 1976.
Beakley, G. C., and H. W. Leach: "Engineering—An Introduction to a Creative Profession," 3d ed., Macmillan Publishing Company, New York, 1977.
Cleland, D. I., and D. F. Kocaoglu: "Engineering Management," McGraw-Hill Book Company, New York, 1981.
Kemper, J. D.: "The Engineer and His Profession," 4th ed., Holt, Rinehart and Winston, New York, 1989.

---

[1] R. Caplan, "By Design," St. Martin's Press, New York, 1982; C. H. Flurschein (ed), "Industrial Design in Engineering," Design Council, London and Springer-Verlag, New York, 1983.

Ramo, S.: "The Management of Innovative Technological Corporations," Wiley-Interscience, New York, 1980.

Roadstrum, W. H.: "Excellence in Engineering," 2d ed., John Wiley & Sons, Inc., New York, 1978.

Roy, R., and D. Wield (eds.) "Product Design and Technological Innovation," Open University Press, Philadelphia, 1985.

CHAPTER
3

DESIGN METHODS

3-1 INTRODUCTION

If you review the steps in the design process given in Sec. 1-3 you quickly conclude that a successful design calls for skills in problem solving and in decision making. At the outset we must clearly define the problem that needs solving. The conceptualization step is perhaps the most crucial, for it is here that we combine the scientific and engineering principles and physical systems that constitute the design. What is done here often determines the ultimate success of the design through the preliminary or embodiment phase and the detail design phase. Methods of unleashing the maximum degree of creativity, tempered with engineering judgement, clearly are desirable. As we move from conceptual design to detail design there is a continual winnowing of ideas. Clearly design involves decision making.

Thus, we can begin to see the need for the development of design methods to assist in the complex process we call engineering design. Design methods are any aids, procedures, or techniques that assist in engineering design. At one extreme are the computer-based "design tools' that are being developed around workstations using expert systems, mathematical modeling and advanced optimization techniques. However, in this chapter we are more concerned with simpler design methods. One of the oldest design methods is

105

sketching or drawing to help visualize a design concept. Another group of methods aimed at systematizing the decision process grew out of World War II developments in operations research. These have been widely adopted in management science, and many techniques have been adapted for engineering design. All have as their intent to bring logical procedures into the design process.

Of the design methods in common use, one group tries to *formalize* certain design procedures and the other tries to *externalize* design thinking. Design methods which introduce formalization attempt to prevent overlooked factors in the design process or the occurrence of errors. These methods also encourage a wider search for appropriate solutions. Design methods that aim at externalizing the design process try to take the thoughts and thinking process out of the designer's head and put them into charts and diagrams. This is important when dealing with complex problems. It is also a necessary part of the teamwork that is so common in design. Putting much of the detail on paper helps to free the mind to pursue more imaginative thinking. While the introduction of design methods has sometimes been challenged as being harmful to creative activity it is far more likely that they lead to more creative solutions than the informal, internal, and often incoherent thinking procedure of the conventional design process.

3-2 CREATIVITY AND PROBLEM SOLVING

Creative thinkers are distinguished by their ability to synthesize new combinations of ideas and concepts into meaningful and useful forms. Engineering creativity is more akin to inventiveness than research. We would all like to be called "creative," yet most of us, in our ignorance of the subject, feel that creativity is reserved for only the chosen few. There is the popular myth that creative ideas arrive with flash-like spontaneity—the flash of lightning and clap of thunder routine. However, students of the creative process[1] assure us that most creative ideas occur by a slow, deliberate process that can be cultivated and enhanced with study and practice. We are all born with an inherent measure of creativity, but the process of maturation takes its toll of our native capacity. A technical education, with its emphasis on precision of thought and correct solutions to mathematical problems, is especially deadly to creativity.

A characteristic of the creative process is that initially the idea is only imperfectly understood. Usually the creative individual senses the total structure of the idea but initially perceives only a limited number of the details. There ensues a slow process of clarification and groping as the entire idea takes shape. The creative process can be viewed as moving from an amorphous idea

---

[1] E. Raudsepp, *Chem. Eng.*, pp. 101–104, Aug. 2, 1976; pp. 95–102, Aug. 16, 1976.

to a well-structured idea, from the chaotic to the organized, from the implicit to the explicit. Engineers, by nature and training, generally value order and explicit detail and abhor chaos and vague generality. Thus, we need to train ourselves to be sensitive and sympathetic to those aspects of the creative process. We need, also, to understand that creative ideas cannot be turned on upon command. Therefore, we need to recognize the conditions and situations that are most conducive to creative thought. We must also recognize that creative ideas are elusive, and we need to be alert to capture and record our creative thoughts.

Listed below are some positive steps you can take to enhance your creative thinking. A considerable literature has been written on creativity,[1] but the steps given here[2] encompass most of what has been suggested:

1. *Develop a creative attitude.* To be creative it is essential to develop confidence that you can provide a creative solution to a problem. Although you may not visualize the complete path through to the final solution at the time you first tackle a problem, *you must have self-confidence*; you must believe that a solution will develop before you are finished. Of course, confidence comes with success, so start small and build your confidence up with small successes.

2. *Unlock your imagination.* You must rekindle the vivid imagination you had as a child. One way to do so is to begin to question again. Ask "why" and "what if," even at the risk of displaying a bit of naïveté. Scholars of the creative process have developed thought games that are designed to provide practice in unlocking your imagination and sharpening your power of observation.[3]

3. *Be persistent.* We already have dispelled the myth that creativity occurs with a lightning strike. On the contrary, it often requires hard work. Most problems will not succumb to the first attack. They must be pursued with persistence. After all, Edison tested over 6000 materials before he discovered the species of bamboo that acted as a successful filament for the incandescent light bulb. It was also Edison who made the famous comment, "Invention is 95 percent perspiration and 5 percent inspiration."

4. *Develop an open mind.* Having an open mind means being receptive to ideas from any and all sources. The solutions to problems are not the property of a particular discipline, nor is there any rule that solutions can

---

[1] J. R. M. Alger and C. V. Hays, "Creative Synthesis in Design," Prentice-Hall, Inc., Englewood Cliffs, N.J., 1964; E. Van Frange, "Professional Creativity," Prentice-Hall, Inc., Englewood Cliffs, N.J., 1959; A. F. Osborne, "Applied Imagination," 3rd ed., Charles Scribner's Sons, New York, 1965; E. Lumsdaine and M. Lumsdaine, "Creative Problem Solving," McGraw-Hill, New York, 1990.

[2] R. J. Bronikowski, *Chem. Eng.*, pp. 103–108, July 31, 1978.

[3] E. Raudsepp, "Creative Growth Games," 2nd ed., Jove Publications, New York, 1982.

come only from persons with college degrees. Ideally, problem solutions should not be concerned with company politics. Because of the NIH factor (not invented here) many creative ideas are not picked up and followed through.

5. *Suspend your judgment.* We have seen that creative ideas develop slowly, but nothing inhibits the creative process more than critical judgment of an emerging idea. Engineers, by nature, tend toward critical attitudes, so special forebearance is required to avoid judgment at an early stage.

6. *Set problem boundaries.* We place great emphasis on proper problem definition as a step toward problem solution. Establishing the boundaries of the problem is an essential part of problem definition. Experience shows that this does not limit creativity, but rather focuses it.

A creative experience often occurs when the individual is not expecting it and after a period when they have been thinking about something else. The secret to creativity is to fill the mind and imagination with the context of the problem and then relax and think of something else. As you read or play a game there is a release of mental energy which your preconscious can use to work on the problem. Frequently there will be a creative "Ah-ha" experience in which the preconscious will hand up into your conscious mind a picture of what the solution might be. Since the preconscious has no vocabulary the communication between the conscious and preconscious will be by pictures or symbols. This is why it is important for engineers to be able to communicate effectively through three-dimensional sketches.

Invention

An invention is something novel and useful. As such, we generally can consider it to be the result of creative thought. A study of a large number of inventions[1] showed that inventions can be classified into seven categories:

1. *The simple or multiple combination.* The most elementary form of invention is a simple combination of two existing inventions to produce a new or improved result.

2. *Labor-saving concept.* This is a higher level of invention sophistication in which an existing process or mechanism is changed in order to save effort, produce more with the same effort, or dispense with a human operator.

3. *Direct solution to a problem.* This category of invention is more typical of what we can consider to be engineering problem solving. The inventor is

---

[1] G. Kivenson, "The Art and Science of Inventing," pp. 14–20, Van Nostrand Reinhold Company, New York, 1977.

confronted with a need and sets out deliberately to design a system that will satisfy the need.

4. *Adaptation of an old principle to an old problem to achieve a new result.* This is a variation of category 3. The problem (need) has been in existence for some time, and the principle of science or engineering that is key to its solution also has been known. The creative step consists in bringing the proper scientific principle to bear on the particular problem so as to achieve the useful result.

5. *Application of a new principle to an old problem.* A problem is rarely solved for all time; instead, its solution is based on the then current limitations of knowledge. As knowledge (new principles) becomes available, its application to old problems may achieve startling results. As an example, the miniaturization of electronic and computer components is creating a revolution in many areas of technology.

6. *Application of a new principle to a new use.* People who are broadly knowledgeable about new scientific and engineering discoveries often are able to apply new principles in completely different diciplinary areas or areas of technology.

7. *Serendipity.* The mythology of invention is full of stories about accidental discoveries that led to great inventions. Lucky breaks do occur, but they are rare. Also, they hardly ever happen to someone who is not already actively pursuing the solution of a problem. Strokes of good fortune seem to be of two types. The first occurs when the inventor is actively engaged in solving a problem but is stymied until a freak occurrence or chance observation provides the needed answer. The second occurs when an inventor suddenly gains a valuable insight or discovers a new principle that is not related to the problem he is pursuing. He then applies the discovery to a new problem and the result is highly successful.

Problem Solving

Problem solving is a process that follows a logical sequence. It begins with a decision maker, the engineer or manager, who identifies the problem, conducts an analysis to find the cause of the problem, and concludes with decision making to decide the course of events that must be followed.

Some psychologists describe the problem-solving process in terms of a simple four-stage model.

Preparation (*stage* 1). The elements of the problem are examined and their inter-relations are studied.

Incubation (*stage* 2). You "sleep on the problem."

Inspiration (*stage* 3). A solution or a path toward the solution suddenly emerges.

Verification (*stage* 4). The inspired solution is checked against the desired result.

This is obviously a simplified model, since we know that many problems are solved more by perspiration than by inspiration. Nevertheless, there is great value in letting a problem lie fallow so as to give the preconscious mind a chance to operate.

It is useful to understand something about how the mind stores and processes information. We can visualize the mind as a three-element computer.

1. The *conscious mind* compares the information and ideas stored in the preconscious mind with external reality.
2. The *preconscious mind* is a vast storehouse of information, ideas, and relations based on past education and experience.
3. The *unconscious mind* acts on the other two elements. It may distort the relation of the conscious and preconscious through its control of symbols and the generation of bias.

The exact details of how the human mind processes information are still the subject of much active research, but it is known that the mind is very inferior to modern computers in its data-processing capacity. It can picture or grasp only about seven or eight things at any instant. Thus, the mind can be characterized as a device with extremely low information-processing capacity combined with a vast subliminal store of information. Those characteristics of the mind have dominated the development of problem-solving methods.

Our attempts at problem solving often are stymied by the mind's low data-processing rate, so that it is impossible to connect with the information stored in the preconscious mind. Thus, an important step in problem solving is to study the problem from all angles and in as many ways as possible to understand it completely. Most problems studied in that way contain more than the seven or eight elements that the mind can visualize at one time. Thus, the elements of the problem must be "chunked together" until the chunks are small enough in number to be conceptualized simultaneously. Obviously, each chunk must be easily decomposed into its relevant parts.

Another important step in problem solving is the generation of divergent ideas and relations. The brainstorming technique described in Sec. 3-3 is one of the generation methods. The objective of this step is to stir up the facts in the preconscious mind so that unusual and creative relationships will be revealed.

It is important to realize that there is no guarantee that the initial approach taken in problem solving will be the successful one. Be willing to change direction when you get stuck. Persisting in an unfruitful line of attack will only lead to desperation and frustration. To keep from making a premature commitment, it is a good idea to keep the initial problem-solving effort approximate and exploratory. Aim first at identifying the right forest, then the right tree, and finally the right branch. In that way you will avoid jumping to conclusions before the problem is thoroughly digested and understood. Also, it is wise to aim for multiple solutions to your problem. By having several solutions in hand, you will be able to optimize your result. Always try to find the principle underlying the solutions. In that way you can generalize and increase your understanding. Finally, it may help to change the medium in which you are working. If you have been developing a solution primarily through verbal means, switch to writing it down, drawing it, or modeling it. Whenever possible, try to get someone else's view of the problem before you are too heavily committed.

There are situations in which group, as opposed to individual, problem solutions are preferred. Groups tend to be superior to individuals in generating a broad spectrum of ideas. They also are more effective than individuals in speed of solution. A group effort reduces the stress of problem solving if the group works together harmoniously. Groups should not be too large; three or four people is the optimum number. On the other hand, a well-trained individual has strong powers of concentration and can pursue the convergent aspects of problem solving much better than a committee. Thus, individuals are better in the problem definition and analysis phases than groups are.

3-3 CREATIVITY METHODS

Brainstorming

The teaching of creativity has almost become a cult. Over a dozen techniques for systematically developing creativity have been proposed.[1] Brainstorming is perhaps the best known operational technique for idea generation. The elements of brainstorming are presented here because the techniques can be used effectively in your student design projects without extensive instruction or indoctrination.

Brainstorming ordinarily is a group activity (typically four to eight people) in which the collective creativity of the group is captured and enhanced. The objective of brainstorming is to generate the greatest number of alternative ideas from the uninhibited responses of the group. Brainstorming is most effective when it is applied to specific rather than general problems. The problem should be limited in scope, open-ended, capable of being handled by verbal rather than graphical or analytical means, and familiar to the participants. Problems having only one correct solution or only a few sensible alternatives do not lend themselves to brainstorming.

There are four fundamental brainstorming principles.

1. *Criticism is not allowed.* Any attempt to analyze, reject, or evaluate ideas is postponed until after the brainstorming session. The idea is to create a supportive environment for free-flowing ideas.

---

[1] W. E. Souder and R. W. Ziegler, *Research Mgt.,* pp. 34–42, July 1977.

2. *All ideas brought forth should be picked up by the other people present.* Individuals should focus only on the positive aspects of ideas presented by others. The group should attempt to create chains of mutual associations that result in a final idea that no one has generated alone. All output of a brainstorming session is to be considered a group result.

3. *Participants should divulge all ideas entering their minds without any constraint.* All members of the group should agree at the outset that a seemingly wild and unrealistic idea may contain an essential element of the ultimate solution.

4. *A key objective is to provide as many ideas as possible within a relatively short time.* It is not unusual for a group to generate 30 to 50 ideas in one-half hour of brainstorming. Obviously, to achieve that output the ideas are described only roughly and without details.

There are some generalized questions that have proved helpful. By posing them to yourself or to the group during a brainstorming session, you can stimulate the flow of ideas.

Combinations. What new ideas can arise from combining purposes or functions?

Substitution. What else? Who else? What other place? What other time?

Modification. What to add? What to subtract? Change color, material, motion, shape?

Elimination. Is it necessary?

Reverse. What would happen if we turn it backward? Turn it upside down? Inside out? Oppositely?

Other use. Is there a new way to use it?

By now you may have decided that brainstorming is a difficult technique to master, but it really isn't. All that is needed is a little practice. Certainly, witholding criticism or argument is foreign to the training and instincts of most engineers. That is why the success of a brainstorming session depends strongly on the ability of the chairman of the group and on the selection of the participants. Ideally, the participants in a session should have different professional backgrounds or experience and not be involved in any strong professional rivalries or competition.

Synectics

Synectics[1] is a technique for creative thinking which draws on analogical thinking, i.e. on the ability to see parallels or connections between apparently

---

[1] W. J. J. Gordon, "Synectics," Harper & Row Pub., New York, 1961.

dissimilar topics. A synectics group should consist of about six people, about half from within the organization and half outsiders. It is important to have a broad distribution of professional or academic backgrounds. As in brainstorming the group aims to build, combine, and develop ideas in a noncritical environment. Unlike in brainstorming, where the group tries to generate a large number of ideas, a synectics group tries to work collectively toward a particular solution. The synectics method appears to work well on a problem that has previously been shown to be real and where a solution has a high probability of implementation. It has been applied extensively to problems of new product development. It is not likely to be as effective in problem identification or matching solutions to the situation.

Synectics groups use the discussion of analogies to tie their spontaneous thinking to the problem. Four types of analogies are used.

1. *Direct analogies.* Most of these are found in biological systems.[1]
2. *Personal analogies.* The designer imagines what it would be like to use one's body to produce the effect that is being sought, e.g. what it would feel like to be a helicopter rotor?
3. *Symbolic analogies.* These are poetic metaphors and similes in which one thing is identified with aspects of another, e.g. the *mouth* of a river, a *tree* of decisions.
4. *Fantasy analogies.* Here we let our imagination run wild and wish for things that don't exist in the real world.

A synectics group meeting starts with the problem statement presented by the client or company management. A discussion follows in which the members purge their minds of obvious solutions that are likely to be simple extensions of what already exists. Then analogies are sought which will transform the given problem into terms that are familiar to the experience of the members of the group. This leads to a conceptualization of the crucial difficulties that prevent a solution of the problem. The chairman of the group then calls for a solution in terms of one of the analogies. The group then develops this in a leisurely way. If the analogies get too abstract the discussion is redirected toward the problem. When a promising idea appears it is developed by the group to the point where rough prototypes could be made and tested.

---

[1] T. W. D'Arcy, "Of Growth and Form," Cambridge Univ. Press, 1961; S. A. Wainwright *et al.*, "Mechanical Design in Organisms," Arnold, London, 1976.

Note that the inventor of the Velcro fastener, George de Mestral, conceived the idea after wondering why cockleburrs stuck to his wool socks. Under the microscope he found that hook-shaped projections on the burrs grasped loops in the wool sock. After a long search he found that a nylon material could be formed into hooks that would retain their shape.

To train an expert synectics group takes about a quarter of the time of six people for a year. A trained synectics group appears to be able to find acceptable solutions to about two major problems and four minor problems a year.

3-4 THE PROBLEM STATEMENT

Obtaining a satisfactory definition of the problem is a task as big as, if not, bigger than, solving the problem. Time spent in defining the problem properly and then writing a complete problem statement invariably pays off in efficient problem solution. The worse case is to arrive at a problem solution only to realize that the solution you have obtained is not for the true problem.

Problem definition is based on identifying on identifying the true needs of the user and formulating them in a set of goals for the problem solution. The problem statement expresses as specifically as possible what is intended to be accomplished to achieve the goals. Design specifications are a major component of the problem statement. The key role of the problem statement in the design process is shown in Fig. 3-1.

The problem statement defines a poorly identified problem area as sharply as possible. Its essential elements are:

- A need statement.
- Goals (aims and objectives)
- Constraints and trade-offs
- Definitions of terms or conditions
- Criteria for evaluating the design

User needs should be categorized as to performance, time, and cost. *Performance* deals with what the design should do when it is completed and in operation. The *time* dimension of need includes all time aspects of the design, especially the schedule for the project. It has to do with such factors as when the design is needed and when it will become obsolete. *Cost* pertains to all monetary aspects of the design.

Maslow[1] developed a hierarchy of human needs that motivate individuals in general.

1. *Physiological needs* such as thirst, hunger, sex, sleep, shelter, and exercise. These constitute the basic needs of the body; and until they are satisfied, they remain the prime influence on the individual's behavior.
2. *Safety and security needs,* which include protection against danger, deprivation, and threat. When the bodily needs are satisfied, the safety and security needs become dominant.

---

[1] A. H. Maslow, *Psych. Rev.,* vol. 50, pp. 370–396, 1943.

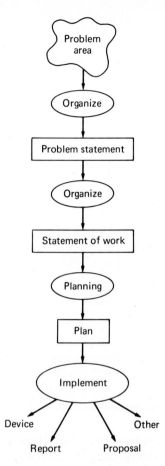

FIGURE 3-1
Key role of the problem statement in design process.

3. *Social needs* for love and esteem by others. These needs include belonging to groups, group identity, and social acceptance.
4. *Psychological needs* for self-esteem and self-respect and for accomplishment and recognition.
5. *Self-fulfillment needs* for the realization of one's full potential through self-development, creativity, and self-expression.

As each need in this hierarchy is satisfied, the emphasis shifts to the next lower need.

Our design problem should be related to the basic human needs, some of which may be so obvious that in our modern technological society they are taken for granted. However, within each basic need there is a hierarchy of problem situations.[1] As the type I problem situations are solved, we move to

---

[1] Based on ideas of Prof. K. Almenas, University of Maryland.

the solution of higher-level problems within each category of basic need. It is characteristic of our advanced affluent society that, as we move toward the solution of type II and III problem situations, the perception of the need by society as a whole becomes less universal.

Basic need	Problem situation		
	I	II	III
Food	Hunger	Vitamin deficiency	Food additives
Shelter	Freezing	Cold	Comfort
Work	Availability	Right to work	Work fulfillment

Problem situation	Analysis of problem	Societal perception of need
I	None required	Complete agreement
II	Definition of problem	Some disagreement in
	Calculation of cost	priorities
	Setting of priorities	
III	Analysis of present and	Strong disagreement
	future costs	on most issues
	Analysis of present and	
	future risks	
	Environmental impact	

Many current design problems deal with type III situations in which there is strong societal disagreement over needs and the accompanying goals. The result is protracted delays and increasing costs.

The *needs analysis* consists of listing the user needs for the design in brief, succinct phrases. Each user need should be identified with the basic need it represents. In addition, there will be technological (performance) needs, time needs, and cost needs. The needs analysis is not complete until we estimate what resources the user will exchange for satisfying his needs. This may take the form of a detailed benefit-cost analysis or a detailed estimate of the cost of manufacturing the design, including profit and marketing costs (see Chap. 9). A less exacting approach is to establish the present cost of alternative ways of meeting the need. This sets an upper bound on the cost for the new design. Finally, authoritative information on costs can be obtained from corporate marketing and sales people or from industry consultants.

The goals of the design project should be expressed in as complete detail as possible at the outset. It is a common mistake to plunge headlong into the problem solution before setting goals; then the tendency is to substitute your solution for the goals. The goals should result from the needs analysis; they should represent what is to be achieved rather than how it is to be done. By setting out the *how* rather than the *what,* we increase the possibilities for creative design solutions.

The goals should be organized into performance, time, and cost goals. *Performance goals* are the technical parameters that define the successful functioning of the design. *Time goals* are the target values for the design schedule. *Cost goals* include all elements of the investment of human and material resources for preparing the design, constructing it, and putting it into place. Generally, the performance goals will dominate over the time and cost goals at the start of a design project. However, as the project moves downstream, it is common to find that time and/or cost goals become dominant.

Every design problem will have certain boundaries or constraints within which the solution must be found. The most basic constraint is that the design must not violate the laws of nature. Most of your engineering education has been devoted to an understanding of those laws, so it is not likely you will trip up here. However, if the design carries you far out of your discipline, say from mechanical design to solid-state electronics, you may be headed for trouble. That is why interdisciplinary design teams are important. More prosaic constraints are introduced by the need for interchangeability of parts (standardization) or the requirement that the dimensions must be within certain bounds, e.g., the width of a railroad vehicle must conform to gage and clearance standards.

Legal constraints on engineering design are becoming increasingly important. Federal and state regulations pertaining to environmental pollution, energy consumption, and public health and safety are examples. Although not legally binding, trade association standards may constitute constraints with a given industry. Many of the legal constraints grew out of social and cultural pressures. Since they usually precede the establishment of laws, many farsighted corporations have ways of monitoring social pressures and incorporating them into their design constraints prior to being forced into compliance by the passage of regulatory laws. Finally, the user may impose constraints on the design that are peculiar to his own desires. If the customer is willing to pay for them, these man-imposed constraints become part of the problem definition.

A very well defined set of design goals are called *design specifications.* For a very routine design problem most of the problem statement will consist of a listing of standard design specifications.

The problem statement of any design project should be reviewed almost continuously. As the design project moves downstream, the goals become more and more detailed. We need to develop criteria for measuring the achievement of those goals. At the beginning of the design process the criteria are called *target specifications.* By the end of the process they have solidified into specifications for manufacturing and purchase. The specifications for a large engineering structure could consist of hundreds of pages of specifications and engineering drawings.

A critical problem is to develop specifications that are sufficiently, but not excessively, detailed to serve as good criteria of when a goal has been achieved. The real purpose of specifications is to screen out unsatisfactory design solutions. Designs that pass all of the tests in the specifications should meet all of the goals set for a satisfactory solution. To be effective, a criterion must discriminate; i.e. there must be a definite yes or no answer to it.

Unfortunately, not all design criteria can be established with mathematical precision. For example, how do we establish specifications for surface appearance or customer appeal? Different scales of measurement are used to describe criteria as we move from qualitative criteria (such as appearance) to quantitative criteria.[1] These scales, in increasing order of objectivity, are:

Nominal scale. A named category, like "shiny".

Ordinal scale. A rank-ordered scale of categories.

Interval scale. A scale of values with end points. The scale has no zero value.

Ratio scale. A linear scale of values that has a zero value.

The nominal scale is basically intuitive and is subject to possible bias. Quantitative scales such as the interval and ratio scales are less subject to bias. Because they are based on measurable quantities, there is less chance for misinterpretation or argument. The degree of objectivity is also determined by who judges the criteria. Increased objectivity and credibility result when the decision of the designer is replaced by the decision of a panel of peers, or even better, a panel of users.

3-5   PRODUCT DESIGN SPECIFICATION

The product design specification (PDS) is a detailed listing of the requirements to be met to produce a successful product or process. Specifications are the formal means of communication between the buyer and the seller. For the usual product the buyer is the consumer and the seller is the design/manufacturing team and the specifications are internal to the producing company, but expressed in whatever warranty is extended to the consumer. A PDS is a document which contains all of the facts related to the product outcome. It should attempt to avoid forcing the design and predicting the outcome, but it also should contain the realistic constraints that are imposed upon the design by the market or the company. The PDS is the basic reference source of the design activity. However, it must be a dynamic document that evolves and changes with the progress of the design.

---

[1] S. F. Love, "Planning and Creating Successful Engineered Design," Van Nostrand Reinhold Company. New York, pp. 73–77, 1980.

This presentation follows that developed by the Institution of Production Engineers of Great Britain.[1] The development of the PDS starts with market research (Sec. 2-9), followed by competitive benchmarking (Sec. 2-10), patent searching, and a thorough examination of the technical literature. The 28 primary elements that go into the PDS are shown in Fig. 3-2.

1. *Performance.* The performance required is an obvious first decision point. This starts with listing the functions to be performed by the product. A function is defined as that action characteristic of a product which meets the basic need or expectation of the user. Every function should be defined in two words—a verb and a noun, e.g. transmit torque, control torque, etc. For each function we must list the required performance. It is important to be realistic in this task and guard against over-specification which will place the product out of the price range of the customer. Also, adding secondary functions to the primary function of a product can increase the complexity and cost, and reduce the reliability of the product. For some products there are performance standards such as ANSI, ASTM, or NEMA standards that describe the necessary product functions in detail. Whenever applicable, these should be included in the PDS.

2. *Environment.* The total environment that the product must operate in should be specified. This includes range of temperature, pressure range, humidity, dirt or dust, corrosive environments, shock loading, vibration noise level, insect or bird damage, degree of abuse by operators, unforeseen hazards.

3. *Service life.* The expected service life and the duty cycle must be established before starting the design.

4. *Maintenance and Logistics.* A product may be designed to be discarded on failure, repaired by the user, repaired at the user's location by a company service representative, or repaired at a repair facility or at the factory. Each of these maintenance concepts can have a different effect on the design of the product. It is important to specify ease of access to components likely to require maintenance. Also, aspects of the logistics support, i.e. spares and spare parts, tools and test equipment, operator and maintenance manuals, and training of company and customer personnel in the operation and maintenance of the product, can be influenced by the design. The speed and ease with which a failed product can be repaired can greatly influence the customer's acceptance of the product.

5. *Target Product Cost.* Target selling costs should be established at the outset. This is part of the benchmarking process. Experience shows that almost all target costs are unrealistically low. Care should be taken to ensure that the target cost is consistent with the costs of the manufacturing

---

[1] "A Guide to Design for Production," Institution of Production Engineers, London, 1984.

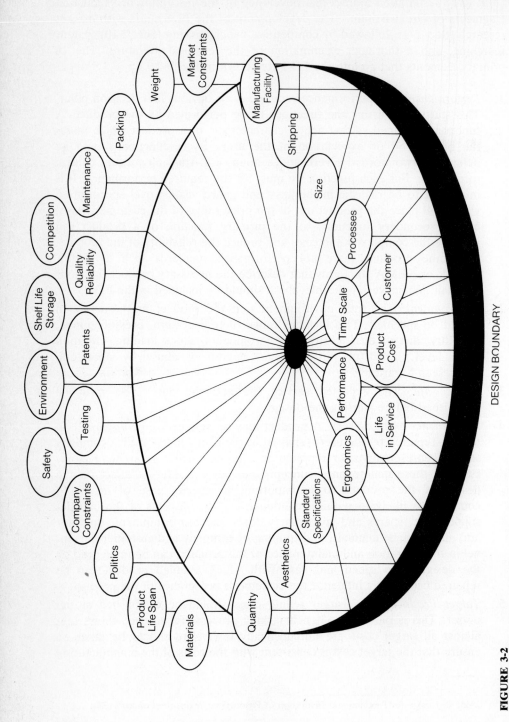

DESIGN BOUNDARY

FIGURE 3-2

Elements of Specifications (*From "A Guide to Design for Production", Institution of Production Engineers, London, 1904.*)

facility available. Many customers are interested in total life cycle cost. Thus, it may be important to establish targets for installation costs and operating and maintenance costs, as well as acquisition cost.

6. *Competition.* The benchmarking process must have included a thorough analysis of the competition. This should include likely future competition as well as existing competitors. If the evolving PDS shows serious deficiencies with what already exists then there must be compelling reasons to proceed with the project.

7. *Shipping.* It is important to know how the product will be delivered to the customer. Such factors as size of boxcar doors or cargo hatches, weight limits on truck trailers, etc. can become design constraints. Provisions must be made for lifting lugs or other design features that facilitate installation. Since we are dealing with a world marketplace, overseas shipping costs must be part of the cost target.

8. *Packing.* With many products provision for packaging in shipping and storage is a factor. It is important to know whether the packaging must be designed with respect to corrosion or shock loading. The cost of packaging will add to product cost and volume.

9. *Quantity.* The estimated number of products is a crucial variable. It greatly determines the cost per unit and has a strong influence on the methods of production and the manufacturing cost.

10. *Manufacturing Facility.* It is important to integrate the design for manufacture with the product design. An important issue is to know whether the product is to be produced in an existing plant or whether a new plant will be built. If it is to be built in an existing plant, how much leeway is there to provide new, more efficient, manufacturing processes? An important issue is to determine whether the part will be made in house or whether parts will be purchased from outside suppliers. This may have an influence on design, tolerances, etc.

11. *Size.* This has been touched on under shipping. Size can be an important constraint when replacing a product already on the market with a new product.

12. *Weight.* Weight is related to size. Also, it is frequently related to cost. Weight can be an important factor in handling on the manufacturing floor, in transportation, and in installation.

13. *Aesthetics, Appearance, and Finish.* The visual appearance of a product is what a customer sees first. The performance comes later. Color, shape, form, and texture of finish should be considered from the outset of design. Expertise in these areas may not reside in the engineering design staff. It may be necessary to call in industrial designers.

14. *Materials.* Material selection is done by the designer for simple situations, but for products with severe service requirements it should be done in conjunction with materials experts. This topic is covered in Chap. 6.

15. *Product Life Span.* At the outset it is important to have some idea of whether the product is likely to remain marketable for 3 years or 30. This factor can affect such issues as the manufacturing investment, sales strategy, and the logistics activity.

16. *Standards and Specifications.* Issues such as the use of SI units or U.S. units should be decided. Existing standards and specifications should be obtained and used.

17. *Ergonomics.* The man–machine interface requirements need to be identified.

18. *Customer.* Information on customer likes, dislikes, preferences, and prejudices should be determined through the benchmarking process. Customer input will depend on whether there are similar products already in the marketplace or whether the product is breaking new ground.

19. *Quality and Reliability.* Quality and reliability are of growing importance in the world marketplace. High risk areas should be identified in the PDS and the risks minimized using formal trade-off techniques in the design process. Risk and reliability are covered in Chap. 12 and product failure and quality in Chap. 13.

20. *Shelf Life in Storage.* Shelf life obviously is important for certain products like batteries or bake goods, but it is important for most products in other ways. Parts which are stored on a construction site for months must be protected against degradation by the elements. Spare parts which can be stored for years must be given the same protection.

21. *Inhouse Processes.* Many parts are given special processing, heat treatment, wear resistant coating, plating, etc. While strictly speaking part of the manufacturing process, these special treatments need consideration at an early stage in design.

22. *Design Schedule.* Adequate time needs to be allocated for the product design process. To short-cut design is to cause problems downstream in manufacturing, quality, or customer acceptance. However, the PDS should contain definite milestones that the design team is required to meet.

23. *Testing and Inspection.* Every product requires testing and inspection to demonstrate that it meets the product specifications. The PDS should specify the tests required and the acceptance testing and any other special quality requirements. Often the product can be designed to facilitate testing.

24. *Safety.* Safety aspects of the design must be considered. Critical parts are defined as those whose failure will result in hazardous or unsafe conditions for persons using or servicing the product. Critical parts should be identified and documented in the PDS so that the designer may give priority to ensuring these parts will be as reliable and safe as possible. Warning labels should be devised and operating manuals should clearly spell out abusive use of the product.

25. *Company Constraints*. Constraints imposed by company practice should be spelled out in the PDS. Examples could be policy requirements to use certain manufacturing facilities, or to limit new plant investment.

26. *Market Constraints*. Feedback from the marketplace cannot be ignored. For example, it may be known that a certain customer will not accept a product using a given producer's diesel engine.

27. *Patents*. All areas of useful information should be consulted prior to launching the design. Knowledge of patents (Sec. 14-7) is particularly important because it may prevent a costly patent infringement suit.

28. *Social and Political Factors*. Legislative action very quickly can create conditions which influence the business climate, and hence the future of the product design. Federal and state regulatory agencies, such as the EPA, FDA, or NRC create design constraints by their actions.

The above is a comprehensive list of the factors that should be considered when writing a PDS. It is necessary to write a separate specification for each factor. The specification should say what a product must *do* not what it must *be*. Whenever possible the specification should be expressed in quantitative terms, and when appropriate it should give limits within which acceptable performance lies. The final list of performance attributes and other factors can be very long. Therefore, it may be appropriate to divide it into two categories, musts and wants. Musts are requirements that must be satisfied, whereas wants are requirements that the customer or designer would like to meet if possible. It is important to remember that the performance attributes should be stated in a way that is independent of any particular solution.

3-6 CONCEPT SELECTION TECHNIQUE

Conceptual design takes the problem statement and generates broad solutions to it in the form of design concepts. It is the phase of design which makes the greatest demands on the designer's creativity. Conceptual design is where engineering science, practical know-how, manufacturing methods, and business aspects need to be brought together. It is the design phase where the most important decisions are made that influence the ultimate success of the design.

The Concept Selection Technique is a methodology for doing conceptual design that was developed by Professor S. Pugh and colleagues at Loughborough University of Technology.[1] The concept selection technique was developed to address the concern that it is impossible to develop and evaluate all possible concepts to satisfy a particular PDS. Thus, to minimize the possibility of the wrong choice of concept a methodology was developed for

[1] S. Pugh, "Concept Selection—a method that works," *Proc. Int. Conf. on Engr. Design*, Rome, 1981; "A Guide to Design for Production," Chap. 4, Institution of Production Engineers, London, 1984.

Concept / Criteria	 1	 2	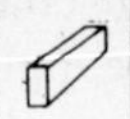 3	 4	 5	 6	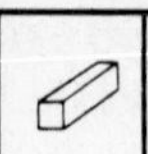 7	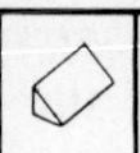 8	 9	 10	 11
A	+	−	+	−	+	−	D	−	+	+	+
B	+	S	+	S	−	−		+	−	+	−
C	−	+	−	−	S	S	A	+	S	−	−
D	−	+	+	−	S	+		S	−	−	S
E	+	−	+	−	S	+	T	S	+	+	+
F	−	−	S	+	+	−		+	−	+	S
Σ^+							U				
Σ^-											
ΣS							M				

FIGURE 3-3
Concept evaluation matrix. (*From "A Guide to Design for Production", Institution of Production Engineers, London, 1984.*)

carrying out concept formulation and evaluation in a progressive and disciplined manner.

The basic rules are as follows.

1. All ideas and embryonic solutions must be relevant to the PDS, i.e. they are solutions to the same problem having the same requirements and constraints.
2. Having developed a number of possible solutions to the problem, they are produced in sketch form to the same level of detail in each case.
3. A concept evaluation matrix is established. This compares the generated design concepts against the criteria for evaluation. Figure 3-3 shows an evaluation matrix. Note that the sketches for each concept are included in the matrix. In making the comparisons between concepts it is important to ensure that they are all at the same generic level.
4. The criteria against which the design concepts will be evaluated are chosen from the detailed requirements of the PDS. It is essential that the criteria are unambiguous and understood by all participants in the evaluation. Also, the PDS must be established before the concept selection begins.
5. A reference or datum is chosen with which all other concepts will be compared. An existing design in the product area is selected as the first datum choice.

6. Each concept/criteria combination is evaluated against the chosen datum. The following symbols are used:

   +: indicates better than, less than, less prone to, etc. relative to the datum.

   −: indicates worse than, more expensive than, more difficult to develop than, more complex than, etc. the datum.

   S: indicates same as datum. Use this when there is any doubt as to whether a concept is better or worse than the datum.

7. An initial comparison of concepts is made using the above scheme. This establishes a score in terms of the number +s, −s, and S's relative to the datum. These scores are for guidance only and must not be added up at this time.

8. Examine the scores for the individual concepts. If certain concepts show unusually high scores try to determine why they score so high. Redo the matrix to see whether on reexamination the same concepts score highly. If so, these are likely to be the best concepts with which to proceed.

9. If a pattern of one or more strong concepts does not emerge, change the datum and reevaluate. If a strong concept still does not emerge it means either that the criteria are ambiguous or that one or more concepts are subsets of the others.

10. If one concept continues to remain the strongest, change the datum and repeat. If the same concept continues to predominate, let this strong concept be the datum and redo the matrix.

11. As strong and weak features of each concept emerge it is important to attempt to make changes that will improve the situation. Often a new concept will emerge. If the effort to eliminate defects in a concept fails then it reinforces the view that it is a weak concept.

The process described above is a group effort, with the team made up of people from different disciplines, backgrounds, and job functions. As is the situation with brainstorming, there is a prohibition on passing judgement on concepts until all ideas are exhausted from the group. However, unlike brainstorming, the concept selection methodology produces a controlled convergence on the problem at hand. The method allows alternate convergent and divergent thinking to occur, since as the process proceeds and a reduction in the number of initial concepts takes place there are new concepts forced to be generated. The continual comparison of alternative concepts makes it impossible to maintain a fixed viewpoint on any one concept. The deliberate attempts to eliminate the poor features of the weaker concepts forces the emergence of new concepts. Since the method is iterative the alternate contraction and expansion of the matrix provides a convergence on the best concept(s).

Frequently the application of the concept selection technique will identify several strong concepts and will cast aside others. To proceed from there into phase II requires further work on these concepts to develop them to a higher

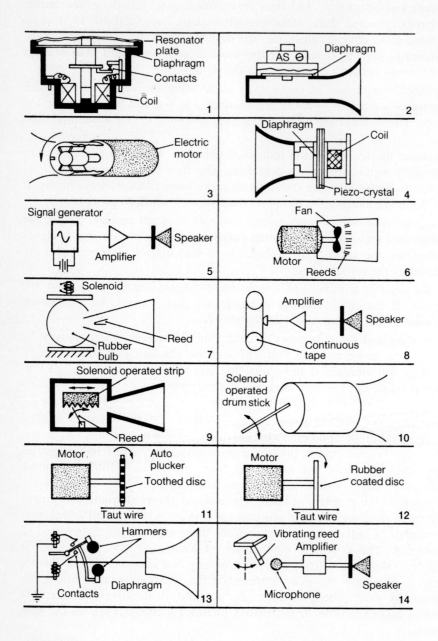

FIGURE 3-4

Sketches of the design concepts developed by a group seeking a better automobile horn. (*From "A Guide to Design for Production", Institution of Production Engineers, London, 1984.*)

Criteria \ Concept	1	2	3	4	5	6	7	8	9	10	11	12	13	14
Ease of achieving 105–125 DbA		S	–		+	–	+	+	–	–	–	–	S	+
Ease of achieving 2000–5000 Hz		S	S	N	+	S	S	+	S	–	–	–	S	+
Resistance to corrosion, erosion and water		–	–	O	S	–	–	S	–	+	–	–	–	S
Resistance to vibration, shock, acceleration	D	S	–	T	S	–	S	–	–	S	–	–	–	–
Resistance to temperature	A	S	–		S	–	–	–	S	S	–	–	S	S
Response time	T	S	–		+	–	–	–	–	S	–	–	–	–
Complexity: number of stages	U	–	+	E	S	+	+	–	–	–	+	+	–	–
Power consumption	M	–	–	V	+	–	–	+	–	–	–	–	S	+
Ease of maintenance		S	+	A	+	+	+	–	–	S	+	+	S	–
Weight		–	–	L	+	–	–	–	S	–	–	–	–	+
Size		–	–	U	S	–	–	–	–	–	–	–	–	–
Number of parts		S	S	A	+	S	S	–	–	+	–	–	S	–
Life in service		S	–	T	+	–	S	–	–	–	–	–	–	–
Manufacturing cost		–	S	E	–	+	+	–	–	S	–	–	–	–
Ease of installation		S	S	D	S	S	+	–	S	–	–	–	S	–
Shelf life		S	S		S	S	–	–	S	S	S	S	S	S
		0+	2+		8+	3+	5+	3+	0+	2+	2+	2+	0+	4+
		6–	9–		1–	9–	7–	12–	11–	8–	13–	13–	8–	9–

FIGURE 3-5
Concept evaluation matrix for the automobile horn problem. (*From "A Guide to Design for Production", Institution of Production Engineers, London, 1984.*)

level and in greater detail than was done in phase I. This should result in greater understanding of the PDS, leading to a refinement and expansion of the criteria for evaluation. The evaluation matrix is reformulated to incorporate the enhanced concepts and the improved criteria and the procedure for judgement and decision is repeated. The result of this reevaluation will either confirm the earlier evaluation or will give rise to a new ordering of the concepts. Like before, it is important to understand the reasons for the priority of concepts and the relative strengths and weaknesses of each. It is not unusual for five or six iterations to be required before a single concept emerges. The time involved to evaluate a single matrix can be a whole day for a complex design situation.

Figure 3-4 shows through sketches the concepts generated by one group working on designs for an automobile horn.[1] Fourteen comparable concepts are illustrated. Figure 3-5 shows the evaluation chart for these concepts. Concept 1, the traditional automobile horn, is chosen as the datum. We note that concept 5 is the overwhelming selection to challenge the existing product design.

---

[1] "A Guide to Design for Production," op. cit.

3-7   METHODS OF CONCEPTUAL DESIGN

Conceptual design is where the design first begins to take place. We saw in the previous section that this is the most critical and demanding stage of design. In that section we presented a design methodology, the concept selection technique, which has shown considerable success in assisting with and systematizing the conceptual design process. In this section we present further ideas and principles for helping the designer produce the best possible conceptual design. This comes chiefly from the works of Professor M. J. French, whose books on the subject[1] should be required reading for all design engineers. The design methods proposed should help by increasing insight into problems, and by diversifying the approach to problems. When faced with an intractable problem it is a great resource to be able to apply a design method which opens up new ways of attacking the problem. Another approach to problem solution is to reduce the size of the mental step required in the design process. The ultimate objective should be to be able to generate a design philosophy appropriate to each particular problem.

Design problems are difficult creative problems. Like other problems of this type they are solved by sheer hard mental effort. Design methods can help the designer to keep working through those periods when creative inspiration will not come. They can help the designer grab onto the glimmer of a solution as it flashes through the imagination. And, since most ideas ultimately prove to be unworkable, design methods can more quickly prove out the usefulness of an idea.

Combinative Ideas

The use of systematic methods to look at combinations of elements that go into a design solution is a good way to assist in the synthesis process and to ensure that good solutions are not overlooked. However, combinative procedures must be used sparingly, otherwise the number of combinations that require examination quickly become excessive. French[2] has proposed that these techniques are good for selecting alternative means of performing essential design functions in all new problems for which no extensive background exists. They are also useful in established design problems where the difficulties are well understood and dominant subproblems have emerged. Thus, there is an economy of effort in concentrating in these areas.

As an example of the combinative approach consider the conceptual design of an automobile. We want to consider the relative advantages of

---

[1] M. J. French, "Conceptual Design for Engineers," Springer-Verlag, New York, 1985; M. J. French, "Invention and Evolution," Cambridge Univ. Press, New York, 1988.

[2] M. J. French, "Conceptual Design for Engineers," Springer-Verlag, New York, 1985, Chap. 2.

TABLE 3-1
Table of Options

Advantage	Combination			
	Ff	Fr	Rf	Rf
Better road holding due to forward center of gravity	×	×		
Better road holding due to front wheel drive	×		×	
No drive shaft, therefore clear floor space	×			×
Less flexure to be accommodated in shafts driving wheels		×		×

locating the engine in the front, F, or in the rear, R, as well as whether the drive to the ground should be through the front wheels, f, or the rear wheels, r. In this simple example we have four possible combinations: Ff, Fr, Rf, and Rr. The chief advantages of each design configuration are given in Table 3-1. We observe that Ff has three of the four advantages, and Rf has only one advantage. We can reject this design concept as being inferior to the other three without further work, and this agrees with historical engineering practice. To narrow the choices further would require additional work using more detailed criteria and analysis.

Although the number of combinations was easy to handle in this simple example, with only four or five design alternatives the number of combinations becomes nearly unmanageable. Thus, procedures for eliminating options become important. Sometimes a short investigation will show that one option is inherently inferior to another no matter what the combination in which it is used. Rejecting the inferior option reduces the complexity. French suggests that the best procedure is to be ruthless in eliminating options to quickly arrive at a small number of preferred solutions. However, it is important to keep these eliminated ideas in mind with the idea of reinstating some of them at a later time as the conceptual design progresses.

Morphological Analysis

This is a common method to widen the search for possible new solutions. The word morphology means the study of shape or form, so that morphological analysis is a way of creating new forms. Morphological analysis is a systematic structured approach to problem definition and solution that uses a simple matrix or morphological box. The intent is to force order out of a fuzzy problem situation and to uncover combinations of factors that would not ordinarily develop from the normal process. Morphological analysis works best

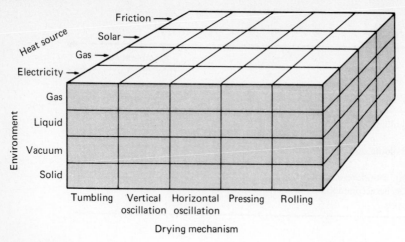

FIGURE 3-6
Morphological analysis of design of clothes dryer. (*After Alger and Hays.*)

when the problem can be readily decomposed into components or sub-problems. Each component should represent a meaningful and identifiable part of the major problem.

The following example[1] will illustrate how morphological analysis can be applied to a design problem. The problem is to develop an innovative design for an automatic clothes dryer that will lead to increased market acceptance. The first step is to identify the two or three major functions that must be performed by a clothes dryer. After a little thought we decide they are:

1. *The source of heat* determines speed and uniformity of drying.
2. *The environment around the clothes* determines smell of the clothes.
3. *The drying mechanism* determines wrinkling of the clothes.

Next, we list the various possible ways of accomplishing these key design functions. We then construct a matrix in which each of the major functions is an axis and each of the methods of performing the functions is a "coordinate" along an axis (Fig. 3-6). Each cell, representing a particular drying mechanism, heat source, and environment, constitutes a potential design concept solution. For this case there are $5 \times 4 \times 4 = 80$ combinations. Some of the combinations may be clearly impractical and can be eliminated readily; others will merit additional study. The chief virtue of the morphological box is that it provides a

---

[1] J. R. M. Alger and C. V. Hays, "Creative Synthesis in Design," Prentice-Hall, Inc., Englewood Cliffs, N.J., 1964.

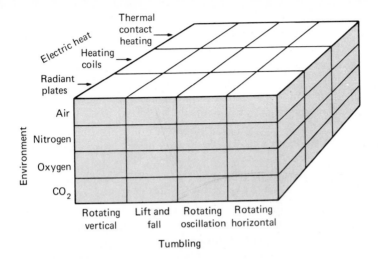

Thermal contact heating

Electric heat

Heating coils

Radiant plates

Air

Nitrogen

Oxygen

CO_2

Environment

Rotating vertical   Lift and fall   Rotating oscillation   Rotating horizontal

Tumbling

FIGURE 3-7
Morphological box for design concepts of electric tumbler clothes dryer.

formal mechanism by which unusual or innovative problem solutions will not be overlooked in the thought process.

The morphological analysis should be repeated for the cells in the matrix that appear most promising. Thus the usual design solution of an electrical heat source, a tumbling drying action, and a gas environment surrounding the clothes can be broken down in greater detail (Fig. 3-7).

Insight

French suggests that insight at the outset of a design comes not from long detailed studies along one set direction but by many rapid calculations which look at the concept in different ways. An important rule is to quantify whenever possible. In an era of ready computer availability, rough "back of the envelope" calculations are not to be sneered at when they provide insight into the design concept. Often an order of magnitude estimate is sufficient to show that an effect can be ignored or that a certain direction of thought must be abandoned. While finite element analysis is important for a detailed stress analysis, the use of this method requires a defined shape and form. Finite elements cannot provide the type of insight that is needed in conceptual design, although they are invaluable tools in later stages of the design process. French[1] gives a number of examples of the use of brief calculations to gain insight.

[1] M. J. French, "Conceptual Design for Engineers," Springer-Verlag, New York, 1985, pp. 77–82.

Optimization of the critical design parameters that determine performance is an important step in conceptual design. Design optimization is considered in detail in Chap. 5.

In gaining insight into a design it is useful to employ certain measures of performance which characterize the elements of the design. Many of these design measures are learned in discipline based engineering courses. Examples are measures like thermodynamic efficiency, strength per unit volume of material, joint efficiency, etc. Where standard measures of performance do not apply it is useful to be able to create new measures from engineering principles.

In many problems important insight can be obtained by examining the effects of scale. For example, if all the dimensions of a pressure vessel (including wall thickness) are increased by the same scale factor then the stresses are unchanged. The ideas of scaling and dimensional analysis are discussed in Sec. 4-3. A related aspect of design insight is proportion. A good designer has a feel for proportion. This type of designer can look at a drawing and tell where certain features are improperly sized.

Two important concepts of design that have been identified by French are matching and disposition. *Matching* refers to creating the proper interface between the separate elements or components of the design so that they perform as an optimized system. Others would call this systems engineering. *Disposition* is concerned with parcelling out some constrained commodity, often space, between a number of functions in the best way. Many disposition problems can be solved by optimization techniques. The concepts of matching and disposition cannot readily be discussed in the abstract, but require detailed examples. The interested reader is referred to French,[1] for a detailed discussion.

Design Guidelines

In a field as poorly researched as design methods it difficult to list with confidence a set of rules of sufficient generality and enough concreteness to be useful and enduring. The guidelines laid out below have been given by French based on his extensive experience and study.

1. Avoid arbitrary decisions. Every choice with which the designer is faced represents an opportunity to improve the design. Sometimes in design an arbitrary decision is made without recognizing it. Other times the designer is faced with a clear decision point but does not know how to take advantage to improve the design.

---

[1] M. J. French, "Conceptual Design for Engineers," Springer-Verlag, New York, 1985, Chaps 5 and 6.

2. Search for alternatives. This imperative has been emphasized throughout this section. It behoves the designer to continually search for and catalog alternative solutions to design situations.

3. Solid models can be of very great help in many design problems because they often suggest ideas and alternative methods. Solid models produced by computer graphics (Sec. 4-5) may provide faster paths to this goal, with enhanced design results.

4. Increase the level of abstraction at which the problem is formulated. Moving the problem to a higher level of abstraction often suggests different solutions. For example, restricting a design study to the level of product features will produce entirely different ideas than if the study was conducted at the higher level of product alternatives.

5. Make tables of design functions and options and use them to develop competing design concepts.

6. In developing some design concept always pursue it to the limits, and then back off. The limits will be set by physical realizability or economic constraints.

7. Aim for clarity of function. There is a tendency in design to add features or devices to overcome problems that arise in the course of the design. The solution is to be willing to stop and start again when difficulties arise. The result will be a clearer and simpler design.

8. Exploit materials and manufacturing methods to the fullest. These topics are covered in Chaps. 6 and 7.

9. Develop a logical chain of reasoning for the design. A good design can be justified by a fairly tight line of reasoning. If a logical reasoning chain cannot be developed then it is difficult to have confidence in the design and more conceptual work is required. It is important to test the logic chain, link by link, to see whether there is some fault in the reasoning. If so, this points out where the design can be improved.

10. Ask questions. The designer should develop an attitude of incredulity. Is this part necessary? What would happen if this component failred? Why did we do it that way?

3-8 DESIGN PRINCIPLES

The methods discussed in the previous sections of this chapter represent the best thinking about how to regularize the design process. However, they are essentially empirical methods. There is no scientific basis for these methods. Rather, they represent a distillation of the best ideas about what works to enhance the design practice.

There is a natural desire to improve upon this situation by developing a theory of design. This would extend intuition and experience by providing a framework for evaluating and extending design concepts. A design theory would make it possible to answer such questions as: Is this a good design?; Why is this design better than others?; How many design parameters (DPs) do I need to satisfy the functional requirements (FRs)?; Shall I abandon the idea or modify the concept? Professor Nam Suh[1] and his colleagues at MIT have developed such a theoretical basis for design which is focused around two design axioms.

An axiom is a proposition which is assumed to be true without proof for the sake of studying the consequences that follow from it. An axiom must be general truth for which no exceptions or counter-examples can be found. Axioms stand accepted, based on weight of evidence, until otherwise shown to be faulty. Suh has proposed two conceptually simple design axioms.

Axiom 1. *The Independence Axiom*
    Maintain the independence of functional requirements (FRs)
Axiom 2. *The Informatio﹍ Axiom*
    Minimize the information content.

The meaning of these two axioms is explored at length in Nam Suh's book.

Fundamental to this theory of design is the idea of functional requirements (FRs) and design parameter (DPs). These concepts have been presented earlier in this text, but Suh gives them sharper meaning. He views the engineering design process as a constant interplay between *what we want to achieve* and *how we want to achieve it.* The former objectives are always stated in the *functional domain,* while the latter (the physical solution) is always generated in the *physical domain.* The design procedure is concerned with linking these two domains at every hierarchical level of the design process, Fig. 3-8. The design objectives are defined in terms of specific requirements called *functional requirements* (FRs). In order to satisfy these functional requirements a physical embodiment must be created in terms of *design parameters* (DPs). As Fig. 3-8 illustrates, the design process consists of mapping the FRs of the functional domain to the DPs of the physical domain to create a product, process, system, or organization that satisfies the perceived societal need. Note that this mapping process is not unique. Therefore, more than one design may result from the generation of the DPs that satisfy the FRs. Thus, the final outcome still depends on the designer's creativity. However, the design axioms provide the principles that the mapping techniques must satisfy to produce a good design, and they offer a basis for comparing and selecting designs.

There are hierarchies of FRs and DPs. Figure 3-9 shows the functional hierarchy for a metalcutting lathe. The hierarchical embodiment of these FRs in the physical domain is shown in Fig 3-10. FRs at the *i*th level cannot be decomposed into the next level of the FR hierarchy without first going over to

---

[1] N. Suh, "The Principles of Design," Oxford University Press, New York, 1990.

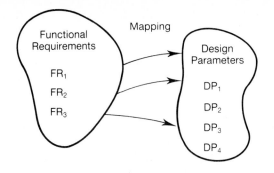

FIGURE 3-8
Suh's concept of design as the process of mapping from functional space to physical space to satisfy the functional requirements (FRs).

the physical domain and developing a solution that satisfies the ith level FRs with all the corresponding DPs. For example, the FR concerning workpiece support and toolholder (Fig. 3-9) cannot be decomposed into the three FRs at the next lower level until it is decided in the physical domain (Fig. 3-10) that a tailstock will be used to satisfy the FRs. An experienced designer will take advantage of the hierarchical structure of FRs and DPs. By identifying the most important FRs at each level of the tree and ignoring the secondary factors from consideration at that level the designer manages to keep the work and information within bounds. Otherwise, the design process becomes too

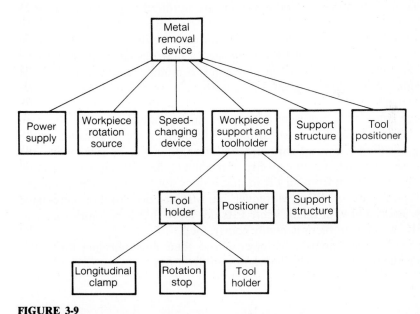

FIGURE 3-9
Hierarchy of functional requirements for a metalcutting lathe. (*From N. P. Suh, "Principles of Design," p. 37, Oxford University Press, New York, 1990.*)

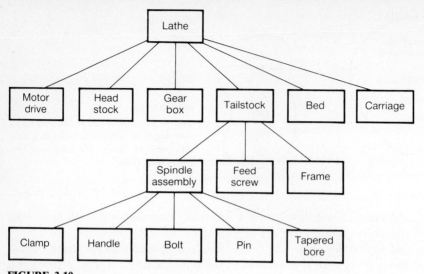

FIGURE 3-10
Hierarchy of lathe design in the physical domain. (*From N. P. Suh, "Principles of Design," p. 38, Oxford University Press, New York, 1990.*)

complex to manage. Remember that according to Axiom 1 each FR must be independent of the other FRs. This may be difficult to do on the first try; it is not unusual to expect that several iterations are required to get a proper set of FRs.

Correspondingly, there can be many design solutions which satisfy a set of FRs. Also, when the set of FRs is changed a new design solution must be found. This new set of DPs must not be simply a modification of the DPs that were acceptable for the original FRs. Rather, a completely new solution should be sought.

Design constraints represent bounds on an acceptable solution. They can be input constraints or system constraints. The former are constraints in the design specifications, like weight, material strength, cost, or size, while the latter represent restrictions on capacity of machines, laws of nature, or shape. A constraint is different from an FR in that it does not have to be independent of other constraints or FRs. Moreover, constraints do not normally have tolerances associated with them, while FRs do. Note that it becomes easier to select a FRs when a problem is highly constrained.

The two design axioms were given above their most succinct form. We now restate them in a somewhat more descriptive form.[1]

---

[1] N. P. Suh, op. cit., Chap. 3

Axiom 1. *The Independence Axiom*

Alternative Statement 1: An optimal design always maintains the independence of the FRs.

Alternative Statement 2: In an acceptable design the DPs and FRs are related in such a way that a specific DP can be adjusted to satisfy its corresponding FR without affecting other functional requirements.

Axiom 2. *The Information Axiom*

Alternative Statement; The best design is a functionally uncoupled design that has the minimum information content.

Consider the design of the door for a conventional kitchen refrigerator. There are two FRs for the vertically hung door.

FR_1: provide access to the food in the refrigerator
FR_2: minimize the energy loss

In this design, when the door is opened to provide access to food (FR_1) the cold air escapes and warm air from the room rushes in. Thus, FR_2 is not satisfied and the vertical door is not a good design because the two FRs are coupled. A better uncoupled design is a door which is hinged horizontally and opens vertically, like in a freezer chest. With this design, when the door is opened to take out the food the cold air does not escape since cold air is heavier than the warm room air. Therefore, with this type of design the FRs are independent and the design satisfies the first axiom. We note that information content is related to complexity. While we aim to prevent functional coupling, physical coupling is desirable because the integration of more than one function in a single part reduces complexity, and according to the second axiom this is beneficial.

Seven important corollaries can be derived from the two basic axioms.[1] We can view these statements as design rules that can be useful in making design decisions.

Corollary 1. *Decoupling of a Coupled Design*

Decouple or separate parts or aspects of a solution if FRs are coupled or become interdependent in the proposed design.

Decoupling does not imply that a part has to be broken into two or more separate physical parts, or that a new element has to be added to the existing design.

Corollary 2. *Minimize FRs*

Minimize the number of FRs and constraints.

---

[1] N. P. Suh, op. cit, pp. 52–54.

Increasing these elements of the design increases the information content. Do not try to produce a design that does more than was intended. A design that fulfills more functions than called for in the FRs will be more costly to operate and maintain, and may have lower reliability.

Corollary 3. *Integration of Physical Parts*
Integrate design features in a single physical part if FRs can be independently satisfied in the proposed solution.

Corollary 4. *Use of Standardization*
Use standardized or interchangeable parts if the use of these parts is consistent with the FRs and constraints.

Corollary 5. *Use of Symmetry*
Use symmetric shapes and/or arrangements if they are consistent with the FRs and constraints. Symmetrical parts require less information to manufacture and to orient in assembly.

Corollary 6. *Largest Tolerance*
Specify the largest allowable tolerance in stating FRs.

Corollary 7. *Uncoupled Design with Less information*
Seek an uncoupled design that requires less information than coupled designs in satisfying a set of FRs. There is always an uncoupled design that involves less information than a coupled design. This corollary follows as a consequence of Axioms 1 and 2, for if this corollary is not true then Axioms 1 and 2 must be invalid. An implication of this corollary is that if a designer proposes an uncoupled design that has more information than a coupled design then the design should be started anew because a better design lies somewhere.

Unfortunately, space does not permit development of the design principles an further detail. The reader is referred to Suh[1] for details of how to determine the independence of FRs, how to measure information content, and for a number of detailed examples of how to apply these techniques in design situations.

3-9 BEHAVIORAL ASPECTS OF DECISION MAKING

Making a decision is a stressful situation for most people. This psychological stress arises from at least two sources.[2] First, the decision maker is concerned about the material and social losses that will result from either course of action

---

[1] N. P. Suh, op. cit.

[2] I. L. Janis and L. Mann, *Am. Scientist,* pp. 657–667, November–December, 1976.

that is chosen. Second, he recognizes that his reputation and self-esteem as a competent decision maker are at stake. Severe psychological stress brought on by decisional conflict can be a major cause of errors in decision making. There are five basic patterns by which people cope with the challenge of decision making.

1. *Unconflicted adherence.* Decide to continue with current action and ignore information about risk of losses.
2. *Unconflicted change.* Uncritically adopt whichever course of action is most strongly recommended.
3. *Defensive avoidance.* Evade conflict by procrastinating, shifting responsibility to someone else, and remaining inattentive to corrective information.
4. *Hypervigilance.* Search frantically for an immediate problem solution.
5. *Vigilance.* Search painstakingly for relevant information that is assimilated in an unbiased manner and appraised carefully before a decision is made.

All of these patterns of decision making, except the last one, are defective.

The quality of a decision does not depend on the particulars of the situation as much as it does on the manner in which the decision-making process is carried out. We will attempt to discuss the basic ingredients in a decision and the contribution made by each.[1] The basic ingredients in every decision are listed in the accompanying table. That a substitution is made for one of them does not necessarily mean that a bad decision will be reached, but it does mean that the foundation for the decision is weakened.

Basic ingredients	Substitute for basics
Facts	Information
Knowledge	Advice
Experience	Experimentation
Analysis	Intuition
Judgment	None

A decision is made on the basis of available facts. Great effort should be given to evaluating possible bias and relevance of the facts. Emphasis should be on preventing arrival at the right answer to the wrong question. It is important to ask the right questions to pinpoint the problem. When you are getting facts from subordinates, it is important to guard against selectivity—the screening out of unfavorable results. The status barrier between a superior and a subordinate can limit communication and transmission of facts. The

---

[1] D. Fuller, *Machine Design*, pp. 64–68, July 22, 1976.

subordinate fears disapproval, and the superior is worried about loss of prestige. Remember that the same set of facts may be open to more than one interpretation. Of course, the interpretation of qualified experts should be respected, but blind faith in expert opinion can lead to trouble.

Facts must be carefully weighed in an attempt to extract the real meaning: knowledge. In the absence of real knowledge, we must seek advice. It is good practice to check your opinions against the counsel of experienced associates. That should not be interpreted as a sign of weakness. Remember, however, that even though you do make wise use of associates, you cannot escape accountability for the results of your decisions. You cannot blame failures on bad advice; for the right to seek advice includes the right to accept or reject it. Many people may contribute to a decision, but the decision maker bears the ultimate responsibility for its outcome. Also, advice must be sought properly if it is to be good advice. Avoid putting the adviser on the spot; make it clear that you accept full responsibility for the final decision.

There is an old adage that there is no substitute for experience, but the experience does not have to be your own. You should try to benefit from the successes and failures of others. Unfortunately, failures rarely are recorded and reported widely. There is also a reluctance to properly record and document the experience base of people in a group. Some insecure people seek to make themselves indispensable by hoarding information that should be generally available. Disputes between departments in an organization often lead to restriction of the experience base. In a well-run organization someone in every department should have total access to the records and experience of every other department.

Before a decision can be made, the facts, the knowledge, and the experience must be brought together and evaluated in the context of the problem. Previous experience will suggest how the present situation differs from other situations that required decisions, and thus precedent will provide guidance. If time does not permit an adequate analysis, then the decision will be made on the basis of intuition, an instinctive feeling as to what is probably right (an educated guess). An important help in the evaluation process is discussion of the problem with peers and associates.

The last and most important ingredient in the decision process is judgment. Good judgment cannot be described, but it is an integration of a person's basic mental processes. Judgment is a highly desirable quality, as evidenced by the fact that it is one of the factors usually included in personal evaluation ratings. Judgment is particularly important because most decisional situations are shades of gray rather than either black or white. An important aspect of good judgment is to understand clearly the realities of the situation.

A decision usually leads to an *action*. A situation requiring action can be thought of as having four aspects:[1] should, actual, must, and want.

---

[1] C. H. Kepner and B. B. Tregoe, "The Rational Manager: A Systematic Approach to Problem Solving and Decision Making," McGraw-Hill Book Company, New York, 1965.

The *should aspect* identifies what ought to be done if there are no obstacles to the action. A should is the expected standard of performance if organizational objectives are to be obtained. The should is compared with the *actual,* the performance that is occurring at the present point in time. The *must* action draws the line between the acceptable and the unacceptable action. A must is a requirement that cannot be compromised. A *want* action is not a firm requirement but is subject to bargaining and negotiation. Want actions are usually ranked and weighted to give an order of priority. They do not set absolute limits but instead express relative desirability.

To summarize this discussion of the behavioral aspects of decision making, we list the sequence of steps that are taken in making a good decision.[1]

1. The objectives of a decision must be established first.
2. The objectives are classified as to importance. (Sort out the musts and the wants.)
3. Alternative actions are developed.
4. The alternatives are evaluated against the objectives.
5. The choice of the alternative that holds the best promise of achieving all of the objectives represents the tentative decision.
6. The tentative decision is explored for future possible adverse consequences.
7. The effects of the final decision are controlled by taking other actions to prevent possible adverse consequences from becoming problems and by making sure that the actions decided on are carried out.

3-10 DECISION THEORY

An important area of activity within the broader subject field of operations research has been the development of a mathematically based theory of decisions.[2] Decision theory is based on utility theory, which develops values, and probability theory, which assesses our stage of knowledge. Decision theory has been applied more to business management situations than to engineering design decisions, but the potential for future applications in design appears strong. The purpose of this section is to acquaint the reader with the basic concepts of decision theory and point out references for future study.

A decision-making model contains the following six basic elements.

1. *Alternative courses of action* can be denoted as $a_1, a_2, \ldots, a_n$. As an example of alternative actions, the designer may wish to choose between

---

[1] Ibid.

[2] J. W. Pratt, H. Raiffa, and R. Schlaifer, "Introduction to Statistical Decision Theory," McGraw-Hill Book Company, New York, 1965; H. Raiffa, "Decision Analysis," Addison-Wesley Publishing Co., Reading, Mass., 1968.

the use of steel (a_1), aluminum (a_2), or glass-reinforced polymer (a_3) in the design of an automotive fender.

2. *States of nature* are the environment of the decision model. Usually, these conditions are out of the control of the decision maker. If the part being designed is to withstand salt corrosion, then the state of nature might be expressed by θ_1 = no salt, θ_2 = weak salt concentration, etc.
3. *Outcome* is the result of a combination of an action and a state of nature.
4. *Objective* is the statement of what the decision maker wants to achieve.
5. *Utility* is the measure of satisfaction or value which the decision maker associates with each outcome.
6. *States of knowledge* is the degree of certainty that can be associated with the states of nature. This is expressed in terms of probabilities.

To carry out the simple design decision of selecting the best material to resist salt corrosion in an automotive fender, we construct a table of the utilities for each outcome. A utility can be thought of as a generalized loss or gain all factors of which (cost of material, cost of manufacturing, corrosion resistance) have been converted to a common scale. We will discuss this complex problem later, but for the present consider that utility has been expressed on a scale of "losses." Table 3-2 shows the loss table for this material selection decision. Note that, alternatively, the utility could be expressed in terms of gains, and then the table would be called the payoff matrix.

Decision-making models usually are classified with respect to the state of knowledge.

1. *Decision under certainty.* Each action results in a known outcome that will occur with a probability of 1. Table 3-2 is based on this type of decision model.
2. *Decision under risk.* Each state of nature has an assigned probability of occurrence.
3. *Decision under uncertainty.* Each action can result in two or more outcomes, but the probabilities for the states of nature are unknown.

TABLE 3-2
Loss table for material selection decision

Courses of action	State of nature			
	θ_1	θ_2	θ_3	θ_4
a_1	1	4	10	15
a_2	3	2	4	6
a_3	5	4	3	2

4. *Decision under conflict.* The states of nature are replaced by courses of action determined by an opponent who is trying to maximize his objective function. This type of decision theory usually is called *game theory*.

Decision Making under Risk

We can extend our design example to the situation of decision making under risk if we assume that the states of nature have the following probability of occurrence.

State of nature	θ_1	θ_2	θ_3	θ_4
Probability of occurrence	0.1	0.3	0.4	0.2

The expected value of an action a_i is given by

$$\text{Expected value of } a_i = E(a_i) = \sum_i P_i a_i$$

Thus, for the three materials in Table 3-1, the expected losses would be:

$$a_1 \text{ (steel)} E(a_1) = 0.1(1) + 0.3(4) + 0.4(10) + 0.2(15) = 8.3$$

$$a_2 \text{ (aluminum) } E(a_2) = 0.1(3) + 0.3(2) + 0.4(4) + 0.2(6) = 3.7$$

$$a_3 \text{ (FRP) } E(a_3) = 0.1(5) + 0.3(4) + 0.4(3) + 0.2(2) = 3.3$$

Therefore, for this example, we would select fiber-reinforced polymer (FRP) as the material that would minimize the loss in utility.

Decision Making under Uncertainty

The assumption in decision making under uncertainty is that the probabilities associated with the possible outcomes are not known. The approach used in this situation is to form a matrix of outcomes, usually expressed in terms of utilities, and base the decision on various decision rules. The *maximum decision rule* states that the decision maker should choose the alternative that maximizes the minimum payoff that can be obtained. This is a pessimistic approach, because it implies that the decision maker should expect the worst to happen. For example, in the loss table shown in Table 3-2 the greatest losses are:

a_1	a_2	a_3
$\theta_4 = 15$	$\theta_4 = 6$	$\theta_1 = 5$

The best choice among these is alternative 3, FRP, so the selection is based on the course of action that maximizes the minimum value (greatest loss) of the outcomes, i.e., maximin.

An opposite extreme in decision rules is the *maximax*. This rule states

TABLE 3-3
Combined decision criterion, $\alpha = 0.3$

Alternative	Optimistic	Pessimistic	Total
Steel	0.3(1)	+0.7(15)	= 10.8
Aluminum	0.3(2)	+0.7(6)	= 5.1
FRP	0.3(2)	+0.7(5)	= 4.1

that the decision maker should select the alternative that maximizes the maximum value of the outcomes. This is an optimistic approach because it assumes the best of all possible worlds. For the payoff matrix in Table 3-2 it would be the alternative with the smallest possible loss, and the decision would be made to select alternative 1, steel.

The use of the maximin decision rule implies that the decision maker is very averse to taking risks. In terms of a utility function, that implies perception of very little utility on any return above the minimum outcome (Fig. 3-11). On the other hand, the decision maker who adopts the maximax approach places little utility on values below the maximum. Neither decision rule is particularly logical.

Since the pessimist is too cautious and the optimist is too audacious, we would like to have an in-between decision rule. It can be had by combining the two rules. By using an index of optimism α, the decision maker can weight the relatively pessimistic and optimistic components. If we weight the decision as three-tenths optimistic, we get the results shown in Table 3-3. On that basis, FRP would be chosen as giving the lowest loss in utility.

Bayesian Decision Making

Perhaps a better approach than classifying problems as to decision making under risk or under uncertainty is to adopt the realistic viewpoint that the probabilities that affect the outcomes usually are not known with much

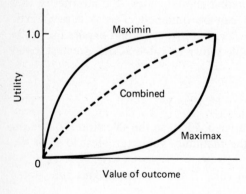

FIGURE 3-11
Utility functions implied by maximin and maximax decision rules.

confidence. Therefore, we should concentrate on achieving the best estimate of those probabilities. In the Bayesian approach to decision making,[1] one attempts to make the best estimate of probabilities, calculates the utilities, and then bases the decision on the outcome with the maximum expected utility. The improved estimation of probabilities is achieved by using *Bayes' theorem,* which is a means of revising prior estimates of probabilities on the basis of new information. Certainly it is logical, when faced with making a decision with uncertainty present, to try to remove the elements of uncertainty by finding out something about the true state of affairs. Thus, the more knowledge we possess the less the uncertainty, but there usually are real limits of cost and time to achieve complete knowledge. The Bayesian approach is a very satisfactory compromise by which we combine additional knowledge with our initial information (prior probabilities) to form revised probabilities (posterior probabilities) that, in turn, provide us with a revised basis on which to make our decision.

Bayes' theorem is expressed by Eq. (3-1). See Chap. 11 for a more detailed discussion of the concepts of probability.

$$P(A_i/B) = \frac{P(A_i)P(B/A_i)}{\sum\limits_{i=1}^{n} P(A_i)P(B/A_i)} \tag{3-1}$$

where

$P(A_i/B) =$ probability of outcome A_i occurring when outcome B
has occurred
$P(A_i) =$ estimate of outcome A_i occurring before we know the event B
$P(B/A_i) =$ conditional probability that outcome B will occur, given
that event A_i has occurred.

Consider the following simple example of the use of Bayes' theorem in decision making. A manufacturer has purchased a batch of machine parts from a supplier. These parts are forgings that are made on either a press or a hammer. The purchaser does not know whether the particular batch was made on the press or the hammer. From previous experience he believes that the level of defective parts produced by the press will be 5 percent and by the hammer forging 10 percent. He also knows that, on the average, 60 percent of the batches are made on the press and 40 percent on the hammer. If the manufacturer selects a random sample of four forgings and finds none defective, what is the posterior probability that the batch was produced by the press?

Let A_1 represent the event of manufacture by the press and A_2 that by the hammer. B is the event that in a random sample of four parts, no defective

---

[1] M. Tribus, "Rational Descriptions, Decisions, and Designs," Pergamon Press, Inc., New York, 1969.

parts were observed. The prior probabilities that the two manufacturing techniques produced the batch are:

$$P(A_1) = 0.6 \qquad P(A_2) = 0.4$$

The probability, for each machine, that no defect will be found in a random sample of four items is

$$P(B/A_1) = (1 - 0.05)^4 = 0.95^4$$
$$P(B/A_2) = (1 - 0.10)^4 = 0.90^4$$

Therefore,

$$P(A_1/B) = \frac{P(A_1)P(B/A_1)}{P(A_1)P(B/A_1) + P(A_2)P(B/A_2)}$$
$$= \frac{0.6(0.95^4)}{0.6(0.95^4) + 0.4(0.90^4)} = 0.651$$

Thus, by using posterior probability, we have increased our knowledge of the probability that the batch of parts was produced by press forging from 0.60 to 0.65.

Utility

All of the decision theory techniques discussed in this section presuppose the ability to determine the utility of each outcome. Utility is the intrinsic worth of an outcome. It is often expressed in monetary terms, but it has broader dimensions than just money. It is equal to the *value in use,* which is not necessarily equal to the *value in exchange* in the marketplace. In a general decision situation, decision makers show two types of preferences for outcomes: 1) a direct preference, as in the statement "I prefer outcome A to outcome C," and 2) an attitude toward risk, as in the statement "I prefer to play it safe and take the outcome that gives me $2000 with certainty to the strategy that gives me a 20 percent chance of losing $10,000 and an 80 percent chance of gaining $8000." Thus, the establishment of utilities is intimately concerned with the attitude toward risk. We have seen in Fig. 3-11 how different the utility functions are for a risk-averse and risk-taking individual.

Although utility is not solely expressed by monetary values, it is sufficient to illustrate the concept in monetary terms. In Table 3-4 are listed the probabilities associated with various outcomes related to the acceptance of two contracts that have been offered to a small R&D laboratory. Using expected values, we would accept contract I.

$$E(I) = 0.6(100,000) + 0.1(15,000) + 0.3(-40,000) = \$62,700$$

$$E(II) = 0.5(60,000) + 0.3(30,000) + 0.2(-10,000) = \$37,000$$

However, because contract I has a 30 percent chance of incurring a fairly large

TABLE 3-4
Probabilities and outcomes to illustrate utility

Contract I		Contract II	
Outcome	Probability	Outcome	Probability
+100,000	0.6	+60,000	0.5
+15,000	0.1	+30,000	0.3
−40,000	0.3	−10,000	0.2

loss (−$40,000), whereas contract II has only a 20 percent chance of a much smaller loss, our attitudes toward risk enter in and utility concepts most likely are important.

To establish the utility function we rank the outcomes in numerical order: +100,000, +60,000, +30,000, +15,000, 0, −10,000, −40,000. The value $0 is introduced to represent the situation in which we take neither contract. Because the scale of the utility function is wholly arbitrary, we set the upper and lower limits as:

$$U(+100,000) = 100 \qquad U(-40,000) = 0$$

Note, however, that in the general case the utility function is not linear between these limits. To establish the utility associated with the outcome of +60,000, $U(+60,000)$, the decision maker asks himself a series of questions, as follows:

Q1: Which would you prefer?
   (a) Gaining $60,000 for certain
   (b) Having a 75 percent chance of gaining $100,000 and a 25 percent chance of losing $40,000
A1: I'd prefer (a) because (b) is too risky.
Q2: Now which would you prefer?
   (a) Gaining $60,000 for certain
   (b) Having a 95 percent chance of gaining $100,000 and a 5 percent chance of losing $40,000.
A2: I'd prefer (b) with those odds.
Q3: How would you feel if the odds in (b) were 90 percent chance of gaining $100,000 and a 10 percent chance of losing $40,000?
A3: It would be a toss-up between (a) and (b) with those chances.

Therefore,

$$U(+60,000) = 0.9U(+100,000) + 0.1U(-40,000)$$
$$= 0.9(1.0) + 0.1(0) = 0.9$$

Thus, the technique is to vary the odds on the choices until the decision maker

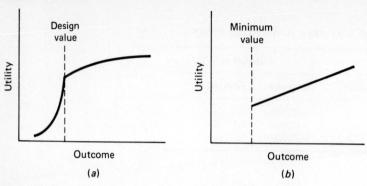

FIGURE 3-12
Common types of utility functions in engineering design.

is indifferent to the choise between (a) and (b). The same procedure is repeated for each of the other values of outcome to establish the utility for those points. A difficulty with this procedure is that many people have difficulty in distinguishing between small differences in probability at the extremes, for example, 0.80 and 0.90 or 0.05 and 0.01.

Nonmonetary values of outcome can be converted to utility in various ways. Clearly, quantitative aspects of a design performance, such as speed, efficiency, or horsepower, can be treated as dollars were in the above example. Qualitative performance indicators can be ranked on an ordinal scale, for example, 0 (worst) to 10 (best), and the desirability evaluated by a questioning procedure similar to the above.

Two common types of utility functions that are found for the dependent variables important in engineering are shown in Fig. 3-12. The utility function shown in Fig. 3-12a is the most common. Above the design value the function shows diminishing marginal return for increasing the value of the outcome. The dependent variable (outcome) has a minimum design value set by specifications, and the utility drops sharply if the outcome falls below that value. The minimum pressure in a city water supply system and the rated life of a turbine engine are examples. For this type of utility function a reasonable design criterion would be to select the design with the maximum probability of exceeding the design value. The utility function sketched in Fig. 3-12b is typical of a high-performance situation. The variable under consideration is very dominant, and we are concerned with maximum performance. Although there is a minimum value below which the design is useless, the probability of going below the minimum value is considered to be very low.

In the usual engineering design situation more than one dependent variable is important to the design. It requires developing a *multiattribute utility function*. The usual approach[1] is to assume that the utility functions for

---

[1] M. Lipson, Value Theory, *in* J. M. English (ed.), "Cost-Effectiveness," John Wiley & Sons, Inc., New York, 1968.

TABLE 3-5
Evaluation Scheme for Design Objectives

11 point scale	Description	5 point scale	Description
0	Totally useless solution	0	Inadequate
1	Very inadequate solution		
2	Weak solution	1	Weak
3	Poor solution		
4	Tolerable solution		
5	Satisfactory solution	2	Satisfactory
6	Good solution with a few drawbacks		
7	Good solution	3	Good
8	Very good solution		
9	Excellent solution (exceeds the requirement)	4	Excellent
10	Ideal solution		

each attribute are independent. Harrington[1] has proposed a desirability function that is a utility function having a scale from 0 to 1. The multiattribute utility is obtained by combining the individual utilities in a geometric mean.

3-11 EVALUATING ALTERNATIVES

An engineer in a design situation must first identify all possible alternatives. Much of this chapter has been devoted to this subject. The next step is evaluating the alternatives in as clear a manner as possible so that a decision can be made.

Assessing Utility

It is necessary to convert the values obtained for various design objectives into a consistent scale of values. Some of the design objectives may be measured on an interval or ratio scale, e.g. 32 miles per gallon or 65,000 psi, while others are expressed on a nominal scale, e.g. poor comfort ride, or on an ordinal scale, i.e. rank order. The simplest way of dealing with design objectives expressed in a variety of ways is to use a point scale. A five point scale (0–4) is used where the knowledge about the objectives is not very detailed. An eleven point scale (0–10) is used when the information is more complete. The values assigned by these scales are given in Table 3-5. It is best if several

---

[1] E. C. Harrington, *Indus. Qual. Control*, pp. 494–498, 1968.

Value scale		Parameter magnitudes			
11 point scale	5 point scale	Fuel consumption g/kWh	Mass per unit power kg/kW	Simplicity of components	Service life km
0	0	400	3.5	extremely	$20 \cdot 10^3$
1		380	3.3	complicated	30
2	1	360	3.1	complicated	40
3		340	2.9		60
4	2	320	2.7	average	80
5		300	2.5		100
6	3	280	2.3	simple	120
7		260	2.1		140
8		240	1.9	extremely simple	200
9	4	220	1.7		300
10		200	1.5		$500 \cdot 10^3$

FIGURE 3-13
Accommodation of both quantitative and qualitative parameters on a value point scale. (*From G. Pahl and W. Beitz, "Engineering Design," Springer-Verlag, New York, 1984, p. 125.*)

knowledgeable people participate in the evaluation. Figure 3-13 shows how both quantitative and qualitative parameters can be compared together on a points scale.

Weighting Factors

Usually each of the design objectives are not of equal importance. Thus, it is appropriate to assign a weighting factor to each objective. A straightforward way of doing this is to list the objectives in rank order of importance. It is helpful to do this as a team effort, since different members of the design team may have different priorities. Priorities might have been established with the client during the problem definition step, or the marketing people might be able to provide customer preferences.

The process of rank ordering can be facilitated by using a digital logic approach. Each design objective is listed and is compared to every other objective, two at a time. In making the comparison the property that is considered the more important of the two is given a 1 and the less important property is given a 0. The total number of possible combinations is $N = n(n-1)/2$, where n is the number of objectives under consideration. Consider the case where there are five design objectives, A, B, C, D, and E. We can use the chart to record the comparisons and arrive at the rank order.

Design objectives	A	B	C	D	E	Row total
A	—	1	0	0	1	2
B	0	—	1	1	1	3
C	1	0	—	0	0	1
D	1	0	1	—	1	3
E	0	0	1	0	—	1
						10

One approach is to let the weighting factor for each objective be proportional to the number of positive responses, i.e. $w_i = m_i/N$. On this basis the weight for objective A would be 0.2, for B it would be 0.3. etc. Note that the sum of the weights for each objective equals unity.

Another approach would be to use the rank order of objectives to place them on a 0–10 scale.

```
B        A       C
D                E
10  9  8  7  6  5  4  3  2  1  0
```

The most important objectives B and D have been given the value of 10, and the others are given values relative to this.

Yet another approach is to distribute 100 points among the design objectives. Points are awarded based on relative ranking, but trade-offs and adjustments between the various objectives are commonly made. Using this method the weighting factors for the five objectives might look like this.

```
D    B    A    C    E
35   30   18   9    8
```

An objectives tree[1] can be used to give a reliable assignment of weighting factors. The highest level, overall objective is given a value of 1.0. At each lower level the objectives are given weights relative to each other, but which also total 1.0. Figure 3-14 illustrates the method. There is a hierarchy of objectives at four levels. Each box in the tree is labeled with the number of the objectives, 0_1, 0_{11}, 0_{111}, etc. Thus, in Fig. 3-14 objectives 0_{12} and 0_{13} are valued equally but each only half as valuable as 0_{11}. The "true weight", given in the right side of each box is calculated as a fraction of the "true weight" of the objective above it. Subobjectives 0_{121}, 0_{122}, and 0_{123} are given values relative to each other of 0.34, 0.33, and 0.33, respectively, but their true values can only

---

[1] N. Cross, "Engineering Design Methods," pp. 45–57, 105–106, John Wiley & Sons, New York, 1989.

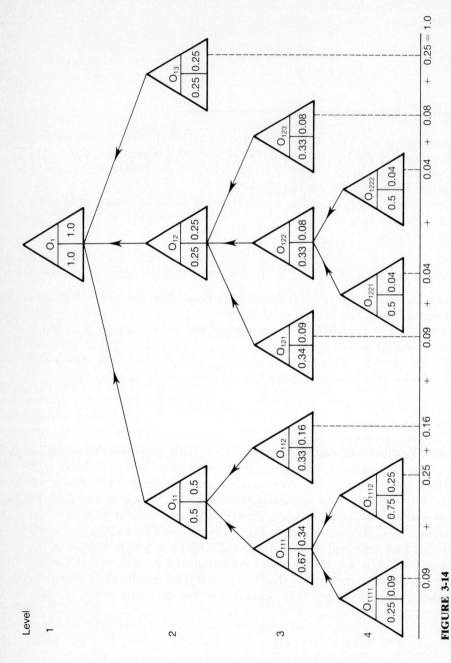

Level
1

2

3

4

FIGURE 3-14
Use of an objective tree for establishing weighting factors. (*From N. Cross, "Engineering Design Methods," John Wiley & Sons, New York, 1989, p. 106.*)

152

add to 0.25. Therefore, the true values are: $O_{121} = 0.34(0.25) = 0.09$; $O_{122} = O_{123} = 0.33(0.25) = 0.08$. Using this method it is easier to assign weighting factors with consistency because it is relatively easy to compare subobjectives in small groups of two or three and with respect to a single higher level objective. It is important to note that the true weights of all of the subobjectives add up to unity.

Decision Matrix

A decision matrix consists of a table listing the design objectives. These are each evaluated according to the methods discussed above for each of the competing design concepts. The score for each design objective is multiplied by the weighting factor for that objective to give the "relative utility". Figure 3-15 shows the decision matrix for the crane hook design discussed in Sec. 1-6.

The simplest comparison between design alternatives is to add up the utility scores and declare the concept with the highest score the winner. Rather than simply ranking the alternatives based on scores, a better way to use the decision matrix is to examine carefully the components that make up the score to see what design factors influenced the result. This may suggest areas for further study or raise questions about the validity of the data or the quality of the individual decisions.

3-12 DECISION TREES

The construction of a decision tree is a useful technique when decisions must be made in succession into the future. Figure 3-16 shows the decision tree concerned with deciding whether an electronics firm should carry out R&D in order to develop a new product. The firm is a large conglomerate that has had extensive experience in electronics manufacture but no direct experience with the product in question. With the preliminary research done so far, the Director of Research estimates that a $4 million ($4M) R&D program conducted over 2 years would provide the knowledge to introduce the product to the marketplace.

A decision point in the decision tree is indicated by a square, and circles designate chance events (states of nature) that are outside the control of the decision maker. The length of line between nodes in the decision tree is not scaled with time, although the tree does depict precedence relations.

The first decision point is whether to proceed with the $4M research program or abandon it before it starts. We assume that the project will be carried out. At the end of the 2-year research effort the research director estimates there is a 50 to 50 chance of being ready to introduce the product. If the product is introduced to the market, it is estimated to have a life of 5 years. If the research is a failure, it is estimated that an investment of an additional $2M would permit the R&D team to complete the work in an additional year. The chances of successfully completing the R&D in a further year are assessed

Objective	Weight factor	Parameter	Built-up plates: welded			Built-up plates riveted			Cast hook		
			Magnitude	Score	Value	Magnitude	Score	Value	Magnitude	Score	Value
Material cost	0.10	¢/lb	25	8	0.8	25	8	0.8	20	9	0.9
Manufacturing cost	0.20	$	1500	7	1.4	1200	9	1.8	2000	4	0.8
Time to produce	0.05	hours	40	7	0.3	25	9	0.4	60	5	0.2
Durability	0.15	experience	high	8	1.2	high	8	1.2	good	6	0.9
Reliability	0.30	experience	good	7	2.1	excellent	9	2.7	fair	5	1.5
Repairability	0.20	experience	good	7	1.4	very good	8	1.6	fair	5	1.0
Overall utility value					7.2			8.5			5.3

FIGURE 3-15
Decision matrix for the crane hook design.

Payoffs

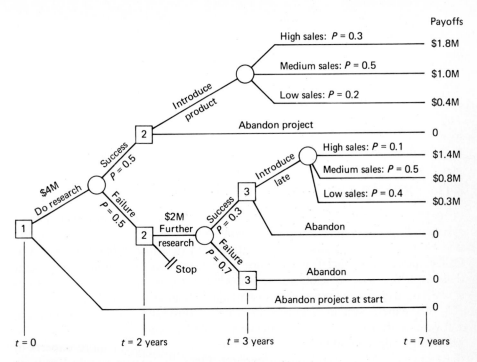

FIGURE 3-16
Decision tree for an R&D project.

at 3 in 10. Management feels that the project should be abandoned if a successful product is not developed in 3 years because there will be too much competition. On the other hand, if the product is ready for the marketplace after 3 years, it is given only a 1 in 10 chance of success.

The payoffs expected at the end are given to the far right at the end of each branch. The dollar amounts should be discounted back to the present time by using techniques of the time value of money (Chap. 8). Alternatively, the payoff could be expressed in terms of utility. As a decision rule we shall use the largest expected value of the payoff. Other decision rules, such as *maximin*, could be used.

The best place to start in this problem is at the end of the branches and work backwards. The expected values for the chance events are:

$$E = 0.3(1.8) + 0.5(1.0) + 0.2(0.4) = 1.12 \quad \text{for the on-time project}$$
$$E = 0.1(1.4) + 0.5(0.8) + 0.4(0.3) = 0.66 \quad \text{for the delayed project at decision point 3}$$
$$E = 0.3(0.66) + 0.7(0) - 2 = -\$1.8M \quad \text{for the delayed project at decision point 2}$$

Thus, carrying the analysis for the delayed project backward to $\boxed{2}$ shows that to continue the project beyond that point results in a large negative expected payoff. The proper decision, therefore, is to abandon the research project if it is not successful in the first 2 years. Further, the calculation of the expected payoff for the on-time project at point $\boxed{1}$ is a large negative value.

$$E = 0.5(1.12) + 0.5(0) - 4.0 = -\$3.44$$

Thus, either the expected payoff is too modest or the R&D costs are too great to be warranted by the payoff. Therefore, based on the estimates of payoff, probabilities, and costs, this R&D project should not have been undertaken.

3-13 EXPERT SYSTEMS

A methodology with great potential for decision making in design is expert systems. Expert systems[1] are called *knowledge-based systems*. They comprise one of the forefront areas of the exciting field of *artificial intelligence*[2] (AI). Other areas of artificial intelligence are automated reasoning, intelligent databases, knowledge acquisition, knowledge bases and knowledge representation, machine learning, natural languages, and vision and sensing.

An expert system acquires knowledge through knowledge-acquisition software tools from a trained specialist called a knowledge engineer. The knowledge engineer obtains his/her knowledge from one or more experts in the technical area, called domain experts. Once the expert system has been constructed this relationship need no longer exist. A prime advantage of expert systems is that they capture the knowledge of experts that may otherwise be lost through death or retirement. Moreover, they can contain the cumulative knowledge of several experts, they are available any time of day or night, and they can be distributed widely throughout an organization. However, it should be quickly added that expert systems are not a substitute for a human expert. Unless a problem is fully understood, which can come only from humans, the expert system project will fail.

The user of an expert system works through a keyboard interface. The input consists of system facts and suppositions of varying degrees of validity. The user interface tends to be highly interactive, following the format of a question and answer session. The expert system returns answers, recommendations, or diagnoses. In a design expert system a graphics interface may be required in order to visualize the object being designed. At present most

---

[1] F. Hayes-Roth, D. Waterman and D. B. Lenat (eds), "Building Expert Systems," Addison-Wesley Pub. Co., Reading, MA, 1983; D. A. Waterman, "A Guide to Expert Systems," Addison-Wesley Pub. Co., Reading, MA, 1986; D. W. Rolston, "Principles of Artificial Intelligence and Expert System Development," McGraw-Hill, 1988.

[2] P. H. Winston, "Artificial Intelligence," Addision-Wesley Pub. Co., Reading, MA, 1977; E. Charniak and D. McDermott, "Introduction to Artificial Intelligence," Addison-Wesley Pub. Co., Reading, MA, 1985.

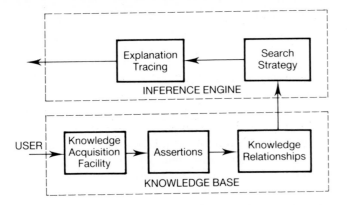

FIGURE 3-17
General structure of an expert system.

expert systems utilize a specialized database, but they are moving rapidly to be able to draw upon generalized databases.

The elements of an expert system are shown in Fig. 3-17. The two major divisions are the knowledge base and the inference engine. The knowledge base is unique to a particular domain but the inference engine may be common to many domains of knowledge. The *knowledge acquisition facility* is the component responsible for entering the knowledge into the database. At its simplest level this facility acts as an editor and knowledge is entered directly in a form acceptable by the language in which the expert system is written. On a more sophisticated level this facility can translate an input in natural language into the representation that the knowledge base can understand. This is an important feature in design, for unless the user is willing to learn the programming language in which the expert system is written it would be impossible to customize the knowledge base.

The *assertions* component, also called the working memory or temporary data store, contains the knowledge about the particular problem being solved. Data are represented by predicate logic, by frames, or by semantic networks. The *knowledge relationships* component contains formulas showing the relationship among several pieces of information. The "if–then" production rule is the most common relationship. This rule has the form:

IF condition THEN action

For example: IF stress level exceeds 85 ksi THEN part will fail. In any knowledge base there is a balance between the assertion of facts (declarative knowledge) and the rules for manipulating those facts (procedural knowledge). Generally, the less knowledge declared the greater the procedural knowledge required, and vice versa.

The inference engine contains the control mechanisms for the expert system. In a production rules expert system the AI reasoning is responsible for choosing which rule to perform next. This constitutes the *search strategy*. Since the probability of one rule following another is less than 100 percent in most cases, it has been necessary to incorporate uncertainty into the rules. With a small number of rules it is practical to search in random order, but if the number of rules is large it is necessary to partition into sublists on some logical basis. *Explanation tracing* is provided to retrace the chain of production rules that led to the development of the system. This greatly enhances the credibility of the expert system.

Special computer languages were developed initially for AI applications like expert systems. LISP (a list programming language) and Prolog (programming in logic) are most commonly used with the early expert systems, but more recently expert systems have employed more general purpose languages like FORTRAN, Pascal, and C. Higher level programming languages for expert systems have been developed. The most common are OPS5, KEE, and ART. As work on expert systems evolved researchers noticed that the knowledge base, dealing with facts and if–then rules, was specific to the problem domain, but the inference engines were very similar from one domain to another similar problem domain. This led to organization of expert systems in ways that separate the knowledge about the problem domain from general knowledge about how to solve problems or how to interact with the user (the inference engine). The latter software is marketed as expert system shells, or expert system building frameworks. They permit the construction of expert systems on computer workstations with little programming being required. In selecting such a design tool for expert systems it is important not to pick a shell tool with more generality than is needed. For best success, try to match the problem characteristics to the tool features, and test the tool early by building a small prototype expert system.

Expert systems are finding rapid application in industry for such tasks as analysis of quality control data, assisting manufacturing personnel to safely operate machinery and accurately tune instruments. A common application is the troubleshooting of equipment that breaks down or the readjustment of equipment that drifts out of calibration. At the present time about 2000 applications have been reported. An area of great promise is real-time process control. The direct application of expert systems to the design process is proceeding more slowly, but it has real potential. Some of the benefits to be expected are:

- The ability to capture valuable expertise and then to put it comfortably into the hands of a novice.
- The ability to improve the consistency of designs within an organization.
- The ability to eliminate errors in problem solving.
- The ability to interface the expert system with advanced software for engineering analysis. By further increasing the analytical capacity the amount

of detail the system can cope with is expanded. An example of this capability is given in Sec. 4-7.

- The ability to search large databases for optimal selection of concepts, components, and materials.
- The ability to search design libraries for similar designs, so that engineers can learn from past experience and avoid duplication.
- The ability to reduce the cost of design while at the same time improving quality.

3-14 EMBODIMENT DESIGN

The majority of this chapter has been concerned with a discussion of methods to assist with the difficult task of conceptual design. The next stage in the design process is embodiment design, where the concept or concepts are invested with bodily form. This is the same phase of the design process which we called *preliminary design* in Chap. 1, but here we use the terminology which is more prevalent in Europe. In this phase the concepts are developed in greater detail. If there is more than one concept then the design of each must be carried to the point where a valid decision can be made.

An important task in embodiment design is to quantify the important design parameters so that an optimum solution can be established. It is often necessary to produce several scale layouts to obtain information about the advantages and disadvantages of different design variants. It is in the embodiment stage that a final check is made on function, spatial compatibility, design aesthetics, and the financial viability of the project. The end product of the embodiment design phase is usually a general set of drawings and specifications describing the design but without detailed dimensions or information on how the design will be manufactured.

3-15 DETAIL DESIGN

This phase of the design process produces the complete description of the design at a level of detail suitable for manufacture. This consists of the arrangement, form, dimensions, tolerances, and surface properties of all individual parts. The materials and manufacturing processes are specified. The way the parts will be assembled is determined. Clearly, a very large number of small but important points need to be decided in detail design. The efficiency and precision with which these are handled determines the speed with which a design reaches the market and has a great deal to do with its cost. The introduction of the computer into the design process has had a growing positive impact on detail design.

Four key factors that influence detail design are: (1) standards, (2) standard components, (3) tolerances, and (4) materials and manufacturing processes.

The importance of designing to standards was discussed in Sec. 1-11. Many companies have compiled their own design and drawing procedures, which incorporate national standards. A well-conceived set of design standards can have an important influence on economical design, while serving as a vehicle for technical communications both within and outside of the organization. It is likely that few companies have achieved the full benefit that could be achieved from design standardization.

The savings to be obtained by the prevention of the design of duplicative parts are obvious. The solution is to institute a system for recognizing similar parts and instilling the discipline in the design engineers to use it. A variety of coding and classification schemes and computer retrieval have been used. A move toward standardization of components not only saves on design time, but it has bigger payoffs in reduced costs in production planning, tooling, inspection, inventory, and accounting.

Tolerances were considered in Sec. 1-8. The relationship between tight tolerances and high manufacturing cost is generally recognized.[1] However, it may not be appreciated that the cost of reducing dimensional tolerances can vary greatly depending upon where the dimension is located on the part. A common difficulty is specifying tolerances from an unsuitable manufacturing datum surface.[1] It is important that the designer carefully determine the influence of precision on the product and then specify realistic tolerances based on functional requirements.

While the first responsibility of the designer is to make sure that the product functions properly, the design responsibility does not stop there. The designer also is responsible for selecting the materials from which the design will be manufactured and, in conjunction with the manufacturing staff, the way it will be produced. There is no question that the cost of manufacturing can be greatly influenced at the design stage. As a result, there has been strong emphasis recently on design for producibility and on simultaneous or concurrent engineering, where the manufacturing issues are made an integral part of design. This requires regular and cooperative interaction between the design engineers and the manufacturing engineers. The subject of material selection for design is discussed in detail in Chap. 6, while the interrelationships between design and manufacturing are covered in Chap. 7.

BIBLIOGRAPHY

Bailey, R. L.: "Disciplined Creativity for Engineers," Ann Arbor Science Publishers, Ann Arbor, Mich., 1978.
Cross, N.: "Engineering Design Methods," John Wiley & Sons, New York, 1989.
French, M. J.: "Conceptual Design for Engineers," 2d ed., Springer-Verlag, New York, 1985.

---

[1] C. Wick and R. F. Veilleux (eds.), "Tool and Manufacturing Engineers Handbook," vol. 4, chap. 4, Soc. of Manuf. Engrs., Dearborn, MI, 1987.

Hill, P. H. *et al.*: "Making Decisions: A Multidisciplinary Approach," Addision-Wesley Pub. Co., Reading, Mass., 1979.

Pugh, S.: "Total Design," Addison-Wesley Pub. Co., Reading, Mass., 1990.

Rubenstein, M. F.: "Patterns of Problem-Solving", Prentice-Hall Inc., Englewood Cliffs, N.J., 1975.

—— and K. Pfeiffer: "Concepts in Problem Solving," Prentice-Hall Inc., Englewood Cliffs, N.J., 1980.

Sandler, B.Z.: "Creative Machine Design," Paragon House Publishers, New York, 1985.

Souder, W.E.: "Management Decision Methods for Managers of Engineering and Research," Van Nostrand Reinhold Co., New York, 1980.

Suh, N.P.: "The Principles of Design," Oxford University Press, New York, 1990.

CHAPTER
4

MODELING
AND
SIMULATION

4-1 THE ROLE OF MODELS IN ENGINEERING DESIGN

A model is an idealization of a real-world situation that aids in the analysis of a problem. You have employed models in much of your education, and especially in the study of engineering you have learned to use and construct models such as the free-body diagram, electric circuit diagram, and the control volume in a thermodynamic system. Figure 4-1 illustrates some of the common types of conceptual models.

A model may be either descriptive or predictive. A *descriptive model* enables us to understand a real-world system or phenomenon; an example is a cutaway model of an aircraft gas turbine. Such a model serves as a device for communicating ideas and information. However, it does not help us to predict the behavior of the system. A *predictive model* is used primarily in engineering design because it helps us to both understand and predict the performance of the system.

We also can classify models as follows:[1] 1) Static-dynamic, 2)

---

[1] W. J. Gajda and W. C. Biles, "Engineering Modeling and Computation," chap. 2, Houghton-Mifflin Company, Boston, 1978.

162

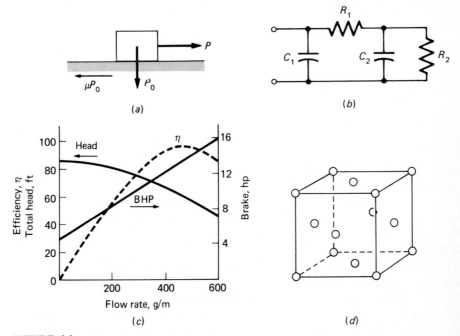

FIGURE 4-1

Examples of common conceptual models. (*a*) Free body diagram; (*b*) electric circuit diagram; (*c*) graphic representation (pump characteristics); (*d*) crystal lattice.

Deterministic-probabilistic, and 3) Iconic–analog–symbolic. A *static model* is one whose properties do not change with time; a model in which time-varying effects are considered is *dynamic*. In the deterministic-probabilistic class of models there is differentiation between models that predict what will happen. A *deterministic model* describes the behavior of a system in which the outcome of an event occurs with certainty. In many real-world situations the outcome of an event is not known with certainty, and these must be treated with *probabilistic models*.

An *iconic model* is one that looks like the real thing. Examples are a scale model of an aircraft for wind tunnel test and an enlarged model of a polymer molecule. Ioconic models are used primarily to describe the static characteristics of a system, and they are used to represent entities rather than phenomena. As geometric representations they may be two-dimensional (maps, photographs, or engineering drawings) or three-dimensional (a balsa wood and paper model airplane or a wood and plastic mockup of a full-size automobile). Three-dimensional models are especially important to communicate a complex design concept, gage customer reaction to design styling, study the human engineering aspects of the design, and check for interferences between parts or components of a large system.

We can distinguish four types of physical or iconic models that are used in engineering design. The *proof of concept model* is a minimally operative model of the basic principle of the design concept. It is usually very elementary and assembled from readily available parts and materials. Sometimes this is known as a "string and chewing gum" model. The *scale model* is dimensionally shrunken or enlarged compared with the physical world. It is often a nonoperating model made from wood or plastic, but it is important for communicating the concept and for visualizing possible space interferences or conflicts.[1] The *experimental model* is a functioning model containing the ideas of the design concept. It is as nearly like the proposed design as possible, but it may be incomplete in appearance. This model is subjected to extensive testing and modification. The *prototype model* is a full-scale working model of the design. It is technically and visually complete. The prototype often is handmade, but in other respects is intended to meet the needs of the user. As we move from proof of concept to prototype the model increases in complexity, completeness, and cost.

Analog models are those that behave like real systems. They are often used to compare something that is unfamiliar with something that is very familiar. Unlike an iconic model, an analog model need look nothing like the real system it represents. It must either obey the same physical principles as the physical system or simulate the behavior of the system. There are many known analogies between physical phenomena,[2] but the one most commonly used is the analogy between easily made electrical measurements and other physical phenomena. An ordinary graph is really an analog model because distances represent the magnitudes of the physical quantities plotted on each axis. Since the graph describes the real functional relation that exists between those quantities, it is a model. Another common class of analog models are process flow charts.

Symbolic models are abstractions of the important quantifiable components of a physical system. A mathematical equation expressing the dependence of the system output parameter on the input parameters is a common symbolic model. A symbol is a shorthand label for a class of objects, a specific object, a state of nature, or simply a number. Symbols are useful because they are convenient, add to simplicity of explanation, and increase the generality of the situation. A symbolic model probably is the most important class of model because it provides the greatest generality in attacking a problem. The use of a symbolic model to solve a problem calls on our analytical, mathematical, and logical powers. A symbolic model is also

[1] L. G. Lamit, "Industrial Model Building," Prentice-Hall, Englewood Cliffs, NJ, 1981.

[2] G. Murphy, D. J. Shippy, and H. L. Luo, "Engineering Analogies," Iowa State University Press, Ames, Iowa, 1963.

important because it leads to quantitative results. We can further distinguish between symbolic models that are *theoretical models,* which are based on established and universally accepted laws of nature, and *empirical models,* which are the best approximate mathematical representations based on existing experimental data.

We have seen that modeling is the representation of a system or part of a system in physical or mathematical form that is suitable for demonstrating the behavior of the system. *Simulation* involves subjecting models to various inputs or environmental conditions to observe how they behave and thus explore the nature of the results that might be obtained from the real-world system. Simulation is manipulation of the model. It may involve system hardware (prototype models) subjected to the actual physical environment, or it may involve mathematical models subjected to mathematical disturbance functions that simulate the expected service conditions.

The solution of models by the straightforward application of mathematical techniques has been the classical approach in much academic engineering education. However, only the simplest (and hence most unrealistic) models can be solved with classic analytic methods. The widespread use of the digital computer, and to a much lesser extent the analog computer, has greatly expanded the scope and usefulness of mathematical modeling. The use of numerical methods for solution and the ease with which iterative procedures can test many specific states of the model have firmly established computer modeling and simulation as a powerful tool of engineering design.

Engineers use models for thinking, communications, prediction, control, and training. Since many engineering problems deal with complex situations, a model often is an aid to visualizing and thinking about the problem. One of the results of an engineering education is that you develop a "menu of models" that are used instinctively as part of your thought process. Models are vital for communicating whether via the printed page, the computer screen, or oral presentation. Generally, we do not really understand a problem thoroughly until we have predictive ability concerning it. Engineers must make decisions concerning alternatives. The ability to simulate the operation of a system with a mathematical model is a great advantage in providing sound information, usually at lower cost and in less time than if experimentation had been required.

Moreover, there are situations in which experimentation is impossible because of cost, safety, or time. While we usually worry about whether the model describes the real situation closely enough, there are situations in which the model constrains and controls the real-world system. For example, a detailed engineering drawing is an iconic model that describes in complete detail how the part is to be built. Finally, we have the growing use of detailed models for the training of operators of complicated systems. Thus, airline pilots train on flight simulators and nuclear power plant operators learn from reactor simulators.

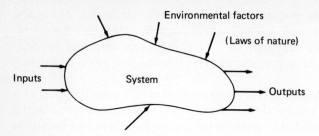

FIGURE 4-2
Characteristics of the modeling process.

4-2 MATHEMATICAL MODELING

In mathematical modeling the components of a system are represented by idealized elements that have the essential characteristics of the real components and whose behavior can be described by mathematical equations. The first step is to devise a conceptual model that represents the real system to be analyzed. You have been exposed to many examples of simple mathematical models in your engineering courses, but modeling is a highly individualized art. A key issue is the assumptions, which determine on the one hand the degree of realism of the model and on the other hand the practicality of the model for achieving a numerical solution. Skill in modeling comes from the ability to devise simple yet meaningful models and to have sufficient breadth of knowledge and experience to know when the model may be leading to unrealistic results.

A generalized picture of a mathematical model for a system or component is shown in Fig. 4-2. The choice of the system that is modeled is an important factor in the success of the model. Engineering systems frequently are very complex. Progress often is better made by breaking the system into simpler components and modeling each of them. In doing that, allowance must be made for the interaction of the components with each other. Techniques for treating large and complex systems by isolating the critical components and modeling them are at the heart of the growing discipline called *systems engineering*.[1]

For example, in the gross sense we might consider the system to be a central-station power plant (Fig. 4-3). But on further reflection we realize that the basic elements of the plant are given by the block diagram in Fig. 4-4. However, each of these components is a complex piece of engineering equipment, so the total power plant system might be better modeled by Fig. 4-5.

---

[1] H. Chestnut, "Systems Engineering Tools," John Wiley & Sons, Inc., New York, 1965; ibid, "Systems Engineering Methods," John Wiley & Sons, Inc., New York, 1967.

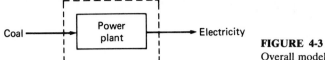

FIGURE 4-3
Overall model for a steam power plant.

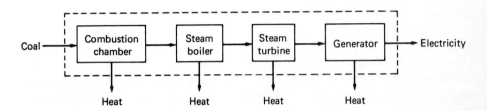

FIGURE 4-4
Block diagram of major components of power plant.

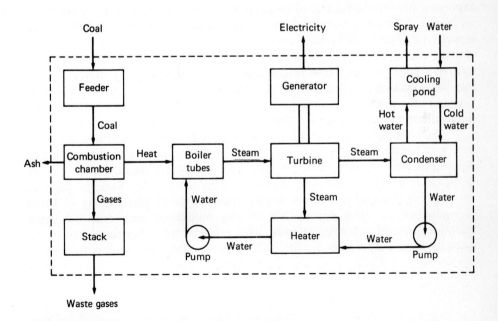

FIGURE 4-5
Detailed systems for a coal-fired power plant.

In developing a model we walk a fine line between simplification and realism. One way to achieve simplification is to minimize the number of physical quantities that must be considered. We do that routinely by neglecting small effects. Thus, we may assume a structural member is completely rigid when its elastic deformation is considered of little consequence to the problem, or we may assume a fluid is Newtonian viscous when it in fact shows a small deviation from ideality. We must be aware that factors that are neglected routinely because they are negligible in one environment may not be of minor consequence in a much different situation. Thus, we neglect surface tension effects when dealing with large objects in a fluid but must consider it when the particles are fine.

Another common assumption is that the environment is infinite in extent and therefore entirely uninfluenced by the system being modeled. In approximate models it also is common practice to assume that the physical and mechanical properties are constants that do not change with time or temperature. Generally, we start with two-dimensional models because they are more mathematically tractable.

Important simplification results when the distributed properties of physical quantities are replaced by their lumped equivalents. A system is said to have *lumped parameters* if it can be analyzed in terms of the behavior of the endpoints of a finite number of discrete elements. Lumped parameters have just single values, whereas distributed parameters have many values spread over a field in space. The mathematical model of a lumped-parameter system is expressed by differential equations, and a distributed-parameter system leads to partial differential equations. Systems that can be represented by linear models, i.e., linear differential equations, are much more easily solved than systems represented by nonlinear models. Thus, a common first step is to assume a linear model. However, since we live in a basically nonlinear world, this simplification often must be abandoned as greater realism is required. Likewise, the usual first step is to develop a deterministic model by neglecting the fluctuations or uncertainties that exist in the input values.

Once the chief components of the system have been identified, the next step is to list the important physical and chemical quantities that describe and determine the behavior of the system. As Fig. 4-2 indicates, they can be grouped into input and output parameters. Next, the various physical quantities are related to one another by the appropriate physical laws.[1] These are modified in ways appropriate to the model to transform the input quantities into the desired output. The relation that transforms the input quantities into output ones is called a *transfer function*. It may take the form of

---

[1] W. Hughes and E. Gaylord, "Basic Equations of Engineering Science," Schaum, New York, 1964; W. G. Reider and H. R. Busby, "Introductory Engineering Modeling," John Wiley & Sons, New York, 1986.

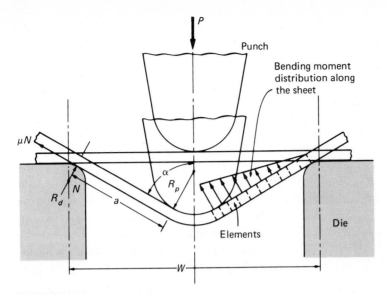

FIGURE 4-6
Schematic details of press brake bending.

algebraic, differential, or integral equations. The solution of these equations, either analytically, numerically, or graphically, is the last step in the modeling process.

Example—A Process Model

This example of mathematical modeling is concerned with press brake bending.[1] The process is used to bend flat sheet into long angle shapes. As Fig. 4-6 illustrates, the process is carried out by placing a flat sheet over a die: a punch attached to the ram of the press brake is then pushed into the sheet to create the bend. The input parameters for the process are:

1. Sheet characteristics
    (*i*) Thickness t
    (*ii*) Width w
    (*iii*) Strain-hardening behavior of metal
2. Tooling characteristics
    (*i*) Punch radius R_p
    (*ii*) Die radius R
    (*iii*) Coefficient of friction μ at the die-sheet interface

---

[1] V. Nagpal, C. F. Billhardt, P. S. Raghupathi, and T. L. Subramanian, "Process Modeling," pp. 287–301, American Society for Metals, Metals Park, Ohio, 1980.

The output parameters are:

1. Radius of curvature R' of the bend after springback
2. Punch load P vs. punch displacement

Several basic assumptions have been made:

1. Because the length of the bend is much greater than the sheet thickness, the deformation occurs under two-dimensional plane strain conditions with zero strain along the bend length. The effect of shear forces in the thickness of the sheet has been neglected, so that equations for pure plastic bending are applicable.
2. The mode of deformation is rigid plastic. Any elastic deformations is neglected.
3. The sheet metal is a rigid plastic material that obeys a linear strain-hardening law. The constitutive equation for the material is taken to be

$$\bar{\sigma} = \sqrt{3}K_0 + \tfrac{3}{2}K_v\bar{\varepsilon} \tag{4-1}$$

where K_0 and K_v are material constants and $\bar{\sigma}$ and $\bar{\varepsilon}$ are the effective stress and strain, respectively.

The equations for the deformation under pure bending of a linearly strain-hardening plastic material are taken from the existing applied mechanics literature.[1] They are the basic engineering science equations. It is typical, as in this case, for the modeler to incorporate the most appropriate fundamental physical laws; although when appropriate basic equations are not available, they may have to be developed before the model can be constructed. Note also that in this case the most useful information was published in German. That illustrates the importance of being aware of world technical developments and having fluency in the major technical languages or access to translations (see Chap. 14). The relation between the bending moment M and the mean radius of curvature r_m is given by Eq. (4-2), the terms of which are the parameters A, η, and X.

$$\frac{M}{K_0 t_0^2} = \eta^2 \left\{ \frac{1}{2} + \frac{1-(X^2/4)(1-e^A)}{X^2} \right.$$
$$\left. + \frac{K_n}{2K_0}\left[\frac{1+(X^2/4)}{X^2}\ln\left(\frac{1+(X/2)}{1-(X/2)}\right) - \frac{1}{X} + \frac{1}{X}\ln\left(\frac{1-(X^2/4)}{\eta^2}\right)\right]\right\} \tag{4.2}$$

and

$$A = \frac{K_v}{2K_0}\ln\left(\frac{1-(X^2/4)}{\eta^2}\right)\ln\left(\frac{1+(X/2)}{1-(X/2)}\right)$$

$$\eta = \frac{r_0 - r_i}{t_0} = \frac{t}{t_0} \qquad X = \frac{2(r_0 - r_i)}{r_0 + r_i} = \frac{t}{r_m}$$

---

[1] F. Proksa, "Plastiches Biegen von Blechen," *Der Stahlbau*, vol. 28, p. 29, 1959 (in German).

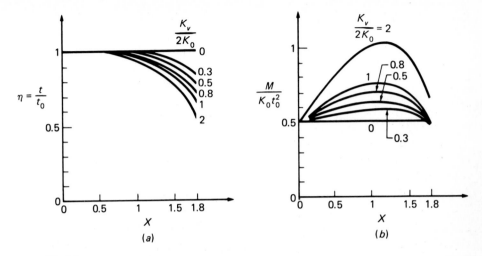

FIGURE 4-7
(*a*) Dependence of change in sheet thickness on relatice curvature; (*b*) dependence of bending moment on relative curvature. (*From Proksa.*)

where t_0 = initial sheet thickness
t = instantaneous sheet thickness
r_0 = radius of the outer surface
r_i = radius of the inner surface

The η term in Eq. (4-2) is a measure of the thickness change in bending. The X term is a measure of the change of curvature. These terms are related for plane-strain bending of a linearly strain-hardening rigid plastic material by Eq. (4-3)

$$\frac{d\eta}{dX} = \frac{\eta}{2X}(e^{-A} - 1) \tag{4-3}$$

The dependence of M and η on X Eqs. (4-2) and (4-3) is plotted in Fig. 4-7.

The bending moment over the width of the sheet is not uniform. M is zero at the die radius, and it increases linearly toward the punch. To determine the stress and strain distributions, the sheet is divided into small elements, each of which is subjected to a pure bending moment of different magnitude. Figure 4-8 shows how the bending moments are established. From the balance of forces in the vertical direction, the normal reaction of the die radius N is given by

$$N = \frac{P}{2(\cos\alpha + \mu\sin\alpha)} \tag{4-4}$$

Now, if the x and y coordinates are taken along the sheet (x) and normal to the sheet (y) and the origin of coordinates is at the die-sheet contact point, the

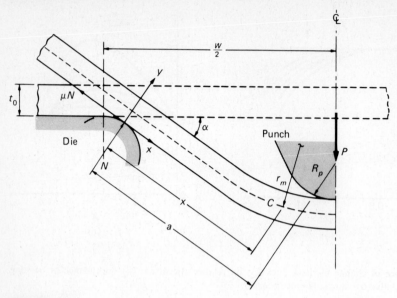

FIGURE 4-8
Deflection of sheet under a load P.

bending moment at any point C on the mid-thickness of the sheet is

$$M_x = N\left[x + \mu\left(y + \frac{t}{2}\right)\right] \qquad \text{for } x < a$$

$$M_x = M_a \qquad \text{for } x = a$$

(4-5)

where $x = a$ is the point at which the sheet starts to wrap over the punch and M_a is the bending moment at $x = a$. With the moment distribution known for a given load P, the mean radius can be determined from Fig. 4-7b and the thickness distribution from Fig. 4-7a. The determination of N in Eq. (4-5) requires knowledge of the angle α. However, α is not known beforehand, so a computer-based iterative procedure must be used to obtain punch displacement (equivalent to α) for each value of punch force.

But this is not the entire story; because when the punch force is removed, the material experiences *elastic recovery*. The sheet "springs backs" so that the radius of curvature R' after release of load is greater than the radius of curvature R before the load was removed. The relation between the two is

$$\frac{1}{R} - \frac{1}{R'} = \frac{12M}{Et^3}(1 - v^2)$$

(4-6)

where v is the Poisson ratio for the sheet.

The numerical procedure used to *simulate* the brake bending process starts with input values for die radius, die width, friction coefficient, and the

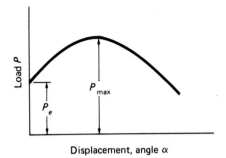

Displacement, angle α

FIGURE 4-9
Typical load vs. displacement curve for air bending.

strain-hardening characteristics of the sheet material. The output data are the strain distribution in the sheet and the load-displacement curve. The strain and the sheet thinning do not increase after the sheet is wrapped over the punch. The maximum strain and thinning are given by

$$\varepsilon_{max} = \ln \frac{2t_0 R_0}{R_0^2 - R_p^2} \tag{4-7}$$

$$\eta_{max} = \frac{t_0}{R_0 + 0.5t_0} \tag{4-8}$$

where $R_0 = R_p + t_0$. To reduce the error in using a linear stress-strain equation (4-1), the total strain range (0 to ε_{max}) is divided into 10 equal segments and a least-squares fit is made to the actual stress-strain curve to obtain values of K_0 and K_v. By using those values, Eq. (4-3) is solved by numerical integration to determine X for an array of η values ranging from 1 to η_{max}. The bending moment for various ratios of K_v/K_0 is determined by using Eq. (4-2).

A typical load vs. displacement curve is shown in Fig. 4-9. Plastic deformation does not occur unless a critical load P_e is exceeded. At some value P_{max} the load reaches a maximum and then decreases for increasing values of displacement. Thus, for a load $P_e \leq P \leq P_{max}$ two solutions for displacement exist. Using the same x and y coordinate system as shown in Fig. 4-8, the computer procedure walks down the bent beam and for each small step in x calculates the bending moment [Eq. (4-5)] and the radius of the bend [Eq. (4-2)]. A value of bend angle is assumed in order to solve for the bend angle vs. load because the normal force, and hence the bending moment at any value of x, is a function of the bend angle. If the angle calculated is close to the assumed angle, the solution is said to be found and the load is incremented to the next value. At a given load P two values of bend angle may exist. Therefore, once the first solution is found, the second solution is found by continuing to step forward in x. As each successful load and bend angle combination is determined, the load, angle, punch displacement, sheet thickness ,and maximum strain value are saved. When all load and angle values have been found, those values are sorted by increasing the value of the bend angle to generate the load vs. displacement curve.

To find the amount of bending necessary to give some desired bend angle after springback, the radius after springback is determined from Eq. (4-6) for two adjacent known load vs. angle values. The exact load and angle are found by interpolation of the known values.

The detailed calculations required to solve this brake bending model have been incorporated into an *interactive computer program*.[1] Thus, without any detailed knowledge of the model or the computation, a manufacturing engineer can obtain a graphic display of the sheet deformation, the load vs. displacement curve, and the strain vs. bend angle curve. Figure 4-10 shows the displays that appear on the graphics terminal of the computer.

An essential step in the development of a model is *validation*. In Fig. 4-11 are shown the bend angle after springback and the punch displacement curve for the theoretical model and experimental results. Quite acceptable agreement is obtained.

4-3   SIMILITUDE AND SCALE MODELS

New processes and designs are not born big; they grow through laboratory, development, and pilot plant stages. Physical models are an important part of the development and design processes. Usually *scale models*, which are less than full size, are employed. Figure 4-12 shows a model for aerodynamics wind tunnel testing. A *pilot plant* reproduces in reduced scale all or most aspects of a chemical, metallurgical, or manufacturing process.

In using physical models it is necessary to understand the conditions under which similitude prevails for both the model and the prototype.[2] *Geometric similarity* is the first type of similitude. The conditions for it are a three-dimensional equivalent of a photographic enlargement or reduction, i.e., identity of shape, equality of corresponding angles or arcs, and a constant proportionality or scale factor relating corresponding linear dimensions.

Model dimension = scale factor × prototype dimension

Sometimes it is convenient and sufficiently accurate to construct a section model of a process. For example, in the study of reactions in a long kiln, it is possible to simulate the kiln by constructing a model with the diameter scaled to the prototype but with a length only a small fraction of the scaled length. In other instances it is not possible to scale all dimensions of the prototype. For

---

[1] V. Nagpal, T. L. Subramanian, and T. Altan, AFML-TR-79-4168, November 1979.

[2] H. L. Langhaar, "Dimensional Analysis and the Theory of Models," John Wiley & Sons, Inc., New York, 1960; R. E. Johnstone and M. W. Thring, "Pilot Plants, Models and Scale-up Method in Chemical Engineering," McGraw-Hill Book Company, New York, 1957; R. C. Pankhurst, "Dimensional Analysis and Scale Factors," Chapman & Hall, London, 1964; D. J. Schuring, "Scale Models in Engineering," Pergamon Press, New York, 1977; E. Szucs, "Similitude and Modeling," Elsevier Scientific Publ. Co., New York, 1977.

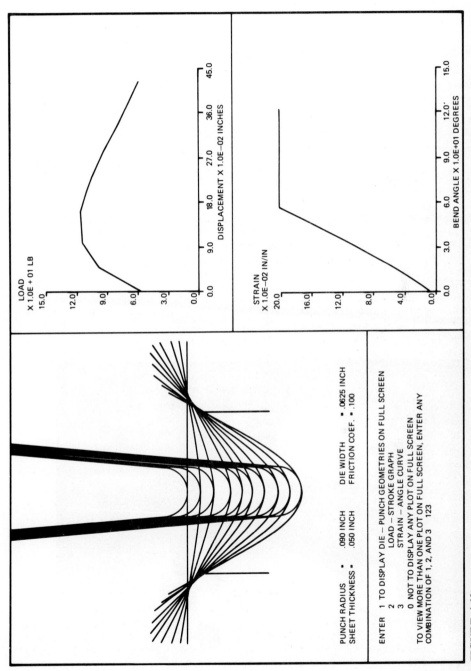

FIGURE 4-10
Brake bending simulation as seen on graphic terminal.

175

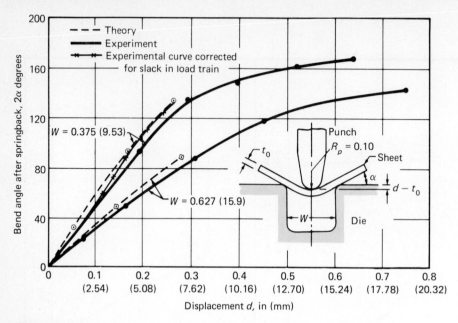

FIGURE 4-11
Bend angle after springback vs. punch displacement. (*From "Process Modeling," p. 297. American Society for Metals, Metals Park, Ohio, 1980.*)

FIGURE 4-12
Wind tunnel test model. (*Courtesy of Glenn L. Martin Wind Tunnel, University of Maryland.*)

example, in modeling a shallow reverberatory furnace, a full-scale model would result in such a shallow bath depth that the model could not be used to study settling and mixing phenomena. Thus, a distorted geometric model with a bath depth deeper than that called for by the scale factor would be used.

Consider the example of a cylindrical bar of length L that is elastically deformed an amount δ by an axial force P.

$$S = \text{scale factor} = \frac{\delta_m}{\delta_p} = \frac{L_m}{L_p} \tag{4-9}$$

where subscript m refers to the model and p to the prototype. The critical variables are deformation δ, load P, cross-sectional area A, elastic modulus E, and length L. These variables can be represented by three dimensionless ratios, which are obtained by manipulating the well-known relation $\delta = PL/AE$. In the more general case, when the functional relation between variables is not known, the dimensional ratios are developed by the procedures of dimensional analysis[1] by using either the method of Lord Rayleigh or the Buckingham π theorem. Dimensional analysis is covered in most undergraduate texts on fluid mechanics and heat transfer.

$$\pi_1 = \frac{\delta}{L} = \frac{\delta_p}{L_p} = \frac{\delta_m}{L_m}$$

$$\pi_2 = \frac{P}{EL^2} = \frac{P_p}{E_p L_p^2} = \frac{P_m}{E_m L_m^2} \tag{4-10}$$

$$\pi_3 = \frac{A}{L^2} = \frac{A_p}{L_p^2} = \frac{A_m}{L_m^2}$$

From the second dimensionless ratio we determine how the load must be adjusted to maintain similarity.

$$\pi_2 = \frac{P_m}{E_m L_m^2} = \frac{P_p}{E_p L_p^2}$$

and

$$P_m = P_p \frac{E_m L_m^2}{E_p L_p^2} = P_p \frac{E_m}{E_p} S^2 \tag{4-11}$$

Suppose we wish to use a plastic model, $E_m = 0.4 \times 10^6$ psi, to model a steel bar, $E_p = 30 \times 10^6$ psi, loaded to 50,000 lb. If the model is built to a 1 to 10 scale ($S = 0.10$), then the proper load is

$$P_m = 50,000 \frac{0.4 \times 10^6}{30 \times 10^6} 0.1^2 = 6.7 \text{ lb} \tag{4-2}$$

[1] P. Bridgman, "Dimensional Analysis," 2d ed., Yale University Press, New Haven, Conn., 1931; E. Isaacson and M. Isaacson, "Dimensional Methods in Engineering and Physics," John Wiley & Sons, Inc., New York, 1975.

We note that an important aspect of dimensional analysis is that dimensional groups are criteria of similarity. In the absence of a theoretical model, dimensional analysis shows how the variables can be arranged to facilitate the experimental determination of a functional relation. Thus, it is necessary to vary, not each factor individually, but only each dimensional group as a whole. We should note, however, that dimensional analysis does not indicate whether a variable is significant. If a significant variable is omitted from the formulation of the problem, an incomplete set of dimensionless groups will be obtained and a proper correlation of the experimental data will not be obtained. If an extraneous variable is included in the initial problem statement, an extra dimensionless group may be derived. The presence of an extra variable should be revealed during the process of data correlation, but it may greatly increase the effort needed in the process. Extensive lists of dimensional groups have been published.[1]

In addition to geometric similarity we have *static similarity*. It implies that, under constant stress, the relative deformation will be in the same proportion as the characteristic geometric dimension. Static similarity was considered in the example of an elastically deformed bar.

Kinematic similarity is based on the ratio of the time proportionality between corresponding events in the model and the prototype. Often the time-scale ratio is combined with the length-scale ratio to express the ratio of velocities at equivalent positions in similar systems. Typical kinematic similarity ratios are:

Acceleration:
$$a_r = \frac{a_m}{a_p} = \frac{L_m T_m^{-2}}{L_p T_p^{-2}} = L_r T_r^{-2} \qquad (4\text{-}13)$$

Velocity:
$$v_r = \frac{v_m}{v_p} = \frac{L_m T_m^{-1}}{L_p T_p^{-1}} = L_r T_r^{-1} \qquad (4\text{-}14)$$

Volume flow rate:
$$Q_r = \frac{Q_m}{Q_p} = \frac{L_m^3 T_m^{-1}}{L_p^3 T_p^{-1}} = L_r^3 T_r^{-1} \qquad (4\text{-}15)$$

In *dynamic similarity* the forces acting at corresponding times and on corresponding locations in the model and the prototype are in a fixed ratio. In fluid-flow situations the forces arise from inertia, viscosity, gravity, pressure, vibration, centrifugal force, or surface tension. In systems in which the forces produce fluid motion, e.g., flow under gravity, dynamic similarity automatically ensures kinematic similarity. In situations in which movement is produced mechanically, it is possible to obtain kinematic similarity without satisfying

---

[1] D. F. Boucher and G. Alves, *Chem. Eng. Prog.*, vol. 55, no. 9, p. 55, 1959; G. D. Fulford and J. P. Catchpole, *Ind. Eng. Chem.*, vol. 58, no. 3, p. 40, 1966; vol. 60, no. 3, p. 71, 1968.

dynamic similarity. Some dynamics similarity ratios are:

$$\text{Inertial force } F_i = \text{mass} \times \text{acceleration}$$

$$= \rho L^3 \frac{L}{T^2} = \rho L^2 v^2 \tag{4-16}$$

$$\text{Viscous force } F_\mu = \tau L^2 = \mu \frac{du}{dy} L^2 = \mu \frac{v}{L} L^2 = \mu L v \tag{4-17}$$

Dimensionless groups involved with dynamic similarity usually are formulated by taking the ratio of the inertia force to the other fluid forces. Thus,

$$\frac{\text{Inertial force}}{\text{Viscous force}} = \frac{F_i}{F_\mu} = \frac{\rho L^2 v^2}{\mu L v} = \frac{\rho L v}{\mu} = N_{\text{Re}} \tag{4-18}$$

The resulting dimensionless groups is the familiar Reynolds number. N_{Re} often is used in similitude considerations of a fluid system. We can write Eq. (4-18) as

$$N_{\text{Re}} = \frac{Lv}{v} = \frac{L_m v_m}{v_m} = \frac{L_p v_p}{v_p} \tag{4-19}$$

where $v = \mu/\rho$ is the kinematic viscosity and L is the linear dimension, usually taken as the diameter of the pipe.

$$v_m = \frac{L_p}{L_m} \frac{v_m}{v_p} v_p = \frac{1}{s} \frac{v_m}{v_p} v_p \tag{4-20}$$

From Eq. (4-20) we can see that if we wish to build a 1 to 10 scale model ($S = 0.1$), the velocity in the model must be increased by a factor of 10 to maintain dynamic similarity. If the higher velocity is difficult to attain, then we may be able to maintain similitude by changing the fluid. At 60°F the kinematic velocity of water is 0.0435 ft²/h compared with 0.568 ft²/h for air at the same temperature. This gives a ratio v_m/v_p of about 1:13. Or, since v of gases increases with temperature, a model employing cold air would result in a 1 to 20 scale factor to simulate air in a furnace at 2500°F.

Thermal similarity requires that the temperature profiles in the model and the prototype must be geometrically similar at corresponding times. In addition, when systems involve bulk movement of material from one point to another, thermal similarity requires that kinematic similarity must also be obtained.

Chemical similarity requires that the rate of a chemical reaction at any location in the model must be proportional to the rate of the same reaction at the corresponding time and location in the prototype. This requires both thermal and kinematic similarity plus a concentration profile within the model that corresponds to the one in the prototype. Chemical simulation usually is difficult to achieve because it requires proportionality of time, temperature, and concentration between the model and the prototype. Since the rate of chemical reaction usually is very sensitive to temperature, chemical models often are operated at the same temperature and concentration conditions as the prototype. Chemical similarity is achieved when the temperatures and concentrations are the same for the model and the prototype at corresponding locations and times.

Similitude and dimensional analysis are necessary requirements for physical modeling. The use of dimensional analysis provides a way of reducing the number of variables needed to describe a system. It also allows data from scale models to be properly interpreted with respect to the design. The use of dimensional analysis can reveal strengths and weaknesses in existing designs and thus point the way to improved new designs.[1]

4-4   SIMULATION

Simulation involves the modeling of a complex situation into a simpler and more convenient form that can be studied in isolation without the troublesome complex side effects that usually accompany a real engineering situation. The purpose of the simulation is to explore the various outputs that might be obtained from the real system by subjecting the model to environments that are in some way representative of the situations it is desired to understand. Simulation invariably involves the use of the computer to perform often laborious computations and to follow the dynamics of the situation. The example of a process model discussed in Sec. 4-2 included a simulation of the brake bending process that utilized a computer program interactive with a graphics display. In this section we shall briefly consider some additional aspects of computer simulation.

Computer simulations involve a time dimension (dynamic behavior), and they fall into three broad categories.

1. The simulation of an engineering system or process by mathematical modeling and computer simulation. An example would be the simulation of a traffic control problem or the solidification of a large steel casting.
2. Simulation gaming (not to be confused with game theory), in which live decision makers use a computer model for complex situations involving military or management decisions. An example would be a game for bidding strategy in the construction industry.
3. Simulation of business or industrial engineering systems, which includes such problems as control of inventory levels, job-shop scheduling, and assembly-line balancing.

---

[1] J. R. Schnittger, *Trans. ASME, Jnl. of Vibration, Acoustics, Stress, and Reliability in Design,* vol. 110, pp. 401–407, 1988.

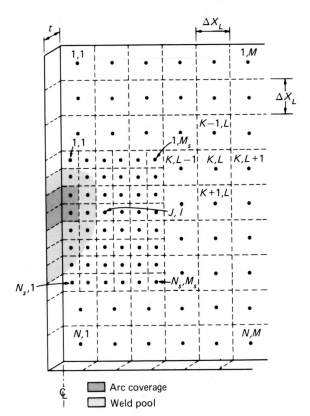

FIGURE 4-13
Example of grid network used in simulation of the pulsed arc welding process. (*From* J. Metals, *p. 21, September 1980.*)

Finite-Difference Method

A typical problem in computer simulation of an engineering process is heat transfer in the pulsed arc welding of a metal sheet.[1] In that welding process the power is cycled from a peak rate of heat input to a reduced level over a fixed pulse time. The formulation of the model for the process employs the finite-difference technique,[2] a method of approximate solution of partial differential equations that is discussed in most courses in heat transfer. Figure 4-13 shows the grid network used for the model. The sheet around the weld pool is divided into a grid of square elements. Because of symmetry, only one-half of the sheet need be considered. The node at the center of each element is assumed to have the average temperature of the element. The

---

[1] H. D. Brody and R. A. Stoehr, *J. Metals,* vol. 32, no. 9, pp. 20–27, 1980.
[2] G. E. Forsythe and W. R. Wasow, "Finite-Difference Methods for Partial Differential Equations," John Wiley & Sons, Inc., New York, 1960; J. A. Adams and D. F. Rogers, "Computer-Aided Heat Transfer Analysis," McGraw-Hill Book Company, New York, 1973.

elements are made smaller (and the nodes are closer together) near the weld pool to achieve better resolution of temperature in that region of special interest.

The rows and columns are numbered according to the following scheme.

Small grid rows: $\qquad 1 < J < N_s$

Small grid columns: $\qquad 1 < I < M_s$

Large grid rows: $\qquad 1 < K < N$ $\hfill$ (4-21)

Large grid columns: $\qquad 1 < L < M$

For example, $T_{K,L}$ refers to the temperature at the node in the Kth row and

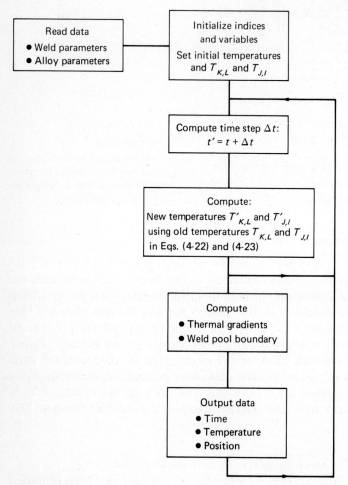

FIGURE 4-14
Simplified flow diagram for finite difference simulation of pulsed arc welding. (*After Brady and Stoehr.*)

Lth column of the large grid. Since arc welding of the sheet is a two-dimensional heat-flow problem, heat is considered to flow into (or out of) each element from its four nearest neighbors. If T' is the temperature after a time step Δt,

$$T'_{K,L} = T_{K,L} + M(T_{K+1,L} + T_{K-1,L} + T_{K,L+1} + T_{K,L-1} - 4T_{K,L}) \quad (4\text{-}22)$$

where $M = \alpha \Delta t / \Delta x^2$. The term $\alpha = K/\rho c_p$ is the thermal diffusivity. The heat from the arc is introduced for the grid elements on the centerline by

$$T'_{J,1} = T_{J,1} + M\left(T_{J+1,1} + T_{J-1,1} + T_{J,2} - 3T_{J,1} + \frac{q}{kt}\right) \quad (4\text{-}23)$$

where q is the heat input from the arc. The relative motion of the arc and the sheet can be considered by changing, through the computer program, the location of the node receiving an external heat input. The initial conditions are established by adjusting the first values of $T_{K,L}$ and $T_{J,I}$ to the initial temperature T_0. The boundary conditions are established by the way the elements are treated. Because of symmetry at the centerline, the left boundary is considered to be adiabatic; no heat enters the centerline element from the face on the centerline, Eq. (4-23). The top and bottom faces of the sheet are assumed to be adiabatic, and that is reflected in Eq. (4-22). In further modification, heat loss by convection or radiation could be handled at those surfaces.

A simplified flow diagram for the computer program used in this computer simulation is shown in Fig. 4-14. Figure 4-15 shows the computer-predicted temperature variation for three different distances from the centerline.

Monte Carlo Method

The Monte Carlo method[1] is a way of generating information for a simulation when events occur in a random way. It uses unrestricted random sampling (it selects items from a population in such a way that each item in the population has an equal probability of being selected) in a computer simulation in which the results are run off repeatedly to develop statistically reliable answers. The technique employs *random numbers* (a collection of random digits) that are generated by the computer.

We shall illustrate the Monte Carlo method with a problem involving the

---

[1] K. D. Toucher, "The Art of Simulation," D. Van Nostrand Company, Inc., Princeton, N.J., 1963; J. M. Hammerslay and D. C. Handscomb, "Monte Carlo Methods," Methuen & Co., London, 1964; H. A. Meyer (ed.), "Symposium on Monte Carlo Methods," John Wiley & Sons, Inc., New York, 1956; R. Y. Rubinstein, "Simulation and the Monte Carlo Method," John Wiley & Sons, Inc., New York, 1981.

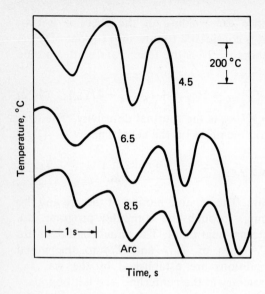

FIGURE 4-15
Variation of temperature with time for pulsed arc welding using finite-difference computer model. (*H. D. Brody and R. A. Stoehr*, J. Metals, *p. 23, September, 1980.*)

statistical distribution of dimensions in an assembled part.[1] The assembly consists of three parts added end to end. Therefore, variations in the lengths of parts 1, 2, and 3 will result in variations in the length of the assembly. We wish to use Monte Carlo simulation to determine the distribution of assembly lengths.

To begin the simulation, we must first have a data base of the distribution of length for a sampling of each part. Fifty samples are measured for each part, and the distribution of length is given in the form of histograms in Fig. 4-16a. To simplify the data analysis, the length of each part is coded. The coded lengths 1 through 5 represent actual measurements of 5.45, 5.46, 5.47, 5.48, and 5.49 mm. Coding is accomplished by subtracting 5.44 from each measurement and dividing the result by 0.01. For example, 14 samples from part 2 have a length of 3 (5.47 mm), since $50(0.28) = 14$.

We note that an observed part length is a random event. To simulate the process, we could construct a roulette wheel with 100 pockets numbered from 00 to 99. For part 1, length 1 ($x_1 = 1$) would correspond to numbers 00 to 05, length 2 would correspond to numbers 06 to 29, etc. as shown in Table 4-1. (We now see the origin of the term Monte Carlo). Note that we have used 100 Monte Carlo numbers (00 to 99) because the total fraction of lengths for a part is 1.00. However, rather than use a roulette wheel to generate these numbers, it is more convenient to use a table of random numbers or to generate random numbers with a computer. In Table 4-1 all lengths for all three parts are represented by Monte Carlo numbers.

---

[1] F. M. Spotts, *Machine Design*, pp. 84–88, Nov. 20, 1980.

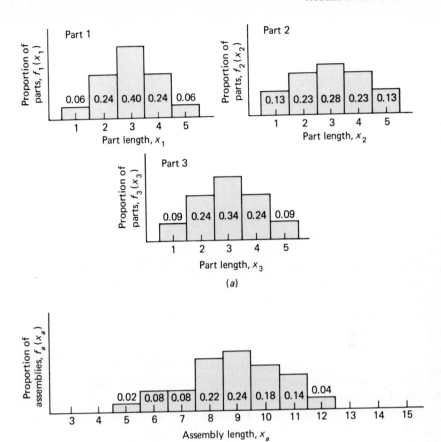

FIGURE 4-16
(*a*) Distribution of lengths of three parts. Part length is coded (see text). (*b*) Distribution of part lengths of 50 assemblies determined from a Monte Carlo simulation. (*M. F. Spotts*, Machine Design, *pp. 84* and *85, Nov. 20, 1980.*)

TABLE 4-1
Assignment of Monte Carlo numbers

	Part 1			Part 2			Part 3	
x_1	$f_1(x_1)$	Monte Carlo No.	x_2	$f_2(x_2)$	Monte Carlo No.	x_3	$f_3(x_3)$	Monte Carlo No.
1	0.06	00 to 05	1	0.13	00 to 12	1	0.09	00 to 08
2	0.24	06 to 29	2	0.23	13 to 35	2	0.24	09 to 32
3	0.40	30 to 69	3	0.28	36 to 63	3	0.34	33 to 66
4	0.24	70 to 93	4	0.23	64 to 86	4	0.24	67 to 90
5	0.06	94 to 99	5	0.13	87 to 99	5	0.09	91 to 99

TABLE 4-2
A sample of random numbers

(1)	(2)	(3)	(4)	(5)	(6)
0095	8935	2939	3092	2496	0359
6657	0755	9685	4017	6581	7292
8875	8369	7868	0190	9278	1709
8899	6702	0586	6428	7985	2979
⋮	⋮	⋮	⋮	⋮	⋮

Now, to simulate the random production of part lengths, we select random numbers from a table or with a computer. Table 4-2 shows a small sample from a table of random numbers. We select the page of numbers by chance and then adopt a purely arbitrary selection scheme. In this case, we use only the first two digits of the four-digit numbers. Digits from columns 1, 2, and 3 are assigned to parts 1, 2, and 3, respectively, for the first 25 assemblies and columns 4, 5, and 6 for the last 25 assemblies. The total of the lengths for the three parts is the length of each assembly x_a. When all 50 assemblies are simulated, the number having length 6, 7, 8, etc. is determined, and the proportion with each length $f_a(x_a)$ is determined. Those numbers are plotted in Fig. 4-16b.

When we examine the histogram in Fig. 4-16b, we note that none of the assemblies in less than 5 units long or greater than 12. Based on the length distribution of the individual parts, it is possible for an assembly $(x_1 + x_2 + x_3 = x_a)$ to be as short as 3 and as long as 15. However, the probability of selecting three sucessive parts from the left tails of the distribution or three sucessive parts from the right tails is very low. Therefore, the assembled lengths tend to bunch more than if the tolerances on the individual parts had simply been added.

Simulation Languages

A number of special-purpose computer languages have been developed for simulation of engineering systems. These are typically used in applications of manufacturing or construction scheduling.[1] The use of these languages can eliminate a large amount of effort compared with starting with a general-purpose language such as FORTRAN.

The general purpose simulation system (GPSS) is a language oriented toward engineering situations in which waiting takes place as parts move through a variety of process steps. This language is used for production-flow

---

[1] N. N. Ekere and R. G. Hannam, *Int. J. Prod. Res.*, vol. 27, no. 4, pp 599–611, 1989.

problems and inventory control. Other general purpose languages are SLAM II, SIMSCRIPT and SIMAN. More specialized manufacturing oriented languages, such as MAP and MAST, have been developed.

Once a simulation model has been constructed it acts as a laboratory in which various design alternatives can be easily tested, compared, and changed. Using a computer-based simulation provides heightened perspective about the design and faithful adherence to detail. With a simulation it is no longer necessary to assume away small but important details so as to be able to use a closed form analytical solution. Such factors as processing time variability, equipment reliability, restrictions on in-process storage, schedule constraints, and complex routing decisions can be accommodated in the simulation model. An important characteristic is that the simulation will give a realistic picture of the dynamic behavior of the system. Moreover, by incorporating advances in computer graphics it is possible to introduce animation into the simulation model display.

4-5 GEOMETRIC MODELING ON THE COMPUTER

In Sec. 1-10 we discussed how the use of interactive computer graphics was revolutionizing engineering design. Here we give a brief insight into the development of computer-generated geometric models. The geometric model is critical because so many subsequent design and manufacturing operations depend on it for a data base. It may, for example, be used to develop a finite-element model for stress analysis. It may be used to produce detailed engineering drawings through automated drafting. More important, computer-aided design can be interfaced with computer-aided manufacturing (CAD/CAM), and the geometric model can be used to generate NC tapes for machining the part on an automated machine tool or for developing process plans for the detailed sequence of steps required to manufacture the part.

The three general classes of pictures that can be created on the CRT of an interactive computer graphics system are 1) wire frame models (stick figures), 2) surface models, and 3) solid models. Most geometric modeling today is done with wire frame models, which represent part shapes with interconnected line elements. These require the least computer time and memory; but although they provide precise information about the location of surface discontinuities on the part, they contain no information about the surfaces themselves nor do they differentiate between the inside and outside of objects.

The development of a wire frame model is shown in Fig. 4-17. The designer uses the CRT screen in much the same manner as a drawing board, but the CAD system contained in the computer software provides many automatic ways of speeding the design process. Most lines making up a wire frame model are straight, but it is possible to produce curves, circles, and conics. Most CAD systems can generate a smooth curve (called a spline)

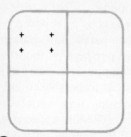

(1) *Screen is split into sections for top, front, side, and isometric views. Four points are specified on the top view representing vertices of that face.*

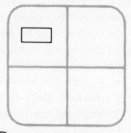

(2) *User commands the system to connect points with straight-line elements outlining the top face of the part.*

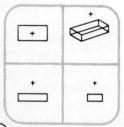

(3) *User commands the system to project the image into the other three views.*

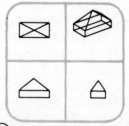

(4) *Face is projected into the second dimension, producing a rectangular block.*

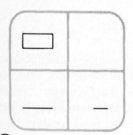

(5) *A point is specified above the block.*

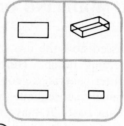

(6) *User commands the system to connect the point to the top corners of the block, creating a pyramid. Model is complete.*

FIGURE 4-17
Development of a wire frame geometric model on a CRT screen. (Machine Design, *p. 99, July 24, 1980.*)

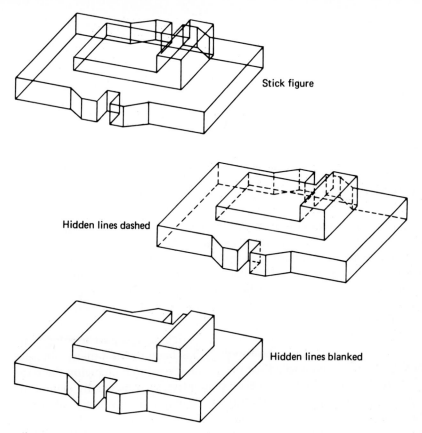

Stick figure

Hidden lines dashed

Hidden lines blanked

FIGURE 4-18
Complications with hidden lines in wire frame models. (Machine Design, *p. 100, July 24, 1980.*)

through a series of arbitrary points specified by the user. Important features that are provided by the CAD software system are the projection of points and lines in one view into other views, the ability to temporarily erase selected lines from the screen without deleting them from the model, and the ability to enlarge or reduce certain areas of the model.[1]

Stick figure or wire frame models often are confusing and difficult to interpret because of extraneous lines. Figure 4-18 shows a stick figure model at

[1] D. L. Ryan, "Computer-aided Graphics and Design," 2d ed., Marcel Dekker, Inc., New York, 1985; D. F. Rogers and J. A. Adams, "Mathematical Elements of Computer Graphics," McGraw-Hill Book Company, New York, 1976; I. D. Faux and M. J. Pratt, "Computational Geometry for Design and Manufacture," Halsted Press, New York, 1979: W. M. Newman and R. F. Sproull, "Principles of Interactive Computer Graphics," 2d ed. McGraw-Hill Book Company, New York, 1979.

the top. With most CAD systems it is possible to clarify the solid geometry of the object by representing the hidden lines as dashed or removing them completely (bottom of Fig. 4-18).

Many of the ambiguities of wire frame models are absent from surface models. *Surface models* define the outside part geometry precisely, but they represent only an envelope of part geometry. That can lead to difficulty in calculating volume-dependent properties such as weight. Surface models are created by connecting various surface elements to user-specified lines. The menu of elements from which to choose includes planes, ruled surfaces, surfaces of revolution, fillets, and sculptured surfaces formed by the intersection of two families of curves.

Solid modeling produces valid and unambiguous representations of solids. The two basic techniques of solid modeling are *constructive solid geometry* (CSG) and *boundary representation* (b-reps). With CSG the solid is constructed in building-block fashion by combining primitive shapes like a block, cylinder, cone or a sphere. The primitives are combined using boolean logic operations such as union, which combines two primitives, or intersection, which defines a volume common to both primitives. Boundary representation is a process in which solids are represented by sets of faces that enclose them completely. Boundary models are useful for complex parts that cannot be modeled conveniently with primitive shapes. It defines the part topology and part geometry separately. After the topology has been defined many different operations can be performed to adjust the geometry without changing the basic topology. Parts having a high degree of symmetry, such as axially symmetric parts, can be modeled quickly with b-reps. A major disadvantage of b-reps modeling is that it cannot guarantee closure.

Sweeping is the creation of a solid shape by moving a line or a plane along a defined trajectory, Fig. 4-19. The sweeping method is fast and easy to use. However, it has the important limitation that it can only create shapes that have translational or rotational symmetry. For example, a block with a hole through it is generated efficiently by sweeping, but a block with a hole only part way through the thickness cannot be generated by sweeping.

Spatial enumeration and *cell decomposition* use combinations of cubes to approximate solid shapes. They are organized as three-dimensional binary trees called octrees. Since modeling irregular surfaces requires almost an infinite number of tiny cubes the method is very memory intensive. Nevertheless, the method is very good at modeling irregular solids. In the variation called cell decomposition the cell may be a distinctive shape, like a finite element mesh. As computer memory and processing power became even more plentiful it is likely that octree methods will become the preferred method of solids modeling.

Feature-based solid modeling, also called parametric modeling, is the cutting edge of CAD. It is aimed at a closer linking of CAD with CAM by embedding a higher level of knowledge in the geometry. It also permits much easier change of the design parameters so that CAD becomes more interactive

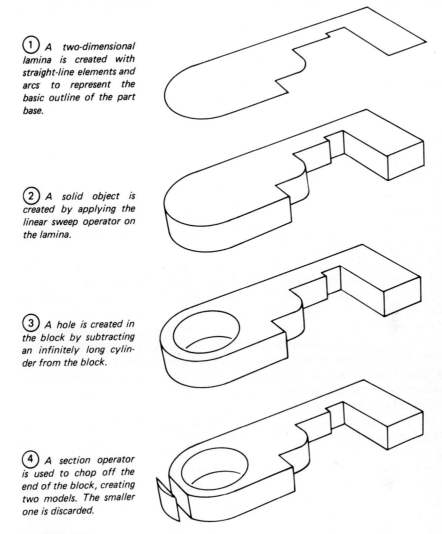

① *A two-dimensional lamina is created with straight-line elements and arcs to represent the basic outline of the part base.*

② *A solid object is created by applying the linear sweep operator on the lamina.*

③ *A hole is created in the block by subtracting an infinitely long cylinder from the block.*

④ *A section operator is used to chop off the end of the block, creating two models. The smaller one is discarded.*

FIGURE 4-19
Development of a solid model. (*Machine Design, p. 104, July 24, 1980.*)

and it is easier to tap the creativity of the designer. Feature based CAD[1] is carried out with a language of features (holes, slots, splines, pockets, etc,) for which the location, dimensions, tolerances, and manufacturing methods can be specified and easily changed. When a relationship is defined for a parametric modeler the system understands that certain geometry is placed in context with

---

[1] S. Drake and S. Sela, *Mech. Engr*, Jan. 1989, pp. 67–73.

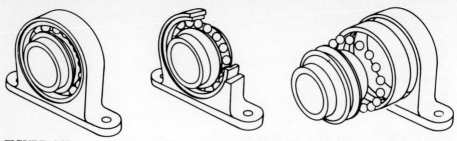

FIGURE 4-20
Interior details of a bearing revealed with solid computer modeling. (*Courtesy Control Data Corp.*)

other geometry. For example, if a fillet is placed between a plane surface and a cylinder, the fillet will change if the cylinder diameter changes. In traditional modelers, the fillet would not change. In feature-based modeling the designer can countersink all the holes in a part with a single command, rather than finding and changing everyone.

An important feature of solid modeling is the ability to cut through the model to reveal interior details, Fig. 4-20. The ability to calculate mass properties is also important. Many computer graphics systems are capable of full color, shading, and shadowing to produce very realistic displays. The evolution of CAD has been from two-dimensional curves, to three-dimensional wireframes, to surfaces, to solids, and more recently to features. The hope is that by incorporating a very easy to use system with a high data rich environment that it will be possible to easily provide the design information that is needed for manufacturing. In this way CAD and CAM will be linked in way that was the dream of 20 years ago.

As an example, one major U.S. car manufacturer has developed a computer-aided styling system that permits designers to sketch automobiles in great detail on video screens. The computer graphics include realistic surface highlights and reflections. This advanced computer modeling eliminates the need for time-consuming clay models. It will trim one year off the four years needed to bring a new product to the market.

4-6 FINITE-ELEMENT ANALYSIS

The finite-element method is a powerful computer-based method of analysis that is finding wide acceptance for the realistic modeling of many engineering problems. The technique was introduced briefly in Sec. 1-8. Here it is presented in some additional detail so you can appreciate its mathematical structure and gain some appreciation of how it is used in computer modeling.[1]

---

[1] R. D. Cook, D. S. Malkus and M. E. Plesha, "Concepts and Applications of Finite Element Analysis," 3d ed. John Wiley & Sons, Inc., New York; L. J. Segerlind, "Applied Finite Element Analysis," John Wiley & Sons, Inc., New York, 1989; J. F. Potts and J. W. Oler, "Finite Element Applications with Microcomputers," Prentice-Hall, Inc., Englewood Cliffs, N.J., 1989.

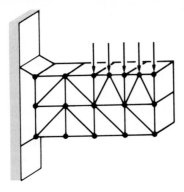

FIGURE 4-21
Simple finite-element representation of a beam.

In finite-element analysis a continuum solid or fluid is considered to be built up of numerous tiny connected elements. Since the elements can be arranged in virtually any fashion, they can be used to model very complex shapes. Thus, it is no longer necessary to find an analytical solution that treats a close "idealized" model and guess at how the deviation from the model affects the prototype. As the finite-element method has developed, it has replaced a great deal of expensive preliminary cut-and-try development with quicker and cheaper computer modeling.

In contrast to the analytical methods that often require the use of higher-level mathematics, the finite-element method is based on simple algebraic equations. However, an FEM solution may require hundreds of simultaneous equations with hundreds of unknown terms. Therefore, the development of the technique required the availability of the high-speed digital computer for solving the equations efficiently by matrix methods. The rapid acceptance of finite-element analysis has been largely due to the increased availability of FEM software through interactive computer systems.

In the finite-element method the loaded structure is modeled with a mesh of separate elements (Fig. 4-21). We shall use triangular elements here for simplicity, but later we shall discuss other important shape elements. The elements are connected to one another at their corners, and the connecting points are called *nodes*. A solution is arrived at by using basic stress and strain equations to compute the deflections in each element by the system of forces transmitted from neighboring elements through the nodal points. The strain is determined from the deflection of the nodal points, and from the strain the stress is determined with the appropriate constitutive equation. However, the problem is more complex than first seen, because the force at each node depends on the force at every other node. The elements behave like a system of springs and deflect until all forces are in equilibrium. That leads to a complex system of simultaneous equations. Matrix algebra is needed to handle the cumbersome systems of equations.

The key piece of information is the *stiffness* matrix for each element. It can be thought of as a kind of spring constant that describes how much the nodal points are displaced under a system of applied forces. In matrix notation

$$\{f\} = [k]\{\delta\} \qquad (4\text{-}24)$$

where $\{f\}$ is the column matrix (vector) of the forces acting on the element, $[k]$ is the stiffness matrix for the element, and $\{\delta\}$ is the column matrix of the deflections of the nodes of the element. The stiffness matrix is constructed from the coordinate locations of the nodal points and the matrix of elastic constants of the material. A triangular element $[k]$ can be constructed from the principles of statics, but more complicated elements require the use of energy principles to derive $[k]$. When all the elements of the systems are assembled, the basic matrix equation is

$$\{F\} = [K]\{\delta\} \qquad (4\text{-}25)$$

where $[K]$ = master stiffness matrix, assembled from the $[k]$ for all the
    elements
  $\{F\}$ = external forces at each node
  $\{\delta\}$ = displacements at each node

The force matrix is known because it consists of numerical values of loads and reactions computed prior to the start of the finite-element analysis. The displacements are the unknowns, and they are solved for by transposing the stiffness matrix in Eq. (4-25). This computer solution gives the displacements at all the nodes.[1] When it is multiplied by the matrix of coordinate locations of the nodes $[B]$ and the matrix of elastic constants $[D]$, it gives the stress at every nodal point.

$$\{\delta\} = [D][B]\{\delta\} \qquad (4\text{-}6)$$

The computer analysis usually is carried through to compute the principal stresses and their directions throughout the part. The printed volume of computer output for complex models is enormous and difficult to handle. A graphics display to display final output data through stress contours, color graphics, etc., is very helpful.

Finite-element analysis was originally developed for two-dimensional (plane-stress) situations. A three-dimensional structure causes orders of magnitude increase in the number of simultaneous equations; but by using higher-order elements and faster computers, these problems are being handled by the FEM. Figure 4-22 shows a few of the elements available for FEM analysis. Figure 4-22a is the basic triangular element. It is the simplest two-dimensional element, and it is also the element most often used. An assemblage of triangles can always represent a two-dimensional domain of any shape. The six-node triangle (b) increases the degrees of freedom available in

---

[1] For example, the deformation in a sheet deformed in brake bending, as in Sec. 4-2, is given by S. I. Oh and S. Kobayashi, *Int. J. Mech. Sci.*, vol. 22, pp. 583–595, 1980.

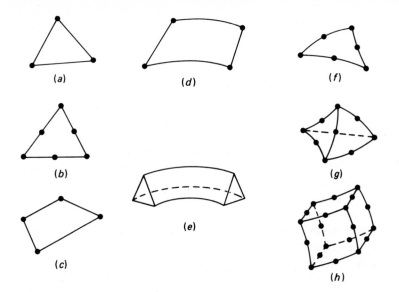

FIGURE 4-22
Some common elements used in FEM analysis.

modeling. The quadrilateral element (c) is a combination of two basic triangles. Its use reduces the number of elements necessary to model some situations. Elements (d) and (e) are three-dimensional but require only two independent variables for their description. These elements are used for problems that possess axial symmetry in cylindrical coordinates. Figure 4-22d is a one-dimensional ring element, and (e) is a two-dimensional triangular ring element. Three-dimensional FEM models are best constructed from isoparametric elements with curved sides. Figure 4-22f is an isoparametric triangle; (g) is a tetrahedron; and (h) is a hexahedron. These elements are most useful when it is desirable to approximate curved boundaries with a minimum number of elements.

A cumbersome part of the FEM solution is the preparation of the input data. The topology of the element mesh must be described in the computer program with the node numbers and the coordinates of the node points, along with the element numbers and the node numbers associated with each element. This bookkeeping task is extremely tedious for an FEM model containing hundreds of nodes. If the data must be prepared by hand and input with cards, the job is very time consuming and subject to errors. To avoid it, many large users of the FEM have developed automatic mesh generation techniques for producing the system topology. Usually these techniques produce meshes containing only one kind of element.

The key to the practical utilization of the FEM is the finite-element model. For sound economics the model should contain the smallest number of elements to produce the needed accuracy. The best procedure is to use an iterative modeling strategy whereby coarse meshes with few elements are increasingly refined with fine meshes in critical areas of the model. Coarse models can be constructed with one-dimensional simple beams and two-dimensional rectangular (plate) elements. Rather than depict the true geometry of the part, the coarse model represents only how the structure reacts to loads. Small holes, ribs, flanges, and similar details are purposely ignored. Computer costs for finite-element analysis increase exponentially with the number of elements. For example, a simple model containing only 50 elements may cost less than $40 in computer time, but the same geometry modeled with 150 elements costs $400. Once the overall structural characteristics (generally deflection to an accuracy no better than 20 percent) have been found by use of the coarse model, the fine-mesh model with more elements is constructed in regions where stress and deflection must be determined more accurately. Plate and beam elements should still be used wherever possible. The use of three-dimensional "brick" elements and the curved isoparametric elements increased costs exponentially over the costs of using the simpler elements.

It is not necessary to develop your own computer program to use the FEM. General-purpose programs are available from a number of commercial sources or can be obtained at nominal cost from the federal government.[1] The extensive documentation that is available makes the programs relatively easy to master. However, one should realize that using a versatile general-purpose program to solve a specialized problem may be far more costly than developing a program specifically for the purpose.

4-7 COMPUTER SIMULATION

The user of computers to carry out extensive simulations involving graphical output has become commonplace. In fact, many people are referring to these as computer experiments, and there is the prediction that this field will eventually grow into a third domain of science, coequal with the traditional domains of theory and experimentation. In doing these simulations finite element methods are by far the predominant method of analysis. A major impetus to the growth of this field is the availability of very high speed computers. For example, the use of a supercomputer to perform an airflow analysis on an automble required only 20 minutes of computational time, while the same analysis using a large mainframe required 10 hours.

Figure 4-23 shows the computer simulation of the deflection of the bottom dome of an aluminum can after it strikes the ground. For the simulated drop test on a supercomputer the computer is fed the shape of the can, the

---

[1] For available programs contact COSMIC Computer Center, University of Georgia, Athens, Ga., 30601.

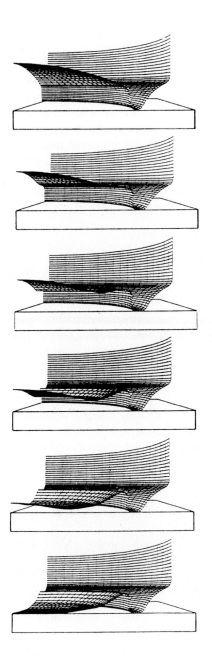

FIGURE 4-23
Computer simulation of bottom dome of aluminum can striking the ground. Only one-quarter of the bottom of the can is included in the simulation. Note that the bottom dome pops downward after the can hits the ground. (*Courtesy of Alcoa.*)

speed at which it hits the ground, the mass and pressure of the liquid inside the can, and the mechanical properties of the aluminum sheet. The finite element analysis combines the equations that describe the complicated interplay of mass, velocity, acceleration, pressure, and mechanical behavior to calculate and display the distortion of the can at one millisecond intervals after impact. Previous to using computer simulation a conventional development approach was used in which experimental designs were fabricated and tested. A typical program would take six months to a year and cost around $100,000, whereas with computer modeling it takes less than two weeks at a cost of about $2000.[1] The same kind of cost saving carries over to larger, more complex systems. A major US auto maker used supercomputer simulation to minimize noise and vibration in the 1986 models. They were able to investigate about 20 different designs for the body and suspension system in the time it would have taken to build three prototypes, at a cost saving of $3 million.

BIBLIOGRAPHY

Doebelin, E. O.: "System Modeling and Response," John Wiley & Sons, Inc., New York, 1980.
Gajda, W. J., and W. E. Biles: "Engineering Modeling and Computation," Houghton-Mifflin Company, Boston, 1978.
Krick, E. V.: "An Introduction to Engineering and Engineering Design," 2d ed., John Wiley & Sons, Inc., New York, 1969.
Mischke, C. R.: "Mathematical Model Building," Iowa State University Press, Ames, Iowa, 1980.
Starfield, A., K. A. Smith, and A. L. Bleloch: "How to Model It," McGraw-Hill Book Co., New York, 1990.
Wellstead, P. E: "Introduction to Physical System Modelling," Academic Press, Inc., New York, 1979.

---

[1] R. Pool, *Science,* vol. 244, pp. 1438–1440, June 23, 1989.

CHAPTER
5

OPTIMIZATION

5-1 INTRODUCTION

We have continually emphasized that design is an iterative process. You start with a poorly defined problem, refine it, develop a model, and arrive at a solution. Usually there is more than one solution, and the first one is not usually the best. Thus, in engineering design we have a situation in which there is a search for the best answer. In other words, *optimization is inherent in the design process*. A mathematical theory of optimization has been developed since 1950, and it has gradually been applied to a variety of engineering design situations. The concurrent development of the digital computer, with its inherent ability for rapid numerical calculation and search, has made the utilization of optimization procedures practical in many design situations.

By the term *optimal design* we mean the best of all feasible designs. Optimization is the process of maximizing a desired quantity or minimizing an undesired one. Optimization theory is the body of mathematics that deals with the properties of maxima and minima and how to find maxima and minima numerically. In the typical design optimization situation the designer has created a general configuration for which the numerical values of the

199

independent variables have not been fixed. An objective function[1] that defines the value of the design in terms of the independent variables is established.

$$U = U(x_1, x_2, \ldots, x_n) \tag{5-1}$$

Typical objective functions could be cost, weight, reliability, and producibility. Inevitably, the objective function is subject to certain constraints. *Constraints* arise from physical laws and limitations or from compatibility conditions on the individual variables. *Functional constraints* ψ, also called equality constraints, specify relations that must exist between the variables.

$$\psi_1 = \psi_1(x_1, x_2, \ldots, x_n) = 0$$
$$\psi_2 = \psi_2(x_1, x_2, \ldots, x_n) = 0$$
$$\vdots \tag{5-2}$$
$$\psi_n = \psi_n(x_1, x_2, \ldots, x_n) = 0$$

For example, if we were optimizing the volume of a rectangular storage tank, where $x_1 = l_1$, $x_2 = l_2$, and $x_3 = l_3$, then the functional constraint would be that $V = l_1 l_2 l_3$. *Regional constraints* ϕ, also called inequality constraints, are imposed by specific details of the problem.

$$\phi_1 = \phi_1(x_1, x_2, \ldots, x_n) \leq L_1$$
$$\phi_2 = \phi_2(x_1, x_2, \ldots, x_n) \leq L_2$$
$$\vdots \tag{5-3}$$
$$\phi_p = \phi_p(x_1, x_2, \ldots, x_n) \leq L_p$$

A type of regional constraint that arises naturally in design situations is based on specifications. *Specifications* are points of interaction with other parts of the system. Often a specification results from an arbitrary decision to carry out a suboptimization of the system.

A common problem in design optimization is that there often is more than one design characteristic that is of value to the user. In formulating the optimization problem, one predominant characteristic is chosen as the objective function and the other characteristics are reduced to the status of constraints. Frequently they show up as rather "hard" or severely defined specifications. Actually, such specifications are usually subject to negotiation (soft specifications) and should be considered as target values until the design progresses to such point that it is possible to determine the penalty that is being paid in trade-offs to achieve the specifications. Siddall[2] has shown how this may be accomplished in design optimization through the use of an interaction curve.

---

[1] Also called the criterion function or the payoff function.
[2] J. N. Siddall and W. K. Michael, *Trans. ASME, J. Mech. Design*, vol. 102, pp. 510–516, 1980.

The following example should help to clarify the above definitions. We wish to design a cylindrical tank to store a fixed volume of liquid V. The tank will be constructed by forming and welding thin steel plate. Therefore, the cost will depend directly on the area of plate that is used.

The design variables are the tank diameter D and its height, h. The surface area of the tank is given by:

$$A = 2(\pi D^2/4) + \pi Dh$$

If C is the cost per unit area of steel plate, then the objective function can be written

$$U = C(\pi D^2/2 + \pi Dh) \tag{5-4}$$

A functional constraint is introduced by the requirement that the tank must hold a specified volume:

$$V = \pi D^2 h/4$$

Regional constraints are introduced by the requirement for the tank to fit in a specified location or to not have unusual dimensions.

$$D_{min} \le D \le D_{max}; \qquad h_{min} \le h \le h_{max}$$

Siddall,[1] who has reviewed the development of optimal design methods, gives the following insightful description of optimization methods.

1. *Optimization by evolution.* There is a close parallel between technological evolution and biological evolution. Most designs in the past have been optimized by an attempt to improve upon an existing similar design. Survival of the resulting variations depends on the natural selection of user acceptance.

2. *Optimization by intuition.* The *art* of engineering is the ability to make good decisions without being able to formulate a justification. Intuition is knowing what to do without knowing why one does it. The gift of intuition seems to be closely related to the unconscious mind. The history of technology is full of examples of engineers who used intuition to make major advances. Although the knowledge and tools available today are so much more powerful, there is no question that intuition continues to play an important role in technological development.

3. *Optimization by trial-and-error modeling.* This refers to the usual situation in modern engineering design where it is recognized that the first feasible design is not necessarily the best. Therefore, the design model is exercised for a few iterations in the hope of finding an improved design. However, this mode of operation is not true optimization. Some refer to *satisficing,* as opposed to optimizing, to mean a technically acceptable job done rapidly

---

[1] J. N. Siddall, *Trans. ASME, J. Mech. Design,* vol. 101, pp. 674–681, 1979.

and presumably economically. Such a design should not be called an optimal design.

4. *Optimization by numerical algorithm.* This is the area of current active development in which mathematically based strategies are used to search for an optimum. The chief types of numerical algorithms are listed in the accompanying table.

Type of algorithm	Example	Reference (see footnotes)
Linear programming	Simplex method	1
Nonlinear programming	Davison-Fletcher-Powell	2
Geometric programming		3
Dynamic programming		4
Variational methods	Ritz	5
Differential calculus	Newton-Raphson	6
Simultaneous mode design	Structural optimization	7
Analytical-graphical methods	Johnson's MOD	8
Monotonicity analysis		9

[1] W. W. Garvin, "Introduction to Linear Programming," McGraw-Hill Book Company, New York, 1960.

[2] M. Avriel, "Nonlinear Programming: Analysis and Methods," Prentice-Hall, Inc., Englewood Cliffs, N.J., 1976.

[3] C. S. Beightler and D. T. Philips: "Applied Geometric Programming," John Wiley & Sons, Inc., New York, 1976.

[4] S. E. Dreyfus and A. M. Law, "The Art and Theory of Dynamic Programming," Academic Press, New York, 1977.

[5] M. H. Denn, "Optimization by Variational Methods," McGraw-Hill Book Company, New York, 1969.

[6] F. B. Hildebrand, "Introduction to Numerical Analysis," McGraw-Hill Book Company, 1956.

[7] L. A. Schmit (ed.), "Structual Optimization Symposium," ASME, New York, 1974.

[8] R. C. Johnson, "Optimum Design of Mechanical Elements," 2d ed., John Wiley & Sons, Inc., New York, 1980.

[9] P. Y. Papalambros and D. J. Wilde, "Principles of Optimal Design," Cambridge University Press, New York, 1988.

There are no "standard techniques" for optimization in engineering design. How well a technique works[1] depends on the nature of the functions represented in the problem. The optimization methods cannot all be covered in a brief chapter, but we shall attempt to discuss a broad range of them as they are applied in engineering design.

---

[1] E. D. Eason and R. G. Fenton, *Trans. ASME, J. Eng. for Ind.*, vol. 96, pp. 196–200, 1974.

5-2 OPTIMIZATION BY DIFFERENTIAL CALCULUS

We are all familiar with the use of the calculus to determine the maximum or minimum values of a function. Figure 5-1 illustrates various types of extrema that can occur. A characteristic property of an extremum is that U is momentarily stationary at each point. For example, as E is approached, U increases; but right at E it stops increasing and soon decreases. The familiar condition for a stationary point is

$$\frac{dU}{dx_1} = 0 \tag{5-5}$$

If the curvature is negative, then the stationary point is a maximum. The point is a minimum if the curvature is positive.

$$\frac{d^2U}{dx_1^2} < 0 \quad \text{indicates a local maximum} \tag{5-6}$$

$$\frac{d^2U}{dx_1^2} > 0 \quad \text{indicates a local minimum} \tag{5-7}$$

Both point B and point E are mathematical maxima. Point B, which is the smaller of the two maxima, is called a local maximum. Point E is the global maximum. Point D is a point of inflection. The slope is zero and the curve is horizontal, but the second derivative is zero. When $d^2U/dx^2 = 0$, higher-order derivatives must be used to find a derivative that becomes nonzero. If the derivative is odd, the point is an inflection point, but if the derivative is even it is a local optimum. Point F is not a minimum point because the objective function is *not continuous* at it; the point F is only a cusp in the objective function.

We can apply this simple optimization technique to the tank problem in

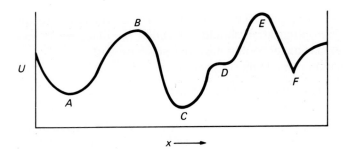

FIGURE 5-1
Different types of extrema in the objective function curve.

Sec. 5-1. The objective function is given by

$$\frac{dU}{dD} = 0 = C\pi D - \frac{4CV}{D^2} \tag{5-8}$$

$$D = \left(\frac{4V}{\pi}\right)^{1/3} = 1.084\, V^{1/3} \tag{5-9}$$

The value of diameter established by Eq. (5-9) results in minimum cost, because the second derivative of Eq. (5-8) is positive. In using Eq. (5-9) we apply the regional constraints that were discussed above.

Lagrange Multipliers

The Lagrange multipliers provide a powerful method for finding optima in multi-variable problems involving functional constraints. We have the objective function $U = U_1(x, y, z)$ subject to the functional constraints $\psi_1 = \psi_1(x, y, z)$ and $\psi_2 = \psi_2(x, y, z)$. We establish a new function, the Lagrange expression (LE)

$$\text{LE} = U_2(x, y, z) + \lambda_1\psi_1(x, y, z) + \lambda_2\psi_2(x, y, z) \tag{5-10}$$

where λ_1 and λ_2 are the Lagrange multipliers. The following conditions must be satisfied at the optimum point.

$$\frac{\partial \text{LE}}{\partial x} = 0 \qquad \frac{\partial \text{LE}}{\partial y} = 0 \qquad \frac{\partial \text{LE}}{\partial z} = 0 \qquad \frac{\partial \text{LE}}{\partial \lambda_1} = 0 \qquad \frac{\partial \text{LE}}{\partial \lambda_2} = 0$$

We will illustrate the determination of the Lagrange multipliers and the optimization method with an example.[1] A total of 300 lineal feet of tubes must be installed in a heat exchanger in order to provide the necessary heat-transfer surface area. The total dollar cost of the installation includes:

1. The cost of the tubes, $700
2. The cost of the shell $= 25D^{2.5}L$
3. The cost of the floor space occupied by the heat exchanger $= 20DL$

The spacing of the tubes is such that 20 tubes will fit in a cross-sectional area of $1\,\text{ft}^2$ inside the shell. The optimization should determine the diameter D and the length L of the heat exchanger to minimize the purchase cost.

The objective function is made up of three costs

$$C = 700 + 25D^{2.5}L + 20DL \tag{5-11}$$

---

[1] W. F. Stoecker, "Design of Thermal Systems," 2d ed., McGraw-Hill Book Company, New York, 1980.

subject to the functional constraint

$$\frac{\pi D^2}{4} L\left(20\frac{\text{tubes}}{\text{ft}^2}\right) = 300 \text{ ft}$$

$$5\pi D^2 L = 300 \tag{5-12}$$

$$\psi_1 = L - \frac{300}{5\pi D^2}$$

The Lagrange expression is

$$\text{LE} = 700 + 25D^{2.5}L + 20DL + \lambda\left(L - \frac{300}{5\pi D^2}\right) \tag{5-13}$$

$$\frac{\partial \text{LE}}{\partial D} = 62.5D^{1.5}L + 20L + 2\lambda\frac{60}{\pi D^3} = 0 \tag{5-14}$$

$$\frac{\partial \text{LE}}{\partial L} = 25D^{2.5} + 20D + \lambda = 0 \tag{5-15}$$

$$\frac{\partial \text{LE}}{\partial \lambda} = L - \frac{300}{5\pi D^2} = 0 \tag{5-16}$$

From Eq. (5-16), $L = 60/\pi D^2$, and from Eq. (5.15), $\lambda = -25D^{2.5} - 20D$. Substituting into Eq. (5-14) yields

$$62.5D^{1.5}\frac{60}{\pi D^2} + 20\frac{60}{\pi D^2} + 2(-25D^{2.5} - 20D)\frac{60}{\pi D^3} = 0$$

$$62.5D^{1.5} + 20 - 50D^{1.5} - 40 = 0$$

$$D^{1.5} = 16 \quad \text{and} \quad D = 1.37 \text{ ft}$$

Substituting the optimum value of diameter into the functional constraint gives $L = 10.2$ ft. The cost for the optimum heat exchanger is \$1540.

This example is simple and could be solved by direct substitution and finding the derivative, but the method of Lagrange multipliers is perfectly general and could be used for a problem with many variables.

5-3 SEARCH METHODS

In engineering we often deal with problems that cannot be expressed by analytical functional relations. Therefore, instead of differentiating an analytical expression to find the optimum conditions, we are interested in devising a search strategy for finding the optimum in the fewest possible experiments. We can identify several classes of search problems. A *deterministic search* is assumed to be free from appreciable experimental error, whereas in a *stochastic search* the existence of random errors must be considered. We can have a search involving only a single variable or the more complicated and more realistic situation involving a search over multiple variables. We can have a simultaneous search, in which the conditions for every experiment are specified and all the observations are completed before any judgment regarding the location of the optima is made, or a sequential search, in which future experiments are based on past outcomes. All well-developed search methods are based on the assumption that the behavior is unimodal (a single peak). Some search problems involve *constrained optimization,* in which certain combinations of variables are forbidden. Linear programming and dynamic programming are techniques that deal well with situations of this class. This section considers only single variable search techniques.

Uniform Search

In the uniform search method the trial points are spaced equally over the allowable range of values. Each point is evaluated in turn in an exhaustive search. Consider the example of our wanting to optimize the yield of a chemical reaction by varying the concentration x of catalyst. Suppose x lies over the range $x = 0$ to $x = 10$. We have four experiments available, and we distribute them at equi-distant spacing over the range $L = 10$ (Fig. 5-2). This divides L into intervals, each of width $L/n + 1$. From inspection of the experimental points we can conclude that the optimum will not lie at $x < 2$ or $x > 6$. Therefore, we know the optimum will lie between $2 < x < 6$. This is known as the *interval of uncertainty.* With four experiments we have narrowed the range of values that require further search to 40 percent of the total range.

Uniform Dichotomous Search

In the uniform dichotomous search procedure, experiments are performed in pairs to establish whether the function is increasing or decreasing. Since the search procedure is uniform, the experiments are spaced evenly over the entire

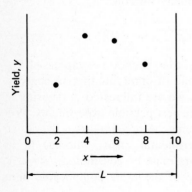

FIGURE 5-2
Example of the uniform search technique.

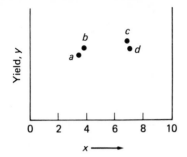

FIGURE 5-3
Example of the uniform dichotomous search technique.

range of values. Using the same example, we place the first pair of experiments at $x = 3.33$ and the other pair at $x = 6.67$. For the n experiments there will be $n/2$ pairs. The range L is divided into $(n/2) + 1$ intervals each of width $L/[(n/2) + 1]$. The two experiments in each pair are separated in x by an amount slightly greater than the experimental error in x. In this case, $x = 3.30$ and 3.36 and $x = 6.64$ and 6.70. In Fig. 5-3, since $y_a < y_b$, the region $0 < x < 3.33$ is excluded. Also, since $y_c > y_d$, the region $6.67 < x < 10$ is excluded, so the optimum lies in the interval $x = 3.33$ to 6.67. Thus, with four experiments we have narrowed the range of values that require further search to 33 percent of the total range.

Sequential Dichotomous Search

The two preceding examples were simultaneous search techniques in which all experiments were planned in advance. This example is a sequential search in which the experiments are done in sequence, each taking advantage of the information gained from the preceding one. Using the same example, we first run a pair of experiments near the center of the range, $x = 5.0$. As before, the two experiments are separated just enough in x to be sure that outcomes are distinguishable. Referring to Fig. 5-4, since $y_b > y_a$, we eliminate $x > 5.0$ from further consideration. The second pair of experiments are run at the center of the remaining interval, $x = 2.5$. Since $y_d > y_c$, we eliminate the region

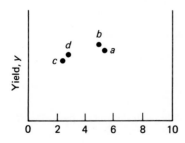

FIGURE 5-4
Example of the sequential dichotomous search technique.

$0 < x < 2.5$ and the optimum is sought in the interval $2.5 < x < 5.0$. With this search technique four experiments have narrowed the range of values that require further search to 25 percent of the total range.

Fibonacci Search

The Fibonacci search is a very efficient sequential technique. It is based on the use of the Fibonacci number series, which is named after a thirteenth-century mathematician. A Fibonacci series is given by $F_n = F_{n-2} + F_{n-1}$, where $F_0 = 1$ and $F_1 = 1$. Thus, the series is

n	0	1	2	3	4	5	6	7	8	9	$\ldots$
F_n	1	1	2	3	5	8	13	21	34	55	

Note that the nth Fibonacci number is the sum of the preceding two numbers.

The Fibonacci search begins by placing experiments at a distance $d_1 = (F_{n-2}/F_n)L$ from each end of the range of values (Fig. 5-5). For $n = 4$, $d_1 = \frac{2}{5}(10) = 4$. Since $y_4 > y_6$, the interval $6 < x < 10$ is eliminated from further consideration. The value of d_2 is obtained by letting $n_2 = n - 1$, so $d_2 = (F_{n-3}/F_{n-1})L$ and $d_2 = \frac{1}{3}(6) = 2$. Therefore, we run experiments at $x = 2$ and $x = 4$ (where we already have a data point). Since $y_4 > y_2$, we eliminate region $0 < x < 2$ and we are left with the region between $x = 2$ and $x = 6$. Since we have used only three data points, we run the last experiment just to the right of $x = 4$ to determine whether the optimum is $2 < x < 4$ or $4 < x < 6$. This experiment eliminates the former interval and we have narrowed the optimum to the region $4 < x < 6$. In this case, with four experiments we have narrowed the range of values that require further search to 20 percent of the total range. Computer programs for the optimization of a single variable by Fibonacci search are available.[1]

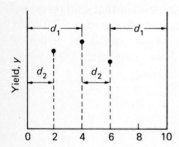

FIGURE 5-5
Example of the Fibonacci search technique.

[1]J. L. Kuester and J. H. Mize, "Optimization Techniques with Fortran," pp. 285–295, McGraw-Hill Book Company, New York, 1973.

Golden Section Search

Although the Fibonacci search is very efficient, it has the disadvantage that it requires an advance decision on the number of experiments before we have any information about the behavior of the function near the maximum. If the function turns out to be very steep near the maximum, we should have chosen more trials. With the sequential dichotomous search we would continue placing pairs of trials in the middle of the interval of uncertainty until the change in value of the objective function from one trial to the next was acceptably small.

The golden section search is a good compromise, because although it is slightly less efficient than the Fibonacci search, it does not require on advance decision on the number of trials. This search technique is based on the fact the ratio of two successive Fibonacci numbers, $F_{n-1}/F_n = 0.618$ for all values of $n > 8$. This same ratio was discovered by Euclid, who called it the golden mean. He defined it as a length divided into two unequal segments such that the ratio of the length of the whole to the larger segment is equal to the ratio of the length of the larger segment to the smaller segment. The ancient Greeks felt 0.618 was the most pleasing ratio of width to length of a rectangle, and they used it in the design of many of their buildings.

In using the golden section search the first two trials are located at $0.618L$ from either end of the range. Note that the same value is used, no matter the ultimate number of trials. Based on the value of y_1 and y_2 we eliminate an interval. With the new reduced interval we take additional experiments at $\pm 0.618L_2$. Actually, only one new experiment is needed, since one of the experiments from the preceding trial lies at exactly the proper location for the new series of trials. This procedure is continued until the maximum is located to within as small an interval as desired.

Comparison of Methods

A measure of the efficiency of a search technique is the reduction ratio, which is defined as the ratio of the original interval of uncertainty to the interval remaining after n trials. It is shown in Table 5-1.

TABLE 5-1
Comparison of univariable search techniques

Search method		Reduction ratio	
		For $n = 4$	For $n = 13$
Uniform	$(n + 1)/2$	2.5	7
Uniform dichotomous	$(n + 2)/2$	3	7.5
Sequential dichotomous	$2^{n/2}$	4	90
Fibonacci	F_n	5	377
Golden search			250

5-4 MULTIVARIABLE SEARCH METHODS

When the objective function is a function of two other variables, the geometric representation is a *response surface* (Fig. 5-6a). It usually is convenient to deal with contour lines produced by the intersection of planes of constant y with the response surface[1] and projected on the $x_1 x_2$ plane.

Lattice Search

In the lattice search, which is an analog to the single-variable exhaustive search, a two-dimensional grid lattice is super-imposed over the contour plot (Fig. 5-7). In the absence of special knowledge about the location of the maximum, the starting point is selected near the center of the region, at point 1. The objective function is evaluated for points 1 through 9. If point 5 turns out to be the largest value, it becomes the central point for the next search. The procedure continues until the location reached is one at which the central point is greater than any of the other eight points. Frequently, a coarse grid is used initially and a finer grid is used after the maximum is approached.

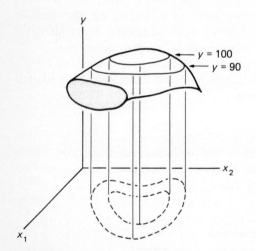
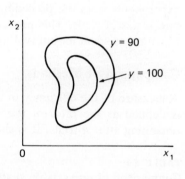

FIGURE 5-6
Representation of two variables by contour lines.

---

[1] A technique for generating contour plots by polynomial curve fitting to the objective function surface has been given by K. A. Afimiwala and R. W. Mayne, *Trans. ASME, J. Mech. Design*, vol. 101, pp. 349–354, 1979.

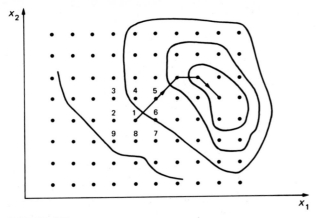

FIGURE 5-7
Procedure for a lattice search. (*From W. F. Stoecker, "Design of Thermal Systems," 2d ed.,*
McGraw-Hill Book Co., 1980; used with permission of McGraw-Hill Book Co.)

Univariate Search

The univariate search is a one-at-a-time method. All of the variables are kept
constant except one, and it is varied to obtain an optimum in the objective
function. That optimal value is then substituted into the function and the
function is optimized with respect to another variable. The objective function
is optimized with respect to each variable in sequence, and an optimal value of
a variable is substituted into the function for the optimization of the succeeding
variables.

Figure 5-8a shows the univariate search procedure. Starting at point 0 we
move along $x_2 = $ const to a maximum by using any one of the single variable
search techniques. Then we move along $x_1 = $ const to a maximum at point 2
and along $x_2 = $ const to a maximum at 3. We repeat the procedure until two
successive moves are less than some specified value. If the response surface
contains a ridge, as in Fig. 5-8b, then the univariate search can fail to find an
optimum. If the initial value is at point 1, it will reach a maximum at $x_1 = $ const
at the ridge, and that will also be a maximum for $x_2 = $ const. A false maximum
is obtained.

Steepest Ascent

The path of steepest ascent (or descent) up the response surface is the *gradient
vector*. Imagine that we are walking at night up a hill. In the dim moonlight we
can see far enough ahead to follow the direction of the local steepest slope.
Thus, we would tend to climb normal to contour lines for short segments and
adjust the direction of climb as the terrain came progressively into view. The
gradient method does essentially that with mathematics. We change the

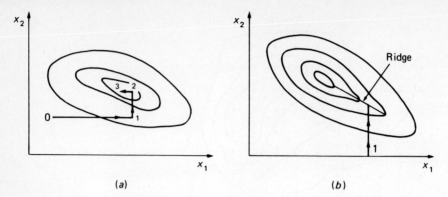

FIGURE 5-8
Univariate search procedure.

direction of the search in the direction of the maximum slope, but we must do so in finite straight segments.

We shall limit our discussion to the situation of two independent variables with the understanding that the steepest ascent method is applicable to many variables. The gradient vector is given by

$$\nabla U = \frac{\partial U}{\partial x_1} \mathbf{i}_1 + \frac{\partial U}{\partial x_2} \mathbf{i}_2 \tag{5-17}$$

To move in the direction of the gradient vector, we take the step lengths δx_1 and δx_2 in proportion to the components of the gradient vector (Fig. 5-9).

$$\frac{\delta x_1}{\delta x_2} = \frac{\partial U / \partial x_1}{\partial U / \partial x_2} \tag{5-18}$$

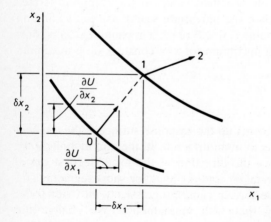

FIGURE 5-9
Definitions in the steepest ascent method.

For the general case of n independent variables, $U = U(x_1, x_2, \ldots, x_n)$

$$\frac{\delta x_i}{\delta x_1} = \frac{\partial U/\partial x_i}{\partial U/\partial x_1} \qquad (5\text{-}19)$$

If the objective function is in analytical form, the partial derivatives can be obtained by calculus. If it is not, a numerical procedure must be used. Starting at the initial point, we take a small finite difference Δx_i in each variable and evaluate the function at each $x_i + \Delta x_i$ in turn, holding all other x_i's at their initial value. $\Delta U_i = U_i - U_0$. The partial derivatives are evaluated by $\Delta U_i/\Delta x_i$, so

$$\frac{\delta x_i}{\delta x_1} \approx \frac{\Delta U_i/\Delta x_i}{\Delta U_1/\Delta x}, \qquad (5\text{-}20)$$

Determining the direction of the gradient vector is standard, but deciding on the length of the step is not as straightforward. Numerous methods for establishing δx_1 have been developed. The most direct is to arbitrarily select δx_1 and compute δx_i from Eq. (5-19) or Eq. (5-20). This method breaks down when one of the partial derivatives becomes zero. The second method is to select the step size such that the objective function improves by a specific amount ΔU.

$$\delta x_1 = \frac{(\partial U/\partial x_1)\Delta U}{(\partial U/\partial x_1)^2 + (\partial U/\partial x_2)^2 + \cdots} \qquad (5\text{-}21)$$

A third method of establishing the step length is to proceed from the initial point in the direction of the gradient vector until an optimal value is reached. Then another gradient vector is calculated and the experiment moves in that direction until an optimal value is obtained. This procedure is repeated until all partial derivatives becomes negligibly small.

An important consideration in working with the gradient method is the scaling of variables. Since the units of each variable may be different and arbitrary in magnitude, the relative scale of the variables is arbitrary. Transforming the scale of a variable changes the shape of the contours of the objective function and the magnitude and direction of the gradient vector. When possible, contours that approach circles are to be preferred.

Nonlinear Optimization Methods

The methods discussed above are not really practical techniques for most real problems. However, multivariable optimization of nonlinear problems has been a field of great activity and many computer-based methods are available. Space permits mention of only a few of the more useful methods. These are of two types, those that require information about derivative values in determining the search direction for optimization (indirect methods), and direct methods that rely solely on the evaluation of the objective function. Because an in-depth understanding requires considerable mathematics for which we do not have space, only a brief word description can be given. The interested student is referred to original sources or the recent text by Arora.[1]

Methods for unconstrained multivariable optimization are discussed first. Newton's method is an indirect technique that employs a second-order approximation of the function.[2] This method has very good convergence properties, but it can be an inefficient method because it requires the calculation of $n(n + 1)/2$ second-order derivatives, where n is the number of design variables. Therefore, methods that require the computation of only first derivatives and use information from previous iterations to speed up convergence have been developed. The DFP (Davidon, Fletcher, and Powell) method is one of the most powerful methods.[3] Updates and improvements on this method were provided by Broyden, Fletcher, Goldfarb, and Shannon in what is called the BFGS method.[4] There is a general feeling among practitioners that the BFGS algorithm is the best one to use.

One of the earliest direct search methods was the method of Hooke and Jeeves.[5] Because it does not use derivatives it is fast and is less sensitive to irregular or discontinuous functions. The method does have problems in highly constrained situations where a search can get stalled when the contour lines and the inequality constraint line have a certain orientation. Other direct search techniques are the simplex method[6] (no relation to the technique in linear programming) and Powell's quadratic convergence method.[7]

Optimization of nonlinear problems with constraints is a more difficult area. The more general approach to this class of problems is called the Generalized Reduced Gradient method (GRG). The idea of GRG is to convert the constrained problem into an unconstrained one by direct substitution. However, with nonlinear constraint equations this is not feasible by direct substitution and the procedures of constrained variation and Lagrange multipliers must be used.[8]

---

[1] J. S. Arora, "Introduction to Optimum Design," McGraw-Hill Book Co., New York, 1989.

[2] J. S. Arora, op. cit., pp. 319–325.

[3] R. Fletcher and M. J. D. Powell, *Computer J.*, vol. 6, pp. 163–180, 1963; Arora, op. cit., pp. 327–330; for a computer program see J. L. Kuester and J. H. Mize, "Optimization Techniques with Fortran," pp. 355–366, McGraw-Hill Book Co., New York, 1973 and J. N. Siddall, "Analytical Decision-Making in Engineering Design," Appendix E, Prentice-Hall, Englewood, NJ, 1972.

[4] For a computer program see R. W. Pike, "Optimization for Engineering Systems," pp. 285–291, Van Nostrand Reinhold, New York, 1986; Arora, op. cit., pp. 330–332..

[5] R. Hooke and T. A. Jeeves, *J. Assoc. Comp. Mach.*, vol. 8, p. 202–229, 1961; for a computer program see Kuester and Mize, op. cit., pp. 309–319 and Siddall, op. cit, Appendix H.

[6] J. A. Nelder and R. Meade, *Computer J.*, vol. 7, pp. 308–318, 1965;

[7] M. J. D. Powell, *Computer J.*, vol. 7, pp. 303–307, 1964.

[8] J. S. Arora, op. cit., pp. 4115–417.

Another approach is to successively linearize the constraints and objective function of a nonlinear problem and solve using the technique of linear programming, Sec. 5-5. Sequential linear programming (SLP) is the name of the method.[1] The most recent, and perhaps the best method of solving nonlinear problems is sequential quadratic programming (SQP).[2]

Computer programs with names like GRG2, NLPQL, OPT, and MINOS are available to perform nonlinear optimization on the mainframe, and programs like GINO, NLP Solver, and OPTISOLVE are available for microcomputers.

5-5 LINEAR PROGRAMMING

The linear programming method is applicable to a large and important class of optimization problems that involve linear objective functions subject to linear constraints.[3] Linear programming problems are common in such engineering fields as water resources, traffic flow control, inventory management and production control. Design problems in aerospace, structural, and mechanical system design are less likely to be linear in nature. However, it is possible to transform a nonlinear programming problem into a sequence of linear programs.[4]

The linear programming method deals with objective functions of the form

$$U = \sum_i k_i x_i \quad \text{to maximize or minimize} \qquad (5\text{-}22)$$

subject to

$$\psi_j = \sum_i a_{ij} x_i \geq r_j \qquad (5\text{-}23)$$

The general linear programming problem consists of a set of m linear equations and/or inequalities involving n variables. We wish to find the nonnegative values of the variables that satisfy the equations and inequalities and also maximize (or minimize) the linear objective function.

We shall illustrate the method of solution of linear programming

---

[1] J. S. Arora, op. cit., pp. 365–371.

[2] J. S. Arora, op. cit., pp. 384–392.

[3] G. H. Hadley, "Linear Programming," Addison-Wesley Publishing Co., Reading, Mass., 1962; S. I. Gass, "Linear Programming," 4th ed., McGraw-Hill Book Company, New York, 1975; ibid., "An Illustrated Guide to Linear Programming," McGraw-Hill Book Company, New York, 1970; B. A. Murtaugh, "Advanced Linear Programming," McGraw-Hill Book Company, New York, 1980.

[4] J. S. Arora, op. cit., pp. 358–384.

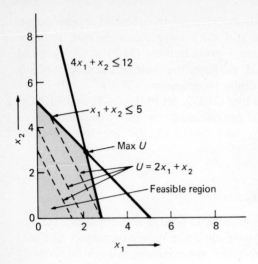

FIGURE 5-10
Graphical solution to LP problem.

problems with the following two-variable example.

$$U = 2x_1 + x_2 \qquad \text{objective function}$$
$$4x_1 + x_2 \leq 12$$
$$x_1 + x_2 \leq 5 \qquad \text{inequality constraints}$$
$$x_1 > 0 \qquad x_2 > 0 \qquad \text{nonnegativity requirement}$$

We can most easily treat this simple problem with a graphical solution (Fig. 5-10). The inequality constraints define the *feasible region* within which a solution will be found. The objective function is defined by a family of parallel dashed lines. As these lines move outward from the origin, they increase in value. The greatest value of U is obtained by the line that just touches the circled point defined by the intersection of the two constraints. Thus, $U_{max} = 7.33$ at $x_1 = 2.33$ and $x_2 = 2.67$. This illustrates a special feature of linear programming; that optimum values occur on the boundary of the feasible region.

To solve the problem by algebra, we convert the inequalities to equalities by introducing so-called slack variables, $s > 0$.

$$4x_1 + x_2 + s_1 = 12$$
$$x_1 + x_2 + s_2 = 5$$

where s_1 and s_2 are slack variables. The variable s_1 represents the unused part of x_1 in terms of how far away we are from the constraint, and s_2 represents the unused part of x_2. Note that if the inequality constraint had been $x_1 + x_2 \geq 5$, we would have written $x_1 + x_2 - s_2 = 5$, that is, negative slack.

For this problem we have $n = 4$ variables (x_1, x_2, s_1, s_2) and two equations $(m = 2)$. U is optimized when at least $n - m$ of the variables are zero. Therefore, we set in turn $n - m = 4 - 2 = 2$ variables equal to zero, which leaves m variables and equations to solve. The number of combinations that must be tried is $n!/m!\,(n-m)!$.

x_1	x_2	s_1	s_2	U
0	12	0	-7	Not allowable, negative s_2
0	5	7	0	5
3	0	0	2	6
5	0	-8	0	Not allowable, negative s_1
0	0	12	5	0
$\frac{7}{3}$	$\frac{8}{3}$	0	0	$7\frac{1}{3}$

Therefore, to optimize a linear system with regional constraints:

1. Define one slack variable for each inequality constraint.
2. The number of unknowns is the number of independent variables plus the number of slack variables.
3. The number of unknowns minus the number of inequality constraints is the number of variables that must be zero.
4. Evaluate U for all combinations of the nonzero variables. The optimum point is the extremum of those combinations.

Simplex Algorithm

The simplex algorithm[1] is an efficient method of searching the many possible combinations for the extremum in the objective function. It is based on the Gauss-Jordan elimination procedure for solving a set of simultaneous linear equations, and it uses a fixed form of data presentation called a *programming tableau.*

We shall consider a slightly larger and more realistic linear programming problem[2] for this example. A small manufacturing plant has two products, A and B. Each product must be worked on by a bank of type 1 machines and then, in succession, by a group of type 2 machines. Product A requires 2 h on type 1 machines and 1 h on type 2 machines. Product B requires 1 h on type 1 machines and 4 h on type 2 machines. A total of 6000 h is available per week on type 1 machines and 10,000 h on type 2 machines. The net profit is $5 per unit for product B and $3.50 per unit for product A. Find the optimal production schedule (mix of A and B products) that will maximize the weekly profit.

---

[1] G. B. Dantzig *in* T. C. Koopmans (ed.), "Activity Analysis of Production and Allocation," John Wiley & Sons, Inc., New York, 1951.

[2] See F. C. Jelen (ed.), "Cost and Optimization Engineering," pp. 264–270, McGraw-Hill Book Company, New York, 1970.

We first must cast the problem into standard linear programming equations. Let x_1 be the number of parts of A produced in a week and x_2 the weekly production of B. The objective function is the weekly profit

$$U = 3.5x_1 + 5x_2 \tag{5-24}$$

The machine capacities introduce regional constraints. The total time required on type 1 machines is

$$2x_1 + x_2 \leq 6000 \tag{5-25}$$

The total time required on type 2 machines is

$$x_1 + 4x_2 \leq 10,000 \tag{5-26}$$

The column headings in the initial tableau are the coefficients of the objective function.

$$U = 3.5x_1 + 5x_2 + 0s_1 + 0s_2 \tag{5-27}$$

The rows consist of the coefficients of the constraint equations.

$$2x_1 + x_2 + s_1 + 0s_2 = 6000 \tag{5-28}$$

$$x_1 + 4x_2 + 0s_1 + s_2 = 10,000 \tag{5-29}$$

The column headed B lists the constants from the constraint equations. We start with the slack variables in the active column, which have a value $s_1 = s_2 = 0$. A feasible solution is to set $x_1 = x_2 = 0$. Next we evaluate the objective function for each variable, $U_j = \sum k_i a_{ij}$ and place those values at the bottom of the tableau.

Initial tableau

Active	k_i	k_j 3.5 x_1	5 x_2	0 s_1	0 s_2	B	θ
s_1	0	2	1	1	0	6,000	6,000
s_2	0	1	4*	0	1	10,000	2,500
$U_j = \sum k_i a_{ij}$		0	0	0	0	0	
$k_j - U_j$		3.5	5	0	0		

The row $k_j - U_j$ indicates the benefit to be derived from adding a unit of each variable. Since the variable x_2 shows the greatest positive value, in the next iteration we substitute x_2 for one of the other active variables. To decide which variable, s_1 or s_2, to take out of the active category, we examine the rows for the row with the lowest contribution. We use the column of θ values, where θ is the value of B for each row divided by the entry in the pivot column for that row. Thus, we drop s_2 from the active variables in the second tableau.

Therefore, variable x_2 enters the active category and variable s_2 leaves.

The variable that enters is on the pivot column and the variable that leaves lies on the pivot row. The intersection of the pivot column and pivot row is the pivot element, in this case 4. The pivot element is marked with an asterisk.

In setting up the next tableau we need to perform some matrix manipulations. The pivot element is made equal to unity by dividing every element in the pivot row by the value of the pivot element. All other elements in the pivot column must be made equal to zero by the appropriate manipulation so that x_2 drops from the other constraint equations. In this case the new s_1 row is obtained by deducting the new x_2 row from the old s_1,

	k_i	x_1	x_2	s_1	s_2	B
s_1 (old)	0	2	1	1	0	6,000
x_2 (new)	5	$\frac{1}{4}$	1	0	$\frac{1}{4}$	2,500

element by element. This interchange of variables is called simplexing. In matrix notation it is expressed as

$$a'_{ij} = a_{ij} - a_{ik}a'_{rj} \qquad i \neq r \tag{5-30}$$

where a_{ij} = old element in the matrix; i denotes row, and j denotes column
a'_{ij} = new element after simplex transformation
$a'_{rj} = a_{rj}/a_{rk}$; r denotes pivot row, and k denotes pivot column

To illustrate, in the initial tableau, $a_{kr} = 4$ and $a'_{21} = \frac{1}{4}$. Let $i = 1, j = 1$

$$a'_{11} = a_{11} - a_{12}a'_{21} = 2 - 1(\tfrac{1}{4}) = \tfrac{7}{4}$$

Let $i = 1, j = 3$

$$a'_{13} = a_{13} - a_{12}a'_{23} = 1 - 1(0) = 1$$

$$a'_{23} = \frac{a_{23}}{a_{22}} = 0/4 = 0$$

Second tableau

	k_j	3.5	5	0	0		
Active	k_i	x_1	x_2	s_1	s_2	B	θ
s_1	0	$\frac{7}{4}$*	0	1	$-\frac{1}{4}$	3,500	2,000
x_2	5	$+\frac{1}{4}$	$+1$	0	$+\frac{1}{4}$	2,500	10,000
U_j		$+\frac{5}{4}$	$+5$	0	$+\frac{5}{4}$	12,500	
$k_j - U_j$		$\frac{9}{4}$	0	0	$-\frac{5}{4}$		

For the third tableau we introduce x_1 into the active variables because it makes the greatest positive contribution to the objective function of all the variables. We drop s_1 because the θ column shows that it will make the least change in the objective function. The pivot element ($r = 1, k = 1$) is $\frac{7}{4}$. We need to make this element unity and to change all other elements in the pivot column to zero. To do so, we multiply each element in the s_1 row by $\frac{4}{7}$ to produce the new x_1. We apply Eq. (5-30) to establish the new coefficients in the x_2 row.

Third tableau

Active	k_j / k_i	3.5 x_1	5 x_2	0 s_1	0 s_2	B
x_1	3.5	1	0	$\frac{4}{7}$	$-\frac{1}{7}$	2,000
x_2	5	0	1	$-\frac{1}{7}$	$\frac{2}{7}$	2,000
U_j		3.5	5	$\frac{9}{7}$	$\frac{13}{14}$	17,000
$k_j - U_j$		0	0	$-\frac{9}{7}$	$-\frac{13}{14}$	

The iteration proceeds until the differences $k_j - U_j$ are less than or equal to zero. Since this is the case for the third tableau, it represents the optimum condition. The constraint equations are:

$$x_1 = 2000 - \tfrac{4}{7}s_1 + \tfrac{1}{7}s_2$$

$$x_2 = 2000 + \tfrac{1}{7}s_1 - \tfrac{2}{7}s_2$$

and

$$U = 17,000 - \tfrac{9}{7}s_1 - \tfrac{13}{14}s_2$$

The optimum solution is obtained for $s_1 = s_2 = 0$ and is found at $x_1 = 2000$ and $x_2 = 2000$, where $U = 17,000$.

The simplex method, as presented, is designed to produce the maximum of the objective function. It can be used to find the minimum of U by changing the sign of the coefficients in the objective function.

Although the simplex method obviously is the hard way to solve a simple linear programming problem, it definitely is the approach to use for a large problem consisting of five or more terms. Because it consists of arithmetic steps, it is well suited for solution with a digital computer.[1] Most computer systems include a linear programming package.

Special Problems in Linear Programming

There are several important practical problems in linear programming for which shorter and quicker solutions than the general simplex method are available.

The *transportation problem* determines the optimum way to distribute a

---

[1] A simplex program is published by Kuester and Mize, op. cit., pp. 9–26.

commodity from a number of dispatch points to a number of destinations. Usually the problem seeks the shipping pattern that will minimize cost.

The *assignment problem* deals with allocation resources of some kind. It may involve assigning people to tasks or assigning machines to the production of specific products. The objective function is either to minimize cost or to maximize production.

Critical-path problems are concerned with finding the shortest path through a network of tasks that must be performed to complete an operation. Another objective is to determine which particular steps in a process are most likely to create delays.

Large-scale linear programs are used routinely to optimize the product mix in petroleum refineries or multiproduct chemical plants. Also LP has been used for modeling and optimizing an industrial steam distribution system.[1]

Karmarkar Algorithm

An alternative to the Simplex method for solving large-scale LP problems was proposed by Karmarkar.[2] The Simplex method searches along the boundary of the feasible region from one vertex (defined by the constraints) to an adjacent vertex until the optimal point is found. For large-scale problems the number of vertices is quite large and the computer time may be excessive. The Karmarkar algorithm works in the interior or the constrained region and seeks out search directions which are much more efficient.

5-6 GEOMETRIC PROGRAMMING

Geometric programming[3] is a nonlinear optimization method that is gaining wide acceptance in engineering design. It is particularly useful when the objective function is the sum of a number of polynomial terms. An example would be an expression such as

$$U = x_1 x_2 + 5x_2 + \frac{12}{\sqrt{x_2}} + 3.4x_1^3$$

A polynomial expression in which all of the terms are positive is called a *posynomial*.

The geometric programming method is best understood in the concrete rather than the general. Consider the insulation of a house where we wish to

---

[1] J. K. Clark and N. E. Helmick, *Chem. Eng.*, pp. 116–128, Mar. 10, 1980.

[2] N. Karmarkar, *Combinatoria*, vol. 4, p. 373, 1984.

[3] R. J. Duffin, E. L. Peterson, and C. Zener, "Geometric Programming," John Wiley & Sons, Inc., New York, 1967; C. Zener, "Engineering Design by Geometric Programming," Wiley-Interscience, New York, 1971.

find the optimum thickness of insulation x that balances the increased cost of insulation against the savings in fuel. The objective function is given by

$$U = 120x + \frac{150,000}{x^4} \qquad (5\text{-}31)$$

$$\text{Total cost} = \text{insulation cost} + \text{fuel cost}.$$

This is an objective function with two terms T and one variable N. By definition, the *degree of difficulty* is $T - (N + 1) = 2 - 2 = 0$. We can rewrite the objective function in the standard form

$$U = u_1 + u_2 = c_1 x^{a_1} + c_2 x^{a_2} \qquad (5\text{-}32)$$

where

$$u_1 = 120x \qquad u_2 = \frac{150,000}{x^4}$$

$$c_1 = 120 \qquad c_2 = 150,000$$

$$a_1 = 1 \qquad a_2 = -4$$

We next develop the function g

$$g = \left(\frac{u_2}{w_1}\right)^{w_1}\left(\frac{u_2}{w_2}\right)^{w_2} = \left(\frac{c_1 x^{a_1}}{w_1}\right)^{w_1}\left(\frac{c_2 x^{a_2}}{w_2}\right)^{w_2} \qquad (5\text{-}33)$$

where $w_1 + w_2 = 1$ and w_i are weighting functions. The value of g^* to minimize U requires that w_1 and w_2 be properly chosen.

$$U^* = 120x + \frac{150,000}{x^4} = g^* = \left(\frac{120x}{w_1}\right)^{w_1}\left(\frac{150,000}{x^4 w_2}\right)^{w_2}$$

where

$$w_1 + w_2 = 1$$

$$a_1 w_1 + a_2 w_2 = w_1 - 4w_2 = 0$$

solving, $w_1 = \frac{4}{5}$ and $w_2 = \frac{1}{5}$

$$U^* = g^* = \left(\frac{120}{\frac{4}{5}}\right)^{4/5} x^{4/5} \left(\frac{150,000}{\frac{1}{5}}\right)^{1/5}\left(\frac{1}{x^4}\right)^{1/5}$$

$$= 150^{4/5}(750,000^{1/5})(x^{4/5})(x^{-4/5})$$

$$= 55.06(14.96)(1) = 824$$

From the theory of geometric programming

$$w_1 = \frac{u_1^*}{u_1^* + u_2^*} = \frac{u_1^*}{U^*} = \frac{-a_2}{a_1 - a_2} \qquad (5\text{-}34)$$

$$\frac{4}{5} = \frac{120x^*}{824} \qquad x^* = 5.49 \text{ in}$$

We note that w_1 is the weighting function that expresses what fraction u_1^* is of the total U^*. The objective function is the sum of the insulation cost u_1 and the fuel cost u_2. The insulation cost is four-fifths of the minimum total cost; the fuel cost is one-fifth. If the cost of insulation increased from $120x$ to $200x$, the total cost at the optimum would be higher and the optimum thickness of insulation would be less but the fraction of the total cost contributed by the insulation would be the same. That is because the relative fraction of the cost is determined by the exponents of x. Thus, when costs are constantly changing (rising!), the distribution of costs among various cost drivers remains constant so long as the functional cost relation remains unchanged.

Consider now an example with three variables.[1] We wish to design a rectangular box to haul V ft^3 of gravel each load. The objective function is

$$U = \frac{40}{LWH} + 10LW + 20LH + 40HW \tag{5-35}$$

where the first term is the cost of transporting a load and the other three terms are the costs of material and labor in building the bottom, sides, and ends of the box. This is a problem with zero degrees of difficulty because $T - (N + 1) = 4 - (3 + 1) = 0$.

$$U^* = g^* = \left(\frac{40}{w_1}\right)^{w_1} \left(\frac{10}{w_2}\right)^{w_2} \left(\frac{20}{w_3}\right)^{w_3} \left(\frac{40}{w_4}\right)^{w_4}$$

and

$$w_1 + w_2 + w_3 + w_4 = 1$$

L:
$$-w_1 + w_2 + w_3 = 0$$

H:
$$-w_1 + w_3 + w_4 = 0$$

W:
$$-w_1 + w_2 + w_4 = 0$$

which results in $w_1 = \frac{2}{5}$; $w_2 = w_3 = w_4 = \frac{1}{5}$

$$U^* = \left(\frac{40}{\frac{2}{5}}\right)^{2/5} \left(\frac{10}{\frac{1}{5}}\right)^{1/5} \left(\frac{20}{\frac{1}{5}}\right)^{1/5} \left(\frac{40}{\frac{1}{5}}\right)^{1/5}$$

$$= 5(20^2 \cdot 10 \cdot 20 \cdot 40)^{1/5} = 5(3,200,000)^{1/5} = 5(20) = 100$$

$$w_2 = \frac{u_2^*}{U^*} = \frac{10LW}{100} = \frac{1}{5} \qquad LW = 2$$

$$w_3 = \frac{u_3^*}{U^*} = \frac{20LH}{100} = \frac{1}{5} \qquad LH = 1$$

$$w_4 = \frac{u_4^*}{U^*} = \frac{40HW}{100} = \frac{1}{5} \qquad HW = \frac{1}{2}$$

---

[1] D. J. Wilde, "Globally Optimum Design," Wiley-Interscience, New York, 1978.

Therefore, to achieve a minimum cost of $100 per load, the optimum dimensions of the box are $L = 2$, $W = 1$, and $H = \frac{1}{2}$.

An important aspect of geometric programming is that when the problem has zero degrees of difficulty, the solution is obtained with a set of linear equations in w_i. If the problem had been solved with calculus, it would have required the solution of simultaneous nonlinear equations. Thus, when the degree of difficulty is zero, geometric programming is an important optimization tool.

Many realistic design problems have at least one degree of difficulty. One approach to a solution is the technique called condensation.[1] We can convert the gravel box problem to a problem with one degree of difficulty if we add runners to the box to make it a gravel sled. The cost of runners is 10 per unit length, so the objective function becomes

$$U^1 = U + 10L = \frac{40}{LWH} + 10LW + 20LH + 40HW + 10L \tag{5-36}$$

We see that

$$T - (N + 1) = 5 - (3 + 1) = 1$$

The cost of this new design is $100 + 10(2) = 120$. We wish to find a new minimum by changing the values of L, W, and H. The technique of condensation combines two of the terms so as to reduce the problem to one of zero degree of difficulty. In selecting the terms to combine, we pick terms with exponents that are not very different. In this case, the second and fifth terms

$$t_2 + t_5 \geq \left(\frac{10LW}{\frac{1}{2}}\right)^{1/2}\left(\frac{10L}{\frac{1}{2}}\right)^{1/2} = 20LW^{1/2} \tag{5-37}$$

where we have used equal weighting factors for each term.

The resulting objective function is

$$U' = \frac{40}{LWH} + 20LW^{1/2} + 20LH + 40HW \tag{5-38}$$

which is a posynomial with zero degrees of difficulty.

$$U'^* = \left(\frac{40}{w_1}\right)^{w_1}\left(\frac{20}{w_2}\right)^{w_2}\left(\frac{20}{w_3}\right)^{w_3}\left(\frac{40}{w_4}\right)^{w_4}$$

and

$$w_1 + w_2 + w_3 + w_4 = 1$$

L:

$$-w_1 + w_2 + w_3 = 0$$

W:

$$-w + \tfrac{1}{2}w_2 + w_4 = 0$$

H:

$$-w_1 + w_3 + w_4 = 0$$

---

[1] Ibid., pp. 80–90.

which results in $w_1 = \frac{3}{8}$, $w_2 = \frac{2}{8}$, $w_3 = \frac{1}{8}$, $w_4 = \frac{2}{8}$

$$U^* = \left(\frac{40}{\frac{3}{8}}\right)^{3/8} \left(\frac{20}{\frac{2}{8}}\right)^{2/8} \left(\frac{20}{\frac{1}{8}}\right)^{1/8} \left(\frac{40}{\frac{2}{8}}\right)^{2/8}$$

$$= \left[\left(\frac{320}{3}\right)^3 (80^2)(160)(160^2)\right]^{1/88} = 115$$

Other techniques that can be used when the degree of difficulty is not zero are partial invariance and dual geometric programming.[1]

5-7 OTHER OPTIMIZATION METHODS

Monotonicity Analysis

Monotonicity analysis is an optimization technique that may be applied to design problems with monotonic properties, i.e. where the objective function and constraints successively increase or decrease with respect to some design variable. This is a situation that is very common in design problems. Engineering designs tend to be strongly defined by physical constraints. When these specifications and restrictions are monotonic in the design variables then monotonicity analysis can often show the designer which constraints are active at the optimum. An active constraint refers to a design requirement which has a direct impact on the location of the optimum. This information can be used to identify the improvements that could be achieved if the feasible domain were modified, which would point out directions for technological improvement.

The ideas of monotonicity analysis were first presented by Wilde.[2] Subsequent work by Wilde and Papalambros has applied the method to many engineering problems[3] and to the development of a computer-based method of solution.[4]

Dynamic Programming

Dynamic programming is a mathematical technique that is well suited for the optimization of staged processes. The word "dynamic" in the name of this technique has no relationship to the usual use of the word to denote changes with respect to time. Dynamic programming is related to the calculus of

[1] Ibid., pp. 91–95.

[2] D. J. Wilde, Trans. ASME, *Jnl. of Engr. for Industry*, vol. 94, pp. 1390–1394, 1975.

[3] P. Papalambros and D. J. Wilde, "Principles of Optimal Design," Cambridge University Press, New York, 1989.

[4] S. Azarm and P. Papalambros, *Trans. ASME, Jnl of Mechanisms, Transmissions, and Automation in Design*, vol. 106, pp. 82–89, 1984.

variations and is not related to linear and nonlinear programming methods. The method is well suited for allocation problems, as when x units of a resource must be distributed among N activities in integer amounts. It has been broadly applied within chemical engineering to problems like the optimal design of chemical reactors. Dynamic programming converts a large complicated optimization problem into a series of interconnected smaller problems, each containing only a few variables. This results in a series of partial optimizations which requires a reduced effort to find the optimum.

Dynamic programming was developed by Richard Bellman[1] in the 1950s. It is a well-developed optimization method.[2]

Johnson's Method

A method of optimum design that is especially suited to the nonlinear problems found in the design of mechanical elements such as gears, roller bearings, and hydrodynamic journal bearings has been developed by R. C. Johnson.[3] The trend in design optimization has been towards methods that require little effort to enter problems into general computer optimization routines. In contrast, Johnson's method often requires significant effort to reduce the system of equations to a form suitable for an optimization study. However, the benefit of the method is that it gives considerable insight into the nature of and possible solutions to the problem. The technique has been applied to a wide range of optimization problems.[4]

5-8 STRUCTURAL AND SHAPE OPTIMIZATION

The design of structures has experienced the greatest degree of application of numerical optimization techniques of all areas of engineering design. Linear programming techniques were combined with plastic design methods in 1956 and nonlinear programming techniques were introduced in the design of elastic structures in 1960. The application of optimization methods to the design of elastic structures has come to be known as "structural synthesis". The

---

[1] R. E. Bellman, "Dynamic Programming," Princeton Univ. Press, Princeton, N.J., 1957.

[2] G. L. Nemhauser, "Introduction to Dynamic Programming," John Wiley & Sons, New York, 1960; E. V. Denardo, "Dynamic Programming Models and Applications," Prentice-Hall Inc., Englewood Cliffs, N.J., 1982; S. Ross, "Introduction to Stochastic Dynamic Programming," Academic Press, New York, 1983.

[3] R. C. Johnson, "Optimal Design of Mechanical Elements," 2d ed., Wiley-Interscience, New York, 1980; R. C. Johnson, *Trans. ASME, J. Mech. Design*, vol. 101, pp. 667–673, 1979.

[4] J. Ellis, *Trans. ASME, J. Mech. Design*, vol. 102, pp. 5224–5531, 1980; G. E. Johnson and M. A. Townsend, *Trans. ASME, J. Mech. Design*, vol. 101, pp. 650–655, 1979.

widespread utilization of finite element analysis (see Sec. 1-10) for structural design has made it relatively easy to couple with optimization methods.[1]

Most engineering design optimization is concerned with the determination of the optimum dimensions or sizes of parts. In this type of structural optimization to the shape of the part has been determined and the goal is to select critical dimensions to satisfy an objective function like minimum weight, minimum cost or minimum deflection at constant weight. The variables which define shape design define the geometry of a two-dimensional plate or a three-dimensional solid structure. In shape optimization the boundary of the structure is variable so parameterization of the geometry is a very important part of the shape design model. The usual approach to shape optimization involves iterative finite element analysis. First an initial geometry is selected. Then the design variables and constraints are selected. The finite element analysis is performed, and at the end of each step the model is changed according to the chosen optimization criteria. However, because the choice of initial shape and design variables is arbitrary the optimization may be severely constrained.

Shape optimization is becoming a more active field with the advent of engineering workstations with high instructions per second. Computer programs have been developed which instantaneously update the change in node coordinates in the finite element model as the designer changes the shape on the computer screen.[2] Another approach is the development of a shape algebra[3] which leads to a systematic method for generating shapes.

5-9 PRACTICAL CONSIDERATIONS OF OPTIMIZATION

We have presented optimization chiefly as a collection of computer-based mathematical techniques. However, of more importance than knowing how to manipulate the optimization tools is knowing where to use them in the design process. In many designs a single design criterion drives the optimization. In consumer products it usually is cost, in aircraft it is weight, and in implantable medical devices it is power consumption. The strategy is to optimize these "bottleneck factors" first. Once the primary requirement has been met as well as possible, there may be time to improve other areas of the design, but if the first is not achieved, the design will fail. In some areas of design there may be no rigid specifications. An engineer who designs a talking, walking teddy bear

---

[1] G. N. Vanderplaats, "Numerical Optimization Techniques for Engineering Design," Chap. 9, McGraw-Hill Book Co., New York, 1984.

[2] R. J. Yang and M. J. Fiedler, "Design Modeling for Large-Scale Three-Dimensional Shape Optimization Problems," General Motors Research Lab., Publication GMR-5689, Oct. 1986.

[3] J. J. Shah, *Trans. ASME, J. Vibration, Acoustics, Stress and Reliability in Design*, vol. 110, pp. 564–570, 1988.

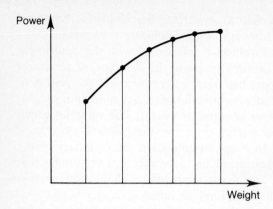

FIGURE 5-11
Trade-off between power and weight for the design of an electric drill (*after Siddall*).

can make almost any trade-off he/she wants between cost, power consumption, realism, and reliability. The designers and market experts will work together to decide the best combination of characteristics for the product, but in the end the four-year old consumers will decide whether it is an optimal design.

However, it frequently occurs in engineering practice that more than one functional characteristic of the design is of value to the user. It is not uncommon that these characteristics oppose each other, i.e. if one is changed to increase user value the other must be changed to decrease value. An example occurs frequently in materials selection, where it is common to find that increasing the strength of a material reduces its fracture toughness.

Thus, an approach is needed to deal with *trade-offs*.[1] For a certain design, a trade-off curve can be established which is a locus of all optimum designs, Fig. 5-11. Each point on the curve is established by maximizing power for a given weight electric drill. The weight in each case can be viewed as a specification. The same curve would be obtained by specifying a maximum power and minimizing the weight. A trade-off curve can be thought of as a plot of optimum designs corresponding to variations in a design specification. To make a decision about trade-off the curve like Fig. 5-11 must be associated with values. Figure 5-12 associates a value curve with design characteristic. These value curves represent the user's feelings about the desirability of changes in power and weight. The best design point on the trade-off curve is the point which gives the maximum combined value.

A related issue is *sensitivity analysis*. With a sensitivity analysis we find which design parameters are most critical to the performance of the design, and what are the critical ranges of those parameters. For some problems the

---

[1] J. N. Siddall, "Optimal Engineering Design," Chap. 5, Marcel Dekker Inc., New York, 1982.

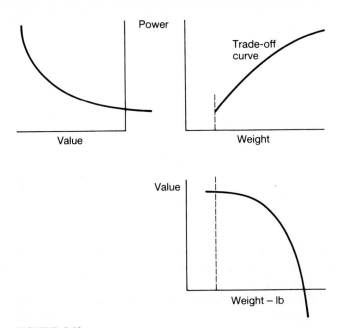

FIGURE 5-12
Value curves associated with trade-off curve (*after Siddall*).

sensitivity analysis can be done analytically, while in others it must be done numerically. To do a sensitivity analysis on an analytically expressed objective function, take the partial derivative with respect to each of the design variables. To put these on a relative basis, each partial derivative should be divided by its actual value.

For example, the cost of producing a chemical in batch runs is given by[1]

$$C^{\text{opt}} = 2\sqrt{K_1 K_2 Q} + K_3 Q$$

The sensitivity coefficients (on an absolute basis) are obtained by taking the partial derivative of C^{opt} with respect to each parameter.

$$\frac{\partial C^{\text{opt}}}{\partial K_1} = \sqrt{\frac{K_2 Q}{K_1}} \qquad \frac{\partial C^{\text{opt}}}{\partial K_2} = \sqrt{\frac{K_1 Q}{K_2}}$$

$$\frac{\partial C^{\text{opt}}}{\partial K_3} = Q \qquad \frac{\partial C^{\text{opt}}}{\partial Q} = \sqrt{\frac{K_1 K_2}{Q}} + K_3$$

[1] T. F. Edgar and D. M. Himmelblau, "Optimization of Chemical Processes," pp. 20–26, McGraw-Hill Book Co., New York, 1988.

If we let $Q = 100,000$, $K_1 = 1.0$, $K_2 = 10,000$, and $K_3 = 40$, then

$$\frac{\partial C^{\text{opt}}}{\partial K_1} = 31,620 \qquad \frac{\partial C^{\text{opt}}}{\partial K_2} = 3.162$$

$$\frac{\partial C^{\text{opt}}}{\partial K_3} = 100,000 \qquad \frac{\partial C^{\text{opt}}}{\partial Q} = 4.316$$

To put these sensitivies on a more meaningful basis we need to determine the relative sensitivities. The relative sensitivity of C^{opt} to K_1 is

$$S_{K_1}^C = \frac{\partial C^{\text{opt}}/C^{\text{opt}}}{\partial K_1/K_1} = \frac{\partial \ln C^{\text{opt}}}{\partial \ln K_1} = \sqrt{\frac{K_2 Q}{K_1}} \cdot \frac{K_1}{C^{\text{opt}}} = 31,620 \cdot \frac{(1.0)}{463,240} = 0.0683$$

Numerical values for the other relative sensitivities are:

$$S_{K_2}^C = 0.0683 \qquad S_{K_3}^C = 0.863 \qquad S_Q^C = 0.932$$

We thus see that changes in the variables Q and K_3 have the greatest influence on C^{opt}.

If the problem could not be obtained with an analytical solution we would determine the sensitivities numerically. First the objective function would be evaluated at the average value of each variable. Then we change each variable, one at a time, by some small amount, e.g. 1%, 5% or 10%, and determine the change in the objective function. This change is repeated in the opposite direction, i.e., minus 10%, to see if the function is symmetrical.

BIBLIOGRAPHY

Arora, J. S.: "Introduction to Optimum Design," McGraw-Hill Book Co, New York, 1989.

Beightler, C. S., D. T. Phillips, and D. J. Wilde: "Foundations of Optimization," 2d ed., Prentice-Hall Inc., Englewood Cliffs, N.J., 1979.

Edgar, T. F., and D. M. Himmelblau: "Optimization of Chemical Processes," McGraw-Hill Book Co., New York, 1988.

Papalambros, P. Y., and D. J. Wilde: "Principles of Optimal Design," Cambridge University Press, New York, 1989.

Pike, R. W.: "Optimization for Engineering Systems," Van Nostrand Reinhold Co., New York, 1986.

Reklaitis, G. V., A. Ravindran, and K. M. Ragsdell: "Engineering Optimization," John Wiley & Sons, New York, 1983.

Siddall, J. N.: "Analytical Decision-Making in Engineering Design," Prentice-Hall Inc., Englewood Cliffs, N.J., 1972.

Siddall, J. N.: "Optimal Engineering Design," Marcel Dekker Inc., New York, 1982.

Vanderplaats, G. N.: "Numerical Optimization Techniques for Engineering Design," McGraw-Hill Book Co., New York, 1984.

CHAPTER
6

MATERIALS SELECTION

6-1 THE PROBLEM OF MATERIALS SELECTION

The selection of the proper material is a key step in the design process because it is the crucial decision that links computer calculations and lines on an engineering drawing with a real or working design. The enormity of this decision process can be appreciated when it is realized that there are over 40,000 currently useful metallic alloys and probably close to that number of nonmetallic engineering materials. An improperly chosen material can lead not only to failure of the part or component but also to unnecessary cost. Selecting the best material for a part involves more than selecting a material that has the properties to provide the necessary service performance; it is also intimately connected with the processing of the material into a finished part (see Chap. 7). Thus, a poorly chosen material can add to manufacturing cost and unnecessarily increase the cost of the part. Also, the properties of the part may be changed by processing, and that may affect the service performance of the part. When we realize that material selection should be based on both material properties (part performance) and material processing (part manufacturing), the number of possible combinations is almost without bound.

Much materials selection is based on past experience. What worked before obviously is a solution, but it is not necessarily the optimum solution. Not too long ago, materials selection was considered a minor part of the design process. Materials were selected from handbooks with limited choice and on the basis of limited property data. Today, however, that is an unacceptable approach for all but the most routine and simple design. In many advanced aerospace and energy applications, materials are subjected to the limits of their properties. In the more consumer-oriented applications, in which performance requirements may not be so severe, the pressures to decrease cost are stronger than ever. In the automotive field, the drive to increase energy efficiency through weight reduction is revolutionizing materials selection. Still another pressure arises from our entering an era of materials shortages; selection of materials on a historical basis will soon be impossible because the material wanted may not be available. Thus, there is a greater need than ever for the selection of materials on a rational basis.

In many manufacturing operations the cost of materials may amount to more than 50 percent of the total cost. The higher the degree of automation, and hence the lower the unit labor cost, the greater the percentage of the total cost that is due to materials. In automobiles, materials cost is about 70 percent of manufacturing cost, and in shipbuilding it is about 45 percent.

6-2 PERFORMANCE CHARACTERISTICS OF MATERIALS

The performance or functional requirements of a material usually are expressed in terms of physical, mechanical, thermal, electrical, or chemical properties. Material properties are the link between the basic structure and composition of the material and the service performance of the part (Fig. 6-1).

We can divide structural engineering materials into metals, ceramics, and polymers. Further division leads to the categories of elastomers, glasses, and composites. Finally, there is the technology driving class of electronic, magnetic, and semiconductor materials. The chief characteristics of metals, ceramics, and polymers are given in Table 6-1.

Not too long ago metals dominated mechanical design so that it was possible to ignore the other classes of materials. Today the range of materials available

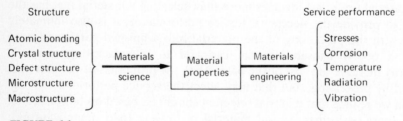

FIGURE 6-1
Materials properties, the link between structure and performance.

TABLE 6-1
Property characteristics of material classes

Metals	Ceramics	Polymers
Strong	Strong	Weak
Stiff	Stiff	Compliant
Tough	Brittle	Durable
Electrically conducting	Electrically insulating	Electrically insulating
High thermal conductivity	Low thermal conductivity	Temperature sensitive

to the engineer is much larger and growing rapidly. It is important to be cognizant of the opportunities for innovation and product improvement that new materials provide.

The ultimate goal of materials science is to predict how to improve the properties of engineering materials by understanding how to control the various aspects of structure. Structure can vary from atomic dimensions to the dimensions of a macroscopic crack in a fillet weld. Figure 6-2 relates various dimensions of structure with typical structural elements. The chief methods of altering structure[1] are through composition control (alloying), heat treatment, and deformation processing. A general background in the way structure controls the properties of solid materials usually is obtained from a course in materials science or fundamentals of engineering materials.[2]

We usually rely on material properties that are reasonably cheap and easy to measure, are fairly reproducible, and are associated with a material response that is well defined and is related to some fundamental response. However, for reasons of technological convenience we often determine something other than the most fundamental material property. Thus, the elastic limit measures the first significant deviation from elastic behavior; but it is tedious to measure, so we substitute the easier and more reproducible 0.2 percent offset yield strenth. That however, requires a carefully machined test specimen, so the yield stress may be approximated by the exceedingly cheap and rapid hardness test.

---

[1] R. W. Cahn, *J. Metals*, pp. 28–37, February 1973.
[2] L. E. Van Vlack, "Elements of Materials Science and Engineering," 6th ed., Addison–Wesley Publishing Co., Reading, MA 1989; D. R. Askeland, "The Science and Engineering of Materials," Brooks/Cole Div. of Wadsworth, Belmont, CA, 1984; W. D. Callister Jr., "Materials Science and Engineering," John Wiley & Sons, New York, 1985; R. A. Flinn and P. K. Trojan, "Engineering Materials and Their Applications," 3d ed, Houghton Mifflin Co., New York, 1986; J. F. Shackelford, "Introduction to Materials Science for Engineers,"Macmillan, New York, 1985; W. F. Smith, "Principles of Materials Science and Engineering," McGraw-Hill Book Co., New York, 1986.

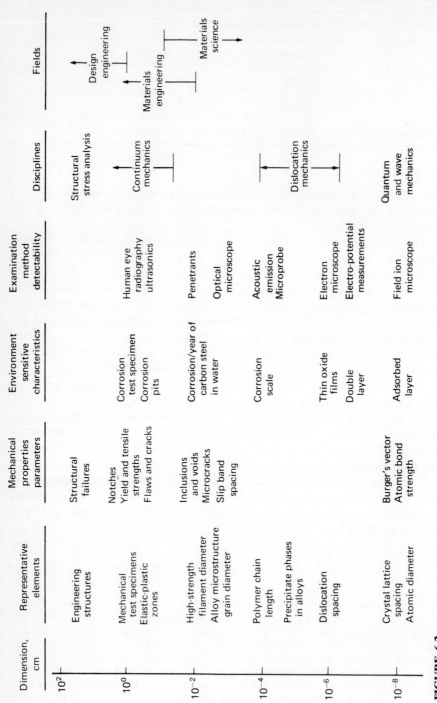

FIGURE 6-2
Dimensions of material structure.

Failure mode	Material property														
	Ultimate tensile strength	Yield strength	Compressive yield strength	Shear yield strength	Fatigue properties	Ductility	Impact energy	Transition temperature	Modulus of elasticity	Creep rate	K_{Ic}	K_{Iscc}	Electrochemical potential	Hardness	Coefficient of expansion
Gross yielding		▓		▓											
Buckling									▓						
Creep										▓					
Brittle fracture							▓	▓			▓				
Fatigue, low cycle					▓	▓									
Fatigue, high cycle	▓				▓										
Contact fatigue			▓												
Fretting			▓										▓		
Corrosion													▓		
Stress-corrosion cracking	▓											▓	▓		
Galvanic corrosion													▓		
Hydrogen embrittlement	▓														
Wear														▓	
Thermal fatigue										▓					▓
Corrosion fatigue					▓								▓		

FIGURE 6-3

Relations between failure modes and mechanical properties. (*C. O. Smith and B. E. Boardman: "Metals Handbook," 9th ed., vol. I, p. 828, American Society for Metals, Metals Park, Ohio, copyright American Society for Metals, 1980.*)

Shaded block at intersection of material property and failure mode indicates that a particular material property is influential in controlling a particular failure mode.

Figure 6-3 shows the relations between some common failure modes and the mechanical properties most closely related to the failures. Note that in most modes of failure, two or more mechanical properties interact to control the material behavior. In addition, the service conditions met by materials in general are more complex than the test conditions used to measure material properties. Thus, the stress level is not likely to be a constant value; instead, it is apt to fluctuate with time in a nonsinusoidal way. Or the service condition consists of a complex superposition of environments, such as a fluctuating stress (fatigue) at high temperature (creep) in a highly oxidizing atmosphere (corrosion). Specialized service simulation tests are developed to "screen materials" for complex service conditions. Finally, the best candidate materials

TABLE 6-2
Material performance characteristics

Physical properties:	*Mechanical properties*:	*Thermal properties*:
Crystal structure	Hardness	Conductivity
Density	Modulus of elasticity	Specific heat
Melting point	Tension	Coef. of expansion
Vapor pressure	Compression	Emissivity
Viscosity	Poisson's ratio	Absorptivity
Porosity	Stress-strain curve	Ablation rate
Permeability	Yield strength	Fire resistance
Reflectivity	Tension	
Transparency	Compression	*Chemical properties*:
Optical properties	Shear	
Dimensional stability	Ultimate strength	Position in
	Tension	electromotive series
Electrical properties:	Shear	Corrosion and degradation
	Bearing	Atmospheric
Conductivity	Fatigue properties	Salt water
Dielectric constant	Smooth	Acids
Coercive force	Notched	Hot gases
Hysteresis	Corrosion fatigue	Ultraviolet
	Rolling contact	Oxidation
Nuclear properties:	Fretting	Thermal stability
	Charpy transition temp.	Biological stability
Half-life	Fracture toughness (K_{Ic})	Stress corrosion
Cross section	High-temperature	Hydrogen embrittlement
Stability	Creep	Hydraulic permeability
	Stress rupture	
	Damping properties	*Fabrication properties*
	Wear properties	Castability
	Galling	Heat treatability
	Abrasion	Hardenability
	Erosion	Formability
	Cavitation	Machinability
	Spalling	Weldability
	Ballistic impact	

must be evaluated in prototype tests or field trials to evaluate their performance under actual service conditions.

The material properties usually are formalized through specifications. There are two types of specification: performance specifications and product specifications. *Performance specifications* delineate the basic functional requirements of the product and set out the basic parameters from which the design can be developed. They are based on the need the product is intended to satisfy and an evaluation of the likely risk and consequences of failure. The *product specifications* define conditions under which the components of the designs are purchased or manufactured. Material properties are an important part of product specifications.

Table 6-2 provides a fairly complete listing of material performance characteristics. It can serve as a checklist in selecting materials to assure that no important properties are overlooked. The crucial subject of producibility or fabrication properties is considered only briefly in Table 6-2. This subject is considered in much greater detail in Chap. 7.

The subject of material properties can quickly become rather complex. A consideration of any one of the properties listed in Table 6-2 can be expanded to include the type of test environment, stress state, or even specimen configuration. Figure 6-4 illustrates the generic tree that is developed by expanding the category of fatigue properties.

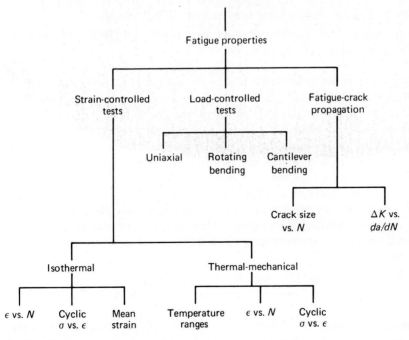

FIGURE 6-4
Generic tree for fatigue properties.

There is a whole set of important factors in materials selection that are not covered by Table 6-2. The economic and practical considerations, which are discussed in detail in Sec. 6-5, are as follows:

1. Availability
   a. Are there multiple sources of supply?
   b. What is likelihood of availability in future?
   c. Is the material available in the forms needed (tubes, wide sheet, etc.)?
2. Size limitations and tolerances on available material
3. Variability in properties
4. Cost. Materials selection comes down to buying properties at best available price

6-3 THE MATERIALS SELECTION PROCESS

The selection of materials on a purely rational basis is far from easy. The problem is not only often made difficult by insufficient or inaccurate property data but is typically one of decision making in the face of multiple constraints without a clear-cut objective function. A problem of materials selection usually involves one of two different situations.

1. Selection of the materials for a new product or new design.
2. Reevaluation of an existing product or design to reduce cost, increase reliability, improve performance, etc.

It generally is not possible to realize the full potential of a new material unless the product is redesigned to exploit both the properties and the manufacturing characteristics of the material. In other words, *a simple substitution of a new material without changing the design rarely provides optimum utilization of the material.* Most often the essence of the materials selection process is *not* that one material competes against another for adoption; rather, it is that the processes associated with the production or fabrication of one material compete with the processes associated with the other. For example, the pressure die casting of a zinc-base alloy may compete with the injection molding of a polymer. Or a steel forging may be replaced by sheet metal because of improvements in welding sheet-metal components into an engineering part.

Materials selection, like any other aspect of engineering design, is a problem-solving process. The steps in the process can be defined as follows:

1. *Analysis of the materials requirements.* Determine the conditions of service and environment that the product must withstand. Translate them into critical material properties.

2. *Screening of candidate materials.* Compare the needed properties (responses) with a large materials property data base to select a few materials that look promising for the application.

3. *Selection of candidate materials.* Analyze candidate materials in terms of trade-offs of product performance, cost, fabricability, and availability to select the best material for the application.

4. *Development of design data.* Determine experimentally the key material properties for the selected material to obtain statistically reliable measures of the material performance under the specific conditions expected to be encountered in service.

We need to say something about the philosophies associated with each step in the material selection process. *A screening property* is any material property for which an absolute lower (or upper) limit can be established for the application. No trade-off beyond that limit is tolerable. The idea in the screening phase is to set one-sided limits that permit a definite yes or no to the question: "Should this material be evaluated further for this application?" In making the final selection, more properties are considered, including such as fabricability and cost. Some of the techniques that can be used in this phase of the selection process will be discussed later in this chapter. *Design data properties* are the properties of the selected material in its fabricated state that must be known with sufficient confidence to permit the design and fabrication of a component that is to function with a specified reliability. The extent to which this phase is pursued depends upon the nature of the problem. In many product areas, service conditions are not severe and specifications established by ASTM may be used without adopting an extensive testing program. In other areas, such as the aerospace and nuclear ones, the need for reliability is so high that extensive effort will be devoted to design property determination.

The materials selection and evaluation process for a complex or advanced design is illustrated in Fig. 6-5. The design process is broken down in the vertical direction into overall design, analysis of components, and testing. Starting with the overall design concept for the system, the individual system components are identified and specifications for their performance are developed. Potential materials that could meet the performance specs are identified (see Sec. 6-4), and they are evaluated on the basis of screening properties. Frequently, actual screening tests are performed to eliminate the unsatisfactory materials. The successful materials in the screening tests now become the candidate materials for evaluation in the candidate design concept. In the material selection phase these materials are further evaluated and tested against a broader and more discriminating set of properties. Trade-offs are made with respect to performance, cost, and availability to arrive at a single or small number of materials for the selected design concept. An extensive testing program may be required to establish variability in properties or to qualify different vendors of material. These detailed property data are fed into the final detailed design. When the design is fabricated into a working concept,

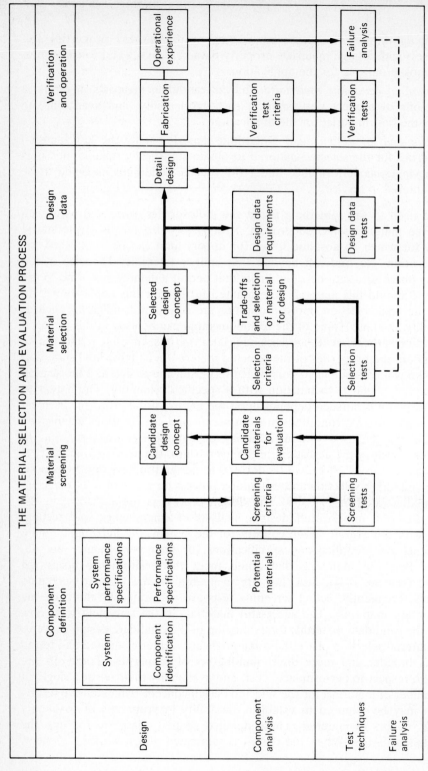

THE MATERIAL SELECTION AND EVALUATION PROCESS

FIGURE 6-5
Materials selection and evaluation process for a complex product. (*Reproduced from "An Approach for Systematic Evaluation of Materials for Structured Application," National Academy Press, Washington, D.C., 1970.*)

additional property testing is required to establish the influence of manufacturing processes on critical design properties. Moreover, critical subassemblies or the entire design may be evaluated in simulated service tests. As designs are put into service, operational experience begins to accumulate. Hopefully, the design will work to perfection, but it is not uncommon in a complex system to experience service failures that are not anticipated by the design analysis or that show up in the simulated service testing. Component failures are a sure indication that the material selection and/or the design was faulty. Failure analysis (see Chap. 13) is a well-developed method that leads to better designs.

While materials selection occurs at every stage of the design process traditionally it has a major role in the later stages. It is suggested that more opportunities for innovative design will occur by considering a broad selection of materials at the conceptual phase of design .At this stage, where all options

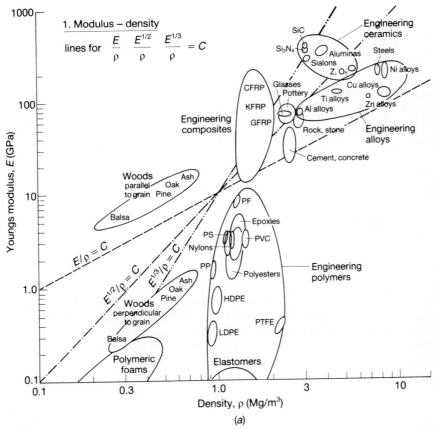

FIGURE 6-6a

Ashby materials selection chart: Young's modulus vs. density. (*M. F. Ashby*, Mater. Sci. and Tech. *vol. 5, p. 521, 1989.*)

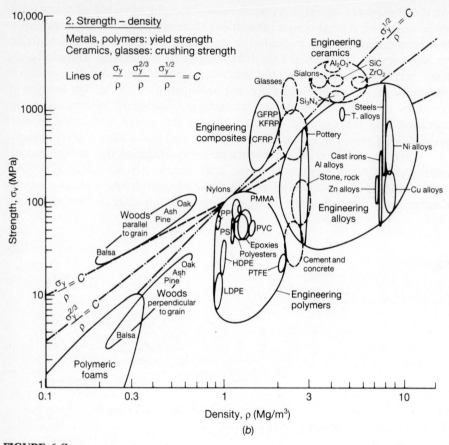

FIGURE 6-6b
Ashby materials selection chart: strength vs. density. (*M. F. Ashby,* Mater Sci. and Eng., *vol. 5, p. 522, 1989.*)

are open, the designer requires approximate data on the broadest possible range of materials. A metal may be the best material for one design concept while a polymer is best for a different concept, even though the two concepts provide the same function. Ashby[1] has created useful material selection charts for this purpose. Figure 6-6*a* plots the elastic modulus of polymers, metals, ceramics and composites against density, while Fig. 6-6*b* shows the same type of plot with strength against density. A common design criterion is to minimize cost or weight. Depending upon the geometry and the loading, different relationships apply (see Sec. 6-7). For simple axial loading the relationship is

---

[1] M. F. Ashby, *Mater. Sci and Tech.*, vol. 5, pp. 517–525, 1989.

E/ρ or σ/ρ. For buckling of a slender column $E^{1/2}/\rho$ applies, and for the bending of a plate the relationship is $E^{1/3}/\rho$. Lines with these slopes are shown on the figures. Thus, if a straight edge is laid parallel to the line $E^{1/2}/\rho = C$ all the materials which lie on the line will perform equally well as a column loaded in compression, while those above the line are better and those below the line are worse.

6-4 SOURCES OF INFORMATION ON MATERIALS PROPERTIES

Most practicing engineers develop a file of trade literature, technical articles, and company reports. Material property data comprise an important part of this personal data system. In addition, many large corporations and government agencies develop their own compendiums of data on materials properties.

The purpose of this section is to provide a guide to material property data that are readily available in the published technical literature. There are several factors to have clearly in mind when using property data in handbooks and other open-literature sources. Usually a single value is given for a property, and it must be assumed that the value is typical. When scatter or variability of results is considerable, the fact may be indicated in a table of property values by a range of values (i.e., the largest and smallest values) or be shown graphically by scatter bands. However, it is rare to find a property data presented in a proper statistical manner by a mean value and the standard deviation (see Chap. 11). Obviously, for critical applications in which reliability is of great importance, it is necessary to determine the frequency distribution of both the material property and the parameter that describes the service behavior. Figure 6-7 shows that when the two frequency distributions overlap, there will be a statistically predictable number of failures.

Table 6-3 gives important references on the properties of materials. These are available in most technical libraries.

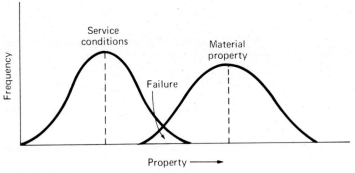

FIGURE 6-7
Overlapping distribution of material property and service requirement.

TABLE 6-3
Sources of material properties

For a quick screening of material properites

C. T. Lynch (ed.), "Handbook of Materials Science," CRC Press, Cleveland, 1975.
  Vol. I "General Properties"
  Vol. II "Metals, Composites and Refractory Materials"
  Vol. III "Non-metallic Materials and Applications"
"Materials Selector," annual publication of *Materials Engineering Magazine,* Penton/IPC, Cleveland, Ohio.
"Materials Handbook," 12th ed., McGraw-Hill Book Co., New York, 1985.
E. A. Brandes (ed.), "Smithells Metals Reference Book," 6th ed., Butterworth, London, 1984.
"Fulmer Materials Optimiser," Fulmer Research Institute, Stoke Poges, England, 1974.

Detailed property data

1. *Metals*
    a. Metals Handbook," 10th ed., Vol. I, "Properties and Selection: Irons and Steels," American Society for Metals, 1990; Vol. II, "Properties and Selection: Nonferrous Alloys and Pure Metals," 1990; 9th ed., Vol. III, "Properties and Selection: Stainless Steels, Tool Materials and Special Purpose Metals," 1980.
    b. "SAE Handbook," Part 1, "Materials, Parts and Components," Society of Automotive Engineers, Warrendale, Pa., published annually.
    c. "Structural Alloys Handbook," vols I and II, Mechanical Properties Data Center, Battelle Memorial Institute, Columbus, Ohio, 1979.
    d. C. F. Walton (ed.), "Gray and Ductile Iron Casting Handbook," Gray and Ductile Iron Founders,' Society, Cleveland, 1971.
    e. "Steel Castings Handbook," 5th ed., Steel Founders Society of America, Rocky River, Ohio, 1980.
    f. R. C. Gibbons (ed.), "Woldman's Engineering Alloys," 6th ed., American Society for Metals, 1979. Use this book to track down information on an alloy if you know only the trade name.
    g. R. B. Ross. "Metallic Materials Specification Handbook," 3d ed., E & FN Spon Ltd., London, 1980. Use this book if you know only the specification designation of an alloy and want more information.
    h. C. Robb, "Metals Databook," The Institute of Metals, Great Britain, 1986.
    i "Metallic Materials and Elements for Aerospace Vehicle Structures," Military Standardization Handbook, MIL-HDBK-5D, U.S. Dept of Defense. Updated every six months and reissued every three years.

2. *Ceramics and inorganic materials*
    a. "Engineering Property Data on Selected Ceramic Materials," Metals and Ceramics Information Center, Battelle Memorial Institute, Columbus, Ohio
      Vol. 1, Nitrides, 1976
      Vol. 2, Carbides, 1979
      Vol. 3, Single oxides, 1983
    b. R. Morrell, "Handbook of Properties of Technical and Engineering Ceramics, "HMSO, London, Part 1, 1985, Part 2, 1987.
    c. C. L. Mantell, "Carbon and Graphite Handbook," Interscience, 1968.
    d. G. W. McLellen and E. B. Shand, "Glass Engineering Handbook," 3d ed., McGraw-Hill Book Co, New York, 1984.

TABLE 6-3 (*Continued*)

3. *Polymers*
    a. C. A. Harper (ed.), "Handbook of Plastics and Elastomers," McGraw-Hill Book Co., New York, 1976.
    b. "Engineering Plastics," Engineered Materials Handbook, Vol. 2, ASM International, Metals Park, Ohio, 1988.
    c. "Modern Plastics Encyclopedia", published yearly by *Modern Plastics,* McGraw-Hill Book Co, New York.
    d. "Plastics Materials Digest," 10th ed., D.A.T.A., San Diego, 1989.
    e. A. K. Bhowmick and H. L. Stevens, "Handbook of Elastomers," Marcel Dekker, Inc., New York, 1988.
    f. "Plastics: Thermoplastics and Thermosets," D. A. T. A., San Diego, 1988.
    g. J. C. Bittence (ed.), "Engineering Plastics and Composites," ASM International, Metals Park OH, 1990—identifies materials by suppliers's trade names and designations.

4. *Composites*
    a. "Composites," Engineered Materials Handbook, Vol. 1, ASM International, Metals Park, Ohio, 1987.
    b. M. M. Schwartz (ed.), "Composite Materials Handbook", McGraw-Hill Book Co., New York, 1984.

5. *Electronic materials*
    a. C. A. Harper (ed.), "Handbook of Materials and Processes for Electronics," McGraw-Hill Book Company, New York, 1970.
    b. "Handbook of Electronic Materials," vols. 1 to 9, Plenum Publishing Corporation, 1971, 1972.

6. *Thermal properties*
    a. Y. S. Touloukian (ed.), "Thermophysical Properties of High Temperature Solid Materials," vols. 1 to 6, Macmillan Publishing Co., New York, 1967.
    b. W. M. Rohsenow and J. P. Harnett (eds.), "Handbook of Heat Transfer," McGraw-Hill Book Company, New York, 1973.

7. *Chemical Properties*
    a. "Metals Handbook," 9th ed., Vol. 13, "Corosion", ASM International, Metals Park, Ohio, 1987.
    b. I. Mellan, "Corrosion Resistant Materials Handbook," 3d ed., Noyes Data Corp., Park Ridge, N.J., 1976.
    c. M. Schumacher (ed.), "Seawater Corrosion Handbook," Noyes Data Corp., Park Ridge, N.J., 1979.
    d. P. A. Schweitzer, (ed.), "Corrosion and Corrosion Protection Handbook," 2d ed, Marcel Dekker Inc., New York, 1989.
    e. P. A. Schweitzer, "Corrosion Resistance Tables," 2d ed, Marcel Dekker Inc., New York, 1986.
    f. "Corrosion Data Survey," 6th ed., National Association of Corrosion Engineers, Houston, Texas, 1986.

Computerized Databases

Because there are over 100,000 commercially available metals, ceramics, polymers, and composites with identifiably different properties and costs there is an obvious advantage to using the computer to store, manipulate, and retrieve this information. If the computer contains data on m properties for n materials then a program written to operate on the $m \times n$ array can carry out sorting and correlation operations. For example, the designer may set minimum limits on properties m_1, m_2, and m_3, and the computer is asked to print out the identification of all materials which meet these criteria. Another task might be to produce a graphical plot of the correlation between two properties, e.g., tensile elongation and impact toughness.

The advantages of computer databases are first the ability to store large amounts of information in a form that specific pieces of information can be retrieved easily for particular applications. Other advantages are the ease with which the data can be kept current, the ability to link related information, and the ease with which the updated information can be made available to users. With numeric data it is possible to perform statistical or other forms of calculations and transfer the data over networks for further analysis in CAD or CAM. While the advantages of computer databases of materials data are very great their evolution has not been very rapid. The chief problem has been the cost associated with building and maintaining a significant database. Other problems are issues of sharing or access to proprietary data and the maintenance of quality control of the data. Computerized databases are available at the Mechanical Properties Data Center at Batelle Columbus Laboratories and for thermophysical, electronic, electrical, and optical properties at the Center for Information and Numerical Data Analysis and Synthesis (CINDAS) at Purdue University. Plans are well advanced for a National Materials Property Data Network supported by industry, technical societies, and users. In a different direction is the growth of materials selection software for personal computers and workstations. MetalSelector, now Mat.DB, includes data on carbon and alloy steels, thermoplastics, tool steels, titanium and aluminum alloys.[1] Mat.DB can search the database by material group, UNS number, common name, manufacturer, specification designation, ranges of chemical composition, product form, heat treated condition, and up to 40 properties. Relative rankings are also given with regard to machinability, weldability, formability, and average processing cost. Similar information is available using IPS/IDES software[2] for over 10,000 grades of plastic material and from Cen BASE/Materials[3]

---

[1] Available from ASM International, Metals Park, Ohio.

[2] Available from D.A.T.A., San Diego, Ca.

[3] Available from Infodex, John Wiley & Sons, New York.

6-5 ECONOMICS OF MATERIALS

Ultimately the decision on a particular design will come down to a trade-off between performance and cost. There is a continuous spectrum of applications, varying from those where performance is paramount (aerospace and defense are good examples) to those where cost clearly predominates (household appliances and low-end consumer electronics are typical examples). In the latter type of application the manufacturer does not have to provide the highest level of performance that is technically feasible. Rather the manufacturer must provide a value to cost ratio that is no worse, and preferably better, than the competition. By *value* we mean the extent to which the performance criteria appropriate to the application are satisfied. *Cost* is what must be paid to achieve that level of value.

Because cost is such an overpowering consideration in many materials selection situations, we need to give this factor additional attention. The basic cost of a material depends upon (1) scarcity, as determined by either the concentration of the metal in the ore or the cost of the feedstock, (2) the cost and amount of energy required to process the material, and (3) the basic supply and demand for the material. In general, large-volume-usage materials like stone and cement have very low prices, while scarce materials, like industrial diamonds, have high prices.

As is true of any commodity, as more work is invested in the processing of a material, the cost increases (value is added). Table 6-4 shows how the price of various steel products increases with processing.

Increases in properties, like yield strength, beyond those of the basic material are produced by changes in structure brought about by compositional changes and additional processing steps. For example, changes in the strength of steel are promoted by expensive alloy additions such as nickel or by heat treatment such as quenching and tempering. However, the cost of an alloy may

TABLE 6-4
Base price of various steel products (February 1980)

Product	Price, ¢/lb
Pig iron	10.1
Billets, blooms, and slabs	14.4
Hot-rolled carbon-steel bars	23.5
Cold-finished carbon1steel bars	40.0
Hot-rolled carbon-steel plate	32.0
Hot-rolled sheet	26.5
Cold-rolled sheet	33.2
Galvanized sheet	37.5

not simply be the weighted average of the cost of the constituent elements that make up the alloy. Often, a high percentage of the cost of an alloy is the need to control one or more impurities to very low levels. That could mean extra refining steps or the use of expensive high-purity raw materials.

Because most engineering materials are produced from depletable mineral resources, there is a continuous upward trend of cost with time. Although some materials, e.g., copper, have fluctuated in price considerably because of temporary oversupply, there is no question that the costs of materials will rise at a rate greater than the costs of goods and services in general. Therefore, wise use of materials will become of increasingly greater importance. The general price level for metals and ceramics can be obtained from the following sources.

Metals; *American Metal Market/Metalworking News, Iron Age*
Ceramics: *Ceramic Industry Magazine*

However, the pricing structure of many engineering materials is quite complex, and true prices can be obtained only through quotations from vendors. The reference sources listed above give only the nominal, baseline price. The actual price depends upon a variety of price extras in addition to the base price (very much as when a new car is purchased). The actual situation varies from material to material, but the situation for steel products is a good illustration.[1]

Price extras are assesed for the following situations.

Metallurgical requirements:
   Grade extra. Each AISI grade has an extra over the cost of the generic type of steel, i.e., hot-rolled bar, hot-rolled plate, etc.
   Chemistry extra. Nonstandard chemical composition for the grade of steel.
   Quality extra, e.g., vacuum melting or degassing.
   Inspection and testing. A charge is made for anything other than routine tensile tests and chemical analysis.
   Special specifications.

Dimensions:
   Size and form. Special shapes, or sizes.
   Length. Precise requirements on length are costly.
   Cutting. Sheared edge, machined edge, flame-cut edge, etc.
   Tolerances. Tighter tolerances on OD or thickness cost extra.

---

[1] R. F. Kern and M. E. Suess, "Steel Selection," John Wiley & Sons, Inc., New York, 1979.

Proccessing:

    Thermal treatment, eg., normalizing or spheroidizing.

    Surface treatment, e.g., pickling or oil dip.

Quantity:

    Purchases in less than heat lots (50 to 300 tons) are an extra.

Pack, Mark, Load:

    Packing, Wrapping, boxing, etc.

    Marking. Other than stamped numbers may be an extra.

    Loading. Special blocking for freight cars, etc.

Metric dimensions:

    U.S. steel producers still make steel to English dimensions, but most of the world is now on the metric system (SI units). Thus, there may be an extra to produce to metric tolerances.

From this detailed listing of price extras we can see how inadvertent decisions of the designer can significantly influence cost. Standard chemical compositions should be used whenever possible, and the number of alloy grades should be standardized[1] to reduce the cost of stocking many grades of steel. Manufacturers whose production rates do not justify purchasing in heat lots should try to limit their material use to grades that are stocked by local steel service centers. Special section sizes and tolerances should be avoided unless a detailed economic analysis shows that the cost extras are really justified.

6-6 METHODS OF MATERIALS SELECTION

There is no method or small number of methods of materials selection that have evolved to a position of prominence . Partly, this is due to the complexity of the comparisons and trade-offs that must be made. Often the properties we are comparing cannot be placed on comparable terms so a clear decision can be made. Partly it is due to the fact that little research and scholarly effort have been devoted to the problem.

    In a general sense a variety of approaches to materials selection are followed by designers and materials engineers. A common path is to examine critically the service of designs in environments similar to the one of the new design. Information on service failures can be very helpful. The results of

---

[1] "Metals Handbook," 8th ed., vol. I, pp. 281–289, American Society for Metals, Metals Park, Ohio, 1961.

accelerated laboratory screening tests or short time experience with a pilot plant can provide valuable input. Often a minimum innovation path is followed and the material is selected on the basis of what worked before or what is used in the competitor's product.

Some of the more common and more analytical methods of materials selection are:

1. Cost vs. performance indices
2. Weighted property indices
3. Value analysis
4. Failure analysis
5. Benefit-cost analysis

The first three methods are described in more detail in this chapter. Failure analysis is considered in Chap. 13, and benefit-cost analysis is discussed in Sec. 8-11.

Because cost is so important in selecting materials, it is logical to consider cost at the outset of the materials selection process. Considerable effort is being given to developing computer-based methods of estimating manufacturing cost that can be employed in the conceptual stage of design. If this is not available it is usually possible to set a target cost and eliminate the materials that are too expensive. Since the final choice is a trade-off between cost and performance (properties), it is logical to attempt to express that relation as carefully as possible. Figure 6-8 shows the costs of substituting lightweight materials to achieve weight saving (fuel economy) in automobiles. The norizontal axis shows the weight reduction made possible by each substitution and the vertical axis shows the cost of the lightweight material relative to its conventional counterpart. In most cases the lightweight materials lie above the breakeven curve where the cost of the substitute part equals the cost of the conventional part, because less material by weight needs to be purchased for the substitute. The exception is high-strength steel substituted for mild steel. Note that this plot does not consider possible savings in processing and assembly for the lightweight material.

Another aspect of cost is the choice between two or more materials with different initial costs and different expected service lives. That is a classical problem in the field of engineering economy. It is discussed in detail in Chap. 8 and illustrated by example in Sec. 6-10.

In the typical problem of materials selection it is necessary to satisfy more than one performance requirement. A simple way to do so is to use the decision matrix described in Sec. 3-11, where the various performance requirements are weighted with respect to significant properties that control performance. Several versions of the weighted property index are presented in Sec. 6-8.

Value analysis is an organized method of finding the least expensive way

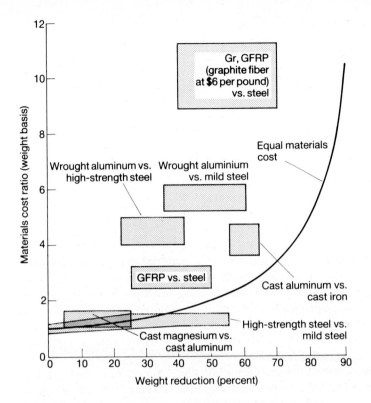

FIGURE 6-8
Cost of substituting a lightweight material in automotive applications. Note: GFRP is glass fiber reinforced polymer. Gr, GFRP is GFRP containing graphite fibers. (*W. D. Compton and N. A. Gjostein,* Scientific American, *vol. 255, p. 98, Oct. 1986.*)

to make a product without compromising quality or reliability. The topic, clearly is more multifaceted than just the materials selection problem. In fact, value analysis is a popular approach to problem solving. However, because it can be used effectively in materials selection, it is presented here in Sec. 6-11.

A rational way to select materials is to determine the way in which actual production parts, or parts similar to a new design, fail in service. Then, on the basis of that knowledge, materials that are unlikely to fail are selected. The general methodology of failure analysis is considered in Chap. 13.

Regardless of how well a material has been characterized and how definitive the performance requirements and the program schedule are, there will always be a degree of uncertainty about the ability of the material to perform. The major risk areas encountered in any complex design and manufacturing program include: 1) program schedule, 2) cost, 3) producibility, 4) performance, 5) maintainability, 6) repairability, and 7) survivability. For high-performance systems such that the consequences of failure can be very severe, material selection based on risk analysis can be very important. Some of the ideas of risk analysis are discussed in Chap. 12.

6-7   COST VS. PERFORMANCE RELATIONS

Since, overall, cost is the most important criterion in selecting a material, it is natural to use it as one of the key factors in the initial screening of materials. Cost is a most useful parameter when it can be related to a critical material property that controls the performance of the design. Such a cost vs. performance index can be a useful parameter for optimizing the selection of a material. The optimization for simple geometric conditions and simple conditions of loading may be easy, but it becomes difficult to construct meaningful indices of performance for the complex situations found in many designs.

It is important to realize that the cost of a material expressed in dollars per pound may not always be the most valid criterion. Often materials provide more of a space-filling function than a load-bearing function, and then dollars per cubic foot is a more appropriate criterion. For example, the cost of plastics usually is justified on a cost per unit volume, rather than a cost per unit weight, basis. It is also important to emphasize that there are many ways to compute costs (see Chap. 9). *Total life-cycle cost* is the most appropriate cost to consider. It consists of the initial material costs plus the cost of manufacturing and installation plus the costs of operation and maintenance. Consideration of factors beyond just the initial materials cost leads to relations like the relation shown in Fig. 6-9. Part costs (curve *A*) can increase with strength because of greater difficulty in fabricating the higher-strength parts. However, the number of parts needed, or the weight per part, will decrease with increasing strength,

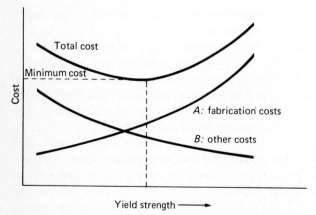

FIGURE 6-9
Relations between cost factors and a material property.

or the service life might increase with strength, etc., to lead to curve B. The total cost is the sum of A and B, and the optimum value of the property occurs at the minimum cost.

Since material cost is directly related to the weight (or volume) of material, the determination of a cost vs. property relation becomes a question of determining the structural equivalency of different materials. Typical problems are determining the relative weight of each material for equal strength and determining the relative weight for equal stiffness.

To illustrate the development of a cost-performance index, consider the simplest case of the yielding of a bar in uniaxial tension.

$$\text{Working stress} = \sigma_w = \frac{\sigma_{ys}}{\text{factor of safety}} = \frac{\text{axial force}}{\text{section area}} = \frac{P}{A} \tag{6-1}$$

If the cross section and the material properties of two bars A and B are denoted by A and σ, respectively, the condition of equal load-carrying ability in both bars is given by

$$A_A \sigma_A = A_B \sigma_B = D_A^2 \sigma_A = D_B^2 \sigma_B \tag{6-2}$$

and

$$\frac{D_B}{D_A} = \left(\frac{\sigma_A}{\sigma_B} \right)^{1/2} \tag{6-3}$$

Since the weight of the bar is

$$W = \rho V = \rho A L = \rho \frac{\pi}{4} D^2 L \tag{6-4}$$

then

$$\frac{W_B}{W_A} = \frac{\sigma_A}{\sigma_B} \frac{\rho_B}{\rho_A} \tag{6-5}$$

If material A costs m_A \$/lb and material B costs m_B \$/lb, then the cost of the bar made from material B to support load P is

$$C_B = W_B m_B \quad \text{and} \quad C_A = W_A m_A \tag{6-6}$$

so that

$$\frac{C_B}{C_A} = \frac{\sigma_A}{\sigma_B} \frac{\rho_B}{\rho_A} \frac{m_B}{m_A} \tag{6-7}$$

Thus, the cost per unit of strength is $C = \rho m / \sigma$. In this example we used yield strength, but in other situations it could be tensile strength, fatigue strength, or creep strength. Table 6-5 compares the cost of the materials to support a load of 100,000 lb in a bar 24 in long.

To derive the relation for structural equivalency for equal stiffness, consider a beam of length L loaded at one end as a cantilever beam. The deflection at the free end δ is given by

$$\delta = \frac{PL^3}{3EI} \tag{6-8}$$

TABLE 6.5
Relative material costs for a bar in uniaxial tension

Material	σ_w, ksi	ρ, lb/in^3	σ_w/ρ	Weight of bar	Cost of bar (1980 prices)
2024 aluminum	40	0.101	396	3.04	$ 3.00
304 stainless steel	30	0.286	105	11.50	$11.27
AISI 1020 steel (annealed)	40	0.284	141	8.50	$ 1.70
AISI 4340 steel (quenched and tempered)	150	0.284	528	2.24	$ 1.01

where E is the modulus of elasticity and I is the moment of inertia of the cross section of the beam. The condition of equal stiffness in two beams of material A and B is given by

$$E_A I_A = E_B I_B \tag{6-9}$$

If we consider a beam with a rectangular cross-sectional area, then

$$I = \frac{wh^3}{12} \tag{6-10}$$

TABLE 6-6
Formulas for cost per unit property

Cross sections and loading conditions	Variable dimension[1]	Cost of unit strength	Cost of unit stiffness
Solid cylindrical bar in tension or compression	Diameter	$\dfrac{\rho m}{\sigma}$	$\dfrac{\rho m}{E}$
Solid cylindrical bar in bending	Diameter	$\dfrac{\rho m}{\sigma^{2/3}}$	$\dfrac{\rho m}{E^{1/2}}$
Solid cylindrical bar in torsion	Diameter	Same as for bending	$\dfrac{\rho m}{G^{1/2}}$
Solid cylindrical bar as slender column	Diameter	—	$\dfrac{\rho m}{E^{1/2}}$
Solid rectangle in bending	Depth	$\dfrac{\rho m}{\sigma^{1/2}}$	$\dfrac{\rho m}{E^{1/3}}$
Thin-walled cylindrical vessel	Wall thickness	$\dfrac{\rho m}{\sigma}$	$\dfrac{\rho m}{E}$

σ = yield strength; E = Young's modulus; G = shear modulus; ρ = density; m = material cost, \$/lb.
[1] Dimension varied to maintain equal strength or stiffness.

where w is the width and h is the depth (thickness) of the beam. For beams of constant width and length, the relative thickness of material [from Eq. (6-9)] is

$$h_B = h_A \left(\frac{E_A}{E_B}\right)^{1/3} \tag{6-11}$$

and the relative cost is given by

$$\frac{C_B}{C_A} = \left(\frac{E_A}{E_B}\right)^{1/3} \frac{\rho_B \, m_B}{\rho_A \, m_A} \tag{6-12}$$

Relationships[1] to determine the relative cost per unit property for either strength or stiffness are given in Table 6-6. The exponents on yield strength and Young's modulus that arise for different types of loading have the effect of bringing materials of different strength and E closer together and therefore make it easier for the designer to select materials of unfavorable σ/ρ if they possess other valuable properties. It is important to realize that material selection is rarely made on the basis of strength alone. Other factors such as corrosion resistance, high-temperature properties, and ease of manufacture also play a major role.

6-8 WEIGHTED PROPERTY INDEX

In most applications it is necessary that a selected material satisfy more than one performance requirement. In other words, compromise is needed in materials selection. We can separate the requirements into three groups (1) go/no-go parameters, (2) nondiscriminating parameters, and (3) discriminating parameters. Go/no-go parameters are those requirements which must meet a certain fixed minimum value. Any merit in exceeding the fixed value will not make up for a deficiency in another parameter. Examples of go/no-go parameters are corrosion resistance or machinability. Nondiscriminating parameters are requirements that must be met if the material is to be used at all. Examples are availability or general level of ductility. Like the previous category these parameters do not permit comparison or quantitative discrimination. Discriminating parameters are those requirements to which quantitative values can be assigned.

The decision matrix that was introduced in Sec. 3-11 is well suited to materials selection with discriminating parameters.[2] In this method each material property is assigned a certain weight depending on its importance to the required service performance. Techniques for assigning weighting factors

---

[1] Derivations may be found in "Metals Handbook," vol. 1, pp. 185–187, American Society for Metals, Metals Park, Ohio, 1961.

[2] M. M. Farag, "Materials and Process Selection in Engineering," Chap. 13, Applied Science Publ., Ltd., London, 1979.

are considered in Sec. 3-11. Since different properties are expressed in different units the best procedure is to normalize these differences by using a scaling factor. The scaling is a simple technique to bring all the different properties within one numerical range. Since different properties have widely different numerical values, each property must be so scaled that the largest value does not exceed. 100.

$$\beta = \text{scaled property} = \frac{\text{numerical value of property}}{\text{largest value under consideration}} 100 \qquad (6\text{-}13)$$

For properties such that it is more desirable to have low values, e.g., density, corrosion loss, cost, and electrical resistance, the scale factor is formulated as follows:

$$\beta = \text{scaled property} = \frac{\text{lowest value under consideration}}{\text{numerical value of property}} 100 \qquad (6\text{-}14)$$

For properties that are not readily expressed in numerical values, e.g., weldability and wear resistance, some kind of subjective rating is required. For example:

	Alternative materials			
Property	A	B	C	D
Weldability	Excellent	Good	Good	Fair
Relative rating	5	3	3	1
Scaled property	100	60	60	20

The material performance index γ is

$$\gamma = \sum \beta_i w_i \qquad (6\text{-}15)$$

where i is summed over all the properties.

Cost can be considered as one of the properties, usually with a high weighting factor. However, if there are a large number of properties to consider, it may be better to emphasize cost by applying it as a moderator to the material performance index.

$$\gamma' = \frac{\gamma}{m\rho} \qquad (6\text{-}16)$$

where m = materials cost, \$/lb
ρ = density, lb/in^3

Example 6-1. The material selection for a cryogenic storage vessel for liquefied propane gas is being evaluated on the basis of 1) low-temperature fracture toughness, 2) fatigue strength, 3) stiffness, 4) thermal expansion, and 5) cost.

Determine the weighting factors for these properties with the aid of a digital logic table. There are $N = 5(5-1)/2 = 10$ possible comparisons of pairs.

Possible design combinations

Property	1 (1) (2)	2 (1) (3)	3 (1) (4)	4 (1) (5)	5 (2) (3)	6 (2) (4)
1. Fracture toughness	1	1	1	1		
2. Fatigue strength	0				0	1
3. Stiffness		0			1	
4. Thermal expansion			0			0
5. Cost				0		

Property	7 (2) (5)	8 (3) (4)	9 (3) (5)	10 (4) (5)	Positive decisions	Weighting factor w_i
1. Fracture toughness					4	0.4
2. Fatigue strength	0				1	0.1
3. Stiffness		0	0		1	0.1
4. Thermal expansion		1		0	1	0.1
5. Cost	1		1	1	3	0.3
					10	1.0

Solution. The weighted property index chart for selecting a material for the cryogenic storage vessel is shown in Table 6-7. Several go no-go screening parameters are included. It is assumed that the aluminum alloy will not be available in the required plate thickness, so that material is dropped from further contention. The material performance index indicates that 9 percent nickel steel is the best choice for this application.

6-9 MATERIALS SELECTION BY EXPERT SYSTEMS

Material selection represents an obvious application for expert systems, see Sec. 3-12. While broad comprehensive knowledge-based systems for materials selection still are under development, expert systems that deal with more restrictive situations currently exist. An example is ESCORT (Expert Software for Corrosion Technology),[1] an expert system for materials selection in the

---

[1] M. J. S. Vancoille, W. F. Bogaerts, and M. J. Rijckaert, Nat. Assoc. Corrosion Engrs 88 Conference, Paper 121, 1988.

TABLE 6-7
Weighed property index chart for selection of a material for a cryogenic storage vessel

| Material | Go no-go screening* | | | | Toughness, 0.4† | Fatigue, 0.1 | Stiffness, 0.1 | Thermal expansion, 0.1 | Cost, 0.3 | Material performance index |
	Corrosion	Weldability	Available in thick plate							
304 Stainless	S	S	S		100	70	93	80	50	79
9 percent Ni Steel	S	S	S		85	100	97	100	83	89
Al alloy	S	S	U		40	70	100	94	100	72
3 percent Ni Steel	S	S	S							

*S = satisfactory U = unsatisfactory

† Property weighting factor

258

chemical process industry. It consists of five knowledge bases: materials, equipment (type of heat exchanger, packed columns, pumps, etc.), environmental characteristics, types of corrosion, and preventive measures.

In starting the expert system the user is shown a **Corrosion Engineering Worksheet** consisting of three windows: (1) Industry, process, or operation; (2) Environment, and (3) Equipment. These windows permit the user to specify the problem in considerable detail. The more detailed the input the more specific will be the final material selection. By specifying the kind of industry or operation in the first window the program will make assumptions concerning certain process conditions. This enables the expert system to ask for additional specific information from the user. Window (2) allows the first selection of the process parameters. Window (3) calls for the specification of the process equipment in which the materials are to be used.

World 1, **World 2**, and **Merge Worlds** are features used while designing process equipment. Frequently the fluids inside and outside of heat transfer tubing are different. The process stream inside the tube might be highly corrosive acid while the outside of the tube might be cooled with water containing chlorides. Using **World 1** the expert system would select all materials which are suitable for the inside of the tubing (acid) and **World 2** would select all materials suitable for the outside of the tubing (water). By using **Merge Worlds** the program will identify the materials compatible with both the inside and outside of the tubing. The command **Input Complete** will cause the system to begin the reasoning process and proceed to the next worksheet. Often the system will produce questions that ask for more information than given in the input to the first Corrosion Engineering Worksheet. **Call NACE-NBS** introduces the option of calling upon external databases, such as the NACE–NBS(NIST) corrosion database. In this way it is not necessary to burden the expert system with too much data. A section called **Requirements** prompts the user to specify the classes of materials he/she prefers. These decisions could be based on previous experience, availability, cost, or codes and specifications.

When the input process has been completed the expert system starts the main reasoning operation using if–then production rules. The system uses two types of rules: shallow rules and deep rules. The shallow rules are based on readily available relationships, like those discussed earlier in this chapter, and on the expertise of individual specialists. The deep rules are even less easily defined, and comprise things like availability, economic factors, company policies, etc. The deep rules further narrow down the material selection. When the expert system completes its task it will present a list of materials that meet the input specifications. The user can ask why a certain material was chosen and the program will show the rules it applied to make this selection. The user can also ask for additional information about a material, e.g. how should it be welded?

The usefulness and power of expert systems in material selection should be obvious from the above brief scenario. Considerable development is expected in this area of knowledge-based engineering in the near future.

6-10 DESIGN EXAMPLE—COST COMPARISON

This design example[1] involving the selection of shipboard piping systems illustrates many of the considerations involved in materials selection for a design application. As is typical, cost is the overriding factor in deciding on the material. This example illustrates the common situation of a trade-off between a material with a lower initial cost and one with a high first cost but longer service life.

The piping system that provides water to a shipboard steam power plant often leads to failure of the condenser tubes because of plugging and biofouling. Expensive alloys are used for the tubes to provide corrosion resistance; but low-cost steel is often used for intake piping, and that leads ultimately to failure of the condenser. Bare or coated steel in contact with seawater provides a surface where marine organisms can attach, chiefly at flow rates below 3 ft/s. Once attached, the organisms grow and ultimately die and are swept into the condenser; they plug tubes, reduce flow, and increase tube failure rates. The problem is aggravated by the higher water temperature in tropical areas.

The countermeasures that have been used to remedy this problem with steel pipes are physical reaming of the pipe and chlorine injection. Reaming is not really a countermeasure, and chlorination, although effective in reducing biofouling, increases the cost of operation and must meet environmental regulations that limit the amount of chlorine discharged into open waters. Another effective biofouling approach is to replace the steel pipe with 90–10 copper-nickel alloy. The alloy is resistant to biofouling growth because of the presence of a very small concentration of copper ions in the corrosion product that forms naturally on the alloy in seawater.

Since seawater is a low-resistance electrolyte, galvanic effects must be taken into account in materials selection.[2] To avoid galvanic corrosion, materials that lie between steel and copper-nickel in the electromotive series should not be selected.

In an application such as the one under consideration there is a pipe size at which installation and operating costs are minimized. Since pipe cost increases and pumping cost (pressure drop) decreases as pipe diameter increases, there is an optimum pipe diameter (and flow rate). For a range of fluids, including water, the economic flow rate occurs between 5 to 8 ft/s. The pressure drop in steel pipe increases significantly as the surface of the pipe is roughened by corrosion and the growth of marine organisms. The friction factor for biofouled steel pipe is 2.5 times that for new steel pipe.

---

[1] A. H. Tuthill and S. A. Fielding, *Nickel Topics,* vol. 27, no. 3, 1974.

[2] M. G. Fontana and N. D. Green, "Corrosion Engineering," 2d ed., McGraw-Hill Book Company, New York, 1979; V. R. Pludek, "Design and Corrosion Control," John Wiley & Sons, Inc., New York, 1977.

Thus, there is a strong case to be made for selecting copper-nickel pipe over steel pipe if the difference in cost can be justified. We need to consider materials cost, installed cost, and life cycle cost. All costs are based on 1974 data, but the comparisons should still be valid.

Galvanized (zinc-coated) schedule 80 carbon steel is the low-first-cost piping material. A 4-in-diameter copper-nickel pipe is 5.3 times the cost of galvanized steel. However, as pipe diameter increases, the premium for 90–10 copper-nickel decreases significantly until at 24 in copper-nickel pipe is only 2.1 times the cost of steel pipe of the same size. But because of the difference in friction factor, a much larger diameter steel pipe is required to carry the same amount of water at the same pressure. For example, an 18-in copper-nickel pipe carries the same flow as a biofouled 24-in steel pipe. When compared on a functionally equivalent basis, the copper-nickel pipe is only 32 percent more expensive than the steel pipe.

The cost of fabricating and installing pipe on shipboard is the largest portion of the cost of a shipboard piping system. Much of the fabrication requires welding of the pipe. Because copper-nickel pipe has a smaller diameter and a thinner wall, fewer man-hours are required to weld it. It is estimated that copper-nickel pipe can be butt-welded in two passes, compared with eight for schedule 80 galvanized steel.

In life cycle cost not only the initial direct material and installation costs but also the effect of the material selection on the costs of maintenance and operations is considered. Obviously, the exact details of this type of cost analysis will depend on the emphasis that is placed on reliable, maintenance-free operation. The analysis performed in this instance showed that copper-nickel was the most economical in life cycle cost unless it could be assumed that the steel piping would not need replacement more than once in 25 years. Experience has shown that to be unlikely. An important element in the analysis is the cost incurred by the ship being out of service because of material failure. If the steel piping leads to a delay of as little as one day per year of sailing, the steel system incurs a cost penalty of 120 percent.

6-11 VALUE ANALYSIS

Value analysis, or value engineering,[1] is an organized system of techniques for identifying and removing unnecessary costs without compromising the quality and reliability of the design. The technique usually is applied to a spectrum of problems much broader than just materials selection, but the framework of

---

[1] C. Fallon, "Value Analysis to Improve Productivity," John Wiley & Sons, Inc., New York, 1971; E. D. Heller, "Value Management," Addison-Wesley Publishing Co., Reading, Mass., 1971; L. D. Miles, "Techniques of Value Analysis in Engineering," 2d ed., McGraw-Hill Book Company, New York, 1972.

value analysis methodology applies admirably to the problem of materials selection.

Value analysis asks the following questions: How can a given function of a design system be performed at minimum cost? What is the value of the contribution that each feature of the design makes to the specific function that the part must fulfill? Value analysis usually is carried out by a team of engineers and managers possessing different backgrounds and viewpoints, so that the problem gets looked at from many aspects. However, a value analysis needs the support and endorsement of top management if it is to be most successful.

The value analysis approach seeks the development of answers to the following questions.

Can we do without the part?

Does the part do more than is required?

Does the part cost more than it is worth?

Is there something that does the job better?

Is there a less costly way to make the part?

Can a standard item be used in place of the part?

Can an outside supplier provide the part at less cost without affecting dependability?

The value analysis job plan consists of those tasks or functions necessary to perform the study (Fig. 6-10). The sequence of tasks has many of the elements described in Chap. 3 for problem solving. A structured plan like the one shown assures that consideration is given to all important aspects, provides for a logical separation of the study into convenient units, and provides a convenient basis for maintaining a written record of progress in the value analysis study.

A key step in value analysis is evaluation of the function of the design or system. It has the same importance that definition of the problem has to more generalized problem solving. In value analysis, function is expressed by two words; a verb and its noun object. The verb answers the question what does it do; the noun answers the question what does it do it to. It might, for example, transmit torque, conduct current, or maintain records. As in preparing a problem statement, the function should be identified so as not to limit the ways in which it could be performed.

The functions of a design or a system should be divided into basic functions and secondary functions. A *basic function* defines a performance feature that must be attained. It answers the question what must it do. A *secondary function* defines performance features of a system or item other than those that must be accomplished. It answers the question what else does it do. For example, the basic function of a coat of paint is to protect a surface from the environment. The secondary function is to improve appearance. The

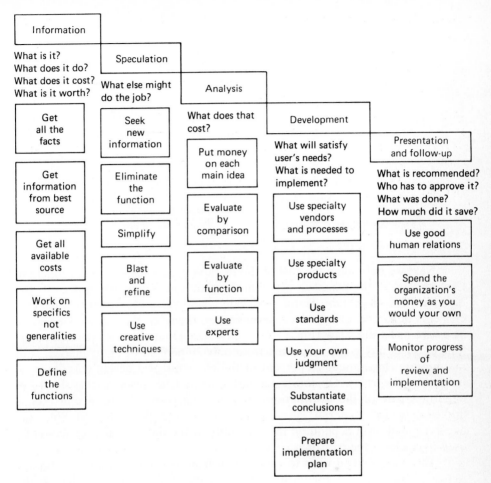

FIGURE 6-10
Components of the value analysis job plan.

ability to identify functions and distinguish between basic and secondary functions is important in value analysis. Value is placed only on basic functions, and it is there we direct our attention.

Once the functions have been established, the next step is to establish a dollar value of worth for each function. The worth of a basic function usually is determined by comparing the present design for attaining the function with other methods of attaining essentially the same function. Considerable skill, knowledge, and judgment are needed to determine worth in terms of dollars. One of the ways to determine the worth of a function is to ask yourself what would be a reasonable amount to pay for the attainment of the basic function if you were to pay for it out of your own funds.

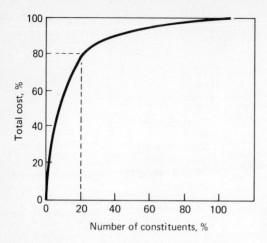

FIGURE 6-11
Pareto's law of distribution of costs.

Next we determine the cost of the method used to carry out the function. Cost applies to the actual design that is used, whereas worth applies to the function. An important feature of value analysis is to identify high-cost elements of the design and focus attention on them. In that regard we should be aware of Pareto's law (Fig. 6-11), which states that about 80 percent of the total effect of any group will come from only 20 percent of the components of that group. Thus, about 20 percent of the elements of a design contribute 80 percent of the costs. Obviously, attention should be given to this small but important part of the distribution. In analyzing costs, the total unit cost is broken down into material, labor, and overhead costs (see Chap. 9). Since the decisions made will depend on the reliability of the data, great care should be used to establish the cost data.

There are several ways to gain insight from cost data. A good technique is to develop the cost for each element in the design for each step in the manufacturing cycle from raw material to finished part. It is also instructive to plot the cost elements from the above analysis as a function of year to determine whether costs are getting out of line owing to out-of-date processing technology. Other types of cost comparison that may be relevant in certain circumstances are cost per pound, cost per dimension (area, volume, or length), and cost per property.

Finally, we come to the determination of the value of the design or system. We have already considered use value (worth) and cost value. Esteem value is concerned with the power of a thing to make us want to possess it. That may be a significant component of value in consumer-oriented products, but it usually plays little role in value analysis. A good way to express value is with the *value index*, the ratio of cost to worth. Large values of value index signal the part of the design that are fruitful for cost reduction.

The steps in the value analysis job plan (Fig. 6-10) are much like those in problem solving. Once the basic information has been gathered, the next step is speculation. The objective of phase II is to generate, by creativity techniques, many alternative means for accomplishing the basic functions identified in phase I. Brainstorming, morphological analysis, etc., are commonly employed. Phase III involves a selection for further analysis and refinement of the most promising alternatives generated in phase II. Each idea is evaluated chiefly in terms of 1) what does it cost and 2) will it attain the basic functions. Phase IV is the development of a complete plan for implementation. Finally, the best solution is selected for implementation, and several alternatives are selected if the first choice is rejected by the approval authority. The last phase involves the critical stage of presenting the best alternatives to persons who have the authority to approve the value analysis proposals.

The following brief example illustrates the value analysis technique. The design under study was the housing for a hydraulic valve (Fig. 6-12). The housing was made of nodular cast iron and was heat-treated to a fully pearlitic matrix to provide the necessary wear resistance for the valve stem. The objective of the value analysis was to reduce the cost of manufacture of the part. Analysis of the costs for each step in the manufacture showed that machining was the preponderant cost driver. It was readily appreciated that machining time (and cost) could be reduced substantially if the valve housings were heat-treated to a fully ferritic matrix. However, wear resistance of the softer condition was not adequate.

Functional analysis showed that the valve housing had two basic functions: 1) to provide an enclosure for the hydraulic fluid and 2) to provide a wear-resistant surface for the piston-like valve stem. In the speculation phase it was realized that those two basic functions were coupled in the original design and that hence the heat treatment of the nodular iron was constrained by the need for high hardness to provide wear resistance. Other suggestions for wear resistance were the use of a tool-steel sleeve in a soft valve housing and the chrome plating of the cylinder bore. The analysis phase further substantiated the cost savings to be achieved by improving the machinability of the nodular iron castings and rejected the new ideas as probably more costly. In the development phase extensive tests were made on the wear characteristics of the valves. These studies showed that wear was significant only during the first 500 cycles of operation as the piston, machined to close tolerances, was running in. This suggested that a low-friction surface coating that would eventually wear away might be a suitable alternative. The tests also pointed

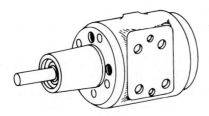

FIGURE 6-12
Hydraulic valve housing.

out that surface finish of the cylinder bore was a critical operational parameter that could not be achieved in dead-soft ferritic nodular iron.

The final solution was to heat-treat the valve housings to a mixed microstructure of ferrite and spheroidized pearlite. This was a compromise metallurgical condition between the hard wear-resistant condition and the soft condition that would have improved machinability but poor surface finish. Wear during the running-in period was accommodated by a phosphate coating on the cylinder bore that provided lubrication during the critical period.

6-12   DESIGN EXAMPLE—MATERIALS SYSTEMS

Engineered systems contain many components, and for each a material must be selected. The automobile is our most familiar engineering system and one that is undergoing a major change in the materials used for its construction. These trends in materials selection reflect the great effort that is being made to decrease the fuel consumption of cars by down-sizing the designs and adopting weight-saving materials. Major shifts in materials selection such as are shown in Table 6-8 can have large economic consequences.

Frequently, complex and severe service conditions can be met economically only by combining several materials into a single component. The surface hardening of gears and other automotive components by carburizing or nitriding[1] is a good example. Here the high hardness, strength, and wear resistance of a high-carbon steel is produced in the surface layers of a ductile and tougher low-carbon steel.

An excellent example of a complex materials system used in a difficult environment is the exhaust valve in an internal-combustion engine.[2] Valve materials must have excellent corrosion- and oxidation-resistance properties to resist "burning" in the temperature range 1350 to 1700°F. They must have 1) sufficient high-temperature fatigue strength and creep resistance to resist failure and 2 suitable hot hardness to resist wear and abrasion.

The critical failure regions in an exhaust valve are shown in Fig. 6-13. Maximum operating temperature occurs in areas A and C. Corrosion and oxidation resistance are especially critical there. The underhead area of the valve, area C, experiences cyclic loading, and because of the mild stress concentrations, fatigue failure may occur at that point. The valve face, area B, operates at a somewhat lower temperature because of heat conduction into the valve seat. However, if an insulating deposit builds up on the valve face, it can

---

[1] Surface Hardening of Steel, "Metals Handbook," 9th ed., Vol. I, "Properties and Selection of Irons and Steels," American Society for Metals, Metals Park, Ohio, 1978.

[2] J. M. Cherrie and E. T. Vitcha, *Metal Prog.*, pp. 58–62, September 1971; J. F. Kocis and W. M. Matlock, ibid., pp. 58–62, August 1975.

TABLE 6-8
Typical distribution of materials in average car

	1979		1988	
	Weight, lb	**Percent of total wt**	**Weight, lb**	**Percent of total wt**
Steel	2000	62	1675	57
Cast iron	515	16	450	15
Aluminum	100	3	150	5
Other metals	65	2	150	5
Plastics	190	5	225	8
Other materials	380	12	300	10
	3250	100	2950	100

lead to burning. Also, the valve seat can be damaged by indentation by abrasive fuel ash deposits. The valve stem is cooler than the valve head. However, wear resistance is needed. Surface wear of the valve stem, area *D*, can lead to scuffing, which will cause the valve to stick open and burn. Wear at the valve tip, area E, where the valve contacts the rocker arm, will cause valve lash and cause the valve to seat with higher than normal forces. Eventually, that will cause failure.

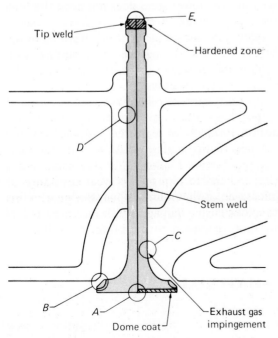

FIGURE 6-13
Typical exhaust valve showing critical regions of failure.

TABLE 6-9
Some properties of value materials

| Alloy | C | Mn | Cr | Ni | Other | Strength at 1350°F | | |
						Tensile strength, psi	Creep strength*	Fatigue limit†
21-2N	0.55	8.25	20.35	2.10	0.3N; bal. Fe	57,700	34,000	26,000
DV2A	0.53	11.50	20.5	—	2W; 1Cb; bal. Fe	78,000	34,000	30,000
Inconel 751	0.10 max	1.0 max	15.5	Bal.	2.3Ti; 1.2Al	82,000	45,000	45,000

* Stress to produce 1 percent elongation in 100 h
† Stress to produce failure in 10^8 cycles

The basic valve material for passenger car application, where $T_{max} = 1300°F$, is an austenitic stainless steel that obtains its good high-temperature properties from a dispersion of precipitates. This alloy, 21-2N, contains 20.35 percent chromium for oxidation and corrosion resistance. It has good PbO corrosion resistance, and its high-temperature fatigue strength is exceeded only by that of the more expensive nickel-base super alloys, Table 6-9.

The entire body of one-piece valves is 21-2N, except for a hard-steel tip at E and a hard chromium plate in area D. However, it is generally more economical to use a two-piece valve in which 21-2N is replaced in the cooler stem portions by a cheaper alloy steel such as SAE 3140 or 4140. Either steel will have sufficient wear resistance, and the lower stem does not need the high oxidation and corrosion resistance of the high-chromium, high-nickel steel. The two materials are joined by welding, as shown in Fig. 6-13. Burning of the valve face, area B, is generally avoided by coating the valve surface with aluminum to produce an Fe–Al alloy or, in severe cases, by hard-facing the valve seat with one of the Co–C–Cr–W Stellite alloys.

With the removal of lead compounds from gasoline has come an easing of the high-temperature corrosion problems that cause burning, but a new problem has arisen. The combustion products of lead-free gasoline lack the lubricity characteristics of those of fuels that contain lead. As a result, the wear rates at the valve seating surfaces have increased significantly. The solution has been to harden the cast-iron cylinder head to improve its wear resistance or install a wear-resistant insert. Unless one or the other is done, the wear debris will weld to the valve seat and cause extensive damage. With increased use of hard-alloy inserts, it is becoming more common to hard-face the valve seat with a Stellite alloy.

6-13 DESIGN EXAMPLE—MATERIALS SUBSTITUTION

This design example illustrates the common problem of substituting a new material for one that has been used for some time. It illustrates that material

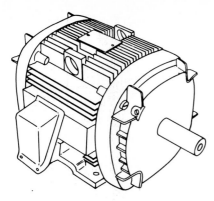

FIGURE 6-14
Horizontal aluminum alloy motor. (*Courtesy of General Electric Company.*)

substitution should not be undertaken unless appropriate design changes are made. Also, it illustrates some of the practical steps that must be taken to ensure that the new material and design will perform adequately in service.

Aluminum alloys have been substituted for gray cast iron[1] in the external supporting parts of integral-horsepower induction motors (Fig. 6-14). The change in materials was brought about by increasing cost and decreasing availability of gray-iron castings. There has been a substantial reduction in gray-iron foundries, partly because of increased costs resulting from the more stringent environmental pollution and safety regulations imposed in recent years by governmental agencies. The availability of aluminum casting has increased owing to new technology and the lesser problem of operating a foundry at a temperature much lower than that required for cast iron.

There are a variety of aluminum casting alloys.[2] Among the service requirements for this application, strength and corrosion resistance were paramount. The need to provide good corrosion resistance to water vapor introduced the requirement to limit the copper content to an amount just sufficient to achieve the necessary strength. Actual alloy selection was dependent on the manufacturing processes used to make the part. That in turn depended chiefly (see Chap. 7) on the shape and the required quantity of parts. Table 6-10 gives details on the alloys selected for this application.

Since the motor frame and end-shield assemblies have been made successfully from gray cast iron for many years, a comparison of the mechanical properties of the aluminum alloys with cast iron is important, Table 6-11.

The strength properties for the aluminum alloys are approximately equal to or exceed those of gray cast iron. If the slightly lower yield strength for alloy

[1] T. C. Johnson and W. R. Morton, IEEE Conference Record 76CH1109-8-IA, Paper PCI-76-14, available from General Electric Company as Report GER-3007.

[2] "Metals Handbook," 9th ed., vol. 2, pp. 140–179, American Society for Metals, Metals Park, Ohio, 1979.

TABLE 6-10

Aluminum alloys used in external parts of motors

Part	Alloy	Composition			Casting process
		Cu	Mg	Si	
Motor frame	356	0.2 max	0.35	7.0	Permanent mold
End shields	356	0.2 max	0.35	7.0	Permanent mold
Fan casing	356	0.2 max	0.35	7.0	Permanent mold
Conduit box	360	0.6 max	0.50	9.5	Die casting

356 cannot be tolerated, it can be increased appreciably by a solution heat treatment and aging (T6 condition) at a slight penalty in cost and corrosion resistance. Since the yield and shear strength of the aluminum alloys and gray cast iron are about equal, the section thickness of aluminum to withstand the loads would be the same. However, since the density of aluminum is about one-third that of cast iron, there will be appreciable weight saving. The complete aluminum motor frame is 40 percent lighter than the equivalent cast-iron design. Moreover, gray cast iron is essentially a brittle material, whereas the cast-aluminum alloys have enough malleability that bent cooling fins can be straightened without breaking them.

One area in which the aluminum alloys are inferior to cast iron is compressive strength. In aluminum, as with most alloys, the compressive strength is about equal to the tensile strength, but in cast iron the compressive strength is several times the tensile strength. That becomes important at the bearing support, where, if the load is unbalanced, the bearing can put an appreciable compressive load on the material surrounding and supporting it. With an aluminum alloy end shield that leads to excessive wear. To minimize the problem, a steel insert ring is set into the aluminum alloy end shield when it is cast. The design eliminates any clearance fit between the steel and

TABLE 6-11

Comparison of typical mechanical properties

Material	Yield strength, ksi	Ultimate tensile strength, ksi	Shear strength, ksi	Elongation in 2 in, percent
Gray cast iron	18	22	20	0.5
Alloy 356 (as cast)	15	26	18	3.5
Alloy 360 (as cast)	25	26	45	3.5
Alloy 356-T61 (solution heat-treated and artifically aged)	28	38		5

aluminum, and the steel insert resists wear from the motion of the bearing just as the cast iron always did.

The greater ease of casting aluminum alloys permits the use of cooling fins thinner and in greater number than in cast iron. Also, the thermal conductivity of aluminum is about three times greater than that of cast iron. Those factors result in more uniform temperature throughout the motor, and the results are longer life and reliability. Because of the higher thermal conductivity and larger surface area of cooling fins, less cooling air is needed. With the air requirements reduced, a smaller fan can be used, and as a result there is a small reduction in noise.

The coefficient of expansion of aluminum is greater than that of cast iron, and that makes it easier to ensure a tight fit of motor frame to core. Only a moderate temperature is needed to expand the aluminum frame sufficiently to insert the core, and on cooling the frame contracts to make a tight bond with the core. That results in a tighter fit between the aluminum frame and the core and better heat transfer to the cooling fins. Complete design calculations need to be made when aluninum is substituted for cast iron to be sure that clearances and interferences from thermal expansion are proper.

Since a motor design that had many years of successful service was being changed in a major way, it was important to subject the redesigned motor to a variety of simulated service tests. The following were used:

- Vibration test
- Navy shock test (MIL-S-901)
- Salt fog test (ASTM B117-57T)
- Axial and transverse strength of end shield
- Strength of integral cast lifting lugs
- Tests for galvanic corrosion between aluminum alloy parts and steel bolts

This example illustrates the importance of considering design and manufacturing in a material-substitution situation. Although the use of aluminum in this situation is very favorable, it must be recognized that this is an application in which the much reduced elastic stiffness of aluminum is not of great importance.

6-14 SUMMARY OF MATERIALS SELECTION

This chapter has shown that there are no magic formulas for materials selection. Rather, the solution of a materials selection problem is every bit as challenging as any other aspect of the design process and follows the same general approach of problem solving. Successful materials selection depends on the answers to the following questions.

1. Have performance requirements and service environments been properly and completely defined?
2. Is there a good correlation between the performance requirements and the material properties used in evaluating the candidate materials?
3. Has the relation between properties and their modification by subsequent manufacturing process been fully considered?
4. Is the material available in the shapes and configurations required and at an acceptable price?

The steps in the materials selection process are:

1. Analysis of material requirements
2. Identification and screening of alternative materials
3. Evaluation of the remaining candidate materials
4. Development of design data for critical applications

Materials selection requires familiarity with available materials, their properties, and their fabrication.

Weighted property techniques using a decision matrix are best for choosing between the competing property requirements in the usual design situation. Failure analysis is an important input to materials selection when a design is modified. The value analysis technique has broad implications in engineering decision making, but it is especially useful for materials selection when a design review on a new product is being conducted or an existing product is being redesigned.

BIBLIOGRAPHY

Charles, J. A., and F. A. A. Crane: "Selection and Use of Engineering Materials," 2d ed; Butterworth London, 1989.

Cornish, E. H.: "Materials and the Designer," Cambridge University Press, Cambridge, 1987.

Farag, M. M: "Selection of Materials and Processes for Engineering Design," Prentice-Hall Inc., Englewood Cliffs, N.J., 1990.

Hanley, D. P.: "Introduction to the Selection of Engineering Materials," Van Nostrand Reinhold Company, New York, 1980.

Kern, R. F., and M. E. Suess: "Steel Selection: A Guide for Improving Performance and Profits," John Wiley & Sons, Inc., New York, 1979.

Lewis, G.: "Selection of Engineering Materials," Prentice-Hall Inc., Englewood Cliffs, N.J., 1990.

Sharp, H. J. (ed.): "Engineering Materials: Selection and Value Analysis," American Elsevier Publishing Co., Inc., New York, 1966.

Verink, E. D. (ed.): "Methods of Materials Selection," Gordon and Breach, Science Publishers, Inc., New York, 1968.

CHAPTER 7

INTERACTION OF MATERIALS, PROCESSING, AND DESIGN

7-1 ROLE OF PROCESSING IN DESIGN

Producing the design is a critical link in the chain of events that starts with a creative idea and ends with a successful product in the marketplace. In modern technology the function of production no longer is a routine activity. Rather, design, materials selection, and processing are intimately related, as shown in Fig. 7.1.

There is a confusion of terminology concerning the engineering function we have called processing. Metallurgical engineers may talk about materials processing to refer to the conversion of semifinished products, like steel blooms or billets, into finished metallurgical products, like cold-rolled sheet or hot-rolled bar. A mechanical, industrial, or manufacturing engineer is more likely to refer to the conversion of the sheet into an automotive body panel as manufacturing. *Processing* is the more generic term, but *manufacturing* seems to be gaining popularity. *Production engineering* is a term used in Europe to describe what we call manufacturing in the United States.

The first half of the twentieth century saw the maturation of manufacturing operations in the western world. Increases in the scale and speed of operations brought about increases in productivity, and manufacturing costs dropped while wages and the standard of living rose. There was a great

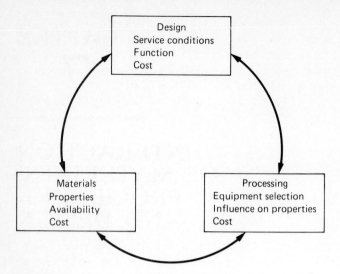

FIGURE 7-1
Interrelations of design, materials, and processing to produce a product.

proliferation of available materials as basic substances were tailor-made to have specific improved properties. One of the major achievements of this era was the development of the production line for mass-producing automobiles, appliances, and other consumer goods. Because of the preeminence in manufacturing that developed in the United States, there has been a tendency in the last half of the century to take the production function for granted. Manufacturing has been downplayed, or even ignored completely, in the education of engineers. Manufacturing positions in industry have been considered routine and not challenging, and as a result they have not attracted their share of the most talented engineering graduates. Fortunately, this situation is improving as the importance of the manufacturing function is being rediscovered and the nature of manufacturing is being changed by automation and computer-aided manufacturing.

Peter Drucker, the prominent social scientist and management expert, has termed the current manufacturing situation "the third industrial revolution." The second industrial revolution began roughly a century ago when machines were first driven directly by fractional-horsepower motors. The use of power, whether generated by falling water or a steam engine, was, of course, the first industrial revolution. However, before machines were direct-driven by electric motors, they were driven by belts and pulleys, which meant that the equipment had to be very close to the source of power. Thus, the second industrial revolution gave flexibility and economy to manufacturing. The third industrial revolution, in which information processing is becoming part of the machine or tool, is converting production from a manual into a knowledge-based operation.

A serious problem has been the tendency to separate design and manufacturing into separate organizational units. Barriers between design and manufacturing can inhibit the close interaction that the two engineering functions should have, as discussed previously under concurrent engineering (Sec. 1-12). When technology is sophisticated and fast-changing, a close coupling between the people in research, design, and manufacturing is very necessary. That has been demonstrated best in the area of solid-state electronic devices. As semiconductor devices replaced vacuum tubes, it became apparent that design and processing could no longer be independent and separable functions. Vacuum-tube technology was essentially a linear situation in which specialists in materials passed on their input to specialists in components who passed on their input to circuit designers who, in turn, communicated with system designers. With the advent of transistors the materials, device construction, and circuit design functions became closely coupled. Then, with the microelectronics revolution of large-scale integrated circuits, the entire operation from materials to system design became interwoven and manufacturing became inseparable from design. The result was a situation of rapid technical advance requiring engineers of great creativity, flexibility, and breadth. The payoff in making the minicomputer a reality has been huge. Nowhere has productivity been enhanced as rapidly as in the microelectronics revolution. That should serve as a model of the great payoff that can be achieved by closer coupling of research, design, and manufacturing.

More conventional manufacturing is divided into 1) process engineering, 2) tool engineering, 3) standards, 4) plant engineering, and 5) administration and control. *Process engineering* is the development of a step-by-step sequence of production. The overall product is subdivided into its components and subassemblies, and the steps required to produce each component are arranged in logical sequence. An important part of process engineering is to specify the related tooling. Vital parameters in process engineering are the rate of production and the cost of manufacturing. *Tool engineering* is concerned with the design of tools, jigs, fixtures, and gages to produce the part. Jigs hold the part, and fixtures guide the tool during processing. Tools do the machining or forming; gages determine whether the dimensions of the part are within specification. *Work standards* are time values associated with each manufacturing operation. Other standards that need to be developed in manufacturing are tool standards and materials standards. *Plant engineering* is concerned with providing the plant facilities (utilities, space, transportation, storage, etc.) needed to carry out the manufacturing process. *Administration and control* is production planning, scheduling, and supervising to assure that materials, tools, machines, and people are available at the right time and in the quantities needed to produce the part.

We ordinarily think of modern manufacturing in terms of the automotive assembly line, but mass production manufacturing systems account for less than 25 percent of metal parts manufactured. In fact, 75 percent of the parts manufactured are produced in lots of fewer than 50 pieces. About 40 percent of the employment in manufacturing is in such job-shop operations. Studies of batch-type metal-cutting production shops have shown that, on the average, a workpiece in such a shop is on a machine tool being productively processed *only 5 percent of the time*. During all the rest of the time the workpiece is in the shop, it is waiting in an inventory of unfinished parts. Moreover, of the very small fraction of time the part is being worked on, it is being cut by the tool only about 30 percent of the time. The remaining 70 percent of the time is taken up by loading and unloading, positioning the workpiece, gaging the part, etc.

Thus, there is a major opportunity for greatly increasing manufacturing productivity in small-lot manufacture. Computer-automated machine tool systems, which include industrial robots and computer software for scheduling and inventory control, have a demonstrated potential for increasing machine utilization time from an average of 5 percent to as much as 90 percent. The

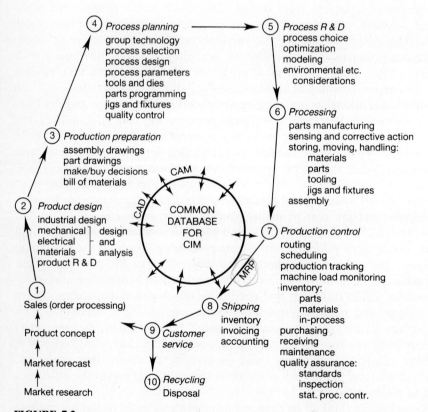

FIGURE 7-2
Spectrum of activities which are encompassed by manufacturing. (*From J. A. Schey, "Introduction to Manufacturing Processes", 2d ed., McGraw-Hill Book Co., 1987. Used with permission of McGraw-Hill Book Co.*)

introduction of computer-controlled machining centers that can perform many operations can greatly increase the productivity of the machine tool. We are moving toward the computer-automated factory in which all steps in parts manufacturing will be optimized by computer software systems, the machine tools will be under computer control, and at least half of the machines will be part of a versatile manufacturing system featuring multiple machining capability and automatic parts handling between work stations. This automated factory will differ from the automotive transfer line in that it will be a flexible manufacturing system capable of producing a wide variety of parts under computer control. This broadbased effort throughout industry to link computers in all aspects of manufacturing is called *computer-integrated manufacturing* (CIM). Figure 7-2 shows the complete spectrum of activities which are encompassed by manufacturing.

7-2 CLASSIFICATION OF MANUFACTURING PROCESSES

It is not a simple unambiguous task to classify the tremendous variety of manufacturing processes. We attempt to come to grips with the problem through the hierarchical classification of business and industry shown in Fig. 7-3. The service industries consist of enterprises, such as banking, education, insurance, and communication, that provide important services to modern society but do not create wealth by converting raw materials. The producing industries acquire raw materials (minerals, natural products, or petroleum) and process them, through the use of energy, machinery, and brainpower, into

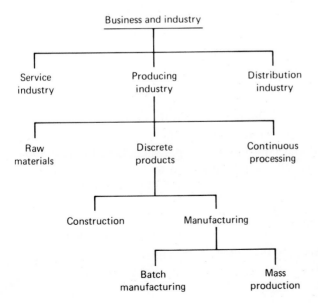

FIGURE 7-3
A hierarchical classification of business and industry.

products that serve the needs of society. The distribution industries, such as merchandising and transportation, make those products available to the general public. A characteristic of modern industrialized society is that an increasingly smaller percentage of the population produces the wealth that makes our affluent society possible. The producing industries account for about 23 percent and the service industries about 68 percent of our gross national product (GNP). Just as the past century saw the United States change from a predominantly agrarian society to a nation in which only 4 percent of the population works in agriculture, so we are becoming a nation in which an ever-decreasing percentage of the work force is engaged in manufacturing. In 1947 about 30 percent of the work force were in manufacturing; in 1980 it was about 22 percent. It is predicted that by the year 2000 less than 10 percent of U.S. workers will be engaged in manufacturing. If that is to come about without destroying our economy through runaway inflation, we will have to greatly increase the productivity of the manufacturing sector.

The producing industries can be divided conveniently into raw materials producers (mining, petroleum, agriculture), producers of discrete products (autos, television, etc.), and industries engaged in continuous processing (gasoline, wood products, steel, chemicals, etc.). Two major divisions of the discrete products are construction (buildings, dams, etc.) and manufacturing. Under manufacturing we recognize batch (low-volume) manufacturing and mass production.

We can classify the great number of processes used in manufacture into the following eight categories.

1. *Solidification* (*casting*) *processes*. Molten metal, plastic, or glass is cast into a mold and solidified into a shape.
2. *Deformation processes*. A material, usually metal, is plastically deformed hot or cold to give it improved properties and change its shape. Typical processes of this type are forging, rolling, extrusion, and wiredrawing. Sheet-metal forming is a special category in which the deformation occurs in a two-dimensional stress state.
3. *Material removal or cutting* (*machining*) *processes*. Material is removed from a workpiece with a sharp tool by a variety of methods such as turning, milling, grinding, shaving, polishing, and lapping.
4. *Polymer processing*. The special properties of polymers have brought about the development of processes, such as injection molding and thermoforming, that are not duplicated in the above categories, which are more oriented to metals processing.
5. *Particulate processing*. This rapidly developing area includes the consolidation of particles of metal, ceramics, or polymers by pressing and sintering, hot compaction, or plastic deformation. It also includes the processing of composite materials.

6. *Joining processing.* Included in joining processing are all categories of welding, brazing, soldering, diffusion bonding, riveting, bolting, and adhesive bonding.

7. *Heat treatment and surface treatment.* This category includes the improvement of mechanical properties by thermal heat treatment processes as well as the improvement of surface properties by diffusion processes like carburizing and nitriding or by alternative means such as sprayed or hot-dip coatings, electroplating, and painting. The category also includes the cleaning of surfaces preparatory to surface treatment.

8. *Assembly processes.* In this, usually the final, step in manufacturing a number of parts are brought together and combined into a subassembly or finished part.

Hopefully the reader will have had a broad-based course on manufacturing processes so the specialized vocabulary of processing and the discussion of various processes have meaning and substance. Realistically, manufacturing processes have been a badly neglected topic in most engineering curricula, since the early 1960s. Although we cannot remedy this deficiency in a single chapter, we list here a complete set of reference books that will provide general information for the beginner and detailed information for the engineer seeking to solve a manufacturing problem.

General Overview of Manufacturing Processes

1. The novice who has no knowledge of manufacturing processes or the more experienced engineer who wants to pull together an uncoordinated background should read seriously "Introduction to Manufacturing Processes", 2d ed. by J. A. Schey, McGraw-Hill Book Company, New York, 1987. This book emphasizes the fundamentals of material behavior and the mechanics of the process and their interaction. The entire scope of manufacturing processes is addressed.

2. The novice without industrial processing experience should follow up the reading of Schey with one or more of the books listed in the bibliography at the end of this chapter. Those books are much more descriptive than Schey and also more voluminous (700 to 1100 pages), but they give a good feel for how the processes are actually carried out.

3. The single most important reference on manufacturing methods is "Tool and Manufacturing Handbook," 4th ed., produced by the Society of Manufacturing Engineers. The five volumes of this handbook provide broad coverage of manufacturing processes.

Selected Guide to Literature on Specific Processes

Solidification processes

1. R. W. Heine, C. R. Loper, Jr., and C. Rosenthal, "Principles of Metals Casting," McGraw-Hill Book Company, New York, 1967. Practice oriented.
2. M. C. Flemings, "Solidification Processing," ibid., 1974. Theoretically oriented.
3. "Metals Handbook," 9th ed., vol. 15, American Society for Metals, Metals Park, Ohio, 1988. Excellent source of practical examples.

Deformation processes

1. W. A. Backofen, "Deformation Processing," Addison-Wesley Publishing Co., Inc., Reading, Mass., 1972. Theoretically oriented.
2. G. W. Rowe, "Principles of Industrial Metalworking Processes," Arnold, London, 1977.
3. T. Z. Blazynski, "Metal Forming: Tool Profiles and Flow," John Wiley–Halsted Press, New York 1976. Chiefly considers wire, rod, tube drawing, and extrusion.
4. "Forging: Equipment, Materials and Practices," Metals and Ceramics Information Center, Battelle Memorial Institute, Columbus, Ohio, 1973.
5. W. L. Roberts, "Cold Rolling of Steel," Marcel Dekker, Inc., New York, 1978.
6. D. P. Koistinen and N. M. Wang (eds.), "Mechanics of Sheet Metal Forming," Plenum Press, New York, 1978.
7. "Metals Handbook," 9th ed., American Society for Metals, Metals Park, Ohio. Vol. 14. "Forming and Forging", 1988.
8. K. Lange (ed), "Handbook of Metalworking", McGraw-Hill, New York, 1985.

Material removal

1. G. Boothroyd and W. A. Knight, "Fundamentals of Machining and Machine Tools", 2d. ed., Marcel Dekker Inc., New York 1989.
2. E. J. A. Armarego and R. H. Brown, "The Machining of Metals," Prentice-Hall, Inc., Englewood Cliffs, N.J. 1969.
3. "Metals Handbook," 9th ed., vol. 16, American Society for Metals, Metals Park Ohio, 1989.
4. "Machining Data Handbook," 3d ed., Machinability Data Center, Metcut Research Associates, Cincinnati, Ohio, 1981.

Polymer processing

1. Z. Tadmor and C. G. Gogos, "Principles of Polymer Processing," John Wiley & Sons, Inc., New York, 1979.
2. J. L. Throne, "Plastics Process Engineering," Marcel Dekker, Inc., New York, 1979.
3. S. Middleman, "Fundamentals of Polymer Processing," McGraw-Hill Book Company, New York, 1977.
4. "Engineered Materials Handbook," vol. 2, Engineering Plastics, ASM International, Metals Park, Ohio, 1988.

Particulate processing

1. J. S. Hirschorn, "Introduction to Powder Metallurgy," American Powder Metallurgy Institute, New York, 1969.
2. H. H. Hausner, "Handbook of Powder Metallugry," Chemical Publishing Co., Inc., New York, 1982.
3. F. V. Lenel, "Powder Metallurgy: Principles and Application," American Powder Metallurgy Institute, New York, 1980.
4. "Metals Handbook," 9th ed. vol. 7, Powder Metallurgy, ASM International, Metals Park, Ohio, 1984.

Joining processes

1. R. A. Lindberg and N. R. Braton, "Welding and Other Joining Processes," Allyn & Bacon, Inc., Boston, 1976.
2. "Welding Handbook," 7th ed., 5 vols, American Welding Society, Miami, Fla. from 1976 on.
3. "Metals Handbook," 9th ed., Vol. 6, "Welding, Brazing, and Soldering," American Society for Metals, Metals Park, Ohio, 1983.
4. C. V. Cagle (ed.), "Handbook of Adhesive Bonding," McGraw-Hill Book Company, New York, 1973.
5. R. O. Parmley, "Standard Handbook of Fastening and Joining," McGraw-Hill Book Co., New York, 1977.

Heat treatment and surface treatment

1. "Metals Handbook," 9th ed., American Society for Metals, Metals Park, Ohio.
   Vol. 4. "Heat Treating," 1981.
   Vol. 5. "Surface Cleaning, Finishing and Coating," 1982.
2. A. K. Graham (ed.), "Electroplating Engineering Handbook," 2d ed., Reinhold Publishing Co., New York, 1962.
3. R. B. Ross, "Handbook of Metal Treatments and Testing," John Wiley–Halsted Press, New York, 1977.

Form	Irons	Steels (carbon, low alloy)	Heat & corr. res. alloys	Aluminum alloys	Copper alloys	Lead alloys	Magnesium alloys	Nickel alloys	Precious metals	Refractory metals	Tin alloys	Titanium alloys	Zinc alloys
Sand castings	■	■	■	■	■	□	■	■			□		□
Shell mold castings	■	□	□	■	■			□					
Full-mold castings	■	■	□	□	□	□		■					
Permanent-mold castings	■	□		■	□	■	■	□			□		□
Die castings				■	□	■	■				□		■
Plaster mold castings				■	■								
Ceramic mold castings	■	■	■	□	□		□	■					□
Investment castings		■	□	■	■		□	■	□				
Centrifugal castings	■	■	■	□	□		□						
Continuous castings		□		■	■d	□							
Open die forgings	□	■	■	□	□		□	□		□		□	
Closed die forgings Blocker type		■	■	□	□		□	□		□		□	
Conventional type		■	■	□	□		□	□		□		□	
Upset forgings		■	■	□	□		□	□		□		□	
Cold headed parts		■	□	■	■	□		□	□				
Stampings, drawn parts		■	□	■	■		□	■	□	□		□	□
Spinnings		■	□	■	■	□	□	■	□			□	□
Screw machine parts	□	■	□	■	■		□	■	□	□		□	□
Powder metallurgy parts	■	■	□	□	■			□	□	■		□	
Electroformed parts	□			□	■	□		■	□		□		□
Cut extrusions		□		■	■	□	■	□		□	□	□	
Sectioned tubing	■	■					■	■		□		■	
Photofabricated parts		■	□	■	■	□	■	■	■	□	□	■	■

a ■ – Materials most frequently used.
□ – also materials currently being used.

b Iron-copper and iron-copper-carbon most frequently used.
c Most frequently used materials are pure nickel and copper.
d Particularly tin-bronze and tin-lead-bronze.

FIGURE 7-4
Processing methods used most frequently with different materials. (*From "Materials Selector,"* Materials Engineering Magazine, *Penton/IPC, Cleveland.*)

TABLE 7-1
Factors that influence the selection of a manufacturing process

Cost of manufacture	Tolerances
Quantity of pieces required	Tooling, jigs, and fixtures
Material	Gages
Geometric shape	Available equipment
Surface finish	Delivery date

Assembly processes

1. G. Boothroyd, C. Poli and L. E. Murch," Automatic Assembly," Marcel Dekker Inc., New York, 1982.
2. E. K. Henriksen, "Jig and Fixture Design Manual," Industrial Press, Inc., New york, 1973.
3. E. Hoffman, "Fundamentals of Tool Design," 2d ed., Soc. of Mfg. Engrs, Dearborn, MI, 1984.

Selecting the best manufacturing process is not an easy task. Rarely can a product be made by only one method, so that several competitive processes generally are available. As in all of engineering design, cost is a very important factor. Therefore, the selection of the optimum process can be made only after the costs of manufacture by the competing processes have been evaluated (see Chap. 9). The evaluations should consider not only the cost of processing the material to a finished product but also the material utilization factor and the effect of the processing method on the material properties and the subsequent performance of the part in service.

Broadly, the selection of the material determines the processing method (Fig. 7-4). Other factors that are important in influencing the selection of the manufacturing process are listed in Table 7-1. These factors are discussed in Sec. 7-4 in relation to the cost of manufacturing.

7-3 DESIGN FOR MANUFACTURE

With increased awareness of the importance of the interaction between design and manufacture a new field aimed at formalizing this relationship is evolving. This is called Design for Manufacture (DFM). The DFM approach examines the product design in all aspects for ways of integrating the product and process concepts so that the best match is made between product and process requirements and that the integrated product/process ensures inherent ease of manufacture. As part of the DFM process the product design is examined to ensure that it is in conformance with the specific requirements of the manufacturing process. Some of these specific design requirements are described in subsequent sections under Design for Casting, Design for Forging, etc. More specific design details are given in the references previously cited and in the book by Trucks.[1]

The major objective of DFM is to ensure that the product (including material selection) and the process are designed together. Therefore, the engineering release package will contain not only part drawings, a part list, and assembly drawings, *but* it also will contain the process plan. This is best achieved by practicing concurrent engineering (Sec. 1-12)

DFM Guidelines

DFM guidelines are statements of good design practice that have been empirically derived from years of experience.[2] Using these guidelines helps narrow the range of possibilities so that the mass of detail that must be considered is within the capability of the designer.

1. MINIMIZE TOTAL NUMBER OF PARTS. Eliminating parts results in great savings. A part that is eliminated costs nothing to make, assemble, move, store, clean, inspect, rework, or service. A part is a good candidate for elimination if there is no need for relative motion, no need for subsequent adjustment between parts, and no need for materials to be different. However, part reduction should not go too far so that it adds cost because the remaining parts become too heavy or complex.

The best way to eliminate parts is to make minimum part count a functional requirement of the design at the conceptual stage of design. Combining two or more parts into an integral design is another approach. Plastic parts are particularly well suited for integral design.[3]

As an example of this principle, a cash register manufacturer designed a new electronic cash register that has 85% fewer parts than the model it replaced, and takes only 25% as much time to assemble. A chief factor was the elimination of all screws by snap-fastened plastic parts. Heavy use was made of CAD and through concurrent engineering the product was introduced in the market just 24 months after development began.

2. DEVELOP A MODULAR DESIGN. A module is a self-contained component with a standard interface with other components in the system. Interchangeable lenses for a camera or plug-in modules for electronic

[1] H. E. Trucks, "Designing for Economical Production," 2d ed., Soc. of Mfg. Engrs, Dearborn, MI, 1987.

[2] H. W. Stoll, *Appl. Mech. Rev.*, vol. 39, no. 9, pp. 1356–1364, 1986.

[3] W. Chow, "Cost Reduction in Product Design," Chap. 5, Van Nostrand Reinhold Co., New York, 1978.

instruments are common examples. Modular design offers the opportunity to standardize by allowing a product to be customized by using different combinations of standard modules. Thus, a modular design is relatively resistant to obsolescence, since a new generation product can utilize most of the old modules. Modular design results in easier service and repair because the defective module can be replaced by a new one. Modular design simplifies final assembly because there are fewer parts to assemble. Products consisting of 4 to 8 modules with 4 to 12 parts per module are preferred for automation assembly. The chief disadvantage of modular design may be cost, because of the extra fittings and interconnections required.

3. MINIMIZE PART VARIATIONS. The risk of quality problems are reduced when part variations (such as the types of screws used) are kept to a minimum. There are few reasons to justify the use of several screw sizes or types of metal in a single part. Reducing the part variation also minimizes the information content required by the production system.

A common way to minimize part variation is to use standard (off the shelf) components. A stock item is always less expensive than a custom-made item. An extension of this concept is to standardize and minimize the number of part numbers supported for common parts and components.

4. DESIGN PARTS TO BE MULTIFUNCTIONAL. A good way to minimize part count is to design such that parts can fulfill more than one function. For example, a part might serve both as a structural member and a spring. The part might be designed to provide a guiding, aligning or self-fixturing feature in assembly.

5. DESIGN PARTS FOR MULTIUSE. It is good sense to use parts in more than one product. For example, the same gear can be used in different products. Design of parts for multiuse provides economy of scale and reduces the cost per piece. A design philosophy built around this guideline attempts to minimize the number of part categories and the number of variations within each category. The group technology approach discussed in Sec. 7-4 is applicable here.

6. DESIGN PARTS FOR EASE OF FABRICATION. Most of this chapter addresses this guideline. As discussed in Chap. 6, the least costly material that satisfies the functional requirements should be chosen. In determining this cost, the cost of manufacture must be considered. However, in many products like automobiles 50 to 60 percent of the total cost is attributable to materials. Since machining processes tend to be costly, manufacturing processes that produce the part to near net shape are preferred whenever possible. Secondary processes such as finish machining and painting should be avoided whenever possible.

7. AVOID SEPARATE FASTENERS The use of screws in assembly is expensive. Snap fits should be used whenever possible. Where screws must be used, quality risks can be reduced by minimizing the number, size, and variations of fasteners and by using standard fasteners.

8. MINIMIZE ASSEMBLY DIRECTION. All parts should be designed so that they can be assembled from one direction. The need to rotate in assembly requires extra time and motion, and may require additional transfer stations and fixtures. The best situation in assembly is when parts are added in a top-down manner to create a *z-axis stack*.

9. MAXIMIZE COMPLIANCE IN ASSEMBLY. Excessive assembly force may be required when parts are not identical or perfectly made. Allowance for this should be made in the product design. Designed-in compliance features include the use of generous tapers, chamfers, and radii. If possible, one of the components of the product can be designed as the part to which other parts are added (part base) and as the assembly fixture. This may require design features which are not needed for the product function.

10. MINIMIZE HANDLING IN ASSEMBLY Parts should be designed to make the required position easy to achieve. Since the number of positions required in assembly equates to increased equipment expense and greater risk of defects, quality parts should be made as symmetrical as function will allow. Orientation can be assisted by design features which help to guide and locate parts in the proper position. Parts that are to be handled by robots should have a flat, smooth top surface for vacuum grippers, or an inner hole for spearing, or a cylindrical outer surface for gripper pickup.

An example of the application of many of these guidelines is the new engine plant announced by Ford Motor Co. The use of flexible manufacturing equipment and modular design will permit production of more than a dozen engine sizes and configuration on a single line. Although most of the engines will share about 75 percent of their parts, important components such as cylinder heads and connecting rods will be specific to an engine. By using DFM concepts the average number of parts in an engine was reduced by 25%. It is estimated that a new V-8 engine design can be brought on line for about $60 million compared with as much as $500 million using older design concepts.

Specific Design Rules

A number of rules for design, more specific than those given in the previous section have been developed.[1]

---

[1] J. G. Bralla (ed.), "Handbook of Product Design for Manufacture," McGraw-Hill Book Co., New York, 1986.

1. Space holes in machined, cast, molded or stamped parts so they can be made in one operation without tooling weakness. This means that there is a limit on how close holes may be spaced due to strength in the thin section between holes.

2. Avoid generalized statements on drawings, like "polish this surface" or "toolmarks not permitted" which are difficult for manufacturing personnel to interpret. Notes on drawings must be specific and unambiguous.

3. Dimensions should be made from specific surfaces or points on the part, not from points in space. This greatly facilitates the making of gages and fixtures.

4. Dimensions should all be from a single datum line rather than from a variety of points to avoid overlap of tolerances.

5. The design should aim for minimum weight consistent with strength and stiffness requirements. While material costs are minimized by this criterion, there also will usually be a reduction in labor and tooling costs.

6. Whenever possible, design to use general purpose tooling rather than special dies, form cutters, etc. An exception is high volume production where special tooling may be more cost effective.

7. Use generous fillets and radii on castings, molded, formed, and machined parts.

8. Parts should be designed so that as many operations as possible can be performed without requiring positioning. This promotes accuracy, and minimizes handling.

Computer-based DFM

For DFM to become widely adopted in a concurrent engineering mode will require the development of effective computer-based design tools. Product designers must usually work with very tight time schedules so that they are reluctant to become involved with DFM analyses. Readily available computer design tools can overcome that problem. Programs that assist the engineer in the design of a part to be produced by a particular manufacturing process have been developed. A good example is Moldflow, a computer simulation of molten polymer running through the gates, runners, and cavity of an injection molding machine. Other programs have been developed for the design of metal castings, for the design of forging dies in three-dimensional plastic deformation, and for the design of sheet metal forming in two dimensions. Some of the commercial packages for finite element analysis are including these types of manufacturing simulation and design.

Another software package known as VSAS (variation simulation analysis software) allows the designer to predict assembly tolerance and manufacturing variation before the prototype is built. The model includes the size and variation of each component and the accuracy of each assembly operation. A Monte Carlo simulation is used to simulate a production run by putting the component together one part at a time. A complete statistical picture of the assembly operation results, including the distribution of critical dimensions, high and low limits, and percentage of out of specification parts. The VSAS software allows the designer to perform "what-if" optimization to investigate the influence of various tolerances and datums.

Perhaps the computer-based design tools closest to the needs of the designer is the DFM software produced by Boothroyd Dewhurst Inc.[1] This was originally based on an analysis for assembly.[2] It has been extended to analysis of injection molded plastic parts, printed circuit board assembly, and will ultimately include machining, forging, sheet forming, and powder metallurgy processing. This software helps the designer both design the part for the particular manufacturing process and at the same time it allows the designer to estimate the cost of producing the part. Thus "what-if" scenarios can be performed at the earliest stage of design where sketches of design concepts are being made and before detail drawings are available. Using the injection molding software the designer selects an appropriate injection molding machine from the database.[3] A materials database containing data on engineering thermoplastics is available. Some of the features of the software are as follows: (a) evaluation of part complexity; (b) calculation of part volume and area; (c) estimation of mold costs based on part size, complexity, tolerances, and appearance factors; (d) determination of filling, packing, cooling, and ejection times; (e) determination of the optimum number of mold cavities for minimum cost; (f) capabilities for instant cost recalculation when changes are made in either the material, part geometry, or production volume; and (g) an interactive cost database to deal with operational variables like machine hourly rates, machine loading limits, etc. Some very spectacular savings have been reported from using this DFM software. In one case the assembly steps were reduced by 78% and the number of parts by 75%.

Another approach toward computer-based DFM is the application of the methods of artificial intelligence. There is considerable activity in capturing empirical knowledge about manufacturing processes in the form of expert systems (Sec. 3-13). Another active area is computer-aided process planning.[4]

Rapid Prototyping

A prototype is the first edition of a design in working form. In effect it is the first stage of manufacture. The great advances in computer graphics and computer modeling have led to a desire to couple into a way of producing

---

[1] *Business Week,* October 2, 1989, pp. 106, 110.

[2] G. Boothroyd and P. Dewhurst, *Machine Design,* vol. 58(4), pp. 72–76, 1984.

[3] G. Boothroyd and P. Dewhurst, *Manf. Engr.,* April 1988, pp. 42–46.

[4] L. Alting and H. Zhang, *Int. J. Prod. Res.,* vol. 27, no. 4, pp. 553–585, 1989.

quickly the computer model as a three-dimensional prototype. S
innovative techniques are under development. They eventually may lead to
"desktop manufacturing", the three-dimensional analog of desktop printing.

One method, called stereolithography, can turn a computer generated
design into a three-dimensional plastic model within hours. With this method a
computer slices the design into cross sections and directs a laser to trace each
section on to the surface of a vat of liquid polymer. The laser radiation causes
the polymer to harden, producing a thin slice of the model. After each pass of
the laser that section drops slightly below the surface so the next layer can be
built on top of it. The second method joins discrete metal powder particles into
a solid body by selective laser sintering.

7-4 ECONOMICS OF MANUFACTURING

For maximum economy the selection of the material for a production part
should be based on processing factors as well as the engineering requirements
of service.[1] With cost of the finished part as the overall criterion, the factors in
selecting a material for production are as listed in Table 7-2.

Generally, the lower the alloy content the lower the cost. Therefore, the
cheapest alloy consistent with required properties is selected. The importance
of using standard alloys whenever possible and avoiding the selection of many
different alloys, so as to reduce inventory costs, have been discussed in
Chap. 6.

When parts have very simple geometric shapes, as straight shafts and
bolts have, the form in which the material is obtained and the method of
manufacture are readily apparent. However, as the part becomes more
complex in shape, it becomes possible to make it from several forms of

TABLE 7-2
Factors in selecting a material for production

Alloy: grade or composition
Form of material: bar, wire, tube, plate, strip, etc.
Size: dimensions and tolerances
Heat-treated condition
Surface finish
Quality level: control of impurities or inclusions
Quantity
Fabricability and/or weldability

---

[1] For detailed examples, see The Selection of Steel for Economy of Manufacture, "Metals Handbook," 8th ed., vol. I, pp. 290–301, American Society for Metals, Metals Park, Ohio, 1961; ibid., 9th ed., vol. III, pp. 838–856, 1980.

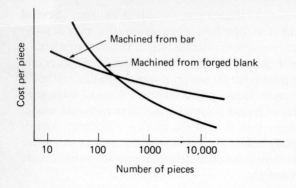

FIGURE 7-5
Comparison of costs with different forms of starting material.

material and by a variety of manufacturing methods. For example, a small gear may be machined from bar stock or, perhaps more economically, from a precision-forged gear blank. The selection of one of several alternatives is based on overall cost of a finished part (see Chap. 9 for details of cost evaluation). Generally, the production quantity is an important factor in cost comparisons (Fig. 7-5). There will be a break-even point beyond which it is more economical to invest in precision-forged preforms in order to produce a gear with a lower unit cost. As the production quantity increases, it becomes easier economically to justify a larger initial investment in tooling or special machinery to lower the unit cost.

The workpiece should be just enough larger than the final part to allow for removal of surface defects. Thus, extra material in a casting may be necessary to permit machining the surface to the needed finish, or a heat-treated steel part may be made oversize to allow for removal of a decarburized layer. Often the manufacturing process dictates the use of extra material, such as sprues and risers in castings or flash in forgings. At other times extra material must be provided for purposes of handling, positioning, or testing the part. Even though extra material removal is costly to some extent, it usually is cheaper to purchase a slightly larger workpiece than to pay for a scrapped part.

Steels, aluminum alloys, and many other engineering materials can be purchased from the supplier in a variety of metallurgical conditions other than the annealed (soft) state. Examples are quenched and tempered steel, solution-treated and cold-worked and aged aluminum alloys, and cold-drawn and stress-relieved brass rod. It may be more economical to have the metallurgical strengthening produced in the workpiece by the supplier of the material than to heat-treat each part separately after it has been manufactured.

Cold-worked metals have a better surface finish than hot-worked metals have. Although cold-worked bars and sheets cost more than the hot-worked products, in some applications it may be more economical to start with a workpiece with the necessary surface finish than to clean and finish each individual manufactured part.

Our discussion of steel pricing in Sec. 6-5 showed that special treatment to produce higher-quality material is a cost extra. Therefore, such quality requirements should not be applied without good reasons. Extra testing in the development phase of the design often may be justified if it results in the use of a material that does not require quality extras.

The ease of processing a material has a major influence on the cost of a part. If the scrap rate is high, because of cracks, poor surface finish, or failure to meet dimensions and tolerances, then either the process should be improved or a material that is easier to work should be selected. Unfortunately, there is an inherent conflict, since high-performance materials (high strength) usually are more difficult to fabricate. Difficulty in working or fabricating a material can often be overcome by a change in the manufacturing process. Thus, an alloy that is difficult to work may be more successfully extruded than drawn, or a high-temperature alloy may be more easily produced into a turbine blade by precision investment casting than by forging. Weldability is a critical material selection parameter, especially for steels. When the carbon content of a steel exceeds 0.2 percent, the susceptibility to weld cracking increases; above 0.3 percent carbon preheating and postheating of the weld are required.

Since the world is running out of mineral resources, the price of engineering materials will continually rise in the future. The percentage of the cost of a manufactured part that is due to cost of materials also is rising. Thus, there is strong economic incentive to conserve material through processing. The predominant manufacturing processes are the machining processes, in which a cutting tool generates the necessary shape by removing successive chips of material. Metal-cutting machine tools have been highly refined, and with computer control they have greatly increased productivity. Practically any shape can be produced by machining processes, but the processes themselves consume a great deal of material as machining chips. The degradation of materials by conventional machining methods is of the order of 30 to 70 percent, and the more complex shapes are at the higher end. Most of the chips are recycled as scrap, but there is a severe economic penalty. As a result, there has been increasing emphasis on "chipless machining" processes by which a part is made to final, or near-net, shape. Precision forging, precision investment casting, and powder-processing techniques are good examples of such processes.

In Fig. 7-6 is shown a simple example of how the interaction between the design and the manufacturing process can have a major influence on material utilization. The part is a simple shaft with a central hub that might later have gear teeth machined in it. To a machine it from solid bar stock would require 18 in^3 of workpiece to get a finished volume of 8 in^3 (Fig. 7-6b). A simpler way to design the part would be as a two-piece assembly in which a machined sleeve was pressed over the shaft (Fig. 7-6c). Much less material would be wasted if the hub were hot-upset-forged on the shaft (Fig. 7-6d) or, even better, produced by cold extrusion.

Each manufacturing process has the capability of producing a part to a certain surface finish and tolerance range without incurring extra cost. Figure

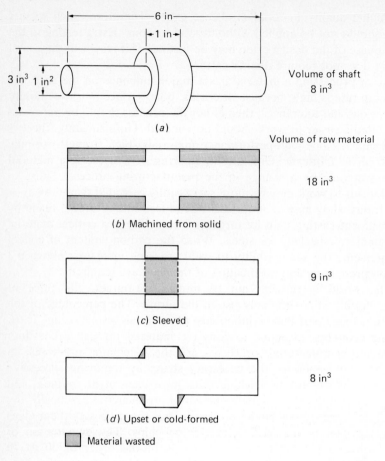

Volume of shaft
8 in³

(a)

Volume of raw material

18 in³

(b) Machined from solid

9 in³

(c) Sleeved

8 in³

(d) Upset or cold-formed

Material wasted

FIGURE 7-6
Interaction of design and manufacturing processes on material utilization.

7-7 shows the general relation between tolerance range and surface roughness. The tolerances apply to a 1-in dimension and are not necessarily scalable to larger or smaller dimensions for all processes. For most economical design, the loosest possible tolerances and coarsest surface finish that will fulfill the function of the design should be specified. Whenever possible, the specified tolerances should be within the range achievable by the manufacturing process that will be used for the part. As Fig. 7-8 shows, processing cost increases nearly exponentially as the requirements for tolerances and surface finish are made more severe.

Different manufacturing processes vary in their limitations on producing complex shapes and parts with minimum dimensions. The classification of shapes (typology) is an active field of research in manufacturing. An important

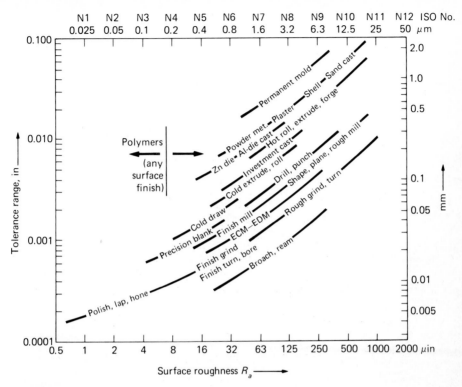

FIGURE 7-7
Approximate values of surface roughness and tolerance on dimensions typically obtained with different manufacturing processes. (*From J. A. Schey, "Introduction to Manufacturing Processes," p. 670, McGraw-Hill Book Company, New York, 1987; used with permission of McGraw-Hill Book Co.*)

application is to *group technology,* which is an approach to increasing the productivity of small batch production. Group technology[1] aims at classifying shapes, because similar shapes tend to be produced by similar processing methods. Once parts belonging to the same group are identified, the production facility can be better organized to produce them. One approach is to organize the machinery to produce parts of a particular group into a *cell,* usually based around a versatile and expensive numerical-control machining center. In this scheme the machine tools, instead of being grouped together,

---

[1] C. C. Gallagher and W. A. Knight, "Group Technology Production Methods in Manufacture," John Wiley & Sons, New York, 1986; R. Wilson and R. Henry, "Introduction to Group Technology in Manufacturing and Engineering," Society of Manufacturing Engineers, Dearborn, Mich., 1980.

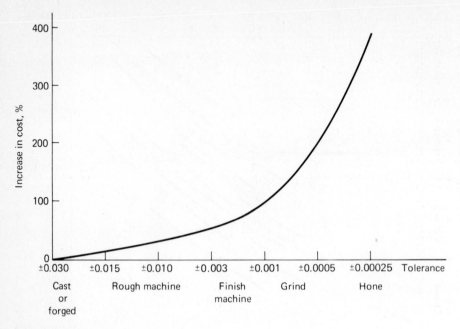

FIGURE 7-8
Influence of tolerance on processing costs.

i.e., all lathes, milling machines, etc., in the same room, are distributed among the cells as needed to produce the class of part shapes assigned to the cell.

Some basic shapes are shown in Fig. 7-9. This is one of many shape classification systems that have been developed. Each shape can be made by a number of different processes. As an example, consider a spool shape, group S3, T2, or F3 in Fig. 7-9. Schey[1] suggests the following alternative processes, which are illustrated in Fig. 7-10: (*a*) machined from a solid bar or tube, (*b*) cast with a horizontal core, (*c*) cast with a vertical core and a ring-shaped insert core, (*d*) forged in the horizontal position followed by machining of the central hole, (*e*) forged in the vertical position with the outer groove machined, (*f*) made from a tube with the flanges welded on. Still other manufacturing processes, not illustrated are:

Forging from a tube by upsetting two flanges
Ring-rolling
Centrifugal casting

---

[1] J. A. Schey, "Introduction to Manufacturing Processes," 2d ed., McGraw-Hill Book Company, New York, p. 862, 1987.

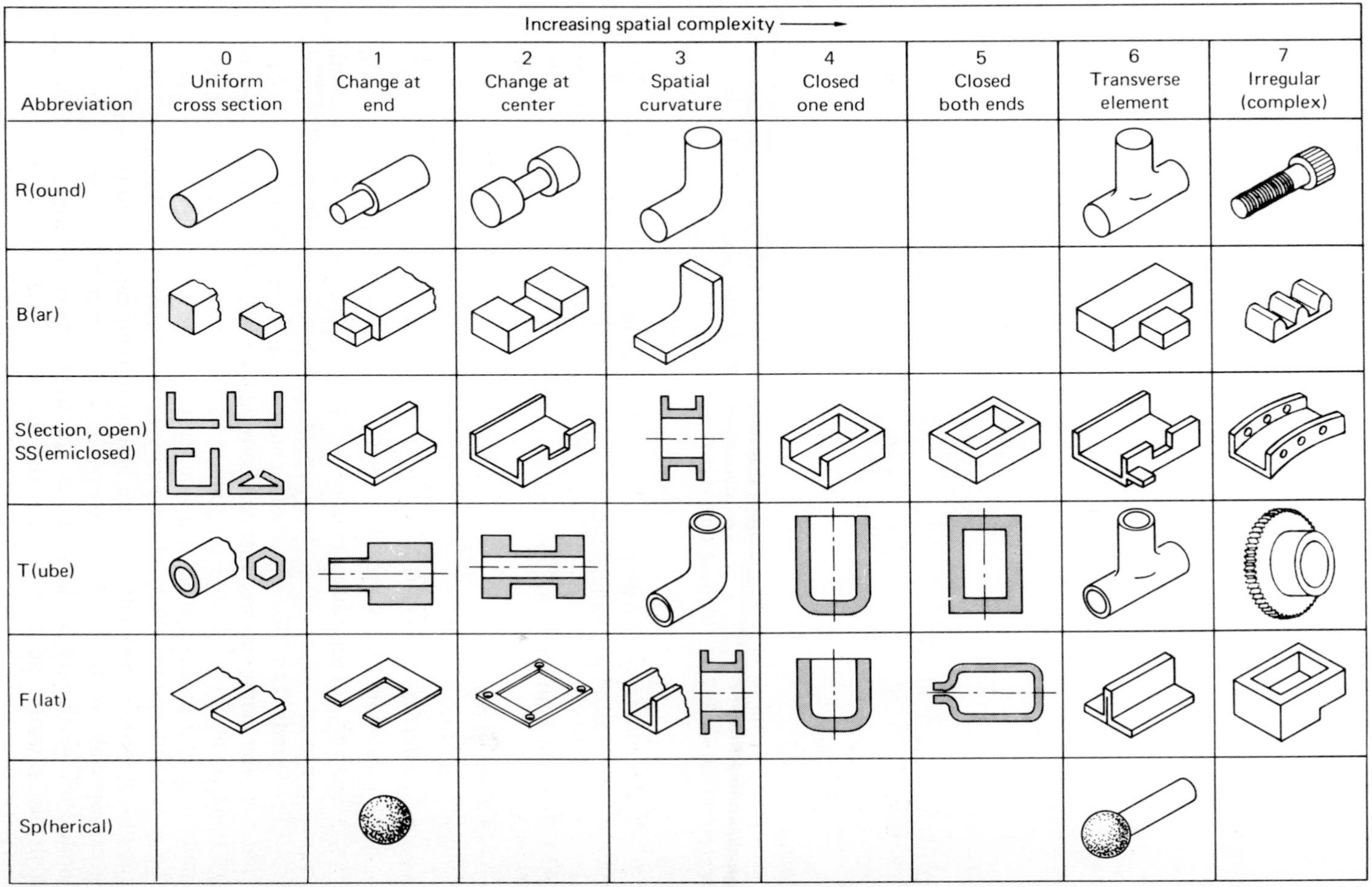
Increasing spatial complexity
Abbreviation
0 Uniform cross section
1 Change at end
2 Change at center
3 Spatial curvature
4 Closed one end
5 Closed both ends
6 Transverse element
7 Irregular (complex)
R(ound)
B(ar)
S(ection, open)
SS(emiclosed)
T(ube)
F(lat)
Sp(herical)

FIGURE 7-9
Some basic shapes in manufacturing. (*From J. A. Schey, "Introduction to Manufacturing Processes," 2d ed. p. 661, McGraw-Hill Book Company, New York, 1987; used with permission of McGraw-Hill Book Company.*)

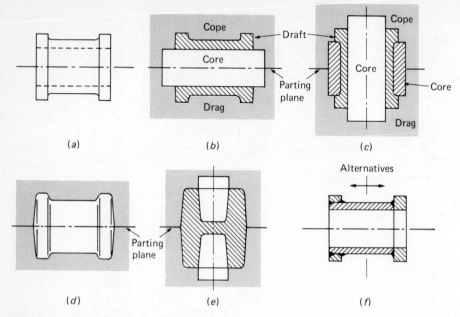

FIGURE 7-10
Some possible methods of making a spool-shaped part. (*From J. A. Schey, "Introduction to Manufacturing Processes," 2d ed., p. 668, McGraw-Hill Book Company, New York, 1987; used with permission of McGraw-Hill Book Company.*)

Powder metallurgy with isostatic compaction in a flexible die or compaction in a multipiece permanent die.

The best manufacturing method would depend on many factors such as:

- Overall part size
- Wall thickness
- Slenderness ratio (L/D)
- Availability of equipment
- Delivery promised by venders

Size limits can be important considerations in selecting a process. Generaly, the maximum size that can be produced by a given process is limited by the size of the available equipment. There may, however, be physical limitations of a process, such as a sand-casting mold that can not stand up to the long solidification times imposed by a very heavy wall thickness. More commonly, processing techniques are limited in their capacity to produce small sizes, especially minimum thicknesses. The wall thickness of a casting may be limited by the fluidity of the metal or the thickness of a forging by the very high die pressures generated when the diameter-to-thickness ratio becomes

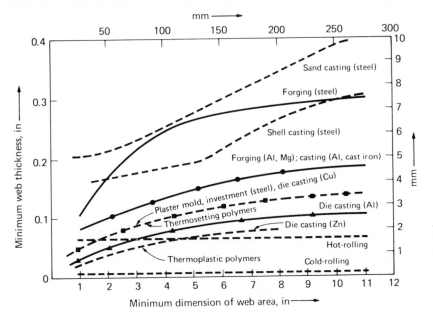

FIGURE 7-11
Minimum thickness normally achieved by different manufacturing processes. (*From J. A. Schey, "Introduction to Manufacturing Processes," 2d. ed., p. 670, McGraw-Hill Book Company, New York, 1987; used with permission of McGraw-Hill Book Company.*)

large. Therefore, very thin, very small, or very large parts usually can be made only under special circumstances and at extra cost. Figure 7-11 gives guidance on the minimum thickness that can normally be achieved with different processes.

The relation between design details and processing is complex and not easily reduced to formulas or simple relationships. Also, there is the tremendous problem of acquiring and processing reliable data on the cost of manufacturing. The U.S. Air Force, as part of its Integrated Computer-Aided Manufacturing Program, is developing a Manufacturing Cost/Design Guide that will be an invaluable first step in the open literature. Because of the obvious complexity and voluminous nature of cost data, it is important to include this type of data base in a computer-aided design system. Figure 7-12 shows in a somewhat simplified way the input and decision points in a computer-aided system for selecting the most economical part design and associated manufacturing process.

7-5 DESIGN FOR CASTINGS

One of the shortest routes from raw material to finished part is casting. In casting, a molten metal is poured into a mold or cavity that ▉proaches the

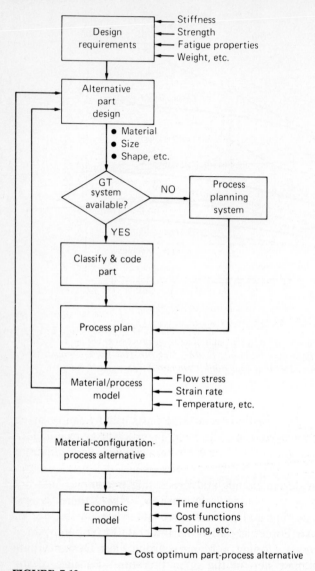

FIGURE 7-12
Flow sheet for a computer-aided process to select the least cost manufacturing process.

shape of the finished part (Fig. 7-13). Heat is extracted through the mold (in this case a sand mold), and the molten metal solidifies into the final solid shape. This seemingly simple process can be quite complex metallurgically, since the metal undergoes a complete excursion from the superheated molten state to the solid state. Solid metal shrinks on solidification. Thus, the casting and mold must be so designed that a supply of molten metal is available to

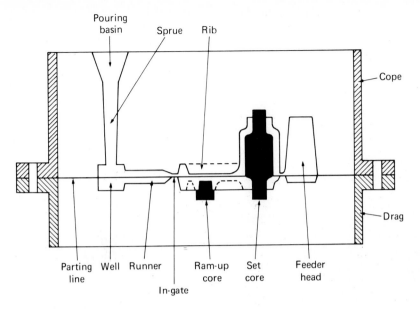

FIGURE 7-13
Parts of a conventional and sand casting process. (*From J. A. Schey, "Introduction to Manufacturing Processes," 2d ed., p. 160, McGraw-Hill Book Company, 1987, used with permission of McGraw-Hill Book Company.*)

compensate for the shrinkage. The supply is furnished by introducing feeder heads (risers) that supply molten metal but must be removed from the final casting (Fig. 7-13). Allowance for shrinkage and thermal contraction must be provided in the design. Also, since the solubility of dissolved gases in the liquid decreases suddenly as the metal solidifies, castings are subject to the formation of gas bubbles and porosity unless proper preventive steps are taken. The grain size and other features of metallurgical structures are influenced by the cooling rate, which in turn is controlled chiefly by the section thickness. Thus, good foundry engineering is needed to produce quality castings. However, by applying modern technology and careful quality control, parts for high-performance service can be produced by casting.

There are a large number of casting processes, which can be classified best with respect to the type of mold that is employed (Fig. 7-14). A brief description of each process is given in Table 7-3. The selection of the proper casting process depends on the following factors.

1. Complexity of the shape:
   a. External and internal shape
   b. Types of core required
   c. Minimum wall thickness

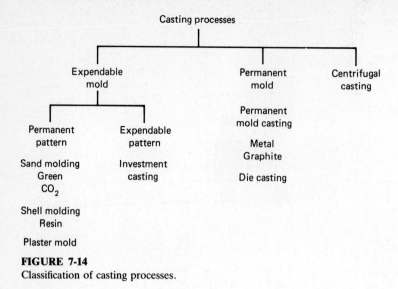

FIGURE 7-14
Classification of casting processes.

2. Cost of the pattern or die
3. Quantity of parts required
4. Tolerances required
5. Surface finish required
6. Strength
7. Weight
8. Overall quality required

In Table 7.4 the various casting processes are compared in respect to their capabilities: size of part, section thickness, and design complexity. Cost aspects of the various processes are the subject of Table 7-5. The dimensional aspects and surface finish produced by each casting process are given in Table 7-6.

Proper attention to design details can minimize casting problems and lead to lower costs.[1] Therefore, close collaboration between the designer and the foundry engineer is important. Some of the more important design considerations for castings are discussed below.

The chief consideration is that the shape of the casting should allow for orderly solidification by which the solidification front progresses from the

---

[1] "Casting Design Handbook," American Society for Metals, 1962; "Metals Handbook," 9th ed.; vol. 15, American Society for Metals, Metals Park, Ohio, 1988.

TABLE 7-3

Advantages and disadvantages of the common casting processes

Form	The process	Advantages	Limitations
Sand castings	Green sand. Moist, bonded sand is packed around a wood or metal pattern, the pattern removed, and molten metal poured into the cavity; when metal solidifies, mold is broken and casting removed.	Almost any metal can be used; almost no limit on size and shape of part; extreme complexity possible; low tool cost; most direct route from pattern to casting	Same machining always necessary; large castings have rough surface finish; close tolerances difficult to achieve; long, thin projections not practical; some alloys develop defects
	Dry sand. Same as above except: core boxes used instead of patterns, sand bonded with a setting binder, and core baked in an oven	Same as above plus ability to handle long, thin projections	Usually limited to smaller parts than possible with green sand
Shell-mold castings	Sand coated with a thermosetting plastic resin is dropped onto a heated metal pattern (which cures resin); shell halves are stripped off and assembled. When poured metal solidifies, shell is broken away from finished casting	Rapid production rate; high dimensional accuracy; smooth surfaces; uniform grain structure; minimized finishing operations	Some metals cannot be cast; requires expensive patterns; equipment, and resin binder; size of part limited
Full-mold castings	Sand casting process using foamed plastic such as polystyrene for pattern and one-piece mold. Pattern is vaporized during casting. One piece or multipiece patterns can be used depending on complexity	Most metals can be cast; almost no limit on shape, size; useful for complex shapes. Plastic patterns easily handled; no draft required; no flash	Pattern costs can be high for low quantities; some limitations imposed by low strength of pattern material
Permanent-mold castings	Mold cavities are machined into metal die blocks designed for repetitive use; molten metal is gravity-fed to cavity (pressure sometimes applied after pouring). Mold consists of two or more parts and is hinged and clamped for easy removal of castings	Good surface finish and grain structure; high dimensional accuracy; repeated use of molds (up to 25,000); rapid production rate; low scrap loss; low porosity	High initial mold costs; shape, size and intricacy limited; high-melting metals such as steel unsuitable

TABLE 7-3 (*Continued*)

Form	The process	Advantages	Limitations
Die castings	Molten metal is poured into closed steel die under pressures varying from 1500 to 25,000 psi; when the metal solidifies, the die is opened and the casting ejected	Extremely smooth surfaces; excellent dimensional accuracy; rapid production rate	High initial die costs; limited to nonferrous metals: size of part limited
Plaster-mold castings	Slurry of special gypsum plaster, water and other ingredients is poured over pattern and allowed to set; pattern is removed and the mold baked. When poured metal cools, mold is broken for removal of casting	High dimensional accuracy; smooth surfaces; almost unlimited intricacy; low porosity	Limited to nonferrous metals; limited to relatively small parts; mold-making time is relatively long
Ceramic-mold castings	Precision technique using stable ceramic powders, binder and gelling agent for mold. Mold can be ceramic or ceramic facing with sand backup	Intricate, close tolerance parts with smooth finishes can be cast	Some limit to maximum size
Investment castings	Refractory slurry is cast around (or dipped on) a pattern formed from wax, plastic, or frozen mercury; when slurry hardens, pattern is melted out and mold is baked. When poured metal solidifies, mold is broken away from casting	High dimensional accuracy; excellent surface finish; almost unlimited intricacy; almost any metal can be used; no flash to remove; no parting line tolerances	Size of part limited; requires expensive patterns and molds; high labor costs
Centrifugal castings	Sand, metal, or graphite mold is rotated in a horizontal or vertical plane (true centrifugal method); molten metal introduced into the revolving mold is thrown to mold wall where it is held by centrifugal force until solidified	Good dimensional accuracy; rapid production rate; good soundness and cleanliness of castings; ability to produce extremely large cylindrical parts	Shape of part limited; spinning equipment expensive

From "Materials Selector," *Materials Engineering Magazine*, Penton/IPC, Cleveland, Ohio.

remotest parts toward the points where molten metal is fed in. Whenever possible, section thickness should be uniform. Large masses of metal lead to hot spots, where freezing is delayed, and a shrinkage cavity is produced when the surrounding metal freezes first.

Figure 7-15 illustrates some design features that can alleviate the shrinkage cavity problem. A transition between two sections of different thicknesses should be made gradually (*a*). As a rule of thumb, the difference in thickness of adjoining sections should not exceed 2 to 1. Wedge-shaped changes in wall thickness should not have a taper exceeding 1 to 4. The thickness of a boss or pad (*b*) should be less than the thickness of the section the boss adjoins, and the transition should be gradual. The local heavy section caused by omitting the outer radius at a corner (*c*) should be eliminated. The radius for good shrinkage control should be from one-half to one-third of the

TABLE 7-4

Comparison of casting processes with regard to size, section thickness, and complexity of design

| Form | Overall size | | Section thickness, in | |
	Max	Min	Max	Min
Sand castings	Green—any weight, dry—5000–6000 lb	1 oz	No limit in floor and pit molds	Al, $\frac{3}{16}$; Cu, $\frac{3}{32}$; Fe, $\frac{3}{32}$; Mg, $\frac{5}{32}$; steel, $\frac{1}{4}-\frac{1}{2}$
Shell-mold castings	Sev. hundred lb; usually <2.5 lb	1 oz	b	$\frac{1}{16}-\frac{3}{4}$
Full-mold castings	40-ton castings have been made	5 lb	No limit	0.1
Permanent-mold castings	1–500 lb 100 lb common in aluminum	Sev. oz	2.0	Iron, $\frac{3}{16}$; Al, $\frac{3}{32}-\frac{1}{8}$; Mg, $\frac{5}{32}$; Cu, $\frac{3}{32}-\frac{5}{16}$
Die castings	Zn, 75 lb; Al, 100 lb; Mg, 45 lb; Cu, 5 lb; Sn, 10 lb; Pb, 15 lb	<1 oz	$\frac{5}{16}$ preferable; usually <0.50	Cu, Mg, 0.05–0.09; Al, 0.03–0.08; Zn, 0.015–0.05; Pb, Sn, 0.03–0.06
Plaster-mold castings	100 lb; usually <15 lb	1 oz	—	0.040–0.060
Ceramic-mold castings	Several tons	3–4 oz	No limit	0.025–0.050
Investment castings	Sev. hundred lb; usually < 10 lb	0.1	3.0	0.025–0.050
Centrifugal castings	Ferrous—50-in diam up to 50 ft length; nonferrous —72-in diam. up to 27-ft length	Ferrous— $1\frac{1}{8}$-in diam., nonferrous —1-in diam.	4.0^c	0.1–0.250^e

TABLE 7-4 (*Continued*)

	Bosses	Undercuts	Inserts	Holes[a]
Sand castings	Yes—small added cost	Yes—small added cost	Yes—small added cost	$\frac{3}{16}-\frac{1}{4}$ in
Shell-mould castings	Yes	Yes	Yes	$\frac{1}{8}-\frac{1}{4}$ in
Full-mold castings	Yes	Yes	Yes	$\frac{1}{4}$ in
Permanent-mold castings	Yes—small added cost	Yes—large added cost, reduced production rate	Yes—no difficulty	$\frac{3}{16}-\frac{1}{4}$ in
Die castings	Yes—small added cost	Yes—large added cost, reduced production rate	Yes—some-what reduced production rate	Cu, $\frac{1}{8}$; Al, Mg, $\frac{3}{32}$; Zn, Sn, Pb, $\frac{1}{32}$ in
Plaster-mold castings	Yes—moderate added cost	Yes—no difficulty		$\frac{1}{2}$ in
Ceramic-mold castings	Yes	Yes	Yes	0.020–0.050 in
Investment castings	Yes—some difficulty	Yes—moderate cost	No	0.020–0.050 in
Centrifugal castings	Possible	No	Yes	Yes

From "Materials Selector," *Materials Engineering Magazine,* Penton/IPC, Cleveland, Ohio.
Note: a) Minimum cored hole diameter; b) avoid section differences where max/min ratio is more than 5:1; c) wall thickness

TABLE 7-5
Cost aspects of various casting processes

Process	Optimum lot size	Tooling costs	Direct labor costs	Finishing costs	Scrap loss
Sand castings	Varies over a wide range	Low	High—much hand labor	High	Moderate
Shell molding	From a few to many, depending on complexity	Low to moderate	Relatively low	Low	Low
Permanent-mold castings	Requires lot size of thousands	Medium	Moderate	Low to moderate	Low
Die casting	Large–1000 to 100,000 pieces	High	Low	Low	Low
Investment casting	Varies widely; adapted to small quantity lots	Low to moderate	High	Low	Low

TABLE 7-6
Dimensional and surface finish aspects of castings

Form	Dimensional tolerances	Draft allowance	Machine finish allowance, in	Surface smoothness, μm rms
Sand castings	Gray iron, $\frac{3}{64}$ in/ft; malleable iron, $\frac{1}{32}$ in/ft; steel, $\frac{1}{16}$ in/ft; Al, Mg, $\frac{1}{32}$ in/ft; Cu, $\frac{3}{32}$ in/ft	1–3°	<table><tr><td></td><td>Iron</td><td>Steel</td><td>Nonfer.</td></tr><tr><td><6 in</td><td>$\frac{3}{32}$</td><td>$\frac{1}{8}$</td><td>$\frac{1}{16}$</td></tr><tr><td>6–12 in</td><td>$\frac{1}{8}$</td><td>$\frac{3}{16}$</td><td>$\frac{1}{16}$</td></tr><tr><td>12–20 in</td><td>$\frac{5}{32}$</td><td>$\frac{1}{4}$</td><td>$\frac{3}{32}$</td></tr><tr><td>20–60 in</td><td>$\frac{3}{16}$</td><td>$\frac{1}{4}$</td><td>$\frac{1}{8}$–4</td></tr></table>	100–1000
Shell-mold castings	0.005 in/in; as little as 0.003 in total possible	$\frac{1}{4}$–$\frac{1}{2}$°	Often none required	50–150
Full-mold castings	$\frac{1}{32}$ in/ft or less	$\frac{1}{2}$° preferred, none required	$\frac{1}{32}$–$\frac{1}{4}$ in	100–1000
Permanent-mold castings	0.015 in/in for first inch; add 0.001–0.002 for each added inch. May be cut to 0.010 in total	2° min on each side; usual range is 2–3°. In recesses, 5° desirable. Standard allowances 0.015–0.020 in/in	$\frac{1}{32}$ for parts up to 4 in; $\frac{1}{16}$ for parts greater than 4 in	100–250
Die castings	0.001–0.007 in first inch; 0.001–0.002 in each additional inch	2° min on each side; usual allowances (in) are Al, Mg, Pb, Sn, 0.010–0.015; Zn, 0.005–0.007; brass, bronze, 0.015–0.020	$\frac{1}{32}$–$\frac{1}{64}$	40–100
Plaster-mold castings	0.005 in/in for first inch; add 0.002 for each additional inch. Common total: 0.005–0.010 in	$\frac{1}{2}$–1°	$\frac{1}{32}$	30–50
Ceramic-mold castings	0.003–0.005 in first inch; 0.002 in each additional inch	1° preferred, none required	0.025	75–150
Investment castings	0.003–0.007 in first inch; 0.002 in each additional inch	Often no draft required; $\frac{1}{2}$–1° for single-parting, noncored dies	0.010–0.025	50–125
Centrifugal castings	Pipe OD, in. Tol. in 3–12 0.06 14–24 0.08 30–36 0.10 42–48 0.12 ID tol about 50 percent greater	$\frac{1}{8}$ in/ft	Ferrous, $\frac{3}{32}$–$\frac{1}{8}$ for small castings, ~ $\frac{1}{4}$ for larger castings; nonferrous, $\frac{1}{16}$ for small castings, $\frac{1}{4}$ for large castings	100–500

From "Materials Selector," *Materials Engineering Magazine,* Penton/IPC, Cleveland, Ohio.

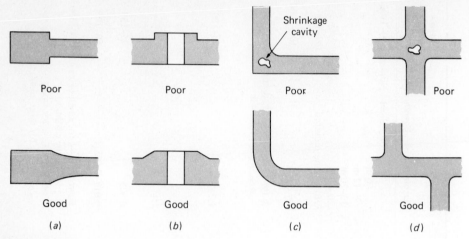

FIGURE 7-15
Some design details to prevent shrinkage cavity.

section thickness. A strong hot spot is produced when two ribs cross each other (*d*); it can be eliminated by offsetting the ribs as shown. A good way to evaluate the hot spot brought about by a large mass of molten metal is to inscribe the largest circle possible in the cross section of the part. The larger the diameter of the circle the greater the thermal mass effect.

Castings must be so designed as to ensure that the pattern can be removed from the mold and the casting from the permanent mold. A draft, or taper, of less than 3° is required on vertical surfaces so the pattern can be removed from the mold. Projecting details or undercuts should be avoided. Allowance for shrinkage and thermal contraction must be provided on the pattern. Molds made with extensive cores cost more money, so castings should be designed to minimize the use of cores. Also, provisions must be made for placing cores in the mold cavity and holding them in place. Locating points for machining are important. When considerable machining is to be done on a casting, a boss should be provided to serve as a reference point for subsequent machining. A finish allowance for machining should be added to the allowances for shrinkage, see Table 7-6.

7-6 DESIGN FOR FORGINGS

Forging processes are among the most important means of producing parts for high-performance applications. Forging is typical of a group of bulk deformation processes in which a solid billet is forced under high pressure to undergo extensive plastic deformation into a final near-to-finished shape. Forging usually is carried out on a hot workpiece, but other deformation processes such as cold extrusion or impact extrusion may be conducted cold, depending upon the material. Because of the extensive plastic deformation that occurs in

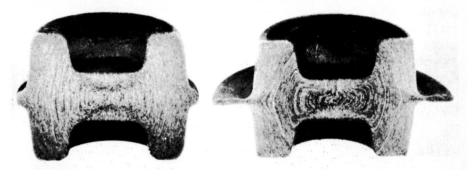

FIGURE 7-16
Flow lines developed in forged web and rib structures.

forging, the metal undergoes metallurgical changes. Any porosity is closed up, and the grain structure and second phases are deformed and elongated in the principal directions of working. As a result of these metallurgical changes, a forging develops a "fiber structure" (Fig. 7-16).

The mechanical fibering due to the preferred alignment of inclusions, voids, segregation, and second-phase particles in the direction of working introduces a directionality to structure-sensitive properties such as ductility, fatigue strength, and fracture toughness. The principal direction of working (such as the long axis of a bar) is defined as the *longitudinal direction*. The *short-transverse direction* is the minimum dimension of the forging, such as the thickness of a platelike shape. The *long-transverse direction* is perpendicular to both the longitudinal and the short-transverse direction. The variation of reduction of area in the tensile test (the most sensitive measure of ductility) with the angle that the specimen axis makes with the forging axis is shown in Fig. 7-17.

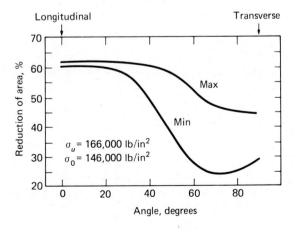

FIGURE 7-17
Relation between reduction of area and orientation within the forging. (*From G. E. Dieter, "Mechanical Metallurgy," McGraw-Hill Book Company, 1976; used with permission of McGraw-Hill Book Company.*)

TABLE 7-7
Advantages and disadvantages of the common forging processes

Form	The process	Advantages	Limitations
Open-die forgings	Compressive forces (produced by hand tools or mechanical hammers) are applied locally to heated metal stock; little or no lateral confinement is involved. Desired shape is achieved by turning and manipulating workpiece between blows	Simple, inexpensive tools; useful for small quantities; wide range of sizes available; good strength characteristics	Limited to simple shapes; difficult to hold close tolerances; machining to final shape necessary; slow production rate; relatively poor utilization of material; high degree of skill required
Closed-die forgings	Compressive forces (produced by a mechanical hammer in a mechanical or hydraulic press) are applied over the entire surface of heated metal stock, forcing metal into a die cavity of desired shape. There are several types of closed-die forgings	Relatively good utilization of material; generally better properties than open-die forgings; good dimensional accuracy; rapid production rate; good reproducibility	High tool cost for small quantities; machining often necessary
	Blocker type. Uses single-impression dies and produces parts with somewhat generalized contours	Low tool costs; high production rates	Machining to final shape necessary; thick webs and large fillets necessary
	Conventional type. Uses preblocked workpiece and multiple-impression dies	Requires much less machining than blocker type; rapid production rates; good utilization of material	Somewhat higher tool cost than blocker type
	Precision type. Uses minimum draft (often 0°)	Close tolerances; machining often unnecessary; excellent material utilization; very thin webs and flanges possible	Requires intricate tooling and elaborate provision for removing forging from tools
Upset forgings	Heated metal stock is gripped by dies (which also form the impression) and pressed into desired shape	Fair amount of intricacy possible; good dimensional accuracy; rapid production rate	Limited to cylindrical shapes; finish not as good as with other forgings; size of part limited; high die costs
Cold-headed parts	Similar to upset forging except metal is cold. Wire up to about 1 mm diam is fed to die in punch press and positioned with one end protruding; this end mushrooms out under force of punch and is formed between die and punch face	Good surface strength; alloys used are generally tough, ductile and crack resistant; excellent surface finish; no scrap loss; rapid production rate	Head volume and shape limited; internal stresses may be left at critical points; size of part limited

From "Materials Selector," *Materials Engineering Magazine*, Penton/IPC, Cleveland, Ohio.

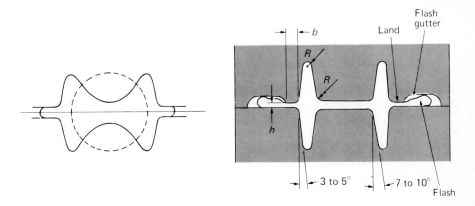

FIGURE 7-18
Schematic of closed-die forging. (*a*) Block die; (*b*) finishing die. (*After Schey.*)

Forgings often are classified as to whether they are made on open, or flat, dies or in closed dies. Open dies are used to impose localized forces for deforming billets progressively into simple shapes, much as the blacksmith does with his hammer and anvil. Closed die forging or impression die forging uses presses or hammers to force the metal to flow into a closed cavity to produce complex shapes to close dimensional tolerances. A wide variety of shapes, sizes, and materials can be utilized in forging. Table 7-7 describes the advantages and disadvantages of the common forging processes. With proper forging die design, grain flow is controlled to give the best properties at the critically stressed regions.

Closed-die forgings rarely are done in a single step. The billet, often a piece of bar stock, must be shaped in blocker dies to place the material properly so it will flow to fill the cavity of the finishing die completely (Fig. 7-18). To ensure complete filling of the die cavity, a slight excess of material is used. It escapes into the flash and is trimmed off from the finished forging.

There are a number of factors that must be considered in the design[1] if the forging is to be made economically and without defects. As with a casting, vertical surfaces of a forging must be tapered to permit removal of the forging from the die cavity. The normal draft angle on external surfaces is 5 to 7°, and for internal surfaces it is 7 to 10°. The maximum flash thickness should not be greater than $\frac{1}{4}$ in nor less than $\frac{1}{32}$ in on average.

---

[1] "Forging Design Handbook," American Society for Metals, Metals Park, Ohio, 1972; "Metals Handbook," ed., vol. 14, 1988; "Forging Industry Handbook," Forging Industry Association, Cleveland, Ohio; A. Thomas, Die Design," Drop Forging Research Association, Sheffield, England, 1980; W. A. Knight and C. Poli; *Machine Design*, Jan. 24 1985; pp. 94–99; C. Poli and W. A. Knight, "Design for Forging Handbook", Univ. of Massachusetts, 1981.

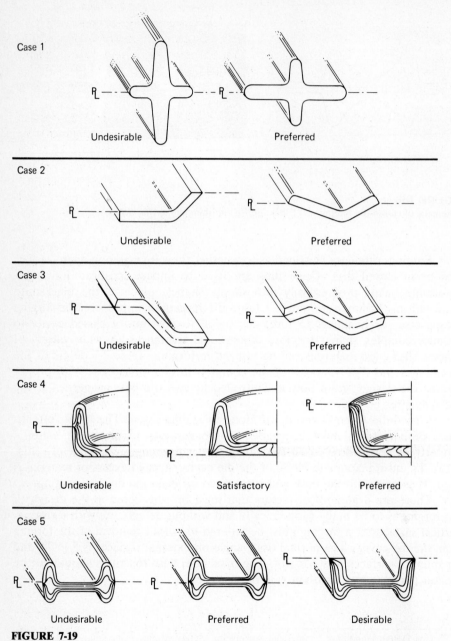

FIGURE 7-19
Examples of desirable and undesirable location of parting line. (*Forging Industry Association Handbook.*)

The parting line, where the die halves meet, is an important design consideration because its location helps to influence grain flow, die costs, and die wear. For optimum economy it should be kept to a single plane if at all possible, since that will make die sinking, forging, and trimming less costly. Because the forging fiber is unavoidably cut through when the flash is trimmed, the parting line is best placed where the minimum stresses arise in the service performance of the forging. Figure 7-19 illustrates a variety of simple shapes and the correct and incorrect parting line locations.

In case 1 the preferred orientation of the part avoids a deep impression that would require high forging pressure for complete filling and might lead to die breakages. Cases 2 and 3 represent nonsymmetrical parts in which the parting line (PL) and the forging plane are no longer coincident. The forging plane always is perpendicular to the direction of ram travel. The preferred locations of the parts in the die orient the parts so that the transverse forces balance out and there is no resultant side thrust on the die. When the PL is not in a single plane, as in cases 2 and 3, die construction is more costly. Sometimes the most economical solution to producing a nonsymmetrical part is to build a die with two cavities in mirror-image positions and in that way balance out side forces without tilting the PL. A good rule in forging die design is to locate the PL near the central height of the part. That avoids deep impressions in either the top or bottom die. However, when parts are dished or hollow, that may not produce the best strength because a centrally located PL interrupts the grain flow, cases 4 and 5. In case 4, the satisfactory location of the PL provides the least expensive design because only the top half of the die requires a machined impression. However, the most desirable grain flow pattern is produced when the parting line is at the top of the dish. In case 5 the location of the PL also is based on grain flow considerations. Placing the PL in the most desirable location often introduces manufacturing problems and is used only when grain flow is an extremely critical factor in design.

Whenever possible in the design of forgings, as in the design of castings, it is desirable to maintain all adjacent sections as uniform as possible. Rapid changes in section thickness should be avoided. Laps and cracks are most likely where metal flow changes because of large differences in the bulk of the sections. To prevent the defects, generous radii must be provided at those locations.

The *machining envelope* is the excess metal that must be removed to bring the forging to the finished size. The ultimate in precision forging is the net-shape forging, in which the machining allowance is zero. Generally, however, allowance must be made for removing surface scale (oxide), correcting for warpage and mismatch (where the upper and lower dies shift parallel to the parting plane), and for dimensional mistakes due to thermal contraction or die wear.

The dimensional tolerances, machining allowances, etc., for closed-die forging of steel are shown in Table 7-8. Cost aspects of various forging processes are considered in Table 7-9.

TABLE 7-8
Dimensional and surface finish aspects of forgings

			Tolerances			
	Thickness, in	Shrinkage and die wear	Draft angle,°	Fillet and corner	Mismatching, in	Machine finish allowance, in
Closed-die forgings	0.2 lb, −0.008, +0.024 1.0 lb, −0.012, +0.036 5.0 lb, −0.019, +0.057 10.0 lb, −0.022, +0.066 50.0 lb, −0.038, +0.114	Shrinkage, 0.003 in/in; die wear; 0.003 in per 2-lb net wt of forging	Drop, press; 7 on outside; 10 on inside; upset, 3 on outside, 5 on inside	0.3 lb, $\frac{3}{32}$ 1.0 lb, $\frac{1}{8}$ 3.0 lb, $\frac{5}{32}$ 10 lb, $\frac{3}{16}$ 30 lb, $\frac{1}{32}$ 100 lb, $\frac{1}{4}$	Up to 1 lb; 0.015 (commercial), 0.010 (close); add 0.002–0.003 for each 6 lb	<3 lb, $\frac{1}{32}$ <40 lb, $\frac{3}{16}$ 100 lb, $\frac{3}{32}$ 200 lb, $\frac{1}{8}$ Allow $\frac{1}{8}$ in on all upset forgings

		Tolerances	
	Shank and shoulder diam, in	Shank length, in	Head diam,, in	Head ht, in
Cold-headed parts	$\frac{1}{16}$–$\frac{3}{16}$ in, 0.002; $\frac{3}{8}$ in, 0.003; $\frac{3}{4}$ in, 0.0045; 1 in, 0.005	Up to 1 in, $\frac{1}{32}$; 1–2 in, $\frac{1}{16}$; 2–6 in, $\frac{3}{2}$; >6 in, $\frac{3}{16}$	$\frac{1}{32}$	0.005

From "Materials Selector," *Materials Engineering Magazine,* Penton/IPC, Cleveland, Ohio.

TABLE 7-9
Cost aspects of forging processes

Process	Optimum lot size	Tooling costs	Direct labor costs	Finishing cost	Scrap loss
Upset forging	Medium to high	High	Low	Medium	Medium
Closed-die forging	Large-lot sizes 10,000 best	High	Medium	Medium	Moderate
Cold heading	Large	Medium	Low	Low	Low

7-7 DESIGN FOR SHEET-METAL FORMING

Sheet metal is widely used for industrial and consumer parts because of its capacity for being bent and formed into intricate shapes. Sheet-metal parts comprise a large fraction of automotive, agricultural machinery, and aircraft components and of consumer appliances. Successful sheet-metal forming depends on the selection of a material with adequate formability, the proper design of the part and the tooling, the surface condition of the sheet, selection and application of lubricants, and the speed of the forming press.

The cold stamping of a strip or sheet of metal with dies can be classified as either a cutting or forming operation.[1] Cutting operations are designed to punch holes or to separate entire parts from sheets by blanking. A blanked shape may be either a finished part or the first stage in a forming operation in which the shape is created by plastic deformation.

The sheared edge that is produced when sheet metal is punched or blanked is neither perfectly smooth nor perpendicular to the sheet surface. Since the die cost depends upon the length and the intricacy of the contour of the blank, simple blank contours should be used whenever possible. It may be less expensive to construct a component from several simple parts than to make an intricate blanked part. Blanks with sharp corners are expensive to produce.

The layout of the blanks on the sheet should be such as to minimize scrap loss. As Fig. 7-20 illustrates, a simple change in design can often greatly improve the material utilization. Notching a blank along one edge results in an unbalanced force that makes it difficult to control dimensions as accurately as with blanking round the entire contour. The usual tolerances on blanked parts are ±0.003 in.

---

[1] For examples see "Metals Handbook," 9th ed., vol. 14, American Society for Metals, Metals Park, Ohio, 1988; G. Sachs, "Principles and Methods of Sheet Metal Fabrication," 2d ed., Reinhold Publishing Co., New York, 1966; F. Strasser, "Functional Design of Metal Stampings," Society of Manufacturing Engineers, Dearborn, Mich., 1971.

Original layout   Improved layout

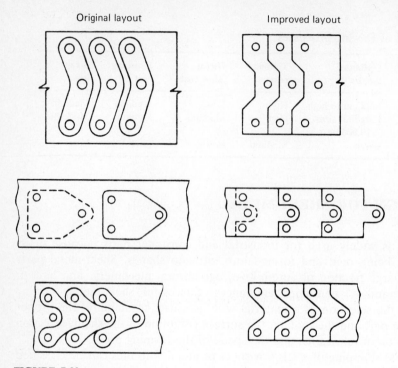

FIGURE 7-20
Changes in design to minimize scrap loss in blanking. (*"Tool and Manufacturing Engineers Handbook," 3d ed.; p. 15–18*).

When holes are punched in metal sheet, only part of the metal thickness is sheared cleanly; that is, a hole with tapered sides is created. If the hole is to be used as a bearing surface, then a subsequent operation will be required to obtain parallel walls. Diameters of punched holes should not be less than the thickness of the sheet or a minimum of 0.025 in. Smaller holes result in excessive punch breakage and should be drilled. The minimum distance between holes, or between a hole and the edge of the sheet, should be at least equal to the sheet thickness. If holes are to be threaded, the sheet thickness must be at least one-half the thread diameter.

Bending is the simplest sheet-forming operation. The greatest formability in bending is obtained when the bend is made across the 'metal grain" (i.e., the line of the bend is perpendicular to the rolling direction of the sheet). The largest possible bend radius should be used, and the bend radius should not be less than the sheet thickness t. The bendability of sheet is usually expressed in multiples of the sheet thickness; thus a $2t$ material has a greater formability than a sheet metal whose minimum bend radius is $4t$. The total length of metal required for bending is the sum of the two legs of the bend plus the bend allowance. The *bend allowance* depends upon how much the metal stretches on bending, which is a function of the angle of bend and the bend radius.

Cost can sometimes be reduced by using sheet metal that is thinner than normal if the strength and rigidity are increased by bending and forming the sheet into ribs, corrugations, and beads.

During forming, the contour of the part matches that of the dies; but upon release of the load, the elastic forces are released. Consequently, the bent material springs back and both the angle of the bend and the bend radius increase. Therefore, to compensate for springback, the metal must be bent to a smaller angle and sharper radius so that when the metal springs back, it is at the desired values. Springback becomes more severe with increasing yield strength and section thickness.

Most sheet forming operations consist of a combination of stretching and deep drawing. In stretching, the limit of deformation is the formation of a localized region of thinning (necking) in the sheet. This behavior is governed by the uniform elongation of the material in a tension test. The greater the capacity of the material to undergo strain hardening the greater its resistance to necking in stretching.

The classic example of deep drawing is the formation of a cup. In deep drawing, the blank is drawn with a punch into a die. The circumference of the blank is decreased when the blank is forced to conform to the smaller diameter of the punch. The resulting circumferential compressive stresses cause the blank to thicken and also to wrinkle unless a sufficient hold-down pressure is applied. However, as the metal is drawn into the die over the die radius, it is bent and then straightened while being subjected to tension. That results in substantial thinning of the sheet in the region between the punch and the die wall. The deformation conditions in deep drawing are substantially different from those in stretching. Success in deep drawing is enhanced by factors that restrict sheet thinning: a die radius about 10 times the sheet thickness, a liberal punch radius, and adequate clearance between the punch and die. Of considerable importance is the crystallographic texture of the sheet. If the texture is such that the slip mechanisms favor deformation in the width direction over slip in the thickness direction of the sheet, then deep drawing is facilitated. This property of the material can be measured in tension test on the sheet from the *plastic strain ratio r*.

$$r = \frac{\text{strain in width direction}}{\text{strain in thickness direction}} = \frac{\varepsilon_w}{\varepsilon_t} \tag{7-1}$$

The best deep-drawing sheet steels have an *r* of about 2.0.

An important tool in developing sheet-forming operations is the Keeler-Goodman forming limit diagram (Fig. 7.21). It is experimentally determined for each sheet material by placing a grid of circles on the sheet before deformation. When the sheet is deformed, the circles distort into ellipses. The major and minor axes of an ellipse represent the two principal strain directions in the stamping. The strains are measured at points of failure for different

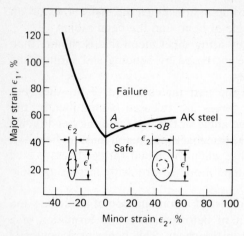

FIGURE 7-21
Keeler-Goodwin forming limit diagram.

stampings with different geometries to fill out the diagram. Strain states above the curve cause failure, and those below do not cause failure. The tension-tension quadrant is essentially stretching, whereas the tension-compression quadrant is closer to deep drawing. As an example of how to use the diagram, suppose point A represents the critical strains in a particular stamping. This failure could be eliminated by moving the strain state to B by increasing the die radius. Alternatively, a material of greater formability could be substituted.

Computer simulation and analysis has greatly reduced the time to design and develop a manufacturing process for a sheet metal part. A computer program called "Sheets", developed by the General Electric R&D Center, generates a finite element analysis and compares the calculated strain distributions with the forming limit diagram of the sheet material. By changing the input constraints of punch speed, punch geometry, and holddown pressure it is possible to quickly optimize the process without building any expensive dies.

7-8 DESIGN FOR MACHINING

Machining operations represent the most versatile and most common manufacturing processes. Practically every part is subjected to some kind of machining operation in its final finishing stages of manufacture. Parts that are machined may have started out as castings or forgings, or they may be machined completely from bar stock or plate.

There is a wide variety of machining processes with which the design engineer should be familiar.[1] We can break them down into three broad

---

[1] For many examples see "Metals Handbook," 9th ed., vol. 16, "American Society for Metals, Metals Park, Ohio, 1989; E. P. DeGarmo, "Materials and Processes in Manufacturing," 5th ed., Macmillan Publishing Co., New York, 1979. For information on specific machine tools see "Machine Tool Specifications," vols. 1 to 4, Society of Manufacturing Engineers, Dearborn, Mich., 1980.

classes: metal-cutting processes, grinding processes, and unconventional or electrical effects. Conventional machining processes, which represent the greatest concentration of effort, can be categorized by whether the tool translates or rotates or is stationary while the workpiece rotates. The classification of machining processes based on this system is shown in Fig. 7-22. The operations and machines that can be used to generate flat surfaces are shown in Fig. 7-23. The operations and machines for machining external cylindrical surfaces are shown in Fig. 7-24, and the common operations and machines for machining internal cylindrical surfaces are shown in Fig. 7-25.

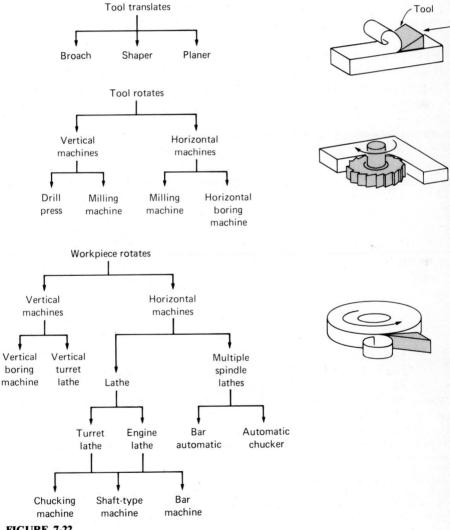

FIGURE 7-22
Classification of metal-cutting processes.

Operation	Block diagram	Most commonly used machines	Machines less frequently used	Machines seldom used
Shaping	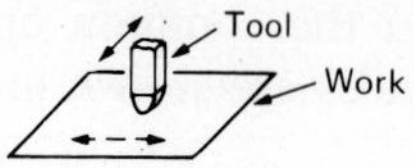 Tool / Work	Horizontal shaper	Vertical shaper	
Planing	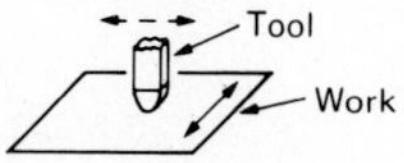 Tool / Work	Planer		
Milling	Slab milling  — Tool / Work; Face milling — Work / Tool 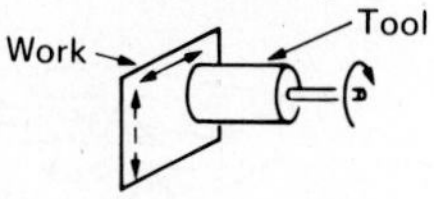	Milling machine		Lathe (with special attachment)
Facing	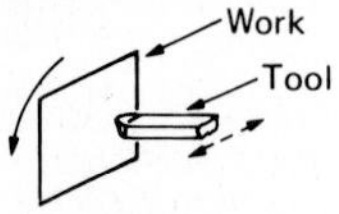 Work / Tool	Lathe	Boring mill	
Broaching	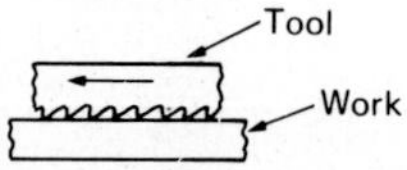 Tool / Work	Broaching machine		
Grinding	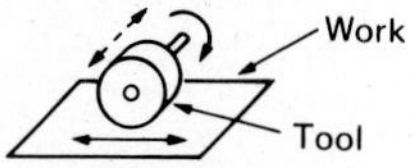 Work / Tool	Surface grinder		Lathe (with special attachment)
Sawing	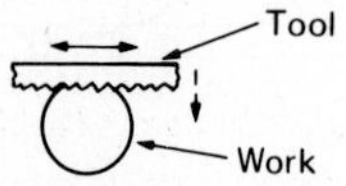 Tool / Work	Cutoff saw	Contour saw	

→ Tool and work motion
- - → Feed only

FIGURE 7-23
Operations and machines for machining flat surfaces. (*From E. P. DeGarmo, "Materials and Processing in Manufacturing," 5th ed., Macmillan Publishing Co., New York, 1979; copyright 1979 by Darvic Associates, Inc.*)

Operation	Block diagram	Most commonly used machines	Machines less frequently used	Machines seldom used
Turning	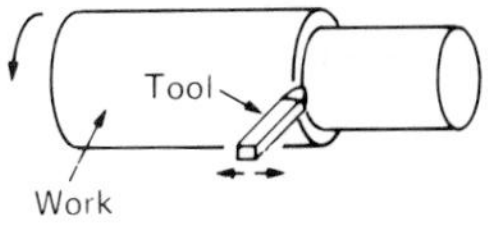 Tool Work	Lathe	Boring mill	Vertical shaper Milling machine
Grinding	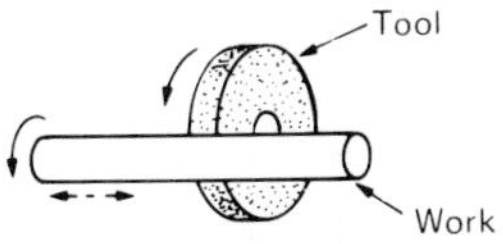 Tool Work	Cylindrical grinder		Lathe (with special attach-ment)
Sawing	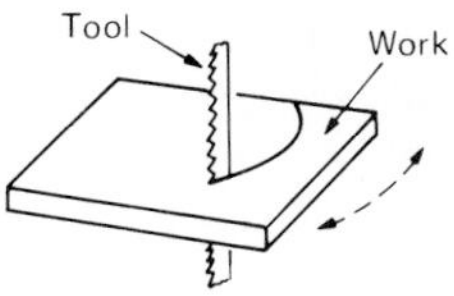 Tool Work	Contour saw		

FIGURE 7-24
Operations and machines for machining external cylindrical surfaces. (*From E. P. DeGarmo "Materials and Processes in Manufacturing", 5th ed., Macmillan Publishing Co., New York, 1979; copyright 1979 by Darvic Associates, Inc.*)

Most metals and alloys can be machined, but they vary a great deal in their ease of machining or machinability. *Machinability* is a complex technological property that is difficult to define precisely. It does, however, involve the ease of chip formation, the ability to achieve a good surface finish, and the achievement of an economical tool life. It usually is evaluated by using a standard machining operation and comparing the tool life at a given surface finish with that for a reference material.

Because machining operations often are finishing operations, a great deal of attention is given to machining tolerances and surface finish. The general relation between surface roughness and minimum dimensional tolerance is given in Fig. 7-26. The relative increase in cost associated with tighter tolerances also is shown. As a general rule, the surface finish should be no finer than is required by the function of the part. The following are examples of various levels of surface finish.

4-μin rms. This is mirrorlike surface that is free from visible marks of any kind. It is used typically in high-quality bearings. It is a very high cost finish that is produced by processes such as lapping and honing or precision grinding.

8-μin rms. This is a scratch-free, close-tolerance finish that would be used for

Operation	Block diagram	Most commonly used machines	Machines less frequently used	Machines seldom used
Drilling	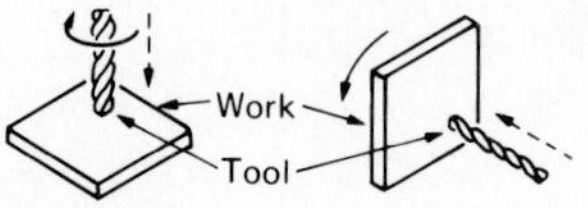	Drill press	Lathe	Milling machine Boring mill Horizontal boring machine
Boring	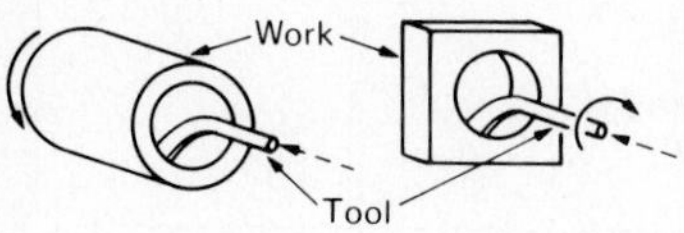	Lathe Boring mill Horizontal boring machine		Milling machine Drill press
Reaming	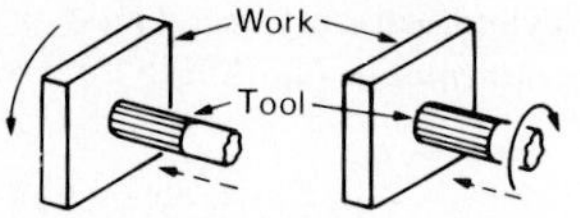	Lathe Drill press Boring mill Horizontal boring machine	Milling machine	
Grinding	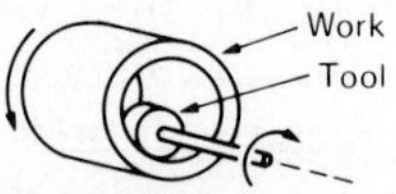	Cylindrical grinder		Lathe (with special attachment)
Sawing	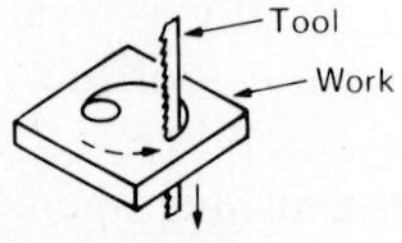	Contour saw		
Broaching	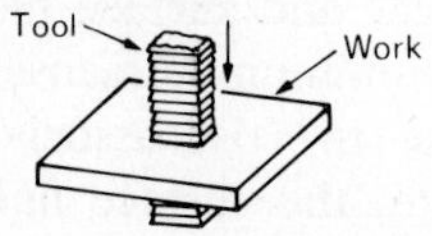	Broaching machine		

FIGURE 7-25
Operations and machines for machining internal cylindrical surfaces. (*From E. P. DeGarmo, "Materials and Processes in Manufacturing," 5th ed., Macmillan Publishing Co., New York, 1979; copyright 1979 by Darvic Associates, Inc.*)

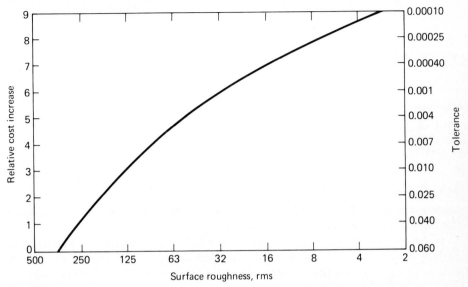

FIGURE 7-26
Relation between dimensional tolerance and surface finish. (*From H. E. Trucks, "Designing for Economical Production," Society of Manufacturing Engineers, Dearborn, Mich., 1974, p. 10.*)

parts like the ID of hydraulic cylinders, pistons, journal bearings, and cam faces. It is produced by diamond turning and boring, or precision grinding.

16-μin rms. This finish is used where the surface finish is of primary importance in the functioning of the part: rapidly rotating shaft bearings, heavily loaded bearings, hydraulic applications, and static sealing rings. It also requires diamond tools or precision grinding.

32-μin rms. This is a fine machine finish produced by careful turning, milling, and drilling operations followed by grinding or broaching. It normally is found in parts subjected to stress concentration and vibration, such as gear teeth and brake drums.

63-μin rms. This is a high-quality, smooth machine finish, as smooth a finish as can be economically produced by turning or milling without subsequent operations. It is suitable for ordinary bearings and ordinary machine parts with fairly close tolerances but where fatigue is not a likely problem.

125-μin rms. This surface finish results from high-quality machining when light cuts, fine feeds, and sharp tools are used in the finishing pass. It can be used for bearing surfaces with light loads and moderately stressed parts, but it should not be used for sliding surfaces.

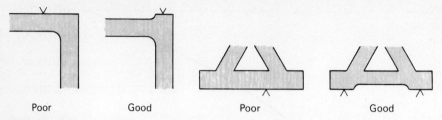

FIGURE 7-27
Examples of design details that minimize the area of the machined surface.

250-μin rms. This surface finish results from ordinary machining operations using medium feeds. It is not objectionable in appearance, and it can be used for the surface of noncritical components.

An important factor for economy in machining is to specify a machined surface only when it is needed for the functioning of the part. Two design examples for reducing the amount of machined area are shown in Fig. 7-27.

In designing a part, the sequence by which the part would be machined must be kept in mind so the design details that make machining easy are incorporated. The workpiece must have a reference surface that is suitable for holding it on the machine tool or in a fixture. A surface with three-point support is better than a large flat surface because the workpiece is then less likely to rock. Sometimes a supporting foot or tab must be added to the rough casting for support purposes, and be removed from the final machined part. When possible, the design should permit all the machining to be done without reclamping the workpiece (Fig. 7-28a). If the part needs to be clamped in a second, different position, one of the already machined surfaces should be used as the reference surface. Whenever possible, the design should be such that existing tools can be used in production. When possible, the radius of the part should be the same as the radius of the tool. Also when possible, design

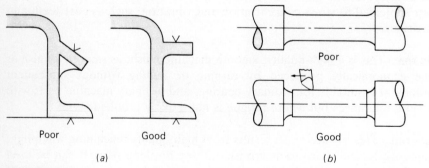

FIGURE 7-28
Design details for improved machining.

should permit the use of the larger tools, which are stronger and can cut at higher speeds. Remember that a cutting tool often requires a runout space because the tool cannot be retracted instantaneously (Fig. 7-28b).

In drilling, the cost of a hole increases proportionately with depth; but when the depth exceeds three times the diameter, the cost increases more rapidly. When a drill is cutting, it should meet equal resistance on all cutting edges. It will if the entry and exit surfaces it encounters are perpendicular to its axis. Holes should not be placed too near the edge of the workpiece. If the workpiece material is weak and brittle, like cast iron, it will break away. Steel, on the other hand, will deflect at the thin section and will spring back afterward to produce a hole that is out of round.

Some additional design guidelines for machining processes are given below.[1] This reference also describes a procedure for the early cost estimation of machined parts.

1. Try to design the part so that it can be machined on one machine tool only.

2. Try to design the part so that machining is not needed on the unexposed surfaces of the workpiece when the part is gripped in the work-holding device.

3. Avoid specifying machined features that the company shop is not equipped to carry out.

4. Design the part so that the workpiece, when gripped in the work-holding device, is sufficiently rigid to withstand the machining forces.

5. Make sure that when the part is machined the tool, tool holder, workpiece, and work-holding device do not interfere with one another.

6. Make sure that auxiliary holes or main bores are cylindrical and have L/D ratios that make it possible to machine them with standard drills or boring bars. For dimensions of standard tools see "Machinery's Handbook".[2]

7. Make sure that auxiliary holes are parallel or normal to the axis of the workpiece and related by a logical drilling pattern.

8. Make sure that the ends of blind holes are conical and, for the case of a tapped blind hole, that the thread does not continue to the bottom of the hole.

FOR COMPONENTS WITH ROTATIONAL SYMMETRY

9. Try to make sure that cylindrical surfaces are concentric and plane surfaces are normal to the part axis.

[1] G. Boothroyd and W. A. Knight, "Fundamentals of Machining and Machine Tools," 2d ed, Chap. 13, Marcel Dekker Inc., New York, 1989.

[2] E. Oberg, F. D. Jones, and H. L. Horton, "Machinery's Handbook," 23d ed., Industrial Press, New York, 1988.

10. For internal corners on the part, specify radii equal to the radius of the rounded tool corner.

11. Avoid internal features in long parts.

12. Avoid parts with very large or very small L/D ratios.

FOR COMPONENTS WITH NONROTATIONAL SYMMETRY

13. Design in a base for work holding and reference.

14. If possible, make sure that the exposed surfaces of the part consist of a series of mutually perpendicular plane surfaces parallel to and normal to the base.

15. Make sure that internal corners normal to the base have a radius equal to the tool radius and that internal corners in machined pockets have as large a radius as possible.

16. If possible, restrict plane surface machining (slots, grooves, etc.) to one surface of the part.

17. Avoid cylindrical bores in long parts.

18. Avoid machined surfaces on long parts by using work material preformed to the required cross section, e.g. extrusions.

19. Avoid extremely long or thin parts.

ASSEMBLY

20. Make sure that it is possible to assemble the parts into the component.

21. Make sure that each operating machined surface on the part has a corresponding machined surface on the mating part.

22. Make sure that internal corners do not interfere with a corresponding external corner on the mating part.

SURFACE FINISH AND ACCURACY

23. Design for the widest tolerances and the roughest surface that will give acceptable performance for operating surfaces.

24. Make sure that surfaces to be finished ground are raised and never intersect to form internal corners.

7-9 DESIGN FOR POWDER METALLURGY

Powder metallurgy (PM) produces parts to final shape and with little or no machining required. In the conventional PM process a finely divided metallic powder is compressed in a die to produce a porous shape. The green compact is sintered in an atmosphere (usually nonoxidizing or reducing) at elevated temperature to close up the porosity of the as-pressed compact. Generally, the sintered part contains from 4 to 10 percent porosity. The porosity can be decreased by coining or restriking the sintered compact. This step also improves dimensional tolerance. Still further densification and strength improvement can be achieved by resintering the part after it has been coined.

Powder metallurgy processing has found greatest acceptance for small parts (under 1 lb) in automotive and appliance applications in which the ability to produce to final shape with a minimum of machining provides a strong economic advantage. The requirements for mechanical strength must not be too severe for pressed and sintered PM parts. Figure 7-29 illustrates a typical PM part and compares the cost for a PM processing route with the cost of a hobbed (machined) gear at an annual production rate of 500,000 pieces.

One of the chief advantages of powder metallurgy is versatility. Metals that can be combined in no other way can be produced by powder metallurgy. Some examples are copper combined with carbon for electrical brushes and cobalt and tungsten carbide for cutting tools. The density (porosity) of the part can be controlled over wide limits to fabricate products with special features such as self-lubricating bearings, metallic filters, and parts with unusual damping properties.

However, the chief trend in powder metallurgy is in the production of full-density, high-strength parts with fewer processing steps than the competing processes of casting or forging. One of the approaches to this is PM hot forging, whereby a sintered PM preform is completely densified and hot-forged to finished shape in a single operation. Another exciting approach is hot isostatic pressing (HIP). In this process, the powder is sealed in a metal or ceramic container that has the shape of the desired part. The container is placed in a special pressure vessel that has the capability of simultaneously heating the container and subjecting it to a hydrostatic argon gas pressure. The powder is compacted, densified, and sintered in one step. The HIP process is particularly suited to producing parts from high-temperature alloys that are difficult to forge and machine. A 35 percent saving in the material needed to make a gas turbine disk has been reported for the HIP process.

Several design rules[1] must be considered to make economical parts by the conventional press and sinter PM process.

1. The design must be such that the part can be ejected from the die. Parts with straight walls are preferred. No draft is required for ejection from a lubricated die. Parts with undercuts or holes at right angles to the direction of pressing cannot be made.
2. In designing the part, consideration should be given to the need for the powder to flow properly into all parts of the die. Therefore, do not design for thin walls, narrow splines, or sharp corners. In general, sidewalls should be thicker than 0.03 in. Abrupt changes in wall thickness should be avoided, since they lead to distortion after sintering.
3. The shape of the part should permit the construction of strong tooling. Dies

---

[1] "Powder Metallurgy Design Manual," Metal Power Industries Federation, Princeton, N.J., 1989.

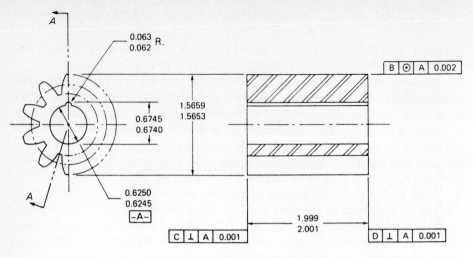

Section A − A
Lubricating oil pump gear, truck

— Comparison of factory costs, materials, and tooling for a truck lubricating oil pump gear

1. Specifications: spur involute form, ten tooth, diametral pitch 8, addendum = 0.157″, pitch diameter 1.250, face width 2.0, 25° pressure angle, I.D. keyed, 0.625″ bore, AGMA Class 3.
2. Material: tensile strength (ult.) 100,000 psi, endurance limit 34,000 psi, hardness 260–302 Bhn or equivalent. Using a cost base of 100 for a hobbed gear, we can illustrate the P/M cost advantage as follows:

Hobbed gear	% of total	P/M gear	% of hobbed total
Material:		Material:	
SAE 1045, including		MPIF FC-0208-S (7.0 g/cc	
2% setup scrap and 46% chips	15.10	density) 5% scrap	9.97
Operations:		Operations:	
Bar chuck, cut off and bore	8.49	Compact (100-ton press)	2.37
Broach keyway	3.17	Sinter	2.56
Hob teeth	47.50	Harden	1.92
Harden	1.92	Grind ends perpendicular to pitch diameter	5.93
Grind ends perpendicular to pitch diameter	5.93	Deburr	0.53
Deburr	0.53	Inspect	0.26
Inspect	0.26	Perishable tools and gages, per piece	8.19
Perishable tools and gages, per piece	17.10		
	100.00		31.73
Cost advantage: 68.27%			

FIGURE 7-29
Comparison of costs for an oil pump gear produced by a PM processing route and by hobbing (machining). (*From S. S. McGee and F. K. Burgess,* Inter. J. Powder Met. Powder Technology, *vol. 12, no. 4, October 1976, p. 315.*)

and punches should have no sharp edges. There should be a reasonable clearance between top and bottom punches during pressing.

4. Since pressure is not transmitted uniformly through a deep bed of powder, the length of a die-pressed part should not exceed about two and a half times the diameter.

5. Keep the part shape simple. The part should be designed with as few levels (diameters) and axial variations as possible.

6. Provide wide dimensional tolerances whenever possible. Wide tolerances mean lower piece-part cost and longer tool life.

7. PM parts may be bonded by assembling in the green (as-pressed) condition and then sintering together to form a bonded assembly.

7-10 DESIGN FOR WELDING

Welding is the most prominent of the processes for joining, into complex parts, components made by some other manufacturing process. Alternative joining techniques are fastening by screws or rivets, soldering, brazing, and adhesive bonding. The many welding processes[1] are classified in Fig. 7-30.

Welding is a process in which two materials are joined permanently by coalescence. It involves some combination of temperature, pressure, and surface condition. A necessary condition for welding is that the two surfaces to be joined must be brought into intimate contact. When fusion takes place in the welding process, the intimate contact is achieved by a molten metal. When melting does not occur, intimate contact is achieved by overcoming the surface roughness of the mating surfaces by producing plastic deformation of the surface asperities and by removing any oxide or other layer that may be on the surfaces.

Probably the oldest welding process is the solid-state one called forge welding. It is the technique, used by the blacksmith, in which two pieces of steel or iron are heated and forced together under point contact. Slag and oxides are squeezed out, and interatomic bonding of the metal results. In the modern version of forge welding, steel pipe is heated by induction or electrical resistance and butt-welded under axial force. The pipe is produced by forming sheet into a cylinder and welding the edges together by forge-seam welding, in which either the sheet is pulled through a conical die or the hot strip is passed between shaped rolls.

As the name implies, cold-welding processes are carried out at room temperature without any external heating of the metal. The surfaces must be

---

[1] For detailed descriptions of these processes and examples of their use, see "Metals Handbook," 9th ed., vol. 6, "Welding, Brazing, and Soldering," American Society for Metals, Metals Park, Ohio, 1983, "Welding Handbook," 7th ed., published in six sections by American Welding Society, Miami, Fla., 1976; R. A. Lindberg and N. R. Braton, "Welding and Other Joining Processes," Allyn & Bacon, Inc., Boston, 1976.

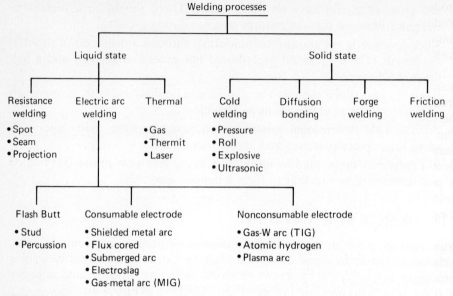

FIGURE 7-30
Classification of welding processes.

clean, and the local pressure must be high to produce substantial cold-working. The harmful effect of interface films is minimized when there is considerable relative movement of the surfaces to be joined. The movement is achieved by passing the metal through a rolling mill or subjecting the interface to tangential ultrasonic vibration. In explosive bonding there is extensive vorticity at the interfaces. Diffusion bonding takes place at a temperature high enough for diffusion to occur readily across the bond zone. Hot roll bonding is a combination of diffusion bonding and roll bonding.

Friction welding (inertia welding) utilizes the frictional heat generated when two bodies slide over each other. In its usual form, one part is held fixed and the other part (usually a shaft or cylinder) is rotated and, at the same time, forced axially against the stationary part. The friction quickly heats the abutting surfaces and, as soon as the proper temperature is reached, the rotation is stopped and the pressure is maintained until the weld is complete. The impurities are squeezed out into a flash, but essentially no melting takes place. The heated zone is very narrow, and therefore dissimilar metals are easily joined.

In the majority of welding applications the interatomic bond is produced by melting. In welding, the workpiece materials and the filler material in the joint have similar compositions and melting points. In soldering and brazing, the filler material has a much different composition that is selected to have a lower melting point than the workpiece materials.

Resistance welding utilizes the heat generated at the interface between two metal parts when a high current is passed through the parts. Spot welding is used extensively to join metal sheets at discrete points (spots). Rather than produce a series of spots, an electrode in the form of a roller often is used to produce a seam weld. If the part to be welded contains small embossed dimples or projections, they are easily softened under the electrode and pushed back to produce the weld nugget.

Other sources of heat for welding are chemical sources or high-energy beams. Gas welding, especially the reaction between oxygen and acetylene to produce an intense flame, has been used for many years. Thermit welding uses the reaction between Fe_2O_3 and Al, which produces Fe and an intense heat. The process is used to weld heavy sections such as rails. Energy from a laser beam is being used to produce welds in sheet metal. Its advantage over an electron beam is that a vacuum is not required. Each form of energy is limited in power, but it can be carefully controlled. Laser beam and electron beam welding lend themselves to welding thin gages of hardened or high-temperature materials.

The thermal energy produced from an electric arc has been utilized extensively in welding. In flash butt welding, an arc is created between two surfaces, typically tubes or bars, which, after they reach temperature, are forced together axially to squeeze out a radial flash. Stud welding is a variant of the process in which a threaded stud is fastened to a flat surface by arc welding. Percussion welding is a process in which two workpieces are brought together at a rapid rate such that, just before the pieces meet, an arc melts both of the colliding surfaces. Percussion welding is particularly good for joining small-diameter wires or dissimilar materials. In all of these flash welding processes, a true electric arc is generated at the welded interface.

Most electric arc welding is done with an arc struck between a consumable electrode (the filler rod) and the workpiece. A coating is applied to the outside surface of the metal electrode to provide a protective atmosphere around the weld pool. The electrode coating also acts as a flux to remove impurities from the molten metal and as an ionizing agent to help stabilize the arc. This is the commonly used shielded metal arc process. Since the electrode coating is brittle, only straight stick electrodes can be used. That restricts the process to a slow hand operation. If the flux coating is placed inside a long tube, the electrode can be coiled, and then the shielded arc process can be made continuous and automatic. In the submerged arc process the consumable electrode is a bare filler wire and the flux is supplied from a separate hopper in a thick layer that covers the arc. In the electroslag process the electrode wire is fed into a pool of molten slag that sits on top of the molten weld pool. Metal transfer is from the electrode to the weld pool through the molten slag. This process is used for welding thick plates and can be automated. In the gas metal arc process the consumable metal electrode is shielded by an inert gas such as argon or helium. Because there is no flux coating, there is no need to remove the slag deposit from the weld bead after each pass.

In nonconsumable electrode welding an inert tungsten electrode is used. Depending on the weld design, a filler rod may be required. In gas tungsten arc welding (TIG welding), argon or helium is used. The process produces high-quality welds in almost any material, especially in thinner-gage sheet. In the atomic hydrogen process molecular hydrogen is passed through an arc maintained by two tungsten electrodes. The hydrogen dissociates to atomic hydrogen in the arc and then reassociates to diatomic hydrogen molecules at the weld surface. This chemial reaction produces an arc with very high temperature. In plasma arc welding two separate gas flows are used. Gas from the central system surrounds the electrode and becomes ionized as it passes through the arc. The flow of this ionized gas is constricted by a small orifice directly below the point of the electrode, which increases the ionization and generates a plasma. The central column of plasma is surrounded by a second cooler sheath of shielding gas. The plasma arc generates the highest temperature of any welding system.

To design a weldment properly, consideration must be given to the selection of materials, the joint design, the selection of the welding process, and the stresses generated by the design. The welding process subjects the workpiece at the joint to a temperature excursion that may exceed the melting point of the material. Heat is applied rapidly and locally. We have a miniature casting in the weld pool, which usually is repeated as successive weld beads are laid down. The base metal next to the weld bead, the heat-affected zone, is subjected to rapid heating and cooling, so that there the original microstructure and properties of the base metal are changed in a nonequilibrium way. Thus, considerable opportunity for defects exists unless the weld processing is properly designed.

Since fusion welding is a melting process, controls appropriate to producing quality castings must be applied. Reactions with the atmosphere are prevented by sealing off the molten pool with an inert gas or a slag or by carrying out the welding in a vacuum chamber. The surfaces of the weld joint should be cleaned of scale or grease before welding is undertaken. The thermal expansion of the weld structure on heating, followed by solidification shrinkage, can lead to high internal tensile stresses that can produce cracking and/or distortion. Rapid cooling of alloy steels in welding can result in brittle martensite formation and consequent crack problems. As a result, it is common to limit welding to carbon steels with less than 0.3 percent carbon or to alloy steels in which the carbon equivalent[1] is less than 0.3 percent carbon. When steels with 0.3 to 0.6 percent carbon must be used because their high strength and high toughness are required, welding without martensite cracking can be performed if the weld joint is preheated before welding and postheated after the weld bead has been deposited. These thermal treatments decrease the rate of cooling of the weld and heat-affected zone, and they thereby reduce the likelihood of martensite formation.

The chief factors in the design of a weldment are 1) the selection of the

---

[1] $C_{eq} = C + \dfrac{Mn}{6} + \dfrac{Cr + Mo + V}{5} + \dfrac{Ni + Cu}{15}$

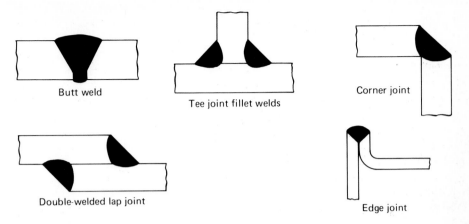

FIGURE 7-31
Basic type of welded joints.

material, 2) the design of the joint, 3) selection of the welding process, and 4) design of the welded joint so it will withstand the applied stresses.

Material selection for welding involves choosing a material with high weldability. *Weldability,* like machinability, is a complex technological property that combines many more basic properties. The melting point of the material, together with the specific heat and latent heat of fusion, will determine the heat input necessary to produce fusion. A high thermal conductivity allows the heat to dissipate and therefore requires a higher *rate* of heat input. Also, metals with higher thermal conductivity result in more rapid cooling and more problems with weld cracking. Greater distortion results from a high thermal expansion, with higher residual stresses and greater danger of weld cracking. There is no absolute rating of weldability of metals because different welding processes impose a variety of conditions that can affect the way a material responds.

The basic types of welded joint are shown in Fig. 7-31. Many variations of these basic designs are possible, depending on the type of edge preparation that is used. A square-edged butt joint requires a minimum of edge preparation. However, an important parameter in controlling weld cracking is the ratio of the width of the weld bead to the depth of the weld. It should be close to unity. Since narrow joints with deep weld pools are susceptible to cracking, the most economical solution is to spend machining money to shape the edges of the plate to produce a joint design with a more acceptable width-to-depth ratio. Ideally, a butt weld should be a full-penetration weld that fills the joint completely throughout its depth. When the gap in a butt joint is wide, a backing strip is used at the bottom of the joint.

Welding electrodes are specified with a code such as E60XX. The last two digits indicate the type of welding application, and the two digits immediately following E indicate the minimum tensile strength of the weld metal in kips per square inch. For example, E7024 has a 70-ksi tensile strength

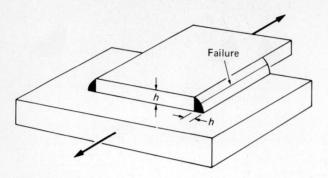

FIGURE 7-32
Fillet weld in a lap joint.

and is intended for ac or dc electric arc welding of steel fillet welds in the horizontal or flat position. The load-carrying ability of a full-penetration butt weld made with E60XX electrodes (50-ksi yield strength) would be

$$\frac{P}{L} = 50t$$

where P/L is joint strength, in kips per inch, and t is the plate thickness. Welds frequently are made with weld metal "reinforcement" that extends above or below the surface of the base metal plate. Some designers believe this increases the strength of the joint and compensates for any weld imperfection. However, such a joint design would serve as a stress concentration under fatigue loading. Therefore, reinforcements on welds should not be used when the welds are subject to fatigue.

Fillet welds are the welds most commonly used in structural design. They are inherently weaker than full-penetration butt welds (Fig. 7-32). A fillet weld fails in shear at the weld throat, given by $0.707h$, and the American Welding Society code allows a shearing yield strength of 30 percent of the tensile strength designation of the electrode. Thus, for an E60XX electrode the load-carrying capacity of a fillet weld is $P/L = 0.30(60)(0.707h)$ kips in.

The selection of the appropriate welding process depends on the required heat input, the availability of equipment, and the economics of the process.[1] Since welding involves the rapid application of heat to a localized area, followed by the rapid removal of the heat, distortion is ever-present. One of the best ways to eliminate welding distortion is to design the welding sequence with thermal distortion in mind. If, because of the geometry, distortion cannot be avoided, then the forces that produce the shrinkage distortion should be balanced with other forces provided by fixtures and clamps. Also, the shrinkage forces can be removed after welding by post-welding annealing and stress-relief operations. It should be kept in mind that distortion arises from

---

[1] R. A. Lindberg and N. R. Braton, op. cit., chap 13.

welding, per se, so that the design should call for only the amount of weld metal that is absolutely required. Overwelding adds not only to the shrinkage forces but also to the costs.

Obviously, the design of weldments[1] calls for much expertise that is beyond the scope of this chapter. However, we can suggest some general design guidelines.

1. Welded designs should reflect the flexibility and economy inherent in the welding process. Do not copy designs based on casting or forging.
2. In the design of welded joints, try to provide for a straight-line force pattern. Avoid the use of welded straps, laps, and stiffeners except as required for strength.
3. Use the minimum number of welds.
4. Whenever possible, weld together parts of equal thickness.
5. Locate the welds at areas in the design where stresses and/or deflections are least critical.
6. Carefully consider the sequence with which parts should be welded together and include that information as part of the design drawing.
7. Make sure that the welder or welding machine has unobstructed access to the joint so that a quality weld can be produced. Whenever possible, the design should provide for welding in the flat or horizontal position, not overhead.

7-11 RESIDUAL STRESSES IN DESIGN

Residual stresses are the system of stresses that can exist in a part when the part is free from external forces, and they are sometimes referred to as internal stresses or locked-in stresses.[2] They arise from nonuniform plastic deformation of a body, chiefly as a result of inhomogeneous changes in volume or shape. For example, consider a metal sheet that is being rolled under conditions such that plastic flow occurs only near the surfaces of the sheet (Fig. 7-33a). The surface fibers of the sheet are cold-worked and tend to elongate while the center of the sheet is unchanged. Since the sheet must remain a continuous whole, its surface and center must undergo strain accommodation. The center fibers tend to restrain the surface fibers from elongating, and the surface fibers seek to stretch the central fibers of the sheet. The result is a residual stress pattern in the sheet that consists of a high compressive stress at the surface and a tensile residual stress at the center (Fig. 7-33b). In general, the sign of the

---

[1] "Design of Weldments," The James F. Lincoln Arc Welding Foundation, Cleveland, Ohio, 1963; O. W. Blodgett, "Design of Welded Structures," ibid., 1966; T. G. F. Gray and J. Spencer, "Rational Welding Design," 2d. ed., Butterworth, London, 1982.

[2] W. B. Young (ed), "Residual Stress in Design, Process and Materials Selection," ASM International, Metals Park, OH, 1987.

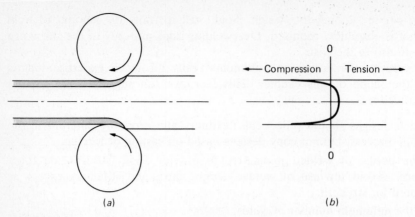

FIGURE 7-33

(a) Inhomogeneous deformation in rolling of sheet; (b) resulting distribution of longitudinal residual stress over thickness of sheet (schematic).

residual stress that is produced by inhomogeneous deformation will be opposite the sign of the plastic strain that produced the residual stress. Thus, for the case of the rolled sheet, the surface fibers that were elongated in the longitudinal direction by rolling are left in a state of compressive residual stress when the external load is removed.

The residual stress system existing in a body must be in static equilibrium. Thus, the total force acting on any plane through the body and the total moment of forces on any plane must be zero. For the longitudinal stress pattern in Fig. 7-33b this means that the area under the curve subjected to compressive residual stresses must balance the area subjected to tensile residual stresses. Actually, the situation is not quite so simple as is pictured in Fig. 7-33. For a complete analysis, the residual stresses acting across the width and thickness of the sheet should be considered, and the state of residual stress at any point is a combined stress derived from the residual stresses in the three principal directions. Frequently, because of symmetry, only the residual stress in one direction need be considered. A complete determination of the state of residual stress in three dimensions is a very considerable undertaking.

Residual stresses are to be considered as only elastic stresses. The maximum value that the residual stress can reach is the elastic limit of the material. A stress in excess of that value, with no external force to oppose it, will relieve itself by plastic deformation until it reaches the value of the yield stress.

Residual and applied stress add algebraically, so long as their sum does not exceed the elastic limit of the material. For example, if the maximum applied stress due to service loads is 60,000 psi tension and the part already contains a tensile residual stress of 40,000 psi, the total stress at the critically stressed region is 100,000 psi. However if the residual stress is a compressive 40,000 psi produced by shot peening, then the actual stress is 20,000 psi.

Any process, whether mechanical, thermal, or chemical, that produces a permanent nonuniform change in shape or volume creates a residual stress pattern. Practically, all cold-working operations develop residual stresses because of non-uniform plastic flow. In surface-working operations, such as shot peening, surface rolling, or polishing, the penetration of the deformation is very shallow. The distended surface layer is held in compression by the less-worked interior. A surface compressive residual stress pattern is highly desirable in reducing the incidence of fatigue failure.

Residual stresses arising from thermal processes may be classified as those due to a thermal gradient alone or to a thermal gradient in conjunction with a phase transformation, as in heat-treating steel. These situations arise most frequently in quenching, casting, and welding.

The situation of greatest practical interest involves the residual stresses developed during the quenching of steel for hardening. In this case, however, the residual stress pattern is due to thermal volume changes plus volume changes resulting from the transformation of austenite to martensite. The simpler situation, in which the stresses are due only to thermal volume changes, will be considered first. This is the situation encountered in the quenching of a metal that does not undergo a phase change on cooling. It is also the situation encountered when steel is quenched from a tempering temperature below the A_1 critical temperature.

The distribution of residual stress over the diameter of a quenched bar in the longitudinal, tangential, and radial directions is shown in Fig. 7-34a for the usual case of a metal that contracts on cooling. Figure 7-34c shows that the opposite residual stress distribution is obtained if the metal expands on cooling (this occurs for only a few situations). The development of the stress pattern shown in Fig. 7-34a can be visualized as follows: The relatively cool surface of the bar tends to contract into a ring that is both shorter and smaller in diameter than it was originally. This tends to extrude the hotter, more plastic center into a cylinder that is longer and thinner than it was originally. If the inner core were free to change shape independently of the outer ring, it would change dimensions to a shorter and thinner cylinder on cooling. However, continuity must be maintained throughout the bar so that the outer ring is drawn in (compressed) in the longitudinal, tangential, and radial directions at the same time the inner core is extended in the same directions. The stress pattern shown in Fig. 7-34 results.

The magnitude of the residual stresses produced by quenching depends on the stress-strain relations for the metal and the degree of strain mismatch produced by the quenching operation. For a given strain mismatch, the higher the elastic modulus of the metal the higher the residual stress. Further, since the residual stress cannot exceed the yield stress, the higher the yield stress the higher the possible residual stress. The yield stress-temperature curve for the metal also is important. If the yield stress decreases rapidly with increasing temperature, the strain mismatch will be small at high temperature because the metal can accommodate to thermally produced volume changes by plastic flow.

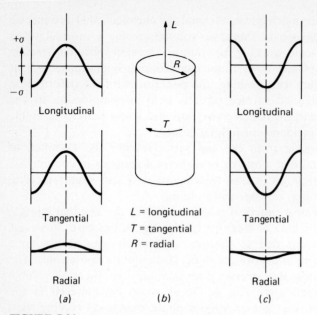

FIGURE 7-34
Residual stress patterns found in quenched bars and due to thermal strains (schematic). (*a*) For metal which contracts on cooling; (*b*) orientation of directions; (*c*) for metal which expands on cooling.

On the other hand, metals that have a high yield strength at elevated temperatures, like superalloys, will develop large residual stresses from quenching.

The following combinations of physical properties will lead to high mismatch strains on quenching:

- Low thermal conductivity k
- High specific heat c
- High coefficient of thermal expansion α
- High density ρ

These factors can be combined into the thermal diffusivity $D_t = k/\rho c$. Low values of thermal diffusivity lead to high strain mismatch. Other factors that produce an increase in the temperature difference between the surface and center of the bar promote high quenching stresses. They are 1) a large diameter of the cylinder, 2) a large temperature difference between the initial temperature and the temperature of the quenching bath, and 3) a high severity of quench.

In the quenching of steels austenite begins to transform to martensite whenever the local temperature of the bar reaches the M_s temperature. Since an increase in volume accompanies the transformation, the metal expands as the martensite reaction proceeds on cooling from the M_s to M_f temperature. This produces a residual stress distribution of the type shown in Fig. 7-34c. The residual stress distribution in a quenched steel bar is the resultant of the competing processes of thermal contraction and volume expansion due to martensite formation. Transformation of austenite to bainite or pearlite also produces a volume expansion, but of lesser magnitude. The resulting stress pattern depends upon the transformation characteristics of the steel, as determined chiefly by composition and hardenability, and the heat-transfer characteristics of the system, and the severity of the quench.

Figure 7-35 illustrates some of the possible residual stress patterns that

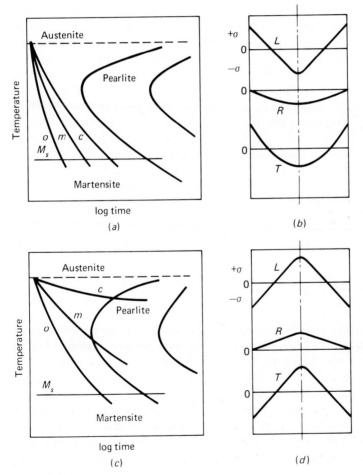

FIGURE 7-35
Transformation characteristics of a steel (*a* and *c*), and resulting residual stress distributions (*b* and *d*).

can be produced by quenching steel bars. On the left side of the figure is a typical isothermal transformation diagram for the decomposition of austenite. The cooling rates of the outside, midradius, and center of the bar are indicated on the diagram by the curves marked o, m, and c. In Fig. 7-35a the quenching rate is rapid enough to convert the entire bar to martensite. By the time the center of the bar reaches the M_s temperature, the transformation has been essentially completed at the surface. The surface layers try to contract against the expanding central core, and the result is tensile residual stresses at the surface and compressive stresses at the center of the bar (Fig. 7-35b). However, if the bar diameter is rather small and the bar has been drastically quenched in brine so that the surface and center transform at about the same time, the surface will arrive at room temperature with compressive residual stresses. If the bar is slack-quenched so that the outside transforms to martensite while the middle and center transform to pearlite (Fig. 7-35c), there is little restraint offered by the hot, soft core during the time when martensite is forming on the surface, and the core readily accommodates to the expansion of the outer layers. The middle and center pearlite regions then contract on cooling in the usual manner and produce a residual stress pattern consisting of compression on the surface and tension at the center (Fig. 7-35d).

To a first approximation the residual stresses in castings are modeled by a quenched cylinder. However, the situation in castings is made more complicated by the fact that the mold offers a mechanical restraint to the shrinking casting. Moreover, the casting design may produce greatly different cooling rates at different locations that are due to variations in section size and the introduction of chills, which produce an artificially rapid cooling rate.

Appreciable residual stresses are developed in welding, even in the absence of a phase transformation. Figure 7-36 shows the residual stresses developed in the longitudinal direction (parallel to the weld joint). As the weld metal and heat-affected zone shrink on cooling, they are restricted by the cool surrounding plate. The weld contains tensile residual stresses, which are balanced by compressive stresses in the adjacent region. Because thermal gradients tend to be steep in welding, the residual stress gradients also tend to be steep.

Chemical processes such as oxidation, corrosion, and electroplating can generate large surface residual stresses if the new surface material retains coherency with the underlying metal surface. Other surface chemical treatments such as carburizing and nitriding cause local volume changes by the diffusion of an atomic species into the surface.

Residual stresses are measured by either destructive or nondestructive methods.[1] The destructive methods relax the locked-in stress by removing a layer from the body. The stress existing before the cut was made is calculated

---

[1] A. A. Denton, *Met. Rev.*, vol. 11, pp. 1–22, 1966; C. O. Ruud, *J. Metals*, pp. 35–40, July 1981.

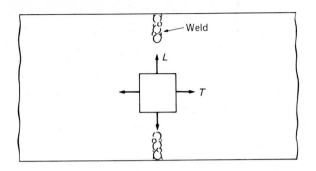

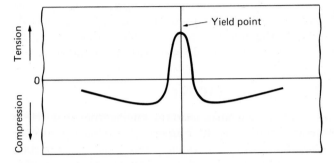

FIGURE 7-36
Longitudinal residual stresses in a butt-welded plate.

from the deformation produced by relaxing the stress. The nondestructive method depends on the fact that the spacing of atomic planes in a crystalline material is altered by stress. This change can be measured very precisely with a diffracted x-ray beam. The x-ray method is nondestructive, but it gives only the value of residual surface stress.

The removal or reduction in the intensity of residual stress is known as stress relief. Stress relief may be accomplished either by heating or by mechanical working operations. Although residual stresses will disappear slowly at room temperature, the process is very greatly accelerated by heating to an elevated temperature. The stress relief that comes from a stress-relief anneal is due to two effects. First, since the residual stress cannot exceed the yield stress, plastic flow will reduce the residual stress to the value of the yield stress at the stress-relief temperature. Only the residual stress in excess of the yield stress at the stress-relief temperature can be eliminated by immediate plastic flow. Generally, most of the residual stress will be relieved by time-dependent stress relaxation. Since the process is extremely temperature-dependent, the time for nearly complete elimination of stress can be greatly reduced by increasing the temperature. Often a compromise must be made between the use of a temperature high enough for the relief of stress in a reasonable length of time and the annealing of the effects of cold-working.

The differential strains that produce high residual stresses also can be eliminated by plastic deformation at room temperature. For example, products such as sheet, plate, and extrusions are often stretched several percent beyond the yield stress to relieve differential strains by yielding. In other cases the residual stress distribution that is characteristic of a particular working operation may be superimposed on the residual stress pattern initially present in the material. A surface that contains tensile residual stresses may have the stress distribution converted into beneficial compressive stresses by a surface-working process like rolling or shot peening. However, it is important in using this method of stress relief to select surface-working conditions that will completely cancel the initial stress distribution. For example, it is conceivable that, if only very light surface rolling were used on a surface that initially contained tensile stresses, only the tensile stresses at the surface would be reduced. Dangerously high tensile stresses could still exist below the surface.

7-12 DESIGN FOR HEAT TREATMENT

One of the reasons why steel is such an important engineering material is that its metallurgical structure, and hence its properties, can be varied over wide limits by heat treatment. Precipitation (aging) reactions are important strengthening heat treatments in aluminum and nickel alloys. Also, annealing heat treatments are important in removing the damaging effects of cold-working so that metal forming can be carried well beyond the point at which fracture would ordinarily occur.

These are only some of the more important examples of the role of thermal treatment in controlling the properties of metals. However, processing by heat treatment requires energy. In addition, it requires a protective atmosphere or protective surface coating to prevent the metal part from oxidizing or otherwise reacting with the furnace atmosphere. Metal parts soften, creep, and eventually sag upon long exposure at elevated temperatures. Therefore, parts may require special fixtures to support them during heat treatment. Since heat treatment is a special processing step, it would be advantageous to eliminate it if possible. Sometimes a cold-worked sheet or bar can be substituted for a heat-treated part, but generally the flexibility and/or superior properties that result from heat treatment will be wanted.

The steel with the best combination of high strength and high toughness is produced by quenching from within the austenite temperature region (1400 to 1650°F depending on composition) and cooling rapidly enough that hard and brittle martensite is formed. The part is then reheated below the austenite region to allow the martensite to break down (temper) into a fine precipitation of carbides in a soft ferrite matrix. The formation of a proper *quenched and tempered* microstructure depends on cooling fast enough that pearlite or other nonmartensitic phases are not formed. This requires a balance between the heat transfer from the part (as determined chiefly by geometry), the cooling power of the quenching medium (water, oil, or air), and the transformation kinetics of the steel (as controlled by the alloy chemistry). These factors are interrelated by the property called hardenability.[1]

In heating for austenitization care should be taken to subject the parts to uniform temperature in the furnace. Long thin parts are especially prone to distortion from nonuniform temperature. Also, parts containing residual stress from previous processing operations may distort on heating.

Quenching is a severe condition to impose upon a piece of steel.[2] In quenching, the part suddenly is cooled at the surface. The part must shrink rapidly because of thermal contraction (steel is at least 0.125 in/ft larger before quenching from the austenitizing temperature), but it also undergoes a volume increase when it transforms to martensite at a comparatively low temperature. As discussed in Sec. 7-11 and Fig. 7-35, this can produce a condition of high residual (internal) stresses. The value of the local tensile stresses may be high enough to produce fractures called quench cracks. Also, local plastic deformation can occur in quenching even if cracks do not form, and that causes warping and distortion.

Problems with quench cracks and distortion chiefly are caused by non-uniformity of temperature distribution and, in turn, by geometry as influenced by the design. Thus, many of the heat-treatment problems can be prevented by proper design. The most important feature is to make the cross sections of the part as uniform as possible. In the ideal design for heat treatment all sections should have equal ability to absorb or give up heat. Unfortunately, designing for uniform thickness or sectional area usually interferes with the functions of the design. A sphere is the ideal geometry for uniform heat transmission, but obviously only a limited number of parts can utilize this shape.

A typical heat-treating problem is illustrated by the gear blank shown in Fig. 7-37. The region A, where the teeth will be machined, is thinner than the hub region B. When the part was heated to the austenitizing temperature, region A increased in temperature faster than region B. As it tried to expand, it was attached to a thicker hub region B whose average temperature was a bit lower. Since the hot metal in A had to go somewhere, local hot upsetting of the surface occurred in region A. On quenching, region A cooled faster, so that it was through the temperature region of martensite formation before the thicker hub region B was. Since there were local temperature variations, there

---

[1] "Metals Handbook," 9th ed., vol. I, pp. 455–525, American Society for Metals, Metals Park, Ohio, 1978; C. A. Siebert, D. V. Doane, and D. H. Breen, "The Hardenability of Steels," ibid., 1977.

[2] "Selecting Steels and Designing Parts for Heat Treatment," ibid., 1969; A. J. Fletcher, "Thermal Stress and Strain Generation in Heat Treatment," Elsevier Applied Science, New York, 1989.

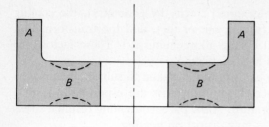

FIGURE 7-37
Gear blank geometry.

were local distortions that resulted in residual stresses. If there were a sharp fillet radius or other type of stress concentration, it could result in quench cracks in the brittle martensite. Thus, nonuniform temperatures, brought on by nonuniform geometry, lead to distortion and possibly fracture. A typical solution is to remove unneeded material in an attempt to produce more uniform heat transfer. In the gear blank of Fig. 7-37 this could be done by redesigning the hub region to remove metal along the dashed lines. In other cases, however, the approach toward uniformity might be to add metal.

Design details that minimize stress concentrations, as would be good design practice to prevent fatigue (Sec. 7-15), also minimize quench cracking. Distortion in heat treatment is minimized by designs that are symmetrical. A single keyway in a shaft is a particularly difficult situation. A part with a special distortion problem may have to be quenched in a special fixture that restrains it from distorting beyond tolerances. Another consideration is to so design the part that the quenching fluid has access to all critical regions that must be hardened. Since the quenching fluid produces a vapor blanket when it hits the hot steel surface, it may be necessary to design for special venting or access holes for the quenching fluid.

A process approach to minimizing the difficulties with heat treatment is called marquenching. In it large thermal gradients are eliminated by initially quenching to a temperature just above the M_s temperature and holding long enough to equalize the temperature of the part but not long enough to transform the austenite (Fig. 7-35). Then the parts are cooled slowly to below the M_f temperature to produce martensite. While seemingly an ideal way to heat-treat steel, marquenching may be difficult to carry out on a practical basis. It generally is more economical to select a steel that can be heat-treated by direct quenching.

7-13 DESIGN FOR ASSEMBLY

The last step in the manufacturing process is to assemble individual components into the final product. Minimizing the cost of assembly is clearly a design function. The decision between manual or automatic assembly depends on such factors as the number of parts per assembly, the number of assemblies required, the design and/or complexity of the components, and the relative costs of capital and labor. The assembly of components into products has been divided as follows:[1]

1. Manual assembly
2. Mechanically aided manual assembly, as when parts feeders are used
3. Automatic assembly using special-purpose synchronous indexing machines and automatic feeders
4. Automatic assembly with special-purpose free-transfer, nonsynchronous machines and automatic feeders
5. Automatic assembly with special-purpose free-transfer, nonsynchronous machines with programmable work heads and parts magazines
6. Automatic assembly with robots and parts magazines

For successful automatic assembly, parts must have proper geometry and be of consistently uniform quality. Clearly, proper attention to those factors in the design phase is important. In automatic assembly,[2] small parts are fed and oriented by vibrating belts and bowls, rotary disks, reciprocating arms, magnetic devices, and tracks. Parts must have sufficient rigidity to withstand feeding forces and selection operations without bending or distorting. Thin, fragile, and brittle components should be avoided. It is a good idea to minimize the number of finished surfaces that must be protected from scratching or damage during automatic assembly. Flanges or projections designed into the part can provide that protection. Fastening by screws is relatively costly in automatic assembly, so designing the part to be assembled by staking, crimping, or welding may be desirable.[3] When possible, the largest and most rigid component of the assembly should be designed to serve as a base or fixture and thereby eliminate the need for and cost of assembly fixtures. If possible, the part should be so designed that all assembly operations can be performed from one plane with straight vertical or horizontal motions.

Since the feeding of small parts generally employs the force of gravity, designers should try to avoid part instability by designing for a low center of gravity. In the design consideration should also be given to the possibility the parts will tangle or nest during feeding or will climb on each other and overlap like shingles on a roof. Moreover, orienting the part in a specific way for

---

[1] K. Swift and A. H. Redford, *Engineering,* pp. 799–802, July 1980.

[2] K. R. Treer, "Automated Assembly," Society of Manufacturing Engineers, Dearborn, Mich., 1979; G. Boothroyd and A. H. Redford, "Mechanical Assembly," McGraw-Hill Book Company, London, 1968; G. Boothroyd, C. Poli, and L. E. Murch, "Automatic Assembly," Marcel Dekker, New York, 1982; M. M. Andreasen, S. Kahler and T. Lund, "Design for Assembly." 2d ed, Springer-Verlag, New York, 1989.

[3] W. Chow, "Cost Reduction in Product Design," Chap. 6, Van Nostrand Reinhold Co., New York, 1978.

assembly can be expensive, so eliminating the need for orientation is a big design plus. Unsymmetrical parts sometimes can be made symmetrical by adding a nonfunctional hole or other design feature. If the part cannot be made symmetrical, then it should be designed to be highly asymmetrical, since very unsymmetrical parts like screws and bolts are easier to orient than parts that are nearly symmetrical.

Another approach is to minimize the number of orientations required at the assembly machine. The orientation produced by a preceding manufacturing step can be retained by placing the parts in racks, magazines, tapes, tubes, arbors, or other transfer devices. For example, thin stampings may be produced in a continuous strip without actually separating the parts from the strip. Coils of strip are then placed in the assembly machine, and the individual parts are separated as required. Another way to eliminate the need for orientation is to produce the parts on the assembly machine. This approach is often taken with simple parts like washers, gaskets, and springs.

Automatic assembly usually requires tighter part tolerances than are required either for the function of the part or for manual assembly. Nonuniformity of manufactured parts in other respects is a problem. Thus, parts can vary in hardness, surface roughness, amount of protruding flash, etc., that cause difficulties in automatic assembly.

The design for assembly (DFA) methodology developed by Boothroyd and Dewhurst[1] has seen wide acceptance. Essentially the DFA method is a step-by-step implementation of the DFM guidelines presented in Sec. 7-3. Guidelines 1, 8, 9, and 10 are especially relevant. The first task is to determine whether the product should be assembled by manual, special-purpose automatic, or programmable automatic methods. This can be answered from knowledge of the production volume, the number of parts, the variety of parts, and the likelihood of design changes without requiring knowledge of the detailed design. The DFA method requires an estimation of the theoretical minimum number of parts in the product. Following this the product is redesigned for ease of assembly by eliminating and combining parts using insights gained in the earlier analysis. To aid designers in implementing these techniques the DFA method has been incorporated in computer-based design tools.[2]

7-14 DESIGN FOR BRITTLE FRACTURE

An important advance in engineering knowledge has been the ability to predict the influence of cracks and cracklike defects on the brittle fracture of materials

---

[1] G. Boothroyd and P. Dewhurst, "Product Design for Assembly Handbook," Boothroyd Dewhurst Inc., Wakefield, RI, 1987.

[2] G. Boothroyd and P. Dewhurst, "The Design for Assembly Software Toolkit," Boothroyd Dewhurst Inc, Wakefield, RI, 1987.

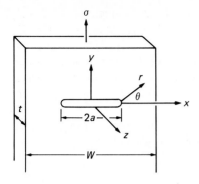

FIGURE 7-38
Model for equations for stress at a point near a crack.

through the science of fracture mechanics.[1] Fracture mechanics had its origin in the ideas of A. A. Griffiths, who showed that the fracture strength of a brittle material, like glass, is inversely proportional to the square root of the crack length. G. R. Irwin proposed that fracture occurs at a fracture stress corresponding to a critical value of crack-extension force G_c, according to

$$\sigma_f = \left(\frac{E G_c}{\pi a} \right)^{1/2} \tag{7-2}$$

where G_c = crack extension force, in. lb/in^2
E = elastic modulus of material, lb/in^2
a = length of the crack, in

An important conceptualization was that the elastic stresses in the vicinity of a crack tip (Fig. 7-38) could be expressed entirely by a stress field parameter K called the stress intensity factor.

The equations for the stress field at the end of the crack can be written

$$\sigma_x = \frac{K}{\sqrt{2\pi r}} \left[\cos \frac{\theta}{2} \left(1 - \sin \frac{\theta}{2} \sin \frac{3\theta}{5} \right) \right]$$

$$\sigma_y = \frac{K}{\sqrt{2\pi r}} \left[\cos \frac{\theta}{2} \left(1 + \sin \frac{\theta}{2} \sin \frac{3\theta}{2} \right) \right] \tag{7-3}$$

$$\tau_{xy} = \frac{K}{\sqrt{2\pi r}} \left(\sin \frac{\theta}{2} \cos \frac{\theta}{2} \cos \frac{3\theta}{2} \right)$$

[1] J. F. Knott, "Fundamentals of Fracture Mechanics," John Wiley & Sons, Inc., New York, 1973; S. T. Rolfe and J. M. Barsom, "Fracture and Fatigue Control in Structures," Prentice-Hall, Inc., Englewood Cliffs, N.J. 1977.

Equations (7-3) show that the elastic normal and elastic shear stresses in the vicinity of the crack tip depend only on the radial distance from the tip r, the orientation θ, and K. Thus, the magnitudes of these stresses at a given point are dependent completely on the stress intensity factor K. However, the value of K depends on the nature of the loading, the configuration of the stressed body, and the mode of crack displacement, i.e., mode I (opening mode), mode II (shearing), or mode III (tearing).

For a center crack of length $2a$ in an infinite thin plate subjected to a uniform tensile stress σ the stress intensity factor K is given by

$$K = \sigma\sqrt{\pi a} = GE \tag{7-4}$$

where K is in ksi$\sqrt{\text{in}}$ or MPa$\sqrt{\text{m}}$. Values of K have been determined for a variety of situations by using the theory of elasticity, often combined with numerical methods and experimental techniques.[1] For a given type of loading, Eq. (7-4) usually is written as

$$K = \alpha\sigma\sqrt{\pi a} \tag{7-5}$$

where α is a parameter that depends on the specimen and crack geometry. For example, for a plate of width w containing a central crack of length $2a$

$$K = \sigma\sqrt{\pi a}\left(\frac{w}{\pi a}\tan\frac{\pi a}{w}\right)^{1/2} \tag{7-6}$$

This dependence of K on geometry may be expressed in other ways, as in Sec. 1-7.

Since the crack tip stresses can be described by the stress intensity factor K, a critical value of K can be used to define the conditions that produce brittle fracture. The tests usually used subject the specimen to the crack opening mode of loading (mode I) under a condition of plane strain at the crack front. The critical value of K that produces fracture is K_{Ic}, the plane-strain fracture toughness. If a_c is the critical crack length at which failure occurs, then

$$K_{Ic} = \alpha\sigma\sqrt{\pi a_c} \tag{7-7}$$

K_{Ic} is a basic material parameter called fracture toughness. If K_{Ic} is known, then it is possible to compute the maximum allowable stress to prevent brittle fracture for a given flaw size. As Fig. 7-39 illustrates, the allowable stress in the presence of a crack of a given size is directly proportional to K_{Ic}, and the

---

[1] A. S. Kobayashi (ed.), "Experimental Techniques in Fracture Mechanics," Society for Experimental Stress Analysis, Westport, Conn., 1973; G. G. Sih, "Handbook of Stress Intensity Factors," Institute of Fracture and Solid Mechanics, Lehigh University, Bethlehem, Pa., 1973; P. C. Paris and G. C. Sih, ASTM Spec. Tech. Publ. 381, pp. 30–83, 1965; Y. Murakami et al., (eds), "Stress Intensity Factors Handbook," (2 vols.), Pergamon Press, New York, 1987.

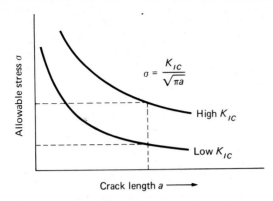

Allowable stress σ

$$\sigma = \frac{K_{IC}}{\sqrt{\pi a}}$$

High K_{IC}

Low K_{IC}

Crack length $a \longrightarrow$

FIGURE 7-39
Relation between fracture toughness and allowable stress and crack size.

allowable crack size for a given stress is proportional to the square of the fracture toughness. Therefore, increasing K_{Ic} has a much larger influence on allowable crack size than on allowable stress. For increasing load in a cracked part, a higher fracture toughness results in a larger allowable crack size or larger allowable stresses at fracture.

Although K_{Ic} is a basic material property, in the same sense as yield strength, it changes with important variables such as temperature and strain rate. The K_{Ic} of materials with a strong temperature and strain-rate dependence usually decreases with decreased temperature and increased strain rate. The K_{Ic} of a given alloy is strongly dependent on such variables as heat treatment, texture, melting practice, impurity level, and inclusion content.

The fracture toughness measured under plane-strain conditions is obtained under maximum constraint or material brittleness. The plane-strain fracture toughness is designated K_{Ic} and is a true material property. Figure 7-40 shows how the measured fracture stress varies with specimen thickness B.

Test specimen

Thin

Medium

Thick

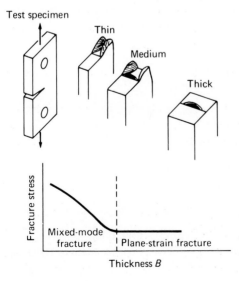

Fracture stress

Mixed-mode fracture

Plane-strain fracture

Thickness B

FIGURE 7-40
Effect of specimen thickness on stress and mode of fracture.

A mixed-mode, ductile brittle fracture with 45° shear lips is obtained for thin specimens. Once the specimen has the critical thickness for the toughness of the material, the fracture surface is flat and the fracture stress is constant with increasing specimen thickness. The minimum thickness to achieve plane-strain conditions and valid K_{Ic} measurement is

$$B = 2.5\left(\frac{K_{Ic}}{\sigma_0}\right)^2 \tag{7-8}$$

where σ_0 is the 0.2 percent offset yield strength. Standardized test specimens and test procedures for fracture toughness testing have been developed by ASTM.[1]

The fracture mechanics concept is strictly correct only for linear elastic materials, i.e., under conditions in which no yielding occurs. However, reference to Eq. (7-3) shows that as r approaches zero the stress at the crack tip approaches infinity. Thus, in all but the most brittle material, local yielding occurs at the crack tip and the elastic solution should be modified to account for crack tip plasticity. However, if the plastic zone size r_y at the crack tip is small relative to the local geometry, for example, r_y/t or $r_y/a < 0.1$, crack tip plasticity has little effect on the stress intensity factor. That limits the strict use of fracture mechanics to high-strength materials. Moreover, the width restriction to obtaining valid measurements of K_{Ic}, as described in Eq. (7-8), makes the use of linear elastic fracture mechanics (LEFM) impractical for low-strength materials. Extensive research to extend LEFM to low-strength materials is underway.

Design Example

We shall illustrate the use of LEFM in design with the example of a disk spinning at high speed.[2] Rotating disks are important components in steam and gas turbines. The maximum stress is a circumferential tensile stress that occurs at the surface of the bore of the disk.[3]

$$\sigma_{max} = \frac{3+v}{4}\rho\omega^2 c^2\left[1 + \frac{1-v}{3+v}\left(\frac{b}{c}\right)^2\right] \tag{7-9}$$

where ρ = mass density
v = Poisson's ratio
$\omega = 2\pi N/60$, the angular velocity

and c and b are as defined in Fig. 7-41. The stress intensity factor for a crack of

---

[1] ASTM Standards, Part 31, Designation E399-70T.

[2] G. C. Sih, *Trans. ASME*, Ser. B., *J. Eng. Ind.*, vol. 98, pp. 1243–1249, 1976.

[3] R. J. Roark and W. C. Young, "Formulas for Stress and Strain," 6th ed. McGraw-Hill Book Company, New York, 1989.

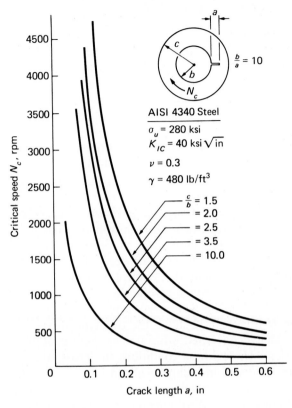

Critical speed N_c, rpm

Crack length a, in

AISI 4340 Steel

σ_u = 280 ksi

K_{IC} = 40 ksi $\sqrt{\text{in}}$

ν = 0.3

γ = 480 lb/ft³

$\frac{c}{b}$ = 1.5

= 2.0

= 2.5

= 3.5

= 10.0

$\frac{b}{a}$ = 10

FIGURE 7-41
Critical speed as a function of
crack length. (*From G. C. Sih,
Trans. ASME, sec. B., vol. 98, p.
1245, 1976.*)

length a located at the bore of a disk, is given by[1] Eq. (7-10) if $b/a > 10$.

$$K_{Ic} = 1.12\sigma_{max}\sqrt{\pi a} \qquad (7\text{-}10)$$

Combining Eqs. (7-9) and (7-10) results in an equation for the critical speed, in rpm, of the disk in terms of the fracture toughness and the crack length.

$$N_c = 13.560 \frac{\sqrt{K_{Ic}}}{a^{1/4}\sqrt{\rho[(3+\nu)c^2 + (1-\nu)b^2]}} \qquad (7\text{-}11)$$

Critical speed is plotted vs. crack length in Fig. 7-41 by using material parameters of a high-strength 4340 steel.

It is important to note that even in this relatively simple example the use of fracture mechanics lends important realism to the calculation. If $a = 0.1$ in and $c/b = 2$, the critical speed, from Fig. 7-41, is 4000 rpm. In the conventional design approach, which ignores the presence of a crack, N_c is determined by setting σ_{max} in Eq. (7-9) equal to the ultimate tensile strength. This oversimplification leads to a critical speed of 8400 rpm.

[1] G. C. Sih, op. cit.

7-15 DESIGN FOR FATIGUE FAILURE

Materials subjected to a repetitive or fluctuating stress will fail at a stress much lower than that required to cause fracture on a single application of load. Failures occurring under conditions of fluctuating stresses or strains are called fatigue failure.[1] Fatigue accounts for the majority of mechanical failures in machinery.

A fatigue failure is a localized failure that starts in a limited region and propagates with increasing cycles of stress or strain until the crack is so large that the part cannot withstand the applied load, whereupon it fractures. Plastic deformation processes are involved in fatigue, but they are highly localized.[2] Therefore, fatigue failure occurs without the warning of gross plastic deformation. Failure usually initiates at regions of local high stress or strain because of abrupt changes in geometry (stress concentration), temperature differentials, residual stresses, or material imperfections. Much basic information has been obtained about the mechanism of fatigue failure, but at present the chief opportunities for preventing fatigue lie at the engineering design level. Fatigue prevention is achieved by proper choice of material, control of residual stress, and minimization of stress concentrations through careful design.

Basic fatigue data are presented in the S-N curve, a plot of stress S vs. the number of cycles to failure N. Figure 7-42 shows the two typical types of behavior. The S-N curve is chiefly concerned with fatigue failure at high numbers of cycles ($N > 10^5$ cycles). Under these conditions the gross stress is elastic, although fatigue failure results from highly localized plastic deformation. As can be seen from Fig. 7-42, the number of cycles of stress that a material can withstand before failure increases with decreasing stress. For most materials, e.g., aluminum alloys, the S-N curve slopes continuously downward with decreasing stress. At any stress level there is some large number of cycles that ultimately causes failure. For steels in the absence of a corrosive environment, however, the S-N curve becomes horizontal at a certain limiting stress. Below that stress, called the fatigue limit, the steel can withstand an infinite number of cycles.

Fatigue Design Criteria

There are several distinct philosophies concerning design for fatigue that must be understood to put this vast subject into proper perspective.

---

[1] G. E. Dieter, "Mechanical Metallurgy," 3d ed., chap. 12, McGraw-Hill Book Company, New York, 1986; N. E. Frost, K. J. Marsh, and L. P. Pook, "Metallic Fatigue," Oxford University Press, London, 1974; S. S. Manson, Avoidance, Control and Repair of Fatigue Damage, ASTM Spec. Tech. Publ. 495, pp. 254–346, 1971. C. C. Osgood, "Fatigue Design," 2d ed., Pergamon Press, New York, 1982.

[2] "Fatigue and Microstructure," American Society for Metals, Metals Park, Ohio, 1979.

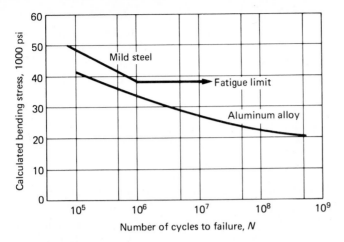

FIGURE 7-42
Typical fatigue curves for ferrous and nonferrous metals.

1. *Infinite-life design.* This design criterion is based on keeping the stresses at some fraction of the fatigue limit of the steel. This is the oldest fatigue design philosophy. It has largely been supplanted by the other criteria discussed below. However, for situations in which the part is subjected to very large cycles of uniform stress it is a valid design criterion.

2. *Safe-life design.* Safe-life design is based on the assumption that the part is initially flaw-free and has a finite life in which to develop a critical crack. In this approach to design one must consider that fatigue life at a constant stress is subject to large amounts of statistical scatter. For example, the Air Force historically designed aircraft to a safe life that was one-fourth of the life demonstrated in full-scale fatigue tests of production aircraft. The factor of 4 was used to account for environmental effects, material property variations, and variations in as-manufactured quality. Bearings are another good example of parts that are designed to a safe-life criterion. For example, the bearing may be rated by specifying the load at which 90 percent of all bearings are expected to withstand a given lifetime. Safe-life design also is common in pressure vessel and jet engine design.

3. *Fail-safe design.* In fail-safe design the view is that fatigue cracks may occur: therefore, the structure is designed so that cracks will not lead to failure before they can be detected and repaired. This design philosophy developed in the aircraft industry, where the weight penalty of using large safety factors could not be tolerated but neither could the danger to life from very small safety factors be tolerated. Fail-safe designs employ multiple-load paths and crack stoppers built into the structure along with rigid regulations and criteria for inspection and detection of cracks.

4. *Damage-tolerant design.* The latest design philosophy is an extension of the fail-safe design philosophy. In damage-tolerant design the assumption is that fatigue cracks will exist in an engineering structure. The techniques of fracture mechanics are used to determine whether the cracks will grow large enough to cause failure before they are sure to be detected during a periodic inspection. The emphasis in this design approach is on using materials with high fracture toughness and slow crack growth. The success of the design approach depends upon having a reliable nondestructive evaluation (NDE) program and in being able to identify the damage critical areas in the design.

Although much progress has been made in designing for fatigue, especially through the merger of fracture mechanics and fatigue, the interaction of many variables that is typical of real fatigue situations makes it inadvisable to depend on a design based solely on analysis. Simulated service testing[1] should be part of all critical fatigue applications. The failure areas not recognized in design will be detected by these tests. Simulating the actual service loads requires great skill and experience. Often it is necessary to accelerate the test, but doing so may produce misleading results. For example, when time is compressed in that way, the full influence of corrosion or fretting is not measured, or the overload stress may appreciably alter the residual stresses. It is common practice to eliminate many small load cycles from the load spectrum, but they may have an important influence on fatigue crack propagation.

The following are some of the most important engineering factors that determine fatigue performance.

Stress cycle
    Repeated or random
    Mean stress
Combined stress state
Stress concentration
    Fatigue notch factor
    Fatigue notch sensitivity
Statistical variation in fatigue life and fatigue limit
Size effect
Surface finish
Surface treatment
Residual stress

---

[1] R. M. Wetzel (ed.), "Fatigue Under Complex Loading," Society of Automotive Engineers, Warrendale, Pa., 1977.

Corrosion fatigue

Fretting fatigue

Cumulative fatigue damage

Space precludes a discussion of these factors. The reader is referred to the basic references that have been cited for details on how the factors control fatigue. There is beginning to be a considerable literature on design methods to prevent fatigue failure.

Fatigue design for infinite life is considered in:

R. C. Juvinall, "Engineering Consideration of Stress, Strain, and Strength," McGraw-Hill Book Company, New York, 1967. Chapters 11 to 16 cover fatigue design in considerable detail.

L. Sors, "Fatigue Design of Machine Components," Pergamon Press, New York, 1971. Translated from the German, this presents a good summary of European fatigue design practice.

C. Ruiz and F. Koenigsberger, "Design for Strength and Production," Gordon & Breach Science Publishers, New York, 1970. Pages 106 to 120 give a concise discussion of fatigue design procedures.

Detailed information on stress concentration factors and the design of machine details to minimize stress can be found in:

R. E. Peterson, "Stress-Concentration Design Factors," John Wiley & Sons, Inc., New York, 1974.

R. B. Heywood, "Designing Against Fatigue of Metals," Reinhold Publishing co., New York, 1967.

C. C. Osgood, "Fatigue Design," 2d ed., Pergamon Press, New York, 1982.

The most complete books on fatigue design, which considers the more modern work on safe-life design and damage-tolerant design are:

H. O. Fuchs and R. I. Stephens, "Metal Fatigue in Engineering," John Wiley & Sons, New York, 1980.

J. E. Bannantine, J. J. Comer and J. L. Handrock, "Fundamentals of Metal Fatigue Analysis," Prentice-Hall Inc., Englewood Cliffs, N.J., 1990.

Conventional fatigue design for infinite life, as well as design for low-cycle fatigue, is covered in:

"Fatigue Design Handbook," 2d ed., Society of Automotive Engineers, Warrendale, PA., 1988.

Damage-tolerant design is considered in:

"Damage Tolerant Design Handbook," Metals & Ceramics Information Center, Battelle, Columbus Laboratories, Columbus, Ohio, 1975.

We will now attempt to bring this huge subject of fatigue into focus by presenting several design examples.

Design Example—Infinite-Life Design[1]

A steel shaft heat-treated to a Brinell hardness of 200 has a major diameter of 1.5 in and a small diameter of 1.0 in. There is a 0.10-in radius at the shoulder between the diameters. The shaft is subjected to completely reversed cycles of stress of pure bending. The fatigue limit determined on polished specimens of 0.2-in diameter is 42,000 psi. The shaft is produced by machining from bar stock. What is the best estimate of the fatigue limit of the shaft?

Since an experimental value for fatigue limit is known, we start with it, recognizing that tests on small unnotched polished specimens represent an unrealistically high value of the fatigue limit of the actual part.[2] The procedure, then, is to factor down the idealized value.

We start with the stress concentration (notch) produced at the shoulder between two diameters of the shaft. A shaft with a fillet in bending is a standard situation covered in all machine design books. If $D = 1.5$, $d = 1.0$, and $r = 0.10$, the important ratios are $D/d = 1.5$ and $r/d = 0.1$. Then, from standard curves, the theoretical stress concentration factor is $K_t = 1.68$. However, K_t is determined for a brittle elastic solid, and most materials exhibit a lesser value of stress concentration when subjected to fatigue. The extent to which the plasticity of the material reduces K_t is given by the fatigue notch sensitivity q.

$$q = \frac{K_f - 1}{K_t - 1} \tag{7-12}$$

where K_t = theoretical stress concentration factor

$$K_f = \text{fatigue notch factor} = \frac{\text{fatigue limit unnotched}}{\text{fatigue limit notched}}$$

For a BHN 200 steel, $q = 0.8$ and $K_f = 1.54$.

---

[1] Based on design procedures described by R. C. Juvinall, "Engineering Consideration of Stress, Strain, and Strength," McGraw-Hill Book Company, New York, 1967.

[2] If fatigue data are not given, they must be determined from the published literature or estimated from other mechanical properties of the material; see H. O. Fuchs and R. I. Stephens, "Metal Fatigue in Engineering," pp. 156–160, John Wiley & Sons, Inc., New York, 1980.

TABLE 7-10
Fatigue reduction factor due to size effect

Diameter, in	C_S
$D \le 0.4$	1.0
$0.4 \le D \le 20$	0.9
$2.0 \le D \le 9.0$	$1 - \dfrac{D - 0.03}{15}$

Returning to the fatigue limit for a small polished specimen, $S_{fl}' = 42,000$ psi, we need to reduce this value because of size effect, surface finish, and type of loading and for statistical scatter

$$S_{fl} = S_{fl}' C_S C_F C_L C_Z \qquad (7\text{-}13)$$

where C_S = factor for size effect
C_F = factor for surface finish
C_L = factor for type of loading
C_Z = factor for statistical scatter

Increasing the specimen size increases the probability of surface defects, and hence the fatigue limit decreases with increasing size. Typical values of C_S are given in Table 7-10. In this example we use $C_S = 0.9$.

Curves for the reduction in fatigue limit due to various surface finishes are available in standard sources.[1] For a standard machined finish in a steel of BHN 200, $C_F = 0.8$.

Laboratory fatigue data (as opposed to simulated service fatigue tests) commonly are determined in a reversed bending loading mode. Other types of loading, e.g., axial and torsional, generate different stress gradients and stress distributions and do not produce the same fatigue limit for the same material. Thus, fatigue data generated in reversed bending must be corrected by a load factor C_L, if the data are to be used in a different loading mode. Table 7-11 gives typical values. Since the bending fatigue data are used for an application involving bending, $C_L = 1.0$.

Fatigue tests show considerable scatter in results. Fatigue limit values are normally distributed with a standard deviation that can be 8 percent of the mean value. If the test or literature value is taken as the mean value of fatigue limit (which in itself is a big assumption), then this value is reduced by a statistical factor[2] according to the reliability level that is desired (Table 7-12).

---

[1] R. C. Juvinall, op. cit., p. 234.
[2] G. Castleberry, *Machine Design*, pp. 108–110, Feb. 23, 1978.

TABLE 7-11
Loading factor for fatigue tests

Loading type	C_L
Bending	1.0
Torsion	0.58
Axial	0.9

TABLE 7-12
Statistical factor for fatigue limit

Reliability, percent	C_Z
50	1.0
99	0.814
98.9	0.752

If we assume a 99 percent reliability level, then $C_Z = 0.814$. Therefore, the unnotched fatigue limit corrected for these factors is

$$S = S'_{fl}C_S C_F C_L C_Z$$
$$= 42,000(0.9)(0.8)(1.0)(0.81) = 24,494 \text{ psi}$$

Since $K_f = 1.54$, the fatigue limit of the notched shaft is estimated to be

$$S_{fl \text{ shaft}} = \frac{24.494}{1.54} = 15,900 \text{ ksi}$$

We note that the working stress is 38 percent of the laboratory value of fatigue limit.

This example is fairly realistic, but it has not included the important situation in which the mean stress is other than zero.[1] This also permits consideration of fatigue strengthening with compressive residual stresses.

Design Example—Safe-life Design

Safe-life design based on failure at a finite number of cycles as influenced by mean stress can be treated by the Haigh diagram as presented by Fuchs.[2] However, a more general and rapidly growing viewpoint is to consider design based on strain-life curves, which are often called low-cycle fatigue curves because much of the data is obtained at less than 10^5 cycles.

When fatigue occurs at a relatively low number of cycles, the stresses that produce failure often exceed the yield strength. Even when the gross stress remains elastic, the localized stress at a notch is inelastic. Under these conditions it is better to carry out fatigue tests under fixed amplitude of strain (strain control) rather than fixed amplitude of stress. Figure 7-43 shows the stress-strain loop that is produced by strain control cycling.

---

[1] See, for example, R. C. Juvinall, op. cit., chap. 14.
[2] H. O. Fuchs, *Trans. ASME*, Ser. D, *J. Basic Eng.*, vol. 87, pp. 333–343, 1965; H. O. Fuchs and R. I. Stephens, op. cit., pp. 148–160.

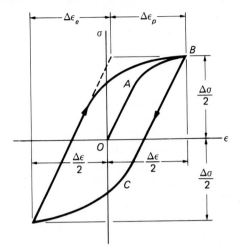

FIGURE 7-43
Typical cyclic stress-strain loop.

The strain-life curve is expressed in terms of the total strain amplitude vs. the number of strain reversals to failure (Fig. 7-44). At given life the total strain is the sum of the elastic and plastic strains.

$$\frac{\Delta\varepsilon}{2} = \frac{\Delta\varepsilon_e}{2} + \frac{\Delta\varepsilon_p}{2} \tag{7-14}$$

Both the elastic and plastic curves are approximated as straight lines. At small strains or long lives the elastic strain predominates, and at large strains or short lives the plastic strain is predominant. The plastic curve has a negative slope of

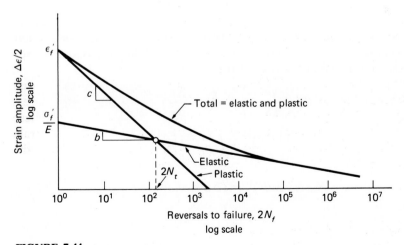

FIGURE 7-44
Typical strain-life curve for mild steel.

c and an intercept at $2N = 1$ of ε'_f. The elastic curve has a negative slope of b and an intercept of σ'_f/E. Substituting into Eq. (7-14) yields

$$\frac{\Delta\varepsilon}{2} = \frac{\sigma'_f}{E}(2N)^b + \varepsilon'_f(2N)^c \tag{7-15}$$

The exponent b ranges from about -0.06 to -0.14 for different materials, and a typical value is -0.1. The exponent c ranges from about -0.5 to -0.7, and -0.6 is a representative value. The term ε'_f in Eq. (7-15). called the fatigue ductility coefficient, is some fraction (0.35 to 1.0) of the true fracture strain measured in the tension test. Likewise, the fatigue strength coefficient σ'_f is approximated by the true fracture stress. Manson[1] has simplified Eq. (7-15) with his method of universal slopes to give

$$\Delta\varepsilon = 3.5\frac{S_u}{E}N^{-0.12} + \varepsilon_f^{0.6}N^{-0.6} \tag{7-16}$$

where S_u = ultimate tensile strength
ε_f = true strain at fracture in tension
E = elastic modulus in the tension test

Equation (7-16) can be used as the first approximation for the strain-life curve for fully reversed fatigue cycles in an unnotched specimen. Strain-life curves give the number of cycles to the formation of a detectable crack with a length between 0.25 and 5 mm, i.e., the life to crack initiation.

An important use of the low-cycle fatigue approach is to predict the life to crack initiation at notches in machine parts where the nominal stresses are elastic but the local stresses and strain at the notch root are inelastic. When there is plastic deformation, both a strain concentration K_ε and a stress concentration K_σ must be considered. Neuber's rule relates these by

$$K_f = (K_\sigma K_\varepsilon)^{1/2} \tag{7-17}$$

The situation is described in Fig. 7-45a, where ΔS and Δe are the elastic stress and strain increments at a location remote from the notch and $\Delta\sigma$ and $\Delta\varepsilon$ are the local stress and strain at the root of the notch.

$$K_\sigma = \frac{\Delta\sigma}{\Delta S} \quad \text{and} \quad K_\varepsilon = \frac{\Delta\varepsilon}{\Delta e}$$

$$K_f = \left(\frac{\Delta\sigma\,\Delta\varepsilon}{\Delta S\,\Delta e}\right)^{1/2} = \left(\frac{\Delta\sigma\,\Delta\varepsilon E}{\Delta S\,\Delta\varepsilon E}\right)^{1/2} \tag{7-18}$$

and

$$K_f(\Delta S\,\Delta e E)^{1/2} = (\Delta\sigma\,\Delta\varepsilon E)^{1/2} \tag{7-19}$$

---

[1] S. S. Manson, *Exp. Mechanics*, vol. 5, no. 7, p. 193, 1965.

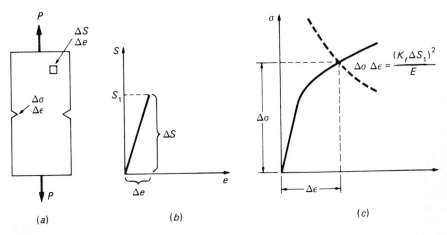

FIGURE 7-45
Notch stress analysis based on Neuber's analysis.

For nominally elastic loading, $\Delta S = \Delta e E$, and

$$K_f \, \Delta S = (\Delta \sigma \, \Delta \varepsilon E)^{1/2} \tag{7-20}$$

Thus, Eq. (7-20) allows stresses and strains remotely measured from the notch to be used to predict notch behavior. Rearranging Eq. (7-20) gives

$$\Delta \sigma \, \Delta \varepsilon = \frac{(K_f \, \Delta S)^2}{E} = \text{const} \tag{7-21}$$

which is the equation of a rectangular hyperbola (Fig. 7-45c). The cyclic stress-strain curve of the material is plotted in Fig. 7-45c, and its intersection with Eq. (7-20) gives the local stress and strain at the notch tip. If the low-cycle fatigue curve (Fig. 7-44) is entered with the value of $\Delta \varepsilon$ at the notch root, an estimate of the fatigue life is obtained.

Actual application of this analysis technique is often much more complex than this simple example.[1] However, the strain-based technique has shown great promise for fatigue analysis of structural components. The technique[2] utilizes the cyclic stress-strain curve and low-cycle fatigue curve. The complex strain-time history that the part experiences is determined experimentally, and this strain signal is separated into individual cycles by a cycle-counting routine such as the rainflow method. The method separates the strain history into

[1] H. O. Fuchs and R. I. Stephens, op. cit., pp. 161–165.

[2] R. W. Landgraf and N. R. LaPointe, Society of Automotive Engineers, Paper 740280, 1974; A. R. Michetti, *Exp. Mechanics*, pp. 69–76, February 1977.

stress-strain hysteresis loops that are comparable with those found in constant-amplitude strain-controlled cycling. Then the notch analysis described above is used in a computer-aided procedure to determine the stresses and strains at the notch for each cycle and the fatigue life is assessed with a cumulative damage rule. Although the design procedures still are evolving, they show great promise of placing fatigue life prediction on a firm basis.

Design Example—Damage-Tolerant Design

Damage-tolerant design starts with the premise that the part contains a fatigue crack of known dimensions and geometry and predicts how many cycles of service are available before the crack will propagate to a catastrophic size that will cause failure. Thus, emphasis is on fatigue crack growth.

Figure 7-46 shows the process of crack propagation from an initial crack of length a_0 to a crack of critical flaw size a_{cr}. The crack growth rate da/dN increases with the cycles of repeated load. An important advance in fatigue design was the realization that there is a generalization in fatigue growth data if the crack growth rate da/dN, is plotted against the stress intensity factor range, $\Delta K = K_{max} - K_{min}$ for the fatigue cycle. Since the stress intensity factor $K = \alpha\sigma\sqrt{\pi a}$ is undefined in compression, K_{min} is taken as zero if σ_{min} is compression in the fatigue cycle.

Figure 7-47 shows a typical plot of rate of crack growth versus ΔK. The typical curve is sigmoidal in shape with three distinct regions. Region I contains the threshold value ΔK_{th} below which there is no observable crack growth. Below K_{th} fatigue cracks behave as nonpropagating cracks. The threshold starts at crack propagation rates of around 10^{-8} in/cycle and at very low values of ΔK, for example $8\,\text{ksi}\sqrt{\text{in}}$ for stress. Region II exhibits essentially a linear relation between $\log da/dN$ and $\log \Delta K$, which results in

$$\frac{da}{dN} = A(\Delta K)^n \tag{7-22}$$

where n = slope of the curve in region II

A = coefficient found by extending the line to $\Delta K = 1\,\text{ksi}\sqrt{\text{in}}$

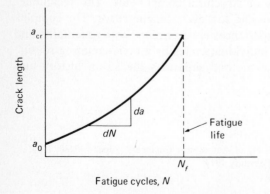

Fatigue cycles, N

FIGURE 7-46
Process of crack propagation (schematic).

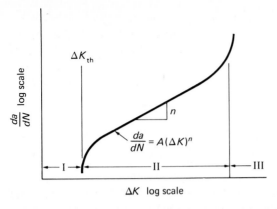

FIGURE 7-47
Schematic fatigue crack growth vs. ΔK curve.

Region III is a region of rapid crack growth that is soon followed by failure. This region is controlled primarily by the fracture toughness K_{Ic}.

The relation between fatigue crack growth and ΔK expressed by Eq. (7-22) ties together fatigue design[1] and linear elastic fracture mechanics (LEFM). The elastic stress intensity factor is applicable to fatigue crack growth even in low-strength, high-ductility materials because the K values needed to cause fatigue crack growth are very low and the plastic zone sizes at the tip are small enough to permit an LEFM approach. By correlating crack growth and stress intensity factor, it is possible to use data generated under constant-amplitude conditions with simple specimens for a broad range of design situations in which K can be calculated. When K is known for the component under relevant loading conditions, the fatigue crack growth life of the *component* can be obtained by integrating Eq. (7-22) between the limits of initial crack size and final crack size. From Sec. 7-14, recall that

$$\Delta K = \alpha \sigma \sqrt{\pi a} \tag{7-23}$$

$$\frac{da}{dN} = A(\Delta K)^n = A(\Delta \sigma \sqrt{\pi a}\,\alpha)^n$$

$$= A(\Delta \sigma)^n (\pi a)^{n/2} \alpha^n \tag{7-24}$$

Now, if the final crack length a_f results in fracture,

$$\sigma_f = \frac{1}{\pi}\left(\frac{K_{Ic}}{\sigma_{max}\alpha}\right)^2 \tag{7-25}$$

$$N_f = \int_0^{N_f} dN = \int_{a_0}^{a_f} \frac{da}{A(\Delta \sigma)^n (\pi a)^{n/2} \alpha^n} \tag{7-26}$$

$$N_f = \frac{a_f^{(-n/2)+1} - a_0^{(-n/2)+1}}{(-n/2 + 1)A(\Delta \sigma)^n (\pi)^{n/2} \alpha^n} \tag{7-27}$$

---

[1] R. P. Wei, *Trans. ASME*, Ser. H., *J. Eng. Materials Tech.*, vol. 100, pp. 113–120, 1978.

Equation (7-27) is the integration of Eq. (7-24) for the special case in which α is not a function of a and $n \neq 2$. Other cases require numerical integration.

Working with Eq. (7-27) shows that the fatigue life is much more dependent on the small initial crack size than on K_{Ic}. Thus, refined NDE techniques to detect small flaws are an important part of a fracture control program.

7-16 DESIGN FOR CORROSION RESISTANCE

Failure of metal components by corrosion is as common as failure due to mechanical causes, such as brittle fracture and fatigue. The National Bureau of Standards estimates that corrosion annually costs the United States $70 billion, of which at least $10 billion could be prevented by better selection of materials and design procedures. Although corrosion failures are minimized by proper materials selection and careful attention to control of metallurgical structure through heat treatment and processing, many corrosion-related failures can be minimized by proper understanding of the interrelation of the fundamental causes of corrosion and design details.[1]

Corrosion of metals is driven by the basic thermodynamic force of a metal to return to the oxide or sulfide form, but it is more related to the electrochemistry of the reactions of a metal in an electrolytic solution. There are eight basic forms of corrosion.[2]

Uniform attack. The most common form of corrosion is uniform attack. It is characterized by a chemical or electrochemical reaction that proceeds uniformly over the entire exposed surface area. The metal becomes thinner and eventually fails.

Galvanic corrosion. The potential difference that exists when two dissimilar metals are immersed in a corrosive or conductive solution is responsible for galvanic corrosion. The less-resistant (anodic) metal is corroded relative to the cathodic metal. Table 7-13 gives a brief galvanic series for some commercial alloys immersed in seawater. Note that the relative position in a galvanic series depends on the electrolytic environment as well as the metal's surface chemistry (passivity).

To minimize galvanic corrosion, use pairs of metals that are close together in the galvanic series and avoid situations in which a small anode metal is connected to a larger surface area of more noble metal. If two metals

---

[1] V. P. Pludek, "Design and Corrosion Control," John Wiley & Sons, Inc., New York, 1977.

[2] M. G. Fontana and N. D. Greene, "Corrosion Engineering," 2d ed., McGraw-Hill Book Company, New York, 1978.

TABLE 7-13
A brief galvanic series for commercial metals and alloys

Noble (cathodic)	Platinum
	Gold
	Titanium
	Silver
	316 Stainless steel
	304 Stainless steel
	410 Stainless steel
	Nickel
	Monel
	Cupronickel
	Cu–Sn bronze
	Copper
	Cast iron
	Steel
Active (anodic)	Aluminum
	Zinc
	Magnesium

far apart in the series must be used in contact, they should be insulated electrically from each other. Do not coat the anodic surface to protect it, because most coatings are susceptible to pinholes. The coated surface would corrode rapidly in contact with a large cathodic area. When a galvanic couple is unavoidable, consider utilizing a third metal that is anodic and sacrificial to both of the other metals.

Crevice corrosion. An intense localized corrosion frequently occurs within crevices and other shielded areas on metal surfaces exposed to corrosive attack. This type of attack usually is associated with small volumes of stagnant liquid at design details such as holes, gasket surfaces, lap joints, and crevices under bolt and rivet heads.

Pitting. Pitting is a form of extremely localized attack that produces holes in the metal. It is an especially insidious form of corrosion because it causes equipment to fail after only a small percentage of the designed-for weight loss.

Intergranular corrosion. Localized attack along the grain boundaries with only slight attack of the grain faces is called intergranular corrosion. It is especially common in austenitic stainless steel that has been sensitized by heating to the temperature range 950 to 1450°F. It can occur either during heat treatment for stress relief or during welding, when it is known as *weld decay*.

Selective leaching. The removal of one element from a solid-solution alloy by corrosion processes is called selective leaching. The most common example of it is the selective removal of zinc from brass (dezincification), but aluminum, iron, cobalt, and chromium also can be removed. When selective leaching occurs, the alloy is left in a weakened, porous condition.

Erosion-corrosion. Deterioration at an accelerated rate is caused by relative movement between a corrosive fluid and a metal surface; it is called erosion-corrosion. Generally the fluid velocity is high and mechanical wear and abrasion may be involved, especially when the fluid contains suspended solids. Erosion destroys protective surface films and enhances chemical attack. Design can play an important role in erosion control in such areas as reducing fluid velocity, eliminating situations in which direct impingement occurs, and minimizing abrupt changes in the direction of flow. Some erosion situations are so aggressive that neither a suitable material nor design can ameliorate the problem. Here the role of design is to provide for easy detection of damage and for quick replacement of damaged components.

A special kind of erosion-corrosion is *cavitation,* which arises from the formation and collapse of vapor bubbles near the metal surface. Rapid bubble collapse can produce shock waves that cause local deformation of the metal surface.

Another special form of erosion-corrosion is fretting corrosion. It occurs between two surfaces under load that are subjected to cycles of relative motion. Fretting produces breakdown of the surface into an oxide debris and results in surface pits and cracks that usually lead to fatigue cracks.

Stress-corrosion cracking. Cracking caused by the simultaneous action of a tensile stress and a specific corrosive medium is called stress-corrosion cracking. The stress may be a result of applied loads or "locked-in" residual stress. Only specific combinations of alloys and chemical environment lead to stress-corrosion cracking. However, many are of common occurrence, such as aluminum alloys and seawater, copper alloys and ammonia, mild steel and caustic soda, and austenitic steel and saltwater.[1] Over 80 combinations of alloys and corrosive environments are known to cause stress-corrosion cracking. Design against stress-corrosion cracking involves selecting an alloy that is not susceptible to cracking in the service environment; but if that is not possible, then the stress level should be kept low. The concepts of fracture mechanics have been applied to SCC.

General precautions. Some of the more obvious design rules for preventing corrosion failure have been discussed above. In addition, tanks and containers

---

[1] B. F. Brown, "Stress Corrosion Cracking Control Measures," Natl. Bur. Standards Monograph 156, GPO, 1977 (AD/A 043 000).

should be designed for easy draining and easy cleaning. Welded rather than riveted tanks will provide less opportunity for crevice corrosion. When possible, design to exclude air; if oxygen is eliminated, corrosion can often be reduced or prevented. Exceptions to that rule are titanium and stainless steel, which are more resistant to acids that contain oxidizers than to those that do not. Many examples of design details for corrosion prevention are given by Pludek[1] and Elliott.[2]

7-17 DESIGNING WITH PLASTICS

Most mechanical design is taught with the implicit assumption that the part will be made from a metal. However, plastics are increasingly finding their way into design applications because of their light weight, attractive appearance, and freedom from corrosion and the ease with which many parts may be manufactured from polymers.

However, polymers are sufficiently different from metals to warrant special attention in design.[3] The chief differences are in mechanical behavior. Polymers have an elastic modulus that is 10 to 100 times lower than that of metals, so design for stiffness and rigidity must be given special attention. Also, the strength properties, yield strength, and tensile strength of most polymers are time-dependent at room temperature. That imposes different ways of looking at allowable strength limits.

Some short-time mechanical properties of metals and polymers at room temperature are compared in Table 7-14. Because many polymers do not have a truly linear initial portion of the stress-strain curve, it is difficult to specify a yield strength; and so the tensile strength usually is reported. Some polymers are brittle at room temperature. That fact is measured rather crudely by the impact test. Note that the units used in testing polymers are foot-pounds per inch. The marked improvement by introducing glass fiber reinforcement to produce a composite structure should be noted. The data in Table 7-14 are aimed at illustrating the differences in properties between metals and polymers. They should not be used for design purposes. Moreover, it should be

---

[1] V. R. Pludek, op. cit.

[2] P. Eliott, Design Details to Minimize Corrosion in "Metals Handbook", 9th ed. Vol. 13, Corrosion, pp. 338–343, ASM International, Metals Park, OH, 1987.

[3] S. Levy and J. H. DuBois, "Plastics Production Design Engineering Handbook," 2d ed., Methuen Inc., New York, 1985; E. Baer (ed.), "Engineering Design for Plastics," Reinhold Publishing Co., New York, 1964; R. D. Beck, "Plastic Product Design," 2d ed., Van Nostrand Reinhold Company, New York, 1980; R. L. E. Brown, "Design and Manufacture of Plastic Parts," John Wiley & Sons, Inc., New York, 1980. E. Miller (ed.), "Plastics Products Design Handbook," Part A: Materials and Components (1981); Part B: Processes and Design for Processes (1983), Marcel Dekker Inc., New York; P. C. Powell, "Engineering with Polymers," Chapman and Hall, New York, 1983.

TABLE 7-14
Typical mechanical properties of metals and polymers

Material	Young's modulus, psi $\times 10^{-6}$	Tensile strength, ksi	Impact ft-lb/in	Specific gravity, g/cc
Aluminum alloys	10	$20 - 60$		2.7
Steel	30	$40 - 200$		7.9
Polyethylene	$0.08 - 0.15$	$3 - 6$	$1 - 12$	0.94
Polystyrene	$0.35 - 0.60$	$5 - 9$	$0.2 - 0.5$	1.1
Polycarbonate	$0.31 - 0.35$	$8 - 10$	$12 - 16$	1.2
Polyacetal	$0.40 - 0.45$	$9 - 10$	$1.2 - 1.8$	1.4
Polyester-glass reinforced	$1.5 - 2.5$	$20 - 30$	$10 - 20$	1.7

recognized that plastics show a wide spread of properties between different polymers and a considerable variation in properties for a given polymer depending on its processing.

The low elastic stiffness of polymers is a problem that often can be surmounted by design. For example, the flexural rigidity of a flat plate is given by[1]

$$D = \frac{Et^3}{12(1 - v^2)} \tag{7(28)}$$

where E = Young's modulus
$\quad v$ = Poisson's ratio
$\quad t$ = the plate thickness

If a polymer has $E = 300,000$ psi and a metal $E = 10,000,000$ psi, the polymer plate will have to be about three times thicker than the metal plate to have the same flexural rigidity. That thickness is probably cost-prohibitive. However, if we increase the thickness of the structure by having two plastic plates of thickness $t/2$ separated by a distance $3t$ (Fig. 7-48b), then the moment of inertia of the split plate is 37 times that of the single plate with the same total thickness.

The separation of plates may be achieved by a thin honeycomb structure or a lightweight foamed plastic core (Fig. 7-49). Using the geometry in Fig. 7-48b, the structural rigidity of the double plate exceeds that of a metal plate of equivalent thickness having an elastic modulus 30 times greater. However, it should be noted that the improved performance of a sandwich panel is limited chiefly to bending resistance. The shear strength of the core must be great enough to resist the shear stresses generated by bending and the interface between the skin and core must be integrally bonded.

---

[1] J. H. Faupel, "Engineering Design," John Wiley & Sons, Inc., New York, 1964.

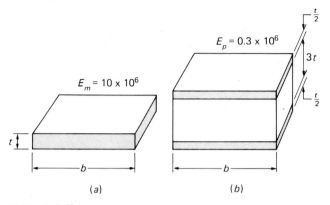

FIGURE 7-48
Design to achieve plates of equivalent flexural rigidity.

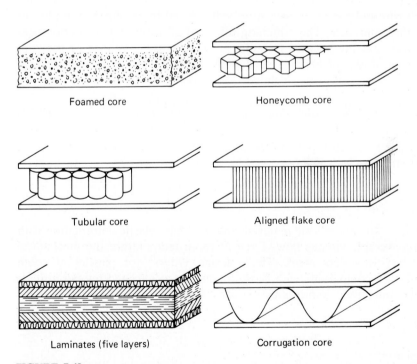

Foamed core

Honeycomb core

Tubular core

Aligned flake core

Laminates (five layers)

Corrugation core

FIGURE 7-49
Examples of sandwich and laminate structures to achieve increased structural rigidity. (*From W. W. Chow, "Cost Reduction in Product Design," p. 211, Van Nostrand Reinhold Company. New York, 1978; reprinted by permission of publisher.*)

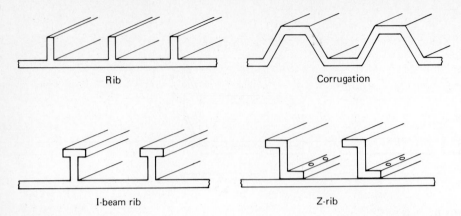

Rib

Corrugation

I-beam rib

Z-rib

FIGURE 7-50
Examples of stiffened skin designs. (*From W. W. Chow, "Cost Reduction in Production Design,"* *p. 211, Van Nostrand Reinhold Company, New York, 1978; reprinted by permission of publisher.*)

Increases in stiffness may be achieved somewhat more cheaply by stiffnened skin approaches that use ribs and corrugations (Fig. 7-50). The ease with which ribs and corrugations can be molded and thermoformed in plastics enhances this approach. The stiffening effect of ribs and corrugations arises from the fact that the geometry results in an increased distance of material from the neutral axis of bending so that the moment of inertia is increased. The stiffening factor may be calculated from geometry[1] and is expressed by

$$\text{SF} = \text{stiffening factor} = \frac{\text{I with ribs}}{\text{I without ribs}} \tag{7-29}$$

When a polymer is substituted for a metal, equivalent stiffness is achieved if the total stiffening factor is unity.

$$\text{Total stiffening factor} = \text{SF}\left(\frac{t_p}{t_s}\right)^3 \frac{E_p}{E_s} \tag{7-30}$$

Polymers are viscoelastic materials that combine elastic deformation with a slow irrecoverable viscous flow. Even at room temperature the mechanical properties are time-dependent. Thus, design criteria for plastics at room temperature are comparable with those for metals at elevated temperature at which creep and stress rupture must be considered.[2]

---

[1] W. W. Chow, "Cost Reduction in Product Design," chap. 4, Van Nostrand Reinhold Company, New York, 1978.

[2] J. G. Williams, "Stress Analysis of Polymers," 2d ed., John Wiley–Halsted Press, New York, 1980.

Although polymers introduce special problems in design, they have many advantages in manufacture. Compression and injection molding of plastics are high-productivity processes that produce parts to final shape. Other processes such as sheet thermoforming, blow molding, and extrusion permit deformation exceeding that normally found with metals. Of course, each process has its design limitations[1] that must be understood. However, when stress levels are not severe, plastics have a versatility in design[2] that is unsurpassed.

BIBLIOGRAPHY

Amstead, B. H., P. F. Ostwald, and M. L. Begelman: "Manufacturing Processes," 8th ed., John Wiley & Sons, Inc., New York, 1987.

Bolz, R. W.: "Production Processes: The Productivity Handbook," Industrial Press Inc., New York, 1981.

Bralla, J. G. (ed.): "Handbook of Product Design for Manufacturing," McGraw-Hill Book Co., New York, 1985.

DeGarmo, E. P.: "Materials and Processes in Manufacturing," 5th ed., Macmillan Publishing Co., New York, 1979.

Kalpakjian, S.: "Manufacturing Engineering and Technology," Addison-Wesley Publ. Co., Reading, MA., 1989.

Lindberg, R. A.: "Processes and Materials of Manufacturing," 2d ed., Allyn & Bacon, Inc., Boston, 1977.

Niebel, B. W., A. B. Draper, and R. A. Wysk: "Modern Manufacturing Process Engineering," McGraw-Hill Book Co., New York, 1989.

Schey, J. A.: "Introduction to Manufacturing Processes," 2d ed., McGraw-Hill Book Company, New York, 1987.

Trucks, H. E. and G. M. Lewis: "Designing for Economical Production," 2d ed., Society of Manufacturing Engineers, Dearborn, Mich., 1987.

---

[1] S. Levy and J. H. DuBois, op. cit., chap. 9; "Engineered Materials Handbook", Vol. 2, Plastics, pp. 277–403, ASM International Metals Park, OH, 1988.

[2] W. W. Chow, op. cit., chap. 5.

CHAPTER
8

ECONOMIC DECISION MAKING

8-1 INTRODUCTION

Throughout this book we have repeatedly emphasized that the engineer is a decision maker and that engineering design basically consists of making a continual stream of decisions. We also have emphasized from the beginning that engineering involves the application of science to real problems of society. In this real-world context, one cannot escape the fact that economics (or costs) may play a role as big as, or bigger than, that of technical considerations in the decision making of design. In fact, it sometimes is said, although a bit facetiously, that an engineer is a person who can do for $1.00 what any fool can do for $2.00.

The major engineering feats that built this nation—the railroads, major dams, and waterways—required a methodology for predicting costs and balancing them against alternative courses of action. In a major engineering project, costs and revenues will occur at various points of time in the future. The methodology for handling this class of problems is known as engineering economy or engineering economic analysis or simply cost engineering. Familiarity with the concepts and approach of engineering economy generally is considered to be part of the kit of tools of all engineers, regardless of discipline. Indeed, an examination on the fundamentals of engineering economy is required for professional engineering registration in all disciplines in all states.

The chief concept in engineering economy is that *money has a time value.* Paying out $1.00 today is more costly than paying out $1.00 a year from now. A $1.00 invested today is worth $1.00 plus interest a year from now. Engineering economy recognizes the fact that use of money is a valuable asset. Money can be rented in the same way one can rent an apartment, but the charge for using it is called interest rather than rent. This time value of money makes it more profitable to push expenses into the future and bring revenues into the present as much as possible.

Before proceeding into the mathematics of engineering economy, it is important to understand where engineering economy sits with regard to related disciplines like economics and accounting. Economics generally deals with broader and more global issues than engineering economy, such as the forces that control the money supply and trade between nations. Engineering economy uses the interest rate established by the economic forces to solve a more specific and detailed problem. However, it usually is a problem concerning alternative costs in the future. The accountant is more concerned with determining exactly, and often in great detail, what costs have been in the past. One might say that the economist is an oracle, the engineering economist is a fortune teller, and the accountant is a historian.

8-2 MATHEMATICS OF TIME VALUE OF MONEY

If we borrow a present sum of money or principal P at a simple interest rate i, the annual cost of interest is $I = Pi$. If the loan is repaid in a lump sum S at the end of n years, the amount required is

$$S = P + nI = P + nPi = P(1 + ni) \qquad (8\text{-}1)$$

where S = future worth (sometimes denoted by F)
  P = present worth
  I = annual cost of interest
  i = annual interest rate
  n = number of years

If we borrow $1000 for 6 years at 10 percent simple interest rate, we must repay at the end of 6 years:

$$S = P(1 + ni) = \$1000[1 + 6(0.10)] = \$1600$$

Therefore, we see that $1000 available today is not equivalent to $1000 available in 6 years. Actually, $1000 in hand today is worth $1600 available in only 6 years at 10 percent simple interest.

We can also see that the *present worth* of $1600 available in 6 years and invested at 10 percent is $1000.

$$P = \frac{S}{1 + ni} = \frac{\$1600}{1 + 0.6} = \$1000$$

In making this calculation we have discounted the future sum back to the present time. In engineering economy the terminology *discounted* refers to bringing dollar values *back in time* to the present.

Compound Interest

However, you are aware from your personal banking experiences that financial transactions usually use compound interest. In *compound interest,* the interest due at the end of a period is not paid out but is instead added to the principal. During the next period, interest is paid on the total sum.

First period: $S_1 = P + Pi = P(1 + i)$
Second period: $S_2 = P(1 + i) + iP(1 + i) = [P(1 + i)](1 + i) = P(1 + i)^2$
Third period: $S_3 = P(1 + i)^2 + iP(1 + i)^2 = P[(1 + i)^2](1 + i) = P(1 + i)^3$
nth period: $S_n = P(1 + i)^n$ (8-2)

Alternatively, we can write this as:

$$S = P(F_{PS}) = P(\text{SCAF})   (8\text{-}3)$$

where F_{PS} is the single-payment compound-amount factor (SCAF) that converts present value P to future worth S.

Example 8-1. How long will it take money to double if it is compounded annually at a rate of 10 percent per year?

Solution. $S = PF_{PS, 10\%, n}$ but $S = 2P$
$2P = PF_{PS, 10\%, n}$

Therefore, the answer clearly is found in a table of single-payment compound-amount factors at the year n for which $F_{PS} = 2.0$. Examining Table B-2 in Appendix B we see that, for $n = 7$, $F_{PS} = 1.949$ and, for $n = 8$, $F_{PS} = 2.144$. Linear exterpolation gives us $F_{PS} = 2.000$ at $n = 7.2$ years. We can generalize the result to establish the financial rule of thumb that the number of years to double an investment is 72 divided by the interest rate (expressed as an integer).

Usually in engineering economy, n is in years and i is an annual interest rate. However, in banking circles the interest may be compounded at periods other than one year. Compounding at the end of shorter periods, such as daily, raises the effective interest rate. If we define r as the nominal annual interest rate and p as the number of interest periods per year, then the interest rate per interest period is $i = r/p$ and the number of interest periods in n years is pn.

Using that notation, Eq. (8-2) becomes:

$$S = P\left[\left(1 + \frac{r}{p}\right)^p\right]^n \tag{8-4}$$

Note that when $p = 1$, the above expression reduces to Eq. (8-2). Standard compound interest tables that are prepared for $p = 1$ can be used for other than annual periods. To do so, use the table for $i = r/p$ and for a number of years equal to $p \times n$. Alternatively, use the interest table corresponding to n years and an effective rate of yearly return equal to $(1 + r/p)^p - 1$.

If the number of interest periods per year p increases without limit, then $i = r/p$ approaches zero.

$$S = P \lim_{p \to \infty} \left(1 + \frac{r}{p}\right)^{pn} \tag{8-5}$$

From calculus, an important limit is $\lim_{x \to 0} (1 + x)^{1/x} = 2.7178 = e$. If we let $x = r/p$, then

$$pn = \frac{p}{r} rn = \frac{1}{x} rn$$

Since $p = r/x$, as $p \to \infty$, $x \to 0$, so Eq. (8-5) is rewritten as

$$S = P\left[\lim_{x \to 0} (1 + x)^{1/x}\right]^{rn} = Pe^{rn} \tag{8-6}$$

Table 8-1 shows the influence of the number of interest periods per year on the effective rate of return.

Engineering economy was developed to deal with sums of money at different times in the future. These situations can rapidly become quite complex, so that a methodology for setting up problems is needed. A good procedure is to place the dollar amounts on a dollar-time diagram, as shown in Fig. 8-1. Receipts (income) are placed above the line, and disbursements (costs) are placed below the line. The length of the arrow should be proportional to the dollar amount.

TABLE 8-1
Influence of compounding period on effective rate of return

Frequency of compounding	No. annual interest periods p	Interest rate for period, %	Effective rate of yearly return, %
Annual	1	12.0	12.0
Semiannual	2	6.0	12.4
Quarterly	4	3.0	12.6
Monthly	12	1.0	12.7
Continuously	∞	0	12.75

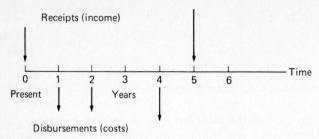

FIGURE 8-1
Dollar-time diagram.

Uniform Annual Series

In many situations we are concerned with a uniform series of receipts or disbursements occurring equally at the end of each period. Examples are the payment of a debt on the installment plan, setting aside a sum S that will be available at a future date for replacement of equipment, and a retirement annuity that consists of series of equal payments instead of a lump sum payment. We will let R be the equal end-of-the-period payment that makes up the uniform annual series. Other authors may use the symbol A for the same thing.

Figure 8-2 shows that if an annual sum R is invested at the end of each year for 3 years, the total sum S at the end of 3 years will be the sum of the compound amount of the individual investments R

$$S = R(1+i)^2 + R(1+i) + R$$

and for the general case of n years.

$$S = R(1+i)^{n-1} + R(1+i)^{n-2} + \cdots + R(1+i)^2 + R(1+i) + R \qquad (8\text{-}7)$$

Multiplying by $1+i$, we get

$$S(1+i) = R(1+i)^n + R(1+i)^{n-1} + \cdots + R(1+i)^3 + R(1+i)^2 + R(1+i)$$

$$(8\text{-}8)$$

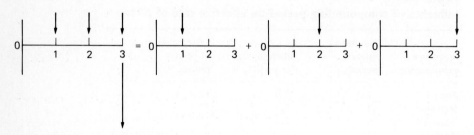

FIGURE 8-2
Equivalence of a uniform annual series.

Subtracting Eq. (8-7) from Eq. (8-8):

$$(1 + i)S = R[(1 + i)^n + \cancel{(1+i)^{n-1}} + \cdots + \cancel{(1+i)^3} + \cancel{(1+i)^2} + \cancel{(1+i)}]$$

$$S = R[\cancel{(1+i)^{n-1}} + \cancel{(1+i)^{n-2}} + \cdots + \cancel{(1+i)^2} + \cancel{(1+i)} + 1]$$

$$iS = R[(1 + i)^n - 1]$$

$$S = R \frac{(1 + i)^n - 1}{i} \tag{8-9}$$

Equation (8-9) gives the future sum of n uniform payments of R when the interest rate is i. This equation may also be written:

$$S = RF_{RS} = R(\text{USCAF}) \tag{8-10}$$

where F_{RS} is the uniform-series compound amount factor (USCAF) that converts a series R to future worth S.

By solving Eq. (8-9) for R, we have the uniform series of end-of-period payments, that, at compound interest i, provide a future sum S.

$$R = S \frac{i}{(1 + i)^n - 1} \tag{8-11}$$

This type of calculation often is used to set aside money in a sinking fund to provide funds for replacing worn-out equipment.

$$R = SF_{SR} = S(\text{SFF}) \tag{8-12}$$

where $F_{SR} = \text{SFF}$ is the sinking fund factor.

By combining Eq. (8-2) with Eq. (8-9), we develop the relation for the present worth of a uniform series of payments R:

$$P = R \frac{(1 + i)^n - 1}{i(1 + i)^n} = RF_{RP} \tag{8-13}$$

Solving Eq. (8-13) for R gives the important relation for capital recovery:

$$R = P \frac{i(1 + i)^n}{(1 + i)^n - 1} = PF_{PR} = P(\text{CRF}) \tag{8-14}$$

where CRF is the capital recovery factor. The R in Eq. (8-14) is the annual payment needed to return the initial capital investment P plus interest on that investment at a rate i over n years.

Capital recovery is an important concept in engineering economy. It is important to understand the difference between capital recovery and sinking fund. Consider the following example:

Example 8-2 What annual investment must be made at 10 percent to provide funds for replacing a $10,000 machine in 20 years?

$$R = SF_{SR, 10\%, 20} = \$10,000(0.01746)$$

$$R = \$174.60 \text{ per year for the sinking fund}$$

What is the annual cost of capital recovery of $10,000 at 10 percent in 20 years?

$$R = PF_{PR,10\%,20} = \$10,000(0.11746)$$

$$R = \$1174.60 \text{ per year for capital recovery}$$

We see that

$$F_{PR} = F_{SR} + i$$

$$0.11746 = 0.01746 + 0.10000$$

Annual cost capital recovery	=	annual cost sinking fund	+	annual interest cost

$$\$1174.60 = \$174.60 + 0.10(\$10,000)$$

With a sinking fund we put away each year a sum of money that, over n years, together with accumulated compound interest, equals the required future amount S. With capital recovery we put away enough money each year to provide for replacement in n years *plus we charge ourselves interest on the invested capital*. The use of capital recovery is a conservative but valid economic strategy. The amount of money invested in capital equipment ($10,000 in the above example) represents an *opportunity cost*, since we are

TABLE 8-2
Summary of S, P, R, and R_b relationships

Item	Conversion	Algebraic relation	Relation by factor	Name of factor
1	P to S	$S = P[(1+i)^n]$	$S = PF_{PS,i,n}$	Compound interest factor
2	S to P	$P = S[(1+i)^{-n}]$	$P = SF_{SP,i,n}$	Present worth factor
3	R to P	$P = R\dfrac{(1+i)^n - 1}{i(1+i)^n}$	$P = RF_{RP,i,n}$	Uniform series present worth factor
4	P to R	$R = P\dfrac{i(1+i)^n}{(1+i)^n - 1}$	$R = PF_{PR,i,n}$	Capital recovery factor
5	R to S	$S = R\dfrac{(1+i)^n - 1}{i}$	$S = RF_{RS,i,n}$ $S = RF_{RP,i,n}F_{PS,i,n}$	Equal payment series future worth factor
6	S to R	$R = S\dfrac{i}{(1+i)^n - 1}$	$R = SF_{SR,i,n}$ $R = SF_{SP,i,n}F_{PR,i,n}$	Sinking fund factor
7	R_b to P	$P = R_b(1+i)\dfrac{(1+i)^n - 1}{i(1+i)^n}$	$P = R_b(1+i)F_{RP,i,n}$	
8	P to R_b	$R_b = \dfrac{P}{1+i}\dfrac{i(1+i)^n}{(1+i)^n - 1}$	$R_b = \dfrac{P}{1+i}F_{PR,i,n}$	
9	R_b to S	$S = R_b(1+i)\dfrac{(1+i)^n - 1}{i}$	$S = R_b(1+i)F_{RS,i,n}$ $S = R_b(1+i)F_{RP,i,n}F_{PS,i,n}$	
10	S to R_b	$R_b = \dfrac{S}{1+i}\dfrac{i}{(1+i)^n - 1}$	$R_b = S\dfrac{1}{1+i}F_{SR,i,n}$ $R_b = S\dfrac{1}{1+i}F_{SP,i,n}F_{PR,i,n}$	

forgoing the revenue that the $10,000 could provide if invested in interest-bearing securities.

In working with a uniform series R it is conventional to assume that R occurs at the end of each period. However, if a beginning-of-period payment R_b is required, it can be determined readily by discounting R one year to the present according to Eq. (8-3), where $S = R$, $P = R_b$, and $n = 1$. Thus,

$$R = R_b(1 + i) \qquad (8\text{-}15)$$

and this can be substituted into Eq. (8-9), (8-11), or (8-13).

A summary of the compound interest relations among S, P, R, and R_b is given in Table 8-2.

8-3 COST COMPARISON

Having covered the usual compound interest relations, we now are in a position to use them to make economic decisions. A typical decision is which of two courses of action is the least expensive when time value of money is considered.

Present Worth Analysis

When the two alternatives have a common time period, a comparison on the basis of present worth is advantageous.

Example 8-3. Two machines each have a useful life of 5 years. If money is worth 10 percent, which machine is more economical?

	A	B
Initial cost	$25,000	$15,000
Yearly maintenance cost	2,000	4,000
Rebuilding at end of third year	—	3,500
Salvage value	3,000	
Annual benefit from better quality production	500	

From the cost diagrams at the top of page 378 we see that the cash flows definitely are different for the two alternatives. To place them on a common basis for comparison, we discount all costs back to the present time.

$$P_A = 25{,}000 + (2000 - 500)F_{RP,10\%,5} - 3000F_{SP,10\%,5}$$
$$= 25{,}000 + 1500(3.791) - 3000(0.621) = \$28{,}823$$

$$P_B = 15{,}000 + 4000F_{RP,10\%,5} + 3500F_{SP,10\%,3}$$
$$= 15{,}000 + 4000(3.791) + 3500(0.751) = \$32{,}793$$

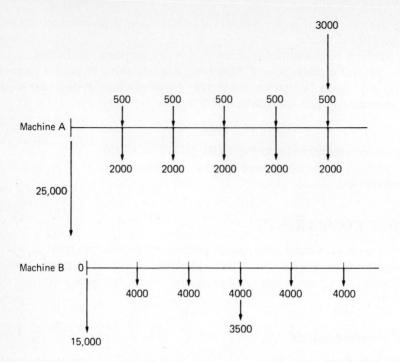

Machine A is the most economical because it has the lowest cost on a present worth basis. In this example we considered both 1) costs plus benefits (savings) due to reduced scrap rate and 2) resale value at the end of the period of useful life. Thus, we really determined the *net present worth* for each alternative. We should also point out that present worth analysis is not limited to the comparison of only two alternatives. We could consider any number of alternatives and select the one with the smallest net present worth of costs.

In Example 8-3, both alternatives had the same life. Thus, the time period was the same and the present worth could be determined without ambiguity. Suppose we want to use present worth analysis for the situation shown below:

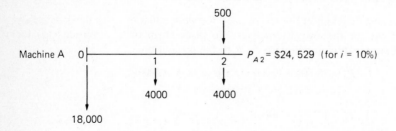

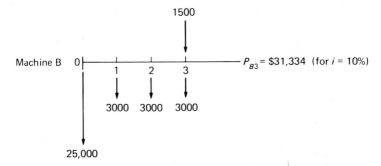

We cannot directly compare P_A and P_B because they are based on different time periods. One way to handle the problem would be to use a common 6-year period, in which we would replace machine A three times and replace machine B twice. This procedure works when a common multiple of the individual periods can be found easily, but a more direct approach is to convert the present worth based on a period n_1 to an equivalent P based on n_2 by[1]

$$P_{n_2} = \frac{P_{n_1} F_{PR,i,n_1}}{F_{PR,i,n_2}} \tag{8-16}$$

For our example, we convert P_B from a 3-year time period to a 2-year period.

$$P_{B_2} = P_{B_3} \frac{F_{PR,10\%,3}}{F_{PR,10\%,2}} = 31{,}334 \frac{0.402}{0.576} = \$21{,}867$$

and $P_{A_2} = \$24{,}529$. Now, when compared on the basis of equal time periods, machine B is the most economical.

Annual Cost Analysis

In the annual cost method, the cash flow over time is converted to an equivalent uniform annual cost or benefit. In this method no special procedures need be used if the time period is different for each alternative, because all comparisons are an annual basis ($n = 1$).

Example	Machine A	Machine B
First cost	$10,000	$18,000
Estimated life	20 years	35 years
Estimated salvage	0	$3000
Annual cost of operation	$4000	$3000

---

[1] For a derivation of Eq. (8-16) see F. C. Jelen, "Cost and Optimization Engineering," pp. 24–25, McGraw-Hill Book Company, New York, 1970.

$$R_A = 10,000F_{PR,10\%,20} + 4000 = 10,000(0.1175) + 4000 = \$5175$$

$$R_B = (18,000 - 3000)F_{PR,10\%,35} + 3000(0.10) + 3000 = \$4855$$

Machine B has the lowest annual cost and is the most economical. Note that in calculating the annual cost of capital recovery for machine B we used the difference between the first cost and the salvage value; for it is only this amount of money that must be recovered. However, although the salvage value is returned to us, we are required to wait until the end of the useful life of the machine to recover it. Therefore, a charge for the annual cost of the interest on the investment tied up in the salvage value is made as part of the annual cost analysis.

Perhaps a more direct way to handle the case of machine B in the above example is to determine the equivalent annual cost based on the cash disbursements minus the annual benefit of the future resale value.

$$R_B = 18,000F_{PR,10\%,35} + 3000 - 3000F_{SR,10\%,35}$$
$$= 18,000(0.1037) + 3000 - 3000(0.0037) = \$4855$$

Capitalized Cost Analysis

Capitalized cost is a special case of present worth analysis. The capitalized cost of a project is the present value of providing for that project in perpetuity ($n = \infty$). The concept was originally developed for use with public works, such as dams and waterworks, that have long lives and provide services that must be maintained indefinitely. Capitalized cost subsequently has been used more broadly in economic decision making because it provides a method that is independent of the time period of the various alternatives.

We can develop the mathematics for capitalized cost quite simply from Eq. (8-16). If we let $n_2 = \infty$ and $n_1 = n$, then

$$P_\infty = P_n \frac{F_{PR,i,n}}{F_{PR,i,\infty}}$$

$$F_{PR,i,n} = \frac{i(1+i)^n}{(1+i)^n - 1} \qquad F_{PR,i,\infty} = \frac{i(1+i)^\infty}{(1+i)^\infty - 1} = i$$

Therefore, the capitalized cost K of a present sum P is given by:

$$K = P_\infty = P\frac{(1+i)^n}{(1+i)^n - 1} = PF_{PK,i,n} \tag{8-17}$$

Since many tables of compound interest factors do not include capitalized cost, we need to note that

$$F_{PK,i,n} = \frac{F_{PR,i,n}}{i} \tag{8-18}$$

In addition, the capitalized cost of an annual payment R is determined as follows:

$$P = RF_{RP}: \qquad\qquad K = PF_{PK} = \frac{PF_{PR}}{i}$$

Substituting for P: $\quad K = \frac{RF_{RP}F_{PR}}{i} = \frac{R}{i} \quad \text{since} \quad F_{RP} = \frac{1}{F_{PR}} \qquad (8\text{-}19)$

The capitalized cost is the present worth of providing for a capital cost in perpetuity; i.e., we assume there will be an infinite number of renewals of the initial capital investment. Consider a bank of condenser tubes that cost \$10,000 and have an average life of 6 years. If $i = 10$ percent, then the capitalized cost is

$$K = PF_{PK,i,n} = P\frac{F_{PR,i,n}}{i} = 10,000\frac{0.2296}{0.10} = \$22,960$$

We note that the excess over the first cost is $22,960 - 10,000 = \$12,960$. If we invest that amount for the 6-year life of the tubes,

$$S = PF_{PS,10\%,6} = 12,960(1.772) = \$22,960$$

Thus, when the tubes need to be replaced, we have generated \$22,960. We take \$10,000 to purchase a new set of tubes (we are neglecting inflation) and invest the difference ($22,960 - 10,000 = 12,960$) at 10 percent for 6 years to generate another \$22,960. We can repeat this process indefinitely. The capital cost is provided for in perpetuity.

Example 8-4. Compare the continuous process and the batch process on the basis of capitalized cost analysis if $i = 10$ percent.

Solution.

	Continuous process	**Batch process**
First cost	\$20,000	\$6000
Useful life	10 years	15 years
Salvage value	0	\$500
Annual power costs	\$1000	\$500
Annual labor costs	\$600	\$4300

Continuous process:

$$K = 20,000\frac{F_{PR,10\%,10}}{0.10} + \frac{1000 + 600}{0.10}$$

$$K = 20,000\frac{0.1627}{0.10} + \frac{1600}{0.10} = \$48,540$$

Batch process:

$$K = 6000\frac{0.1315}{0.10} - 500\frac{1}{(1+0.10)^{15}}\frac{0.1315}{0.10} + \frac{4800}{0.10}$$
$$= 7890 - 500(0.2394)(1.315) + 48,000 = \$55,733$$

Note that the $500 salvage value is a negative cost occurring in the fifteenth year. We bring this to the present value and then multiply by $F_{PK} = F_{PR}/i$.

Each of the three methods of analysis will give the same result when applied to the same problem. The best method to use depends chiefly on whom you need to convince with your analysis and which technique you feel they will be more comfortable with.

8-4 DEPRECIATION

Capital equipment suffers a loss in value with time. This may occur by wear, deterioration, or obsolescence, which is a loss of economic efficiency because of technological advances. Therefore, a company must lay aside enough money each year to accumulate a fund to replace the obsolete or worn-out equipment. This allowance for loss of value is called depreciation. Another important aspect of depreciation is that the federal government permits depreciation to be deducted from gross profits as a cost of doing business. In a capital-intensive business, depreciation can have a strong influence on the amount of taxes that must be paid.

Taxable income = total income − allowable expenses − depreciation

The basic questions to be answered about depreciation are: 1) what is the time period over which depreciation can be taken and 2) how should the total depreciation charge be spread over the life of the asset? Obviously, the depreciation charge in any given year will be greater if the depreciation period is short (a rapid write-off).

The Economic Recovery Act of 1981 introduced the *accelerated cost recovery system* (ACRS) as the prime capital-recovery method in the United States. This was modified in the 1986 Tax Reform Act to MACRS. The statute sets depreciation recovery periods based on the expected useful life. Some examples are:

- special manufacturing devices; some motor vehicles 3 yrs
- computers; trucks; semiconductor mfg. equipment 5 yrs
- office furniture; railroad track; agricultural buildings 7 yrs
- durable-goods mfg. equipment; petroleum refining 10 yrs
- sewage treatment plants; telephone systems 15 yrs

Residential rental property is recovered in 27.5 years and nonresidential rental property in 31.5 years. Land is a nondepreciable asset, since it is never used up.

We shall consider four methods of spreading the depreciation over the recovery period n: 1) straight-line depreciation, 2) declining balance, 3) sum-of-the-years digits, and 4) the MACRS procedure. Only MACRS and the straight-line method currently are acceptable under the U.S. tax laws but the others methods are useful in classical engineering economic analyses. Moreover, depreciation schedules established prior to 1981 with these methods must remain in force.

Straight-Line Depreciation

In straight-line depreciation an equal amount of money is set aside yearly. The annual depreciation charge D is

$$D = \frac{\text{initial cost} - \text{salvage value}}{n} = \frac{C_i - C_s}{n} \tag{8-20}$$

The *book value* is the initial cost minus the sum of the depreciation charges that have been made. For straight-line depreciation, the book value B at the end of the jth year is

$$B = C_i - \frac{j}{n}(C_i - C_s) \tag{8-21}$$

Declining-Balance Depreciation

The declining-balance method provides an accelerated write-off in the early years. The depreciation charge for the jth year D_j is a fixed fraction F_{DB} of the book value at the beginning of the jth year (or the end of year $j - 1$). For the book value to equal the salvage value after n years,

$$F_{DB} = 1 - \sqrt[n]{\frac{C_s}{C_i}} \tag{8-22}$$

and the book value at the beginning of the jth year is

$$B_{j-1} = C_i(1 - F_{DB})^{j-1} \tag{8-23}$$

Therefore, the depreciation in the jth year is

$$D_j = B_{j-1}F_{DB} = C_i(1 - F_{DB})^{j-1}F_{DB} \tag{8-24}$$

The most rapid write-off occurs for double declining-balance depreciation. In this case $F_{DDB} = 2/n$ and $B_{j-1} = C_i(1 - 2/n)^{j-1}$. Then

$$D_j = C_i\left(1 - \frac{2}{n}\right)^{j-1}\frac{2}{n}$$

Since the DDB depreciation may not reduce the book value to the salvage value at year n, it may be necessary to switch to straight-line depreciation in later years.

Sum-of-Years Digits Depreciation

The sum-of-years digits (SOYD) depreciation is an accelerated method. The annual depreciation charge is computed by adding up all of the integers from 1 to n and then taking a fraction of that each year, $F_{SOYD,j}$.

For example, if $n = 5$, then the sum of the years is $(1 + 2 + 3 + 4 + 5 = 15)$ and $F_{SOYD,2} = 4/15$, while $F_{SOYD,4} = 2/15$. The denominator is the sum of the digits; the numerator is the digit corresponding to the jth year when the digits are arranged in *reverse order*.

Modified Accelerated Cost Recovery System (MACRS)

In MACRS the annual depreciation is computed using the relation

$$D = qC_i \tag{8-25}$$

where q is the recovery rate obtained from Table 8-3 and C_i is the initial cost. In MACRS the value of the asset is completely depreciated even though there may be a true salvage value. The recovery rates are based on starting out with a declining-balance method and switching to the straight-line method when it

TABLE 8-3
Recovery rates q used in MACRS method

Year	Recovery rate, q, %				
	$n = 3$	$n = 5$	$n = 7$	$n = 10$	$n = 15$
1	33.3	20.0	14.3	10.0	5.0
2	44.5	32.0	24.5	18.0	9.5
3	14.8	19.2	17.5	14.4	8.6
4	7.4	11.5	12.5	11.5	7.7
5		11.5	8.9	9.2	6.9
6		5.8	8.9	7.4	6.2
7			8.9	6.6	5.9
8			4.5	6.6	5.9
9				6.5	5.9
10				6.5	5.9
11				3.3	5.9
12–15					5.9
16					3.0

n = recovery period, years.

TABLE 8-4
Comparison of depreciation methods

| Year | $C_i = \$6000$, $C_s = \$1000$, $n = 5$ | | | |
	Straight line	Declining balance	Sum-of-years digits	MACRS
1	1000	1807	1667	1200
2	1000	1263	1333	1920
3	1000	882	1000	1152
4	1000	616	667	690
5	1000	431	333	690
6	—	—	—	348

offers a faster write-off. MACRS uses a half-year convention which assumes that all property is placed in service at the midpoint of the initial year. Thus, only 50 percent of the first year depreciation applies for tax purposes and requires that a half year of depreciation be taken in year $n + 1$.

Table 8-4 compares the annual depreciation charges for these four methods of calculation.

8-5 TAXES

Taxes are an important factor to be considered in engineering economic decisions. The chief type of taxes that are imposed on a business firm are:

1. *Property taxes.* Based on the value of the property owned by the corporation (land, buildings, equipment, inventory). These taxes do not vary with profits and usually are not too large.
2. *Sales taxes.* Imposed on sales of products. Sales taxes usually are paid by the retail purchaser, so they generally are not relevant to engineering economy studies.
3. *Excise taxes.* Imposed on the manufacture of certain products like tobacco and alcohol. Also usually passed on to the consumer.
4. *Income taxes.* Imposed on corporate profits or personal income. Gains resulting from the sale of capital property also are subject to income tax.

Generally, federal income taxes have the most significant impact on engineering economic decisions. Although we cannot delve into the complexities of tax laws, it is important to incorporate the broad aspects of income taxes into our analysis.

The 1986 Tax Reform Act established the following federal income tax schedule for corporations and individuals: the first $50,000 of taxable income is taxed at 15 percent, the next $25,000 at 25 percent, with income over $75,000

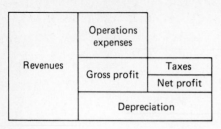

FIGURE 8-3
Distribution of corporate revenues.

taxed at 34 percent. For incomes between \$100,001 and \$350,000 there is a tax surcharge of 5 percent. All income above \$350,000 is taxed at the 34 percent rate. Many studies use an estimate of 34 percent for the federal. Most states and some cities also have an income tax. For simplicity in economic studies an effective tax rate is often used. This commonly varies from 35 percent to 50 percent. Since state taxes are deductible from federal taxes, the effective tax rate is given by

$$\text{effective tax rate} = \text{state rate} + (1 - \text{state rate})(\text{federal rate}) \quad (8\text{-}26)$$

The chief effect of corporate income taxes is to reduce the rate of return on a project or venture.

$$\text{After-tax rate of return} = \text{before tax rate of return} \times (1 - \text{income tax rate})$$

$$r = i(1 - t) \quad (8\text{-}27)$$

Note that this relation is true only when there are no depreciable assets. For the usual case when we have depreciation, capital gains or losses, or investment tax credits, Eq. (8-27) is a rough approximation. The importance of depreciation in reducing taxes is shown in Fig. 8-3. The depreciation charge appreciably reduces the gross profit, and thereby the taxes. However, since depreciation is retained in the corporation, it is available for advancement of the enterprise.

Consider a depreciable capital investment $C_d = C_i - C_s$. At the end of each year depreciation amounting to $D_f C_d$ is available to reduce the taxes by an amount $D_f C_d t$

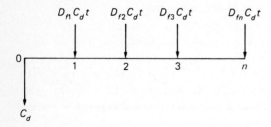

Note that the fractional depreciation charge each year D_f may vary from year to year depending on the method used to establish the depreciation schedule

See for example Table 8-4. The present value of this series of costs and benefits is

$$P = C_d - C_d t \left[\frac{D_{f_1}}{1 + r} + \frac{D_{f_2}}{(1 + r)^2} + \frac{D_{f_3}}{(1 + r)^3} + \cdots + \frac{D_{f_n}}{(1 + r)^n} \right] \quad (8.\text{-}28)$$

The exact evaluation of the term in brackets will depend on the depreciation method selected.

Example 8-5. A manufacturing company of modest size is considering an investment in energy-efficient electric motors to reduce its large annual energy cost. The initial cost would be $12,000, and over a 10-year period it is estimated that the firm would save $2200 annually in electricity costs. The salvage value of the motors is estimated at $2000. Determine the after-tax rate of return.

Solution. First we will establish the before-tax rate of return. We need to determine the cash flow for each year. Cash flow, in this context, is the net profit or savings for each year. We shall use straight-line depreciation to determine the depreciation charge. Table 8-5 shows the cash flow results. The before-tax rate of return is the interest rate at which the before-tax cash flow savings just equals the purchase cost of the motors.

$$12,000 = 2200(F_{RP.i,10}) + 2000(F_{SP.i,10})$$

We find the rate of return by trying different values of i in the compound interest tables. For $i = 14$ percent,

$$12,000 = 2200(5.2161) + 2000(0.2697)$$

$$= 11,475 + 539 = 12,014$$

Therefore, the before-tax rate of return is very slightly more than 14 percent. To find the after-tax rate of return, we use the after-tax cash flow in Table 8-5. From Eq. (8-27) we estimate the after-tax rate of return to be 7 percent.

$$12,000 = 1600 F_{RP,i,10} + 2000 F_{SP,i,10}$$

For $i = 6\%$: $12,000 = 1600(7.3601) + 2000(0.5584)$

$$= 11,776 + 1117 = 12,893 \quad i \text{ too low}$$

TABLE 8-5
Cash flow calculations for example

Year	Before-tax cash flow	Depreciation	Taxable income	50% income tax	After-tax cash flow
0	−12,000				−12,000
1 to 9	2,200	1000	1200	−600	1,600
10	2,200	1000	1200	−600	1,600
	2,000				2,000

For $i = 8\%$:
$$12{,}000 = 1600(6.7101) + 2000(0.4632)$$
$$= 10{,}736 + 926 = 11{,}662 \qquad i \text{ too high}$$

$$i = 6\% + 2\% \frac{12{,}893 - 12{,}000}{12{,}893 - 11662} = 6\% + 2\%(0.72)$$

$$i = 6 + 1.44 = 7.44\%$$

The method of including taxes illustrated above is the most straightforward approach, but it is not the quickest or shortest path to an answer. We can rewrite Eq. (8-28) as

$$P = C_d(1 - t\psi) \tag{8-29}$$

where ψ represents the term in brackets in Eq. (8-28). Jelen[1] presents expressions for annual cost and capitalized cost calculations based on straight-line depreciation and sum-of-years digits. The following example illustrates the approach.

Example 8-6. (Selection of alternative material based on capitalized cost.) We wish to decide whether carbon-steel or special alloy-steel tubes should be used in a heat exchanger. One material has a higher initial cost but offers longer service life, lower annual maintenance costs, and a saving in product due to less downtime. The comparison is based on capitalized cost using an after-tax rate of return $r = 10$ percent and a tax rate $t = 50$ percent. Depreciation is based on the sum-of-the-years digits method.

Solution.

	Carbon steel	Special alloy steel
Initial cost	$10,000	$95,000
Salvage value	—	20,000
Estimated service life, years	4	10
Tube cleaning inside, yearly	3,000	—
Tube cleaning outside, every third year	4,000	—
Tube repair, every fourth year	1,500	—
Tube maintenance, yearly	—	1,000
Annual savings from less product loss because of less downtime	—	−6,000

Capitalized cost of carbon-steel tubes

Initial cost:
$$K_1 = PF_{PK,r,n} = C_d F_{PK,r,n}(1 - tF_{SDP,r,n'})$$
$$= (10{,}000 - 0)(3.1547)[1 - 0.5(0.83013)] = 18{,}452$$

[1] F. C. Jelen, op. cit., pp. 58–59.

Note: $F_{\text{SDP},r,n'}$ is ψ evaluated for sum-of-years digits, where n' is IRS depreciation life.

Yearly cost, starting at the beginning of the year:

$$K_2 = \frac{R_b}{i} F_{PK,r,1}(1 - F_{\text{SDP},r,1})$$

$$= \frac{3000}{0.10}(1.100)[1 - 0.5(0.90909)] \qquad 17,998$$

Tube cleaning outside, beginning of third year:

$$K_3 = C_{bx}\frac{1 - tF_{\text{SDP},r,1}}{(1 + r)^{x-1}} F_{PK,r,n}$$

$$= \frac{4000}{1.10^{3-1}}[1 - 0.5(0.90909)](3.1547) \qquad 5,688$$

Tube repair, beginning of fourth year:

$$K_4 = C_{bx}\frac{1 - tF_{\text{SDP},r,1}}{(1 + r)^{x-1}} F_{PK,r,n}$$

$$= \frac{1500}{1.10^{4-1}}[1 - 0.5(0.90909)](3.1547) \qquad \frac{1,939}{\$44,077}$$

Capitalized cost of special alloy steel

Initial cost:

$$K_1 = (95,000 - 20,000)(1.6275)[1 - 0.5(0.70099)] \qquad 79,280$$

Salvage value: K_2 $\qquad\qquad$ 20,000

Note annual saving, treated as a beginning-of-year saving:

$$1000 - 6000 = -5000 \text{ net savings}$$

$$K_3 = \frac{R_b}{i} F_{PK,r,1}(1 - tF_{\text{SDP},r,1})$$

$$= \frac{-5000}{0.10}(1.100)[1 - 0.5(0.90909)] \qquad \frac{-29,997}{\$69,283}$$

This shows that the higher purchase cost of the special steel does not justify use of the steel with an after-tax return of 10 percent.

The expenditures that a business incurs are divided for tax purposes into two broad categories. Those for facilities and production equipment with lives in excess of one year are called capital expenditures, they are said to be "capitalized" in the accounting records of the business. Other expenses for running the business, such as labor and material costs, direct and indirect costs, and facilities and equipment with a life of one year or less, are ordinary business expenses. Usually they total more than the capital expenses. In the accounting records, they are said to be "expensed." The ordinary expenses are directly subtracted from the gross income to determine the taxable income, but only the annual depreciation charge can be subtracted for the capitalized expenses.

When a capital asset is sold, a capital gain or loss is established by subtracting the book value of the asset from its selling price. Frequently in our modern history capital gains have received special treatment by being taxed at a rate lower than for ordinary income. However, under the 1986 Tax Act capital gains and ordinary income are taxed at the same rate.

Investment in capital is a vital step in the innovation process that leads to increased national wealth. Therefore, the federal government frequently uses the tax system to stimulate capital investment. This most often takes the form of a tax credit, usually 7 percent but varying with time from 4 to 10 percent. This means that 7 percent of the purchase price of qualifying equipment can be deducted from the taxes that the firm owes the U.S. government. Moreover, the depreciation charge for the equipment is based on its full cost.

8-6 PROFITABILITY OF INVESTMENTS

One of the principal uses for engineering economy is to determine the profitability of proposed projects or investments. The decision to invest in a project generally is based on three different sets of criteria.

Profitability. Determined by techniques of engineering economy to be discussed in this section.

Financial analysis. How to obtain the necessary funds and what it will cost. Funds for investment come from three broad sources: 1) retained earnings of the corporation, 2) long-term commercial borrowing from banks, insurance companies, and pension funds, and 3) the equity market through the sale of stock.

Analysis of intangibles. Legal, political, or social consideration or issues of a corporate image or desire for goodwill often outweigh financial considerations in deciding on which project to pursue. For example, a corporation may decide to invest in the modernization of an old plant because of its responsibility to continue employment for its employees when investment in a new plant 1000 miles away would be economically more attractive.

However, in our free-enterprise system a major goal of a business firm is to show a profit. It does so by committing its funds to ventures that appear to be profitable. If investors do not receive a sufficiently attractive profit, they will find other uses for their money, and the growth—even the survival—of the firm will be threatened.

Four methods of evaluating profitability are commonly used. Accounting rate of return and payback period are simple techniques that are readily understood, but they do not take time value of money into consideration. Net present value and discounted cash flow are the most common profitability measures in which time value of money is considered. Before discussing them, however, we need to look a bit more closely at the concept of cash flow.

Cash flow measures the flow of funds into or out of a project. Funds flowing in constitute positive cash flow; funds flowing out are negative cash flow. The cash flow for a typical plant construction project is shown in Fig. 8-4. From an accounting point of view, cash flow is defined as:

$$\text{Cash flow} = \text{net annual cash income} + \text{depreciation}$$

You might consider cash income as "real dollars" and the depreciation a book-keeping adjustment to allow for capital expenditures. Table 8-6 shows how cash flow can be determined in a simple situation.

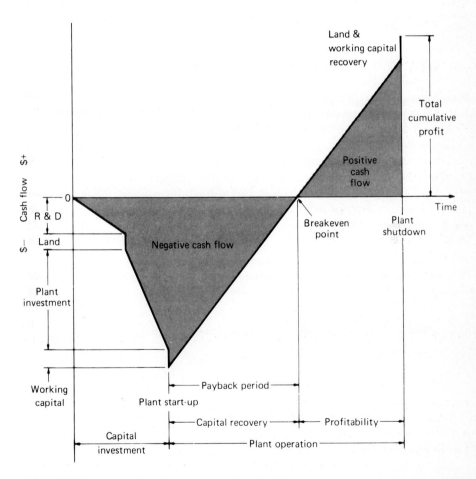

FIGURE 8-4
Typical costs in the life cycle of a capital plant investment.

TABLE 8-6
Calculation of cash flow

(1) Revenue (over 1-year period)	$500,000
(2) Operating costs	360,000
(3) (1) − (2) = gross earnings	140,000
(4) Annual depreciation charge	60,000
(5) (3) − (4) = taxable income	80,000
(6) (5) × 0.35 = income tax	28,000
(7) (5) − (6) = net profit after taxes	52,000
Net cash flow (after taxes)	
(7) + (4) = 52,000 + 60,000	112,000

Rate of Return

The rate of return on the investment (ROI) is the simplest measure of profitability. It is calculated from a strict accounting point of view without consideration of the time value of money. It is a simple ratio of some measure of profit or cash income to the capital investment. There are a number of ways to assess the rate of return on the capital investment. ROI may be based on 1) net annual profit before taxes, 2) net annual profit after taxes, 3) annual cash income before taxes, or 4) annual cash income after taxes. These ratios, usually expressed as percents, can be computed for each year or on the average profit or income over the life of the project. In addition, capital investment sometimes is expressed as the average investment. Thus, although the ROI is a simple concept, it is important in any given situation to understand clearly how it has been determined.

Example 8-7. An initial capital investment is $360,000 and has a 6-year life and a $60,000 salvage value. Working capital is $40,000. Total net profit after taxes over 6 years is estimated at $167,000. Find the ROI.

Solution. Using straight-line depreciation, the annual charge is

$$D = \frac{C_i - C_s}{n} \frac{360,000 - 60,000}{6} = \$50,000$$

$$\text{Average annual net profit} = \frac{167,000}{6} = \$28,000$$

ROI on initial capital investment

$$= \frac{28,000}{360,000 + 40,000} = 0.07$$

Payback Period

The payback period is the period of time necessary for the cash flow to fully recover the initial total capital investment (Fig. 8-4). Although the payback method uses cash flow, it does not include a consideration of the time value of money. Emphasis is on rapid recovery of the investment. Also, in using the method, no account is taken of cash flows or profits recovered after the payback period. Consider the following example.

	Cash flow	
Year	Project A	Project B
0	$ − 100,000	$ − 100,000
1	50,000	0
2	30,000	10,000
3	20,000	20,000
4	10,000	30,000
5	0	40,000
6	0	50,000
7	0	60,000
	$10,000	$110,000
Payback period	3 years	5 years

By the payback period criterion, project A is the most desirable because it recovers the initial capital investment in 3 years. However, project B, which returns a cumulative cash flow of $110,000, obviously is the most profitable overall.

Net Present Worth

In Sec. 8-3, as one of the techniques of cost comparison, we introduced the criterion of net present worth (NPW).

Net present worth = present worth of benefits − present worth of costs

By this technique the expected cash flows (both + and −) through the life of the project are discounted to time zero at an interest rate representing the minimum acceptable return on capital. The project with the greatest positive value of NPW is preferred. NPW depends upon the project life, so strictly speaking the net present worths of two projects should not be compared if the projects have different service lives.

Obviously, the value of NPW will be dependent upon the interest rate used for the calculation. Low values of interest rate will tend to make NPW more positive, for a given set of cash flows, and large values of interest will push NPW in a negative direction. There will be some value of i for which the sum of the discounted cash flows equals zero; NPW = 0. This value of i is called the discounted cash flow rate of return.

Discounted Cash Flow

Discounted cash flow (DCF) is the rate of return for which the net present worth equals zero. We can express this criterion in a variety of ways.

$$PW \text{ of benefits} - PW \text{ of costs} = 0$$

$$\frac{PW \text{ of benefits}}{PW \text{ of costs}} = 1$$

$$PW \text{ of benefits} = PW \text{ of costs}$$

If, for example, the DCF rate of return is 20 percent, it implies that 20 percent per year will be earned on the investment in the project, in addition to which the project will generate sufficient funds to repay the original investment. Depreciation is considered implicitly in NPW and DCF calculations through the definition of cash flow.

Because the decision on profitability is expressed as a percentage rate of return in the DCF method, it is more readily understood and accepted by engineers and businessmen than the NPW method, which produces a sum of money as an answer. In the NPW method it is necessary to select an interest rate for use in the calculations, and that may be a difficult and controversial thing to do. But by using the DCF method, we compute a rate of return, called the internal rate of return, from the cash flows. One situation in which NPW has an advantage is that individual values of NPW for a series of subprojects may be added to give the NPW for the complete project. That cannot be done with the rate of return developed from DCF analysis.

Example 8-8. A machine has a first cost of $10,000 and a salvage value of $2000 after a 5-year life. Annual benefits (savings) from its use are $5000, and the annual cost of operation is $1800. The tax rate is 50 percent. Find the DCF rate of return.

Solution. Using straight-line depreciation, the annual depreciation charge is

$$D = \frac{C_i - C_s}{n} = \frac{10,000 - 2,000}{5} = \$1600$$

The annual cash flow after taxes is the sum of the net receipts and depreciation.

$$A_{CF} = (5000 - 1800)(1 - 0.50) + 1600(0.50)$$
$$= 1600 + 800 = \$2400$$

Year	Cash flow
0	−10,000
1	2,400
2	2,400
3	2,400
4	2,400
5	2,400 + 2,000 (C_s)

$$NPW = 0 = -10{,}000 + 2400F_{RP,i,5} + 2000F_{SP,i,5}$$

If $i = 10$ percent, NPW $= +340$; if $i = 12$ percent, NPW $= -214$. Thus, we have the DCF rate of return bracketed, and

$$i = 10\% + (12\% - 10\%)\frac{340}{340 + 214}$$

$$= 10 + 2 + \frac{340}{564} = 10 + 12 = 11.2\%$$

It is an important rule of engineering economy that *each increment* of investment capital must be justified on the basis of earning the minimum required rate of return.

Example 8-9. A company has the option of investing in one of the two machines described in the following table. Which investment is justified?

	Machine A	Machine B
Initial cost C_i	$10,000	$15,000
Useful life	5 years	10 years
Salvage value C_s	$2,000	0
Annual benefits	$5,000	$7,000
Annual costs	$1,800	$4,300

Solution. Assume a 50 percent tax rate and a minimum attractive rate of return of 6 percent. The conditions for machine A are identical with those in Example 8-8, for which $i = 11.2$ percent. Calculation of the DCF rate of return for machine B shows it is slightly in excess of the minimum rate of 6 percent. However, that is not the proper question. Rather, we should ask whether the *increment of investment* ($15,000 - $10,000) is justified. In addition, because machine B has twice the useful life of machine A, we should place them both on the same time basis.

Cash flow

Year	Machine A	Machine B	Difference, B − A
0	−10,000	−15,000	−5,000
1	2,400	2,100	−300
2	2,400	2,100	−300
3	2,400	2,100	−300
4	2,400	2,100	−300
5	2,400 − 10,000 + 2,000	2,100	−300 + 8,000
6	2,400	2,100	−300
7	2,400	2,100	−300
8	2,400	2,100	−300
9	2,400	2,100	−300
10	2,400 + 2,000	2,100	−300 − 2,000

$$NPW = 0 = -5000 - 300F_{RP,i,10} + 8000F_{SP,i,5} - 2000F_{SP,i,10}$$

But, even at $i = \frac{1}{4}$ percent, NPW = -2009, and there is no way that the extra investment in machine B can be justified economically.

When only costs—not income (or savings)—are known, we can still use the DCF method for incremental investments, but not for a single project. We assume that the lowest capital investment is justified without being able to determine the DCF rate of return, and we then determine whether the additional investment is justified.

Example 8-10. On the basis of the data in the following table, determine which machine should be purchased.

	Machine A	Machine B
First cost	$300	$4000
Useful life	6 years	9 years
Salvage value	$500	0
Annual operating cost	$2000	$1600

Solution. This solution will be based on cash flow before taxes. To place the machines on a common time frame, we use a common life of 18 years.

Year	Machine A	Machine B	Difference, B − A
0	−3000	−4000	−1000
1 to 5	−2000	−1600	+400
6	−2000 − 2500	−1600	+400 + 2500
7 to 8	−2000	−1600	+400
9	−2000	−1600 − 4000	+400 − 4000
10, 11	−2000	−1600	+400
12	−2000 − 2500	−1600	+400 + 2500
13 to 17	−2000	−1600	+400
18	−2000 + 500	−1600	+400 − 500

$$\text{NPW} = 0 = -1000 + 400F_{RP,i,18} + 2500F_{SP,i,6} + 2500F_{SP,i,12}$$
$$- 400F_{SP,i,9} - 500F_{SP,i,18}$$

Trial and error shows that $i \approx 47$ percent, which clearly justifies purchase of machine B.

We have presented information on the four most common techniques for evaluating the profitability of an investment. The rate-of-return method has the advantage of being simple and easy to use. However, it ignores the time value of money and the consideration of cash flow. The payback period also is a simple method, and it is particularly attractive for industries undergoing rapid technological change. Like the rate-of-return method, it ignores time value of money, and it places an undue emphasis on projects that achieve a quick payoff. The net present worth method takes both cash flow and time value of money into account. However, it suffers from the problem of ambiguity in setting the required rate of return, and it may present problems when projects with different service lives are compared. Discounted cash flow rate of return has the advantage of producing an answer that is the real internal rate of return. The method readily permits comparison between alternatives, but it is assumed that all cash flows generated by the project can be reinvested to yield a comparable rate of return.

8-7 OTHER ASPECTS OF PROFITABILITY

Innumerable factors[1] affect the profitability of a project in addition to the mathematical expressions discussed in Sec. 8-6. The purpose of this section is to round out our consideration of the crucial subject of profitability.

We need to realize that profit and profitability are not quite the same concept. Profit is measured by accountants, and its value in any one year can be manipulated in many ways. Profitability is inherently a long-term parameter of economic decision making. As such, it should not be influenced much by short-term variations in profits. In recent years there has been a strong trend toward undue emphasis on quick profits and short payoff periods that work to the detriment of long-term investment in high-technology projects.

Estimation of profitability requires the prediction of future cash flows, which in turn requires reliable estimates of sales volume and sales price by the marketing staff and of material price and availability. The quadrupling of crude oil price in the early 1970s was a dramatic (and traumatic) example of how changes in raw material costs can greatly influence profitability predictions. Similarly, trends in operating costs must be looked at carefully, especially with respect to whether it is more profitable to reduce operating costs through increased investment, as with automation.

The estimated investment in machinery and facilities that is required for the project is usually the most accurate component of the profitability evaluation. (This topic is considered in more detail in Chap. 9.) The depreciation method used influences how the expense is distributed over the years of a project, and that in turn determines what the cash flow will be. However, a more fundamental aspect of depreciation is the effect of writing off a capital investment over a long time period. As a result, costs are underestimated and selling prices are set too low. A long-term write-off combined with inflation results in insufficient cash flow to permit reinvestment. Inflation creates hidden expenses like inadequate allowance for depreciation. When depreciation methods do not allow for inflated replacement costs, those

---

[1] F. C. Jelen, *Hydrocarbon Processing*, pp. 111–115, January, 1976.

costs must be absorbed on an after-tax basis. Profit and profitability are overstated in an inflationary period.

A number of technical decisions are intimately related to the investment policy and profitability. At the design stage it may be possible to ensure a level of product superiority that is more than needed by the current market. Later, when competitors enter the market, the superiority would prove useful, but it is not achieved without an initial cost to profitability. Economics generally favor building as large a production unit as the market can absorb. However, this increased profitability is achieved at some risk to maintaining continuity of production should the unit be down for repairs. Thus, there often is a trade-off between the increased reliability of having a number of small units over which to spread the risk and a single large unit with somewhat higher profitability.

The profitability of a particular product line can be influenced by decisions of cost allocation. Such factors as overhead, utility costs, transfer prices between divisions of a large corporation, or scrap value often require arbitrary decisions for allocation between various products. Thus, the situation often favors certain products and discriminates against others because of cost allocation policies. Sometimes corporations take a position of milking an established product line with a limited future (a "cash cow") in order to stimulate the growth of a new but promising product line. Another profit decision is whether to charge a particular item as a current expense or capitalize it to make a future expense. In a period of inflation there is strong pressure to increase present profitability by deferring costs into the future by capitalizing them. It is argued that a fixed dollar amount deferred into the future will have less consequence in terms of future dollars.

The role of the government in influencing profitability is very great. In the broader sense the government creates the general economic climate through its policies on money supply, taxation, and foreign affairs. It provides subsidies to stimulate selected parts of the economy. Its regulating powers have had an increasing influence on profitability in such areas as pollution control, occupational health and safety, consumer protection and product safety, use of federal lands, antitrust, minimum wages, and working hours.

Since profitability analysis deals with future predictions, there is inevitable uncertainty. Incorporation of uncertainty or risk is possible with advanced techniques. Unfortunately, the assignment of the probability to deal with risk is in itself a subjective decision. Thus, although risk analysis (Sec. 8-10) is an important technique, one should always realize its true origins.

8-8   INFLATION

Since engineering economy deals with decisions based on future flows of money, it is important to consider inflation in the total analysis. Surprisingly, however, inflation is considered substantively in few engineering economy

TABLE 8-7
Consumer price index (CPI)[1]

Year	CPI	Percent change	Year	CPI	Percent change
1967	100.0	2.9	1978	199.6	9.0
1968	104.2	4.2	1979	226.1	13.3
1969	109.8	5.4	1980	254.2	12.4
1970	116.3	5.9	1981	276.8	8.9
1971	121.3	4.3	1982	287.3	3.8
1972	125.3	3.3	1983	298.2	3.8
1973	136.3	8.8	1984	309.9	3.9
1974	152.9	12.2	1985	321.6	3.8
1975	163.6	7.0	1986	325.2	1.1
1976	171.4	4.8	1987	339.5	4.4
1977	183.1	6.8	1988	352.7	3.9

[1] Monthly Labor Review, U.S. Dept. of Labor.

texts.[1] Perhaps one reason is that the average annual increase in the consumer price index from 1926 to 1986 has been 3.1 percent. However, the annualized growth rate of inflation in the United States was 8.7 percent from 1975 to 1982.

We are all painfully aware that inflation is the situation in which prices of goods and services are increasing so that a given amount of money buys less and less as time goes by. Interest rates and inflation are directly related. The basic interest rate is about 2 to 3 percent higher than the inflation rate. Thus, in a period of high inflation, not only does the dollar purchase less each month but the cost of borrowing money also rises.

Price changes may or may not be considered in an economic analysis. For meaningful results, costs and benefits must be computed in comparable units. It would not be sensible to calculate costs in 1980 dollars and benefits in 1990 dollars. However, we could have a situation in which future benefits fluctuate with the inflation of the dollar. In that somewhat rare situation, inflation would have no effect on the before-tax economic analysis. However, as will be seen later, adjusting future before-tax benefits for inflation (constant-value money) will not avoid the impact of inflation in an after-tax analysis.

Changes in price are determined by the U.S. Department of Labor and published as various price indices. Table 8-7 lists the consumer price index over the past 22 years. The wholesale industrial price index would be more appropriate for many engineering economy problems. Note that 1967 has been selected as the baseline year at which the CPI = 100. The increase in prices

[1] B. W. Jones, "Inflation in Engineering Economic Analysis," John Wiley & Sons, New York, 1982.

also can be expressed as the annual compound rate at which prices rise. For example, over the 10-year period 1967 to 1976

$$S = PF_{PS,i,10}$$

$$171.1 = 100.0F_{PS,d,10}$$

$$F_{PS,d,10} = 1.71$$

and on interpolating from the compound interest tables we find that prices increased at a compound rate of $d = 5.48$ percent.

When considering inflation in economic analysis, it is useful to define two situations: current money and constant-value money. Current money refers to ordinary money units, which decline in purchasing power with time. To work with *current money,* we derive an equivalent interest rate f that combines the interest rate after taxes r and the inflation rate d. To develop the rate, consider a future payment S that is discounted to the present (current time). But with inflation we must also discount the future value for the inflation rate d.

$$P = \frac{S}{(1+r)^n(1+d)^n} = \frac{S}{(1+r+d+rd)^n} = \frac{S}{(1+f)^n}$$

and

$$f = r + d + rd \approx r + d \tag{8-30}$$

Constant-value money means that money units, i.e., dollars, represent equal purchasing power at any future time. To make the correction, current dollars are discounted each year at the inflation rate d:

$$\text{Constant-value money} = \text{current money} \left(\frac{1}{1+d}\right)^n \tag{8-31}$$

Example 8-11. A project requires an investment of $10,000 and is expected to return, in future, or "then current," dollars, $2500 at the end of year 1, $3000 at the end of year 2, and $7000 at the end of year 3. The monetary interest rate is 10 percent, and the inflation rate is 6 percent per year. Find the net present worth of this investment opportunity.

Solution. The equivalent interest rate is $0.10 + 0.06 + (0.10)(0.06) = 0.166$; for simplicity we shall use $f = 0.17$.

Current-dollars approach

Year	Cash flow	$F_{SP,17,n}$	Present worth
0	−10,000	1.00	−10,000
1	2,500	0.8547	2,137
2	3,000	0.7305	2,191
3	7,000	0.6244	4,971
		NPW =	−711

Constant-value-dollars approach

Year	Cash flow*	$F_{SP,10,n}$	Present worth
0	−10,000	1.00	−10,000
1	2,358	0.9091	2,144
2	2,670	0.8264	2,206
3	5,877	0.7513	4,415
		NPW =	−1,235

* Adjusted for $d = 0.06$.

The difference in the NPW's found by the two treatments is due to using an approximate combined discount rate instead of the more accurate value of $f = 0.166$. However, the approximation is justified in view of the uncertainty in predicting the rate of inflation. It should be noted that, for this example, the NPW is +$10 if inflation is ignored. That emphasizes the fact that neglecting the influence of inflation overemphasizes the profitability.

When profitability is measured by the DCF rate of return r, the inclusion of the inflation rate d results in an effective rate of return r' based on constant-value money.[1]

$$1 + r = (1 + r')(1 + d)$$
$$r' = r - d - r'd \approx r - d \tag{8-32}$$

To a first approximation, the DCF rate of return is reduced by an amount equivalent to the average inflation rate.

Interest rates are quoted to investors in current money r, but investors generally expect to cover any inflationary trends and still receive an acceptable return. In other words, investors hope to obtain a constant-value interest rate r'. Therefore, the current interest rate tends to fluctuate with the inflation rate. If the calculation is to be made with constant-value money, discounting should be done with the normal after-tax interest rate r. If calculations are in terms of current money, then the discount rate should be $f \approx r + d$.

Note that tax allowance for depreciation has a reduced benefit when constant money is used for profitability evaluations. By law, depreciation is defined in terms of current money. Therefore, under high inflation when constant-money conditions are appropriate, a full tax credit for depreciation is not achieved.

Another effect of inflation is that it increases the cash flow because the prices received for goods and services rise as the value of money falls. Even when constant-value-money is used, the yearly cash flows should display the current money situation.

---

[1] F. A. Holland and F. A. Watson, *Chem. Eng.*, pp. 87–91, Feb. 14, 1977.

Detailed relations for capitalized cost that include an inflation factor have been developed by Jelen.[1] They can also be used to introduce inflation calculation into annual cost calculations.

8-9 SENSITIVITY AND BREAK-EVEN ANALYSIS

A *sensitivity analysis* determines the influence of each factor in the problem on the final result, and therefore it determines which factors are most critical in the economic decision. Since there is a considerable degree of uncertainty in predicting future events like sales volume, salvage value, and rate of inflation, it is important to see how much the economic analysis depends on the magnitude of the estimates. One factor is varied over a reasonable range and the others are held at their mean (expected) value. The amount of computation involved in a sensitivity analysis of an engineering economy problem can be very considerable, but the use of computers has made sensitivity analysis in this context a much more practical endeavor.[2]

A *break-even analysis* often is used when there is particular uncertainty about one of the factors in an economic study. The break-even point is the value for the factor at which the project is just marginally justified.

Example 8-12. Consider a $20,000 investment with a 5-year life. The salvage value is $4000, and the minimal acceptable return is 8 percent. The investment produces annual benefits of $10,000 at an operating cost of $3000. Suppose there is considerable uncertainty as to whether the new machinery will survive 5 years of continuous use. Find the break-even point, in terms of life, at which the project just becomes economically viable.

Solution. Using the annual cost method,

$$\$10,000 - 3000 - (20,000 - 4000)F_{PR,8,n} - 4000(0.08) = 0$$

$$F_{PR,8,n} = \frac{6680}{16,000} = 0.417$$

and interpolating in the interest tables gives us $n = 2.8$ years. Thus, if the machine does not last 2.8 years, the investment cannot be justified.

Break-even analysis frequently is used in problems dealing with staged construction. The usual problem is to decide whether to invest more money initially in unused capacity or to add the needed capacity of a later date when needed, but at higher unit costs.

---

[1] F. C. Jelen, op. cit., pp. 122–128; *Chem. Eng.*, pp. 123–128, Jan. 27, 1958.

[2] J. C. Agarwal and I. V. Klumpar, *Chem. Eng.*, pp. 66–72, Sept. 29, 1975.

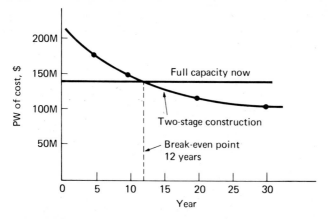

FIGURE 8-5
Break-even plot for Example 8-13.

Example 8-13. A new synthetic fuel plant will cost $100 million for the first stage and $120 million for the second stage at n years in the future. If it is built to full capacity now, it will cost $140 million. All facilities are expected to last 40 years. Salvage value is neglected. Find the preferable course of action.

Solution. The annual cost of operation and maintenance is assumed the same for a two-stage construction and full-capacity construction. We shall use a present worth (PW) calculation with a 10 percent interest rate. For full-capacity construction now, PW = $140 million ($140M). For two-stage construction

$$PW = \$100M + \$120M \ F_{SP,10,n}$$

$$n = \ 5 \text{ years: } PW = 100 + 120(0.6201) = \$174M$$

$$n = 10 \text{ years: } PW = 100 + 120(0.3855) = \$146M$$

$$n = 20 \text{ years: } PW = 100 + 120(0.1486) = \$118M$$

$$n = 30 \text{ years: } PW = 100 + 120(0.0573) = \$107M$$

These results are plotted in Fig. 8-5. The break-even point (12 years) is the point at which the two alternatives have equivalent cost. If the full capacity will be needed before 12 years, then full capacity built now would be the preferred course of action.

8-10 UNCERTAINTY IN ECONOMIC ANALYSIS

In the preceding section we discussed the fact that engineering economy deals chiefly with decisions based on future estimates of costs and benefits. Since none of us has a completely clear crystal ball, such estimates are likely to contain considerable uncertainty. In all of the examples presented so far in this chapter we have used a single value that was the implied best estimate of the future.

Now that we are willing to recognize that estimates of the future may not be very precise, there are some ways by which we can guard against the imprecision. The simplest procedure is to supplement your estimated most likely value with an optimistic value and a pessimistic value. The three estimates are combined into a mean value by

$$\text{Mean value} = \frac{\text{optimistic value} + 4(\text{most likely value}) + \text{pessimistic value}}{6}$$

(8-33)

In Eq. (8-33) the distribution of values is assumed to be represented by a beta frequency distribution. The mean value determined from the equation is used in the economic analysis.

The next level of advance would be to associate a probability with certain factors in the economic analysis. In a sense, by this approach we are transferring the uncertainty from the value itself to the selection of the probability.

Example 8-14. The expected life of a piece of mining equipment is highly uncertain. The machine costs $40,000 and is expected to have $5000 salvage value. The new machinery will save $10,000 per year, but it will cost $3000 annually for operations and maintenance. The service life is estimated to be:

3 years, with probability $= 0.3$

4 years, with probability $= 0.4$

5 years, with probability $= 0.5$

value. The new machinery will save $10,000 per year, but it will cost $3000 annually for operations and maintenance. The service life is estimated to be:

Solution. For 3-year life:

Net annual cost $= (10,000 - 3000) - (40,000 - 5000)F_{PR,10,3} - 5000(0.10)$

$\qquad = 7000 - 35,000(0.4021) - 500 = -8573$

For 4-year life:

Net annual cost $= 7000 - 35,000(0.3155) - 500 = -4542$

For 5-year life:

Net annual cost $= 7000 - 35,000(0.2638) - 500 = -2733$

Expected value of net annual cost $= E(\text{AC}) = \sum \text{AC} \times P(\text{AC})$

$\qquad = -8573(0.3) + [-4542(0.4)] + [-2733(0.3)] = -5207$

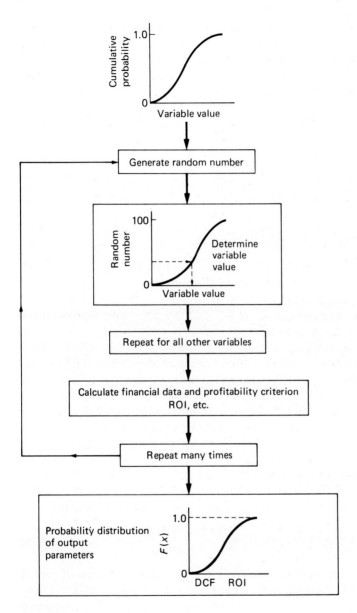

FIGURE 8-6
Steps in economic risk analysis.

Risk analysis[1] is a probabilistic approach to economic analysis in which Monte Carlo simulation is employed (see Sec. 4-4). Each variable in the analysis is assigned a cumulative probability distribution (Fig. 8-6). The probability distributions are constructed from historical data, econometric analysis, or any theoretical models that might be applicable. Once the probability distributions have been constructed, the next step is to establish a value of each variable by the use of random numbers. After all variables have been assigned values in this way, they are substituted into the economic model of the problem to determine cash flow and calculate the profitability criterion, such as DCF ROI. Then new random numbers are generated and other numerical values are assigned to the problem variables. The process is repeated to give a new value of ROI. The procedure is repeated hundreds of times so that a probability distribution of the profitability criteria can be developed.

8-11   BENEFIT-COST ANALYSIS

An important class of engineering decisions involves the selection of the preferred system design, material, purchased subsystem, etc., when economic resources are constrained. The methods of making cost comparisons and profitability analysis described in Secs. 8-3 and 8-6 are important decision-making tools in this type of situation.

Frequently, comparisons are based on a *benefit-cost ratio*, which relates the desired benefits to the capital investment required to produce the benefits. This method of selecting alternatives is most commonly used by governmental agencies for determining the desirability of public works projects. A project is considered viable when the net benefits associated with its implementation exceed its associated costs. *Benefits* are advantages to the public (or owner), expressed in terms of dollars. If the project involved disadvantages to the owner, these *disbenefits* must be subtracted from the benefits. The costs to be considered include the expenditures for construction, operation, and maintenance, less salvage. Both benefits, disbenefits, and costs must be expressed in common dollar terms by using the present worth or annual cost concept.

$$\text{Benefit/cost ratio (BCR)} = \frac{\text{benefits} - \text{disbenefits}}{\text{costs}} \tag{8-34}$$

A design or project for which BCR < 1 does not cover the cost of capital to create the design. Generally, only projects for which BCR > 1 are acceptable. The benefits used in the BCR would be factors like improved component performance, increased payload through reduced weight, and

---

[1] D. B. Hertz, *Harvard Bus. Rev.*, vol. 42, pp. 95–106, 1964; J. B. Malloy, *Chem. Eng. Prog.*, vol. 67, pp. 60–77, 1971.

increased availability of equipment. Benefits are defined as the advantages minus any disadvantages, i.e., the net benefits. Likewise, the costs are the total costs minus any savings. The costs should represent the initial capital cost as well as costs of operation and maintenance; see Chap. 9.

Very often in problems of choosing between several alternatives the incremental or marginal benefits and costs associated with changes beyond a base level or reference design should be used. The alternatives are ranked with respect to cost, and the lowest-cost situation is taken as the initial reference. This is compared with the next higher cost alternative by computing the incremental benefit and incremental cost. If $\Delta B/\Delta C < 1$, then alternative 2 is rejected because the first alternative is superior. Alternative 1 now is compared with alternative 3. If $\Delta B/\Delta C > 1$, then alternative 1 is rejected and alternative 3 becomes the current best solution. Alternative 3 is compared with number 4, and if $\Delta B/\Delta C < 1$, then alternative 3 is the best choice. We should note that this may not be the alternative with the largest overall benefit/cost ratio.

When used in a strictly engineering context to aid in the selection of alternative materials, the benefit/cost ratio is a useful decision-making tool. However, it often is used with regard to public projects financed with tax moneys and intended to serve the overall public good. There is a psychological advantage to the BCR concept over the discounted cash flow rate of return in that it avoids the connotation that the government is profiting from public moneys. Here questions that go beyond economic efficiency become part of the decision process. Many of the broader issues are difficult to quantify in monetary terms. Of even greater difficulty is the problem of relating monetary cost to the real values of society.

Consider the case of a hydroelectric facility. The dam produces electricity, but it also will provide flood control and recreational boating. The value of each of the outputs should be included in the benefits. The costs include the expenditures for construction, operation, and maintenance. However, there may be social costs like the loss of virgin timberland or a scenic vista. Great controversy surrounds the assignment of costs to environmental and aesthetic issues.

Although benefit-cost analysis is a widely used methodology, it is not without problems. The assumption is that costs and benefits are relatively independent. Basically, it is a deterministic method that does not deal with uncertainty in a major way. As with most techniques, it is best not to try to push it too far. Although the quantitative ratios provided by Eq. (8-34) should be used to the greatest extent possible, they should not preempt the utilization of common sense and good judgment.

BIBLIOGRAPHY

Blank, L. T., and A. J. Tarquin: "Engineering Economy," 3d ed., McGraw-Hill Book Co., New York, 1989.

Canada, J. R.: "Intermediate Economic Analysis for Management and Engineering," Prentice-Hall, Inc., Englewood Cliffs, N.J., 1971.

English, M. J.: "Project Evaluation: A Unified Approach to the Analysis of Capital Investments," Macmillan Publ. Co., New York, 1984.

Fabrycky, W. J., and G. J. Thuesen: "Economic Design Analysis," 2d ed., Prentice-Hall, Inc., Englewood Cliffs, N.J., 1980.

Grant, E. L., W. G. Ireson, and R. S. Leavenworth: "Principles of Engineering Economy," 6th ed., The Ronald Press Co., New York, 1976.

Holland, F. A., F. A. Watson, and J. K. Wilkinson: "Introduction to Process Economics," John Wiley & Sons, Inc., New York, 1974.

Jelen, F. C., and J. H. Black (eds.): "Cost and Optimization Engineering," 2d ed., McGraw-Hill Book Company, New York, 1983.

Kurtz, M. (ed.): "Handbook of Engineering Economics," McGraw-Hill Book Co., New York, 1984.

Newnan, D. G.: "Engineering Economic Analysis," Engineering Press, San Jose, Calif., 1976.

Riggs, J. L., and T. West: "Engineering Economics," 3d ed., McGraw-Hill Book Company, New York, 1986.

CHAPTER 9

COST
EVALUATION

9-1 INTRODUCTION

An engineering design is not complete until we have a good idea of the cost required to build the design or manufacture the product. Generally the lowest-cost design will be successful in a free marketplace. We saw that in Chap. 5, where the most common objective function was cost and the most frequent optimization procedure was to find the particular parameters of the design that would minimize the cost.

An understanding of the elements that make up cost is vital, because competition between companies and between nations is fiercer than ever before. The world is becoming a single gigantic marketplace in which newly developing countries with very low labor costs are acquiring technology and competing successfully with the well-established industrialized nations. To maintain markets requires a detailed knowledge of costs and an understanding of how new technology can impact to lower costs. A very good example of how technological advances can dramatically lower costs is in the field of microelectronics and large-scale integrated circuits (Fig. 9-1).

Broadly speaking, two classes of situations are included in cost evaluation:

1. Estimation of the cost of building a plant or installing a process within a plant to produce a product or line of products.

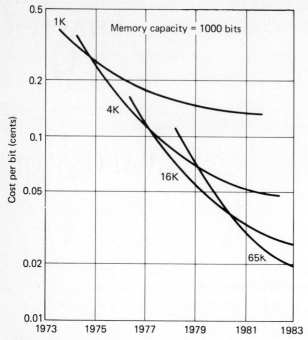

FIGURE 9-1
Historic cost trend of integrated circuits for memory applications.

2. Estimation of the cost of manufacturing a part based on a particular sequence of manufacturing steps.

Some of the uses to which cost estimates are put are the following:

1. To provide information to be used in establishing the selling price of a product or a quotation for a good or service.

2. To determine the most economical method, process, or material for manufacturing a product.

3. To be used as a basis for a cost-reduction program.

4. To determine standards of production performance that may be used to control costs.

5. To provide input concerning the profitability of a new product.

It can be appreciated that cost evaluation inevitably becomes a very detailed and "nitty-gritty" activity. This type of information rarely is published in the technical literature, partly because it does not make interesting reading but more important, because costs are highly proprietary information. Therefore, the emphasis in this chapter will be on the identification of the elements of costs and on some of the more generally accepted cost evaluation methods. It should be realized that cost estimation within a particular industrial or governmental organization will follow highly specialized and standardized procedures peculiar to the organization. However, the general concepts of cost evaluation described here will still be valid.

9-2 CATEGORIES OF COSTS

There are two broad categories of costs:

1. *Nonrecurring costs.* These are one-time costs, which we usually call capital costs. They are divided further into fixed capital costs, which include depreciable facilities such as plant building or manufacturing equipment and tools, and nondepreciated capital costs, such as land.
2. *Recurring costs.* These costs are direct functions of the manufacturing operation and occur over and over again. They usually are called operating costs or manufacturing costs.

Another classification division is into fixed and variable costs. *Fixed costs* are independent of the rate of production of goods; *variable costs* change with the production rate. A cost often is called a *direct cost* when it can be directly assigned to a particular cost center, product line, or part. *Indirect costs* cannot be directly assigned to a product but must be "spread" over an entire factory. The general categories of fixed and variable costs are given below.

Fixed Costs

1. Indirect plant cost
   (*a*) Investment costs
   Depreciation on capital investment
   Interest on capital investment and inventory
   Property taxes
   Insurance
   (*b*) Overhead costs (burden)
   Technical services (engineering)
   Nontechnical services (office personnel, security, etc.)
   General supplies
   Rental of equipment
2. Management expenses
   (*a*) Share of corporate executive staff
   (*b*) Legal staff
   (*c*) Share of corporate research and development staff
3. Selling expenses
   (*a*) Sales force
   (*b*) Delivery and warehouse costs
   (*c*) Technical service staff

Variable Costs

1. Materials
2. Direct labor (including fringe benefits)
3. Direct production supervision
4. Maintenance costs
5. Power and utilities
6. Quality-control staff
7. Royalty payments
8. Packaging and storage costs
9. Scrap losses and spoilage

Fixed costs such as marketing and sales costs, legal expense, security costs, financial expense, and administrative costs are often lumped into an overall category known as general and administrative expenses (G&A expenses). The above list of fixed and variable costs is meant to be illustrative of the chief categories of costs, but it is not exhaustive.[1]

The way the elements of cost build up to establish a selling price is shown in Fig. 9-2. The chief cost elements of direct material and direct labor determine the *prime cost*. To it must be added indirect manufacturing costs such as light, power, maintenance, supplies, and factory indirect labor. The manufacturing cost is made up of the factory cost plus general fixed expenses such as depreciation, engineering, taxes, office staff, and purchasing. The total cost is the manufacturing cost plus the sales expense. Finally, the selling price is established by adding a profit to the total cost.

Another important cost category is *working capital*, the funds that must be provided in addition to fixed capital and land investment to get a project started and provide for subsequent obligations as they come due. It consists of raw material on hand, semifinished product in the process of manufacture, finished product in inventory, accounts receivable, and cash needed for day-to-day operation. The working capital is tied up during the life of the plant, but it is considered to be fully recoverable at the end of the life of the project.

A concept that provides a rough estimate of the investment cost for a new product is the turnover ratio:

$$\text{turnover} = \frac{\text{annual sales}}{\text{total investment}} \tag{9-1}$$

---

[1] For an expanded list of fixed and variable costs see R. H. Perry and C. H. Chilton, "Chemical Engineers' Handbook," 5th ed., pp. 25–13 and 25–27, McGraw-Hill Book Company, New York, 1973.

Profit

Sales
expense

General
expense

Factory
expense

Manufacturing
cost

Total
cost

Selling
price

Direct
material

Prime
cost

Factory
cost

Direct
labor

FIGURE 9-2
Elements of cost that establish the selling price.

In the chemical industry the turnover ratio for many products is near 1.0; in the steel industry it is around 0.6. Suppose we wanted a quick estimate of the investment for a plant' producing 20,000 tons/year of a chemical product that sells for 30¢/lb. Since the total annual sales are

$$0.30 \text{ \$/lb} \times 2000 \text{ lb/ton} \times 2 \times 10^4 \text{ ton/year} = 12 \times 10^6 \text{ \$/year}$$

the total plant investment is of the same magnitude for a turnover ratio of 1.0.

The fact that the variable costs depend on the rate or volume of production and fixed costs do not, leads to the idea of a break-even point (Fig. 9-3). The determination of the production lot size to exceed the break-even point and produce a profit is an important consideration. There are many things to be considered, but a common decision associated with economic lot size is how to allocate production among different machines, plants, or

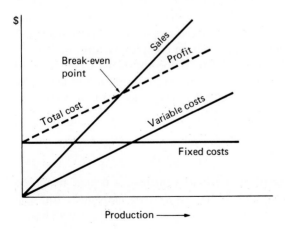

FIGURE 9-3
Break-even curve showing relation between fixed and variable costs and profit before taxes.

processes of various efficiencies or cost structure to make a product at minimum cost.

9-3 METHODS OF DEVELOPING COST ESTIMATES

The methods used to develop cost evaluations fall into three categories: methods engineering, costs by analogy, and statistical analysis of historical data.

Methods Engineering (Industrial Engineering Approach)

In the methods engineering approach the separate elements of work are identified in great detail and summed into the total cost per part. A typical, but simplified example, is the production of a simple fitting from a steel forging.

Cost per part

Operations	Material	Labor	Overhead	Total
Steel forging	37.00			37.00
Set up on milling machine		0.20	0.80	1.00
Mill edges		0.65	2.60	3.25
Set up on drill press		0.35	1.56	1.91
Drill 8 holes		0.90	4.05	4.95
Clean and paint		0.30	0.90	1.20
	37.00	2.40	9.91	$49.31

Material cost, 75 percent; direct labor, 5 percent; overhead, 20 percent

It illustrates that overhead charges (fixed costs) often are many times more than the direct labor costs. The exact ratio will depend on numerous factors (see Sec. 9-8). Also, in this case material cost is high because the production sequence starts with a finished forging rather than a workpiece cut from bar stock. Using the methods engineering approach on complex systems requires a great deal of effort and computation. However, there are strong trends toward putting material and processing costs into a computer data base and using the computer to search out the optimum processing sequence and calculate the costs.

Analogy

In cost estimation by analogy, the future costs of a project or design are based on past costs of a similar project or design, with due allowance for cost escalation and size differences. The method therefore requires a backlog of experience or published cost data. As will be shown subsequently, this method of cost evaluation commonly is used for feasibility studies of chemical plants and process equipment. When cost evaluation by analogy is used, future costs must be based on the same state of the art. For example, it would be valid to use cost data on a DC-8 transport aircraft to estimate costs for a larger DC-10, but it would not be correct to use the same data to predict the cost of a supersonic transport because many discontinuities in technology are involved. One major difference is the change from predominantly aluminum to titanium airframe construction in the SST.

Another concern with determining cost by analogy is to be sure that costs are being evaluated on the same basis. Equipment costs often are quoted FOB (free on board) the manufacturer's plant location, so that delivery cost must be added to the cost estimate. Costs sometimes are given for the equipment not only delivered to the plant site but also installed in place, although it is more usual for costs to be given FOB some shipping point. For example, the purchase cost of a 10-hp electric motor in 1970 was about $1400, but the cost installed in a plant was about $5000.

Statistical Approach

In the statistical approach to cost estimation, techniques such as regression analysis are used to establish relations between system cost and initial parameters of the system: weight, speed, power, etc. This approach, involving as it does cost estimation at a high level of aggregation, is the direct opposite of the methods engineering approach. For example, the cost of developing a turbofan aircraft engine might be given by

$$C = 0.13937x_1^{0.7435}x_2^{0.0775}$$

where C is in millions of dollars, x_1 is the maximum engine thrust, in pounds, and x_2 is the number of engines produced. Cost data expressed in this empirical form can be useful in trade-off studies.

Levels of Cost Evaluation

The American Association of Cost Engineers lists five types of cost evaluations based on the level of detail and accuracy that is involved:

Kind of estimate	Probable error, %
Order of magnitude (ratio)	±40
Study (scope)	±30
Preliminary (budget authorization possible)	±20
Definitive (complete data on project short of detailed drawings and specifications)	±10
Detailed (complete engineering drawings and specifications)	±5

The order-of-magnitude estimate is not based on complete flow sheets or equipment lists. It is developed by analogy using scale-up ratios and cost escalation factors for known costs of similar designs. Depending on the size of the project, an order-of-magnitude estimate takes from a few hours to a few days to complete. The main purpose of such an estimate is to determine whether there is enough likelihood of profit to proceed with further engineering work.

The study estimate proceeds from incomplete but developing detailed engineering design information. The purpose of a study estimate is to provide management with a "ball park figure" so it can decide whether further engineering work is justified.

The preliminary estimate is prepared from well-defined engineering design information such as flow sheets and detailed equipment lists. The accuracy of the cost estimate (±20 percent) is precise enough to allow preliminary budget approval.

The definitive estimate works from complete design information so that the data are sufficiently established to provide numbers for project control.

The detailed estimate adds a complete set of detailed engineering drawings and project specifications to a level of detail suitable to going out for bids or submitting a purchase order.

A number of factors must be given special attention in construction cost estimates:

1. *Allowances for extras.* Every project will incur extra costs depending on its complexity, site location and conditions, and mistakes on and omissions from the construction drawings.

2. *Cost escalation.* In an inflationary economy, if the time for the construction project is long or if the project becomes delayed, the cost of labor and material may have increased considerably before completion. Escalation clauses in labor contracts and purchase orders must be carefully examined.

3. *Design modification.* A complex plant or process often requires changes in the design after start-up so the system can achieve the expected performance. Hopefully, with good engineering design these additions and/or revisions will be minor.

4. *Contingency.* A fund must be set aside to provide for unforeseen cost eventualities. The more detail and work that have gone into a cost estimate the less the contingency fund will be. However, it is not unusual for the contingency fund of a complex project to be between 10 and 20 percent of the total cost of the project.

5. *Expiration date.* Every cost estimate should show a date (3 to 6 months hence) after which the estimate is no longer valid.

9-4   COST INDEXES

Because the purchasing power of money decreases with time, *all published cost data are out of date.* To compensate for this shortcoming, cost indexes are used

TABLE 9-1
Engineering cost indexes

Type of cost	Source of cost index
General price indexes	
Consumer (CPI)	Bureau of Labor Statistics
Producer (wholesale)	U.S. Department of Labor
Construction cost indexes	
General construction	*Engineering News Record*
Chemical plants	*Chemical Engineering*
Petroleum refinery	*Oil and Gas Journal*
Industrial equipment	Marshall and Swift index
Plant maintenance	*Factory*

to convert past costs to present costs.

$$C_{t_2} = C_{t_1} \frac{I_{t_2}}{I_{t_1}} \tag{9-2}$$

In Table 9-1 are listed the most common cost indexes and the source of the data.

The *Chemical Engineering* plant cost index is an overall index for the cost of constructing chemical process plants. In 1959 the CE index was 100, and in 1986 it was 318. The index has four major components: equipment and machinery, construction labor, buildings (materials and labor), and engineering design. The equipment and machinery subindex is further broken down into fabricated equipment, process machinery, pipes, valves, and fittings, process instruments and control, pumps and compressors, electrical equipment, and structural supports.

The Marshall and Swift cost index (formerly Marshall and Stevens) is an overall index for the cost of industrial equipment. The index is subdivided for specific industries like cement, chemical, paper, and steam power. In 1926 the Marshall and Swift index was 100, and in 1986 it was 797.

You should be aware of some of the pitfalls inherent in using cost indexes. First, you need to be sure that the index you plan to use pertains to the problem you must solve. For example, the cost components in the *Engineering News Record* index would not apply to estimating the cost of an ore beneficiation plant.[1] Of more basic concern is the fact that the cost indexes reflect the costs of past technology and design procedures. When new radical

---

[1] Details of the composition of the cost indexes are given in the following sources: Chemical Engineering Index: C. H. Chilton, *Chem. Eng.*, vol. 73, no. 9, p. 184, 1966; J. Matley, *Chem. Eng.*, vol. 89. no. 8, p. 153, 1982; Engineering News Record: *Eng. News Record*, 142, no. 11, p. 161, 1949; Marshall and Swift Index: R. W. Stevens, *Chem. Eng.*, p. 124, November 1947; Petroleum Index: W. L. Nelson, *Oil Gas J.*, vol. 65, p. 97, May 15, 1967.

technology is introduced, it usually changes the cost trends very sharply. For example, through improvements in technology the cost of building a catalytic cracking unit for high-octane gasoline in 1969 was 40 percent less than the cost of a plant of similar capacity in 1946. This cost effect of improved technology, called a learning curve, will be discussed in Sec. 9-10. Finally, it should be noted that published cost indexes reflect national averages and may not accurately reflect local conditions. Information on labor rates and productivity in various sectors of the United States is available from the Department of Labor. Methods of converting cost indexes to another country are available.[1]

Major items of engineering equipment, such as steam turbines, supertankers, and nuclear reactors, require several years to build. With the current trend of inexorable inflation it is important to consider that in the cost estimate. A common way is to use a *cost escalation factor* to estimate the cost of equipment at the date it is ready to be shipped to the customer. A typical cost escalation formula is

$$E = C_0 \times F \times M \times \frac{(BLS)_A}{(BLS)_B} + C_0 \times F \times (1 - M) \times \frac{(BLS)_A'}{(BLS)_B'} \qquad (9\text{-}3)$$

This formula is based on Bureau of Labor Statistics (BLS) cost indexes. Considered separately are the escalation of material costs through the BLS Producer Price Index $(BLS)_A$ and of wages through the BLS Labor Index $(BLS)_A'$. The use of this formula is shown by an example.[2]

Example 9-1. An order is placed for a major equipment item in June 1975 for delivery in December 1976. The ·uoted cost in June is $500,000. The manufacturer stipulates that 54 percent of .he cost is subject to escalation over which he has no control; therefore, $F = 0.54$. These costs typically are purchased components, materials, etc. The materials-to-labor ratio is 70 percent to 30 percent; therefore, $M = 0.70$. Using the appropriate BLS indexes for material and labor costs:

$(BLS)_A = 199$ at December 1976 $(BLS)_A' = 5.62$ at December 1976

$(BLS)_B = 188.6$ at June 1975 $(BLS)_B' = 5.03$ at June 1975

$$E = 500,000(0.54)(0.7)\left(\frac{199}{188.6}\right) + 500,000(0.54)(0.30)\left(\frac{5.62}{5.03}\right) = 289,920$$

This is the escalated cost in June 1976 for the fraction of the original cost subject to escalation. The fraction not subject to escalation is

$$5000,000(1 - 0.54) = 230,000$$

Therefore, the total cost of the equipment at the time of shipment will be $289,920 + 230,000 = $519,920.

---

[1] A. V. Bridgwater, *Chem. Eng.*, pp. 119–121, November 5, 1979.

[2] A. Pikulik and H. E. Draz, *Chem. Eng.*, pp. 107–21, October 10, 1977.

9-5 COST-CAPACITY FACTORS

The cost of most capital equipment is not directly proportional to the size or capacity of the equipment. For example, doubling the horsepower of a motor increases the cost by only about one-half. The economy of scale is an important factor in engineering design. The cost-capacity relation usually is expressed by

$$C_1 = C_0\left(\frac{Q_1}{Q_0}\right)^x \tag{9-4}$$

where C_1 and C_0 are the capital costs associated with the capacity or size Q_1 and Q_0. The exponent x varies from about 0.4 to 0.8, and it is approximately 0.6 for many items of process equipment. For that reason, the relation in Eq. (9-4) often is referred to as the "six-tenths rule." Values of x for different types of equipment are given in Table 9-2.

The cost-capacity relation also may be used to estimate the effect of plant size on capital cost. In addition, cost indexes and capacity may be combined as shown in the following example.

Example 9-2. A 200-ton per day sulfuric acid plant cost $900,000 in 1966. How much would a 400 ton/day plant cost in 1976?

Solution. Using the CE plant cost index for the appropriate years and the six-tenth rule,

$$C_{1976} = \$900,000\left(\frac{400}{200}\right)^{0.6}\left(\frac{192}{107}\right) = \$2,450,000$$

In making cost estimations for chemical and metallurgical plants and other manufacturing plants, there are two different situations to consider. A

TABLE 9-2
Typical values of size exponent for equipment

Equipment	Size range	Capacity unit	Exponent x
Blower, single stage	1000–9000	ft^3/min	0.64
Centrifugal pumps, S/S	15–40	hp	0.78
Dust collector, cyclone	2–7000	ft^3/min	0.61
Heat exchanger, shell & tube, S/S	50–100	ft^2	0.51
Motor, 440-V, fan-cooled	1–20	hp	0.59
Pressure vessel, unfired			
carbon steel	6000–30,000	lb	0.68
Tank, horizontal, carbon-steel	7000–16,000	lb	0.67
Transformer, 3-phase	9–45	kW	0.47

Source: R. H. Perry and C. H. Chilton, "Chemical Engineers' Handbook," 5th ed., p. 25–18, McGraw-Hill Book Company, New York, 1973.

grass-roots plant is a complete plant erected on a new (green field) site. The cost of the plant must include land acquisition, site preparation, and auxiliary facilities, such as rail line, loading dock, and warehouses, in addition to the cost of the processing equipment. A *battery limits plant* is an existing plant to which process equipment has been added. The battery limits circumscribe the new equipment to be added but do not include auxiliary facilities and site preparation, which already exist. When using published cost data, you need to know for which type of situation the costs have been determined.

9-6 FACTOR METHODS OF COST ESTIMATION

The need in the chemical industry for rapid preliminary estimates of cost led to the development of methods by which the total project cost is built up by applying various factors to the cost quotations for major items of equipment. The equipment costs to which the factors are applied are called the *base cost*. Lang proposed this method in the aggregate,

$$C_i = fC_b \qquad (9\text{-}5)$$

where C_i = installed plant cost
$\quad C_b$ = base cost
$\quad f$ = Lang factor, which depends on the nature of the plant
$\quad f$ = 3.1 for a solids-processing plant
$\quad f$ = 3.6 for a solid-fluid-processing plant
$\quad f$ = 4.7 for a fluid-processing plant

These factors show that the installed cost of a plant is many times the purchase cost of the major items of equipment, but they give no information on details.[1] In the following example the factor method is extended to a higher level of detail.

> **Example 9-3.**[2] The purchased cost of a fabric filter gas cleaning system (baghouse) to handle 30,000 std ft^3/min is $192,000. Needed auxiliary equipment (cooler, duct work, etc.) is $134,000. The base cost is developed as:
>
> | | |
> |---|---|
> | Primary equipment | 192,000 |
> | Auxiliary equipment | 134,000 |
> | | 326,000 |

[1] For more detailed breakdown of factors see D. R. Woods, "Financial Decision Making in the Process Industry," p. 181, Prentice-Hall, Inc., Englewood Cliffs, N.J., 1975.

[2] W. M. Vatavuk and R. B. Neveril, *Chem. Eng.*, pp. 157–162, November 3, 1980.

			Comment
Instruments and controls $(0.10 \times 326{,}000 \times 1.5)$	$= 48{,}900$		0.10 std% for instrument; 1.5 adjustment for continuous operation
Taxes $(0.03 \times 326{,}000)$	$= 9{,}780$		
Freight $(0.05 \times 326{,}000 \times 0.6)$	$= 9{,}780$		0.05 std% for freight; 0.6 adjustment for delivery near major city

Base cost $394,460

Find the installed cost of the system.

Solution. Note that this system consists of standard factors for items like instruments and controls plus an adjustment factor for deviations from the normal conditions. The cost factors for the fabric filter and for other common types of air pollution control equipment are listed in Table 9-3.

In Table 9-4 are listed the adjustments to the average costs based on the degree of complexity or difficulty in such areas as handling and erection, site preparation, engineering, and supervision. Continuing with our example, we find that installation presents no unusual problems because the filter is installed in an existing plant. Thus, the installation multiplier factor is 0.72. When considering indirect costs, we realize that the equipment is standard and not too massive, so engineering and supervision can be reduced from 0.10 to 0.05. The other factors are standard, so the indirect cost multiplier factor is 0.40.

Thus, the estimate for the entire filter system is:

Base cost	$394,460
Installation $0.72 \times 394{,}460 =$	284,011
Indirect cost $0.40 \times 394{,}460 =$	157,784
	$836,000 rounded off

The accuracy of this factored cost estimate is ± 20 percent, so the installed cost could range from $1,003,000 to $669,000.

Cost factors for process piping, electrical work, instrumentation, field erection, etc., are available in the literature.[1] Very detailed cost data for process plants and equipment have been presented by Guthrie.[2] In this approach detailed cost information is presented, in graphical form, for

---

[1] C. A. Miller, *Am. Assoc. Cost Eng. Bull.*, p. 92, September 1965; D. R. Woods, op. cit., pp. 186–192.

[2] K. M. Guthrie, *Chem. Eng.*, pp. 114–142, March 24, 1969, "Process Plant Estimating, Evaluation, and Control," Craftsman Book Co, Saloma Beach, Calif., 1974.

TABLE 9-3
Average cost factors for air pollution control equipment

Direct costs	Electrostatic precipitator	Venturi scrubber	Fabric filter	Thermal and catalytic incinerator	Carbon adsorber	Gas absorber	Refrigerator	Flare
Purchased equipment:								
Primary	A							
Auxiliary	B							
Instruments and controls	$0.10\,(A+B)$		Same for all equipment					
Taxes	$0.03(A+B)$							
Freight	$0.05\,(A+B)$							
Base price	X (total of above)							
Installation:								
Foundation and supports	0.04	0.06	0.04	0.08	0.08	0.12	0.08	0.12
Handling and erection	0.50	0.40	0.50	0.14	0.14	0.40	0.14	0.40
Electrical	0.08	0.01	0.08	0.04	0.04	0.01	0.08	0.01
Piping	0.01	0.05	0.01	0.02	0.02	0.30	0.02	0.02
Insulation	0.02	0.03	0.07	0.01	0.01	0.01	0.10	0.01
Painting	0.02	0.01	0.02	0.01	0.01	0.01	0.01	0.01
Site preparation*				As required				
Facilities and buildings*				As required				

Direct costs	Electrostatic precipitator	Venturi scrubber	Fabric filter	Thermal and catalytic incinerator	Carbon adsorber	Gas absorber	Refrigerator	Flare
Multiplier for direct installation costs	0.67	0.56	0.72	0.30	0.30	0.85	0.43	0.57
Indirect costs								
Installation:								
Engineering and supervision	0.20	0.10	0.10	0.10	0.10	0.10	0.10	0.10
Construction and field	0.20	0.10	0.20	0.05	0.05	0.10	0.05	0.10
Construction fee	0.10	0.10	0.10	0.10	0.10	0.10	0.10	0.10
Start-up	0.01	0.01	0.01	0.02	0.02	0.01	0.02	0.01
Performance test	0.01	0.01	0.01	0.01	0.01	0.01	0.01	0.01
Model study	0.02	—	—	—	—	—	—	—
Contingencies	0.03	0.03	0.03	0.03	0.03	0.03	0.03	0.03
Multiplier for indirect installation costs	0.57	0.35	0.45	0.31	0.31	0.35	0.31	0.35

Reprinted by special permission from *Chemical Engineering*, November 3, 1980, p. 158.

TABLE 9-4
Adjustments to average cost factors in Table 9-3

Cost category	Adjustment	Cost category	Adjustment
Instrumentation		**Facilities and buildings**	
1. Simple, continuous, manual	0.5 to 1.0	1. Outdoor process; utilities at site	0
2. Intermittent control; instrumentation modulates flow	1.0 to 1.5	2. Outdoor process, with some weather enclosures; utilities brought to site; access roads, fencing and minimum lighting	1
3. Hazardous process; includes safety backup	3	3. Buildings with heating and cooling, sanitation facilities; shops and office; may include railroad sidings, truck depot and parking	2
Freight			
1. Major metropolitan areas in continental U.S.	0.2 to 1.0	**Engineering and supervision**	
2. Remote areas in continental U.S.	1.5	1. Small-capacity standard equipment; duplicate of typical system; turnkey quotation	0.5
3. Alaska, Hawaii and other countries	2	2. Custom equipment; automatic control	1 to 2
Handling and erection		3. Prototype equipment; large system	3
1. Delivered cost includes assembly; supports, base, skids provided; small-to-moderate system	0.2 to 0.5	**Construction and field**	
2. Modular equipment; ducts and piping less than 200 ft long for compact site; moderate-size system	1	1. Small-capacity system	0.5
		2. Medium-capacity system	1
3. System large and scattered, requiring long ducts and piping; fabrication on-site, with extensive welding and erection	1 to 1.5	3. Large-capacity system	1.5
		Construction fee	
		1. Turnkey, erection and installation included in equipment cost	0.5
4. Retrofitting existing plant; includes equipment removal and site renovation; moderate-to-large system	2	2. Single contractor for entire installation	1
		3. Many contractors; supervision by primary contractor	2
Site preparation			
1. Within battery limits of existing plant; minimum clearing and grading	0	**Contingency**	
2. Outside battery limits; extensive leveling and removal of structures; includes land survey and study	1	1. Developed process	1
		2. Prototype or experimental process subject to change	3 to 5
3. Extensive excavation, filling and leveling; may include draining and setting piles	2	3. Guarantee of efficiencies and operating specifications requires pilot tests; deferment of payment until final certification, and penalty for failure to meet completion date or efficiency	5 to 10

Reprinted by special permission from *Chemical Engineering,* November 3, 1980, p. 159.

equipment such as process furnaces, heat exchangers, vessels, and pumping units. Separate cost factors are presented to allow for variation in material of construction, design features, and operating pressure. Other detailed cost data are presented for process piping, plant structures, and off-site facilities. An example of the use of Guthrie's method plus extensive cost tables is given by Bassel.[1]

9-7 MANUFACTURING COSTS

Manufacturing costs or operating costs are equal to the direct production costs + fixed charges + plant overhead costs. The components of direct production costs (variable costs) are as follows:

1. Raw materials. Be sure to include a credit for recycled scrap or a by-product that is sold elsewhere. Note that most raw material prices are negotiated. Prices listed in trade magazines are only guidelines.
2. Operating labor. The most accurate method of determining labor cost is to set up a manning table and find the actual number of people required to run the process line. Be sure to account for standby personnel. Costs of operating labor include fringe benefits. As an order-of-magnitude estimate of labor costs

$$\frac{\text{Operating man-hours}}{\text{Tons of product}} = K \frac{\text{no. of process steps}}{(\text{tons/day})^{0.76}}$$

where $K = 23$ for a batch operation
17 for average labor requirements
10 for automated process

3. Direct supervisory and clerical labor
4. Utilities
5. Maintenance and repairs; typically they vary from 3 to 10 percent of plant investment per year
6. Operating supplies; 10 to 20 percent of costs for maintenance and supplies
7. Laboratory and/or quality-control charges
8. Patents and royalties

Fixed charges are made up of the following components:

1. Depreciation
2. Local taxes

---

[1] W. D. Bassel, "Preliminary Chemical Engineering Plant Design," pp. 254–258, Elsevier Scientific Publishing Co., Inc., New York, 1976.

3. Insurance

4. Floor space costs

5. Interest on investment

Plant overhead costs include general plant upkeep; payroll overhead; medical, safety, and security services; food and recreation for employees; shipping and receiving; salvage; laboratories; and storage.

The total product cost equals the manufacturing cost plus the general and administrative expenses referred to in Sec. 9-2.

Manufacturing costs may be calculated on the basis of dollars per unit of production or dollars per hour or dollars per year. It is important to determine the manufacturing costs for both full and reduced levels of production. Manufacturing costs generally are not a linear function of production. The best source of information on manufacturing costs is a similar or identical production operation within the same company. The most serious source of error is in overlooking some element of cost. To avoid the error, it is a good idea to use a standard form for cost evaluation. The use of minicomputers to speed up the cost evaluation process and enhance the attention to detail is a major step in this field.[1] For machining processes the data provided by Ostwald[2] is useful.

Example 9-4. Determine the cost of hardening a carburized pinion gear made of 8720-H steel and weighing 76.25 lb. The furnace is a roller-hearth type that is electrically heated with three zones of 190-, 100-, and 37-kW rating. It is equipped for protective atmosphere and requires one man for operation, and it operates at capacity (26.7 parts/h) with no idle periods. (This example is from *Metal Progress*, July 15, 1954 with update on costs to 1980.)

Solution. The direct production costs (variable costs), based on dollars per hour, are as follows:

1. Raw materials

$$\text{Atmosphere } \$7.80 \text{ per h} \times \frac{2500 \text{ chf used}}{8000 \text{ cfh generated}} \qquad\qquad 2.437$$

$$\text{Alloy trays, 24 per furnace } \frac{\$90.00}{5000\text{-h life}} \times 24 \qquad\qquad 0.432$$

Supplies—quench oil loss and inert gas to purge furnace $\qquad\qquad$ 0.270

[1] J. G. Goldberg, *Manuf. Engr.*, Feb., 1987, pp. 37–41.

[2] P. F. Ostwald, "AM Cost Estimator, 1987–88 ed.," McGraw-Hill Book Co., N.Y., 1987.

2. Direct labor

$$1 \text{ operator} \times \frac{\$4.25 \text{ per h}}{0.92 \text{ efficiency}} \qquad\qquad 4.619$$

Fringe benefits at 30 percent   1.386
Premium pay—prorated annual overtime   0.150
     $6.155

3. Direct supervisory labor
     $0.36 \times \$6.155$ $2.216

4. Utilities
     Electricity cost, 215 kWh $\times$ $0.035 $7.525

5. Maintenance and supplies

$$\frac{\$2512 \text{ labor} + \$1210.39 \text{ supplies}}{5760 \text{ productive hours per year}} \qquad\qquad \$0.646$$

6. Scrap and rework

$$\frac{\$127.25 \text{ scrap} + \$419.33 \text{ rework}}{5760} \qquad\qquad \$0.095$$

Total variable costs per hour $19.776

Fixed costs, initially based on dollars per year, are as follows:

1. Furnace depreciation
     8.33 percent $\times$ $129,375 (12-year life)   10,776.94

2. Property tax   945.60

3. Insurance

$$\$0.095 \text{ per hundred} \times \frac{\$129,375}{100} \qquad\qquad 122.83$$

4. Building expense (space charge)
     Depreciation   12,000
     Taxes   500
     Insurance   90
     Fire protection   16,200
     Maintenance   42,750
     Heat   16,921
     Light   6,250
     $94,711

$$\text{Cost per square foot of floor area} = \frac{94,711}{36,250} = \$2.613$$

Furnace area
936 ft^2 $\times$ $2.613 $2,445.77

5. Interest on investment

$$\frac{\text{Building}}{\text{value}} \times \frac{\text{furnace floor area}}{\text{building floor area}} \times \frac{\text{interest}}{\text{rate}}$$

$$\$200,000 \times \frac{936 \text{ ft}^2}{36,250 \text{ ft}^2} \times 0.08 = \$413.13$$

Furnace cost × interest rate
$$\$129,375 \times 0.08 = \$10,350$$

Total interest	10,763.13
Total fixed costs (annual)	25,054.27 \$/year
Fixed costs per hour $= \dfrac{25,054.27}{5,760}$	4.349 \$/h

Plant overhead allocated to the heat-treatment department is \$97,200, and it is further apportioned to the furnace on the basis of the direct labor involved.

$$\$97,200 \times \frac{1 \text{ direct labor}}{62 \text{ employees in HT dept.}} \qquad \$1,567.74$$

$$\text{Overhead cost per hour} = \frac{1567.74}{5,760} \qquad 0.272$$

Summary

Variable costs	19.776
Fixed costs	4.349
Overhead costs	0.272
Total costs, per hour	\$24.397

$$\text{Cost per part} = \frac{24.397}{26.7} \qquad \$0.914$$

$$\text{Cost per pound} = \frac{24.397}{26.7 \times 76.25} \qquad \$0.0119$$

We must realize that overhead can be allocated according to a number of different bases. The most usual bases are direct labor hours, materials cost, number of parts produced, and space utilized. In Example 9-4 the plant overhead cost was unusually low because the allocation was based on direct labor, which was very low for the heat-treating operation. However, it is not unusual to have factory overhead rates of 150 to 200 percent of direct labor. The subject of overhead is discussed more fully in the next section.

9-8   OVERHEAD COSTS

Perhaps no aspect of cost evaluation creates more confusion and frustration in the young engineer than overhead cost. Many engineers consider overhead to be a tax on their creativity and efforts, rather than the necessary and legitimate cost it is. However, as we have seen above, overhead can be computed in a

variety of ways. Therefore, you should know something about how accountants assign overhead charges.

An overhead cost is any cost not specifically or directly associated with
the production of identifiable goods or services. The two main categories of
overhead costs are factory or plant overhead (discussed in Sec. 9-7) and
corporate overhead. Just as factory overhead includes the costs of manufacturing that are not related to direct labor and material, so corporate overhead is
based on the costs of running the company that are outside the manufacturing
or production activities. Since many manufacturing companies operate more
than one plant, it is important to be able to determine factory overhead for
each plant and to lump the other overhead costs into corporate overhead.
Typical cost contributions to corporate overhead are the salaries and fringe
benefits of corporate executives, sales personnel, accounting and finance, legal
staff, R & D, corporate engineering and design staff, and the operation of the
corporate headquarters building.

Example 9-5 A modest-sized corporation operates three plants with direct labor
and factory overhead as follows:

Cost	Plant A	Plant B	Plant C	Total
Direct labor	$750,000	400,000	500,000	1,650,000
Factory overhead	900,000	600,000	850,000	2,350,000
Total	1,650,000	1,000,000	1,350,000	4,000,000

In addition, the cost of management, engineering, sales, accounting, etc. is
$1,900,000. Find the corporate overhead rate based on direct labor.

Solution. Corporate overhead rate $= \dfrac{1,900,000}{4,000,000}(100) = 47\%$

In the next example of overhead costs, we consider the use of factory
overhead in determining the cost of performing a manufacturing operation.

Example 9-6. A batch of 100 parts requires 0.75 h of direct labor each in the
gear-cutting operation. If the cost of direct labor is $7.50 per h and the factory
overhead is 160 percent, determine the total cost of processing a batch.

Solution. The direct labor cost is

$$(100 \text{ parts})(0.75 \text{ h/part})(\$7.50/\text{h}) = \$562.50$$

The factory overhead charge is

$$562.50(1.60) = \$900$$

The cost of gear cutting for the batch of 100 parts is

$$562.50 + 900 = \$1462.50$$

The most common basis for allocating factory overhead is direct labor hours. The total overhead charges to a cost center are estimated. The estimated hours of direct labor are divided into the sum to give the overhead rate, in dollars per hour. As production proceeds, the direct labor hours actually used are totaled. The total direct labor hours (DLH) is multiplied by the overhead rate to establish the amount of overhead to be charged to the product.

Example 9-7. It is estimated that the factory overhead costs for the powder metallurgy operation of a large manufacturing plant will be $180,000 over the next year. It is further estimated that 30,000 DLH will be used in production in that shop over the same time period. Therefore, the factory overhead for this cost center will be $180,000/30,000 = $6 per DLH. During the past month 900 units of PM gears were started and completed. They required 580 direct labor hours and resulted in the charges for direct labor and direct materials shown below. Find the amount of overhead to be charged to the product.

Solution.

	Total	Per unit
Direct labor	$7,400	$8.22
Direct materials	$9,800	$10.89
Overhead		
580 DLH × $6 per DLH	$3,480	$3.87
Total cost	$20,680	$22.98

We note if the work force were suddenly cut to produce only 15,000 DLH without a corresponding reduction in the overhead costs, the overhead rate would double.

The allocation of overhead on the basis of direct labor hours sometimes can cause confusion as to the real costs when process improvements result in an increase in manufacturing productivity. Consider the following example.

Example 9-8. A change from a high-speed steel cutting tool to a new coated WC tool results in a halving of the time for a machining operation. The data for the old tool and the new coated carbide tool are shown below in columns 1 and 2, respectively. Because the cost of overhead is based on the DLH, the cost of overhead apparently is reduced along with the cost of direct labor. The apparent savings per piece is $72.00 − $36.00 = $36.00. However, a little reflection will show that the cost elements that make up the overhead (foreman, toolroom, maintenance, inspection, etc.) will be constant per piece produced; and if the rate of production doubles, so should the overhead charge. The true costs are given in column 3. Thus, the actual savings per piece is $72.00 − $56.00 = $16.00.

	(1) **Old tool**	(2) **New tool** **(apparent** **cost)**	(3) **New tool** **(true cost)**
Machining time, DLH	4.0	2.0	2.0
Direct labor rate, per hour	$8.00	$8.00	$8.00
Direct labor cost	$32.00	$16.00	$16.00
Overhead rate, per hour	$10.00	$10.00	$20.00
Cost of overhead	$40.00	$20.00	$40.00
Cost of direct labor and overhead	$72.00	$36.00	$56.00

Operating budgets frequently are based on direct labor hours. Thus, in Example 9-8, a manufacturing engineer who reduced DLH could be penalized unless recognition was given to the increased maintenance costs, etc. from the true overhead costs of a more productive situation.

Direct labor hours may not be the best basis for allocation in many manufacturing situations. Consider a plant whose major cost centers are a machine shop, a paint line, and an assembly department. We see that it is reasonable for each cost center to have a different overhead rate in units appropriate to the function that is performed.

Cost center	Est. factory overhead	Est. number of units	Overhead rate
Machine shop	$250,000	40,000 machine hours	$6.25 per machine hour
Paint line	80,000	15,000 gal of paint	$5.33 per gallon of paint
Assembly dept.	60,000	10,000 DLH	$6.00 per DLH

The above examples show that the allocation of overhead on the basis of DLH may not be the best way to do it. This is particularly true of automated production systems where overhead has become the dominant project cost. In such situations overhead rates are often between 500 and 800 percent of the direct labor cost. In the limit, the overhead rate for an unmanned manufacturing operation would be infinity. There is a trend to base the overhead allocation on the actual machine hours used in production.

There is a danger that improper overhead allocation can lead top management to make the wrong decisions.[1] A company sold one of its product lines because overhead allocations made it appear unprofitable. When the product line was sold profits decreased rather than increased. Moreover, since

---

[1] R. Strauss, *Chem. Engr,* Mar. 16, 1987, pp. 103–106.

the overhead from the disposed of product had to be reallocated to the other existing products several of them now appeared unprofitable.

9-9   STANDARD COSTS

In the most common forms of cost accounting actual or recorded costs or standard or predetermined costs are used. In the preceding examples the costs used were actual costs, in that costs of direct labor and material were charged to the part or the product and factory overhead was allocated on the basis of actual direct labor hours. However, the nomenclature *actual cost* is misleading. Usually average labor rates are used rather than the actual ones and arbitrary allocations are made for use of capital equipment and for general and administrative costs. Also, actual costs are compiled long after the job is completed. This approach is chiefly aimed at financial accountability rather than cost control.

The standard cost approach to cost accounting[1] was developed to provide meaningful benchmarks that could be used to monitor day-to-day performance and signal when deviations from predetermined levels were occurring. Standard costs are based on the proposition that there is a certain amount of material in a part and a given amount of labor goes into the part's manufacture. In a given period of time costs tend to vary around some average cost per unit or per hour. The system of standard costs consists of two parts: 1) a base standard and 2) a current standard. The *base standard* is determined infrequently, e.g., once a year, and the *current standard* represents the latest cost. The difference between the two is the *cost variance*.

Example 9-9 (Standard material cost). The standard cost for a casting (Ax3472-14) has been set at $2.25 based on analysis of the purchase price plus transportation cost over the past year. Analysis of real costs showed the price varied from $2.05 to $2.34, with a weighted mean of $2.25. The casting is machined to produce a component. Data from the production line show that there is a certain scrap loss (due to porosity, runout in machining, etc.), so that the standard is set at 510 castings to produce 500 finished units. Thus, the standard material cost is $510 \times \$2.25 = \$1147.50$

In the next production batch of 500 units, 507 castings are used at a cost of $2.27 for an actual cost of $1150.89. This results in an unfavorable (+) variance of $3.39. We see that this is caused by a $0.02 increase per unit in purchased castings and a decrease in the scrap rate, so 507 castings instead of 510 are required to produce 500 completed units.

---

[1] J. P. Simmini, "Cost Accounting Concepts for Nonfinancial Executives and Managers," American Management Association, New York, 1976; R. Strauss, *Chem. Engr.*, April 27, 1987, pp. 66–70; May 25, 1987, pp. 87, 88.

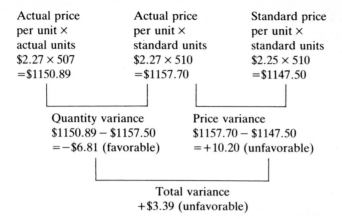

Actual price per unit × actual units	Actual price per unit × standard units	Standard price per unit × standard units
$2.27 × 507	$2.27 × 510	$2.25 × 510
=$1150.89	=$1157.70	=$1147.50

Quantity variance
$1150.89 − $1157.50
=−$6.81 (favorable)

Price variance
$1157.70 − $1147.50
=+10.20 (unfavorable)

Total variance
+$3.39 (unfavorable)

The same information on cost variance could be developed by determining the price variance first.

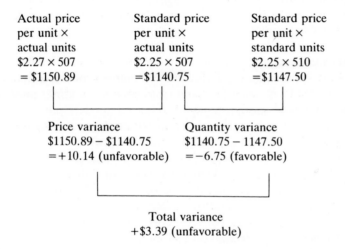

Actual price per unit × actual units	Standard price per unit × actual units	Standard price per unit × standard units
$2.27 × 507	$2.25 × 507	$2.25 × 510
= $1150.89	=$1140.75	=$1147.50

Price variance
$1150.89 − $1140.75
=+10.14 (unfavorable)

Quantity variance
$1140.75 − 1147.50
=−6.75 (favorable)

Total variance
+$3.39 (unfavorable)

We note that there are slight differences in the values of the individual variances depending upon whether the quantity variance or the price variance was determined first, but the total variance is the same in each instance.

Example 9-10 (Standard labor costs). Production workers of various levels of skill and wage rates are used in a cost center that does job shop machining.

Class of operator	Wage rate, $/h	Number	Accumulated average
A	$10.50	1	$10.50
B	8.25	3	8.81
C	6.00	5	7.25
D	4.00	2	6.66

The weighted average of the various wage rates, $6.66 per h, is the standard labor rate. If the labor force in the job shop used 325 hours to machine a batch of 500 castings, the direct labor cost would $325 \times 6.66 = \$2164.50$. However, the distribution of effort by wage rate turned out to be

Wage rate, $/h	Hours required	Actual cost	Cumulative cost
10.50	40	420	$420
8.25	68	561	981
6.00	27	162	1143
4.00	190	760	1903
	325		

Direct labor variance = 1903 − 2164 = −$261

The favorable variance of $261 would be credited to the labor variance account. This situation arose because we assigned the bulk of the machining to the lowest-grade operators because it was of a routine nature. That is reflected by the large favorable variance.

The development of standard costs requires a considerable amount of work. As Fig. 9-4 illustrates, it involves a number of departments in the company. The bill of materials and the process routing must be well established. Time and motion study is required to determine the standard labor hours and the standard time to set up the machines for production of the part. However, the time and expense are well worthwhile in most manufacturing situations in which cost control is vital for profitability. With a system of standard costs, increasing or decreasing costs of labor and material are reflected in the variance accounts. Armed with that information, management can take steps to reduce costs and/or increase prices.

We need to differentiate between cost control and cost reduction. *Cost control* is a management information system that breaks costs into meaningful components and tracks the costs with time. *Cost reduction* primarily is an engineering function aimed at changing the design, the material, or the process in a way that will reduce cost without compromising the product function or quality. The value analysis technique discussed in Sec. 6-11 is an excellent framework in which to carry out cost reduction.

9-10 LEARNING CURVE

It is a common observation in a manufacturing situation that the workers, as they gain experience in their jobs, can produce more product in a given unit of time. That is due to an increase in the workers' level of skill, to improved production methods that evolve with time, and to better management practices involving scheduling and other aspects of production planning. The extent and rate of improvement also depend on such factors as the nature of the

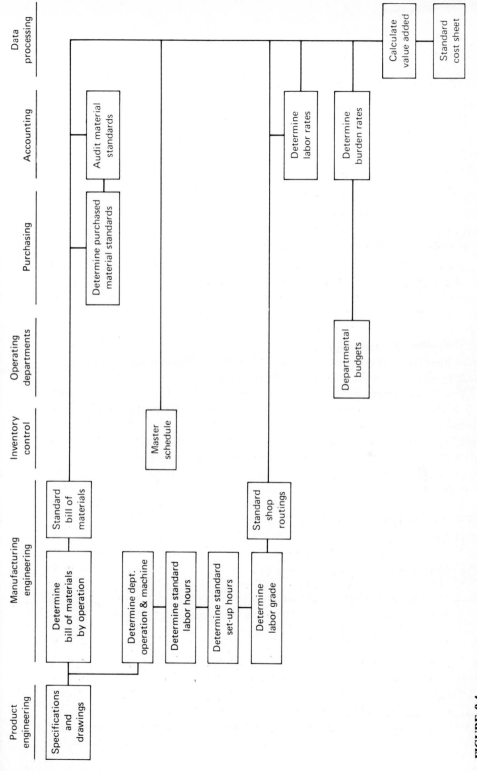

FIGURE 9-4
Steps in the development of standard costs.

435

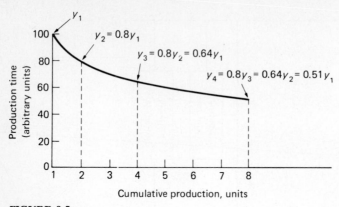

FIGURE 9-5
An 80 percent learning curve.

production process, the standardization of the product design, the length of the production run, and the degree of harmony in worker/management relationships. Usually, the processes that are more people-dominated show greater improvement than those that are dominated by machinery, as in a chemical process plant or a central station generating plant.

The improvement phenomenon usually is expressed by a learning curve, also called a product improvement curve. Figure 9-5 shows the characteristic features of an 80 percent learning curve. Each time the cumulative production doubles ($x_1 = 1$, $x_2 = 2$, $x_3 = 4$, $x_4 = 8$, etc.) the production time (or production cost) is 80 percent of what it was before the doubling occurred. For a 60 percent learning curve the production time would be 60 percent of the time before the doubling. Thus, there is a constant percentage reduction for every doubled[1] production. Such an obviously exponential curve will become linear when plotted on log-log coordinates (Fig. 9-6).

The learning curve is expressed by

$$y = kx^n \tag{9-6}$$

where y = production effort, h/unit or $/unit
$\quad k$ = effort required to manufacture the first unit of production
$\quad x$ = cummulative total of units produced
$\quad n$ = negative slope

---

[1] The learning curve could be constructed for a tripling curve of production or any amount, but it is customary to base it on a doubling.

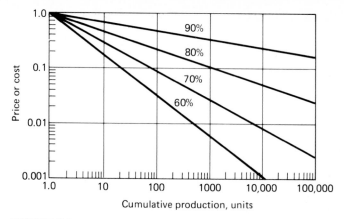

FIGURE 9-6
Standard learning curves.

The value of n can be found as follows: For an 80 percent learning curve $y_2 = 0.8y_1$ for $x_2 = 2x_1$; then,

$$\frac{y_2}{y_1} = \left(\frac{x_2}{x_1}\right)^n$$

$$\frac{0.8y_1}{y_1} = \left(\frac{2x_1}{x_1}\right)^n$$

$$n \log 2 = \log 0.8$$

$$n = \frac{-0.0969}{0.3010} = -0.322$$

Other values of n for different percentage learning curves are given in Table 9-5. Note that the learning curve percentage is $P = 2^n$.

TABLE 9-5
Exponent values for typical learning curve percentages

Learning curve percentage P	n
65	−0.624
70	−0.515
75	−0.415
80	−0.322
85	−0.234
90	−0.074

Example 9-11. The first of a group of 80 machines takes 150 h to build and assemble. If you expect a 75 percent learning curve, how much time would it take to complete the fortieth machine and the last machine?

Solution.

$$y = kx^n$$

For $P = 75\%$, $n = -0.415$, and $k = 150$,

$$y = 150(x^{-0.415})$$

For $x = 40$

$$y_{40} = 150(40^{-0.415}) = 32.4 \text{ h}$$

For $x = 80$

$$y_{80} = 150(80^{-0.415}) = 24.3 \text{ h}$$

The learning curve can be expressed with the production time in hours per unit *or* as the cumulative average hours for x units. The distinction between these two ways of expressing the output is shown in the accompanying table. We note that, for a given number of units of output, the cumulative average is greater than the unit values. Note, however, that the learning improvement percentage (80 percent) that applies to the unit values does not apply to the cumulative values. Similarly, if the unit values are derived from cumulative values, the constant percentage does not apply. In constructing learning curves from historical data it is more likely to find records of cumulative total hours than the hours to build each unit.

The total hours required to manufacture a cumulative total of x_e units is given by

$$T = yx_e \tag{9-7}$$

Based on an 80 percent learning curve

x units	y, h/unit	Cumulative total hours	y, cumulative average h/unit
1	100.00	100.00	100.00
2	80.00	180.00	90.00
3	70.22	250.22	83.41
4	64.00	314.22	78.55
5	59.56	373.78	74.76
6	56.16	429.94	71.66
7	53.44	483.38	69.05
8	51.19	534.57	66.82

and since $y = kx^n$

$$T = kx_e^n x_e = k(x_e)^{1+n} \tag{9-8}$$

The total hours to manufacture a group of consecutively produced units when the first unit in the group is x_f and the x_e is given, is[1]

$$T_1 = y_e x_e - y_f(x_f - 1)$$
$$= kx_e^n x_e - kx_f^n(x_f - 1)$$
$$= kx_e^{1+n} - kx_f^n(x_f - 1) \tag{9-9}$$

The unit time required to manufacture the xth unit is given by

$$U \approx (1 + n)kx^n \qquad \text{if } x > 10 \tag{9-10}$$

9-11 HOW TO PRICE A PRODUCT

There are two aspects to pricing a product. The first is to be sure that *all costs* are included, not just the obvious costs of labor, materials, and overhead. The second aspect concerns the business strategy of setting the price based on the volume-price relation and the estimate of the market potential for the product. These topics are discussed briefly in this section.

Figure 9-2 shows the elements of cost that add together to establish the selling price of a product. These cost factors are considered in the following example in somewhat greater detail.

Example 9-12. Consider the oil pump gear shown in Fig. 7-29. The basic design document is the bill of materials. The following information is obtained from that document.

1. Part identification number
2. Title or description of the part
3. Quantity of part required per unit of product
4. Material specification and/or raw material stock number
5. Component code classification; if a purchased item, vendor control number
6. Other products in which part may be used
7. Cross reference to similar part numbers

The total part cost $C_T(Q)$ for Q units of production is given by

$$C_T(Q) = [C_m + C_{dl}(1 + OH_F)](1 + OH_{SGA}) \tag{9-11}$$

where C_m = material cost
  C_{dl} = direct labor cost
  OH_F = factory overhead
  OH_{SGA} = cost of selling and general and administrative costs

Rather than use an equation, the cost build-up is best developed in tabular form. The processing steps for the hobbed gear shown in Fig. 7-29 will be used. The cost estimate is based on a lot size of 500 gears.

---

[1] E. M. Malstrom and R. L. Shell, *Manuf. Eng.*, pp. 70–75, May 1979.

Dept. no.	Operation	h/lot	Rate	Labor cost
125	Deliver bar stock	1.3	$4.20	5.46
110	Machine shop	423.0	7.00	2961.00
114	Heat treatment	3.5	6.30	22.05
120	Inspection	48.8	7.80	380.64
115	Toolroom	95.7	7.10	679.47
	Subtotal, manufacturing labor			$4048.62
	Fringe benefits on labor at 31 percent			1255.07
	Total manufacturing labor			$5303.69

Item	Description	Per unit	Per lot
1	Manufacturing labor (from above)	$10.61	$5503.69
2	Factory overhead at 82 percent of item 1	8.70	4349.02
3	Subtotal	$19.30	$9652.71
4	General & administrative at 20 percent of item 3	3.86	1930.54
5	Material cost and purchased parts	0.45	225.00
6	Overhead on item 5 at 15%	0.06	33.75
7	Special engineering	—	—
8	Overhead on item 7 at 50%	—	—
9	Research and development	—	—
10	Contingency	—	—
11	Subtotal	$23.68	$11,842.00
12	Selling expense at 20% of item 11	4.74	2,368.84
13	Total cost	$28.42	$14,210.84
14	Profit at 15%	4.26	2,131.63
15	Selling price	$32.68	$16,342.47

Items 7 to 10 are included for completeness even though they do not apply to this particular example. We note that the selling price is about three times the cost of labor plus materials. The costs in this example are fictitious but reasonable for 1980. They could vary widely depending upon the particular condition.

Two other methods of establishing selling price are used.[1] In the *operation method* the standard cost is used for each operation, following the process route sheet. The labor cost, overhead cost, and material cost are added for each operation, and the manufacturing cost is arrived at by summing the costs for the operations. In the *variable-cost method* costs are broken into the costs that vary with production volume and the fixed costs that are 1) product-related and 2) non-product-related.

[1] P. F. Ostwald, "Cost Estimating for Engineering and Management," pp. 280–288, Prentice-Hall, Inc., Englewood Cliffs, N.J., 1974.

The actual setting of price is at the heart of business practice, but describing it briefly is not easy. A key issue in establishing a price is the reaction of the competition and the customer. When competition is fierce, the sales force may push hard for price reductions, claiming they can more than make up the lost revenues by increased volume. Special offers such as promotional discounts, premiums, extras, trade-ins, volume discounts, and trade-in allowances are often used.

Frequently prices are set by a group or committee of people who understand the products' market, cost structure, and technical aspects. Price setting usually is performed by the corporate management, even in companies that are highly decentralized. Since there are severe penalties for collusion of several companies of an industry in establishing prices, certain rigid policies may be adopted to avoid any semblance of price fixing.

9-12 LIFE CYCLE COSTING

Life cycle costing is based on all the design-associated costs from inception in the research, development, and engineering (RD&E) stage until the product is retired from service and scrapped. The typical problem, which we already have considered in Chap. 8, is whether it is more economical to spend more money in the initial purchase to obtain a product with lower operating and maintenance costs or less costly to purchase a product with lower first costs but higher operating costs. However, life cycle costing[1] goes into the analysis in much greater detail in an attempt to evaluate all relevant costs, both present and future.

The life cycle of a system or a design generally is divided into the following phases.

1. RD&E
2. Production
3. Operating and support
4. Disposal

The life cycle costing concept (LCC) has found strong advocates in the area of military procurement, where it is used for comparison of competing weapons systems.[2] It also is important when various approaches to logistic support are considered or decisions about replacing aging equipment are made.

---

[1] M. R. Selden, "Life Cycle Costing," Westview Press, Boulder, Colo., 1979; A. J. Dell'Isola and S. J. Kirk, "Life Cycle Costing for Design Professionals," McGraw-Hill Book Company, New York, 1981; R. J. Brown and R. R. Yanuck, "Introduction to Life Cycle Costing," The Fairmont Press, Atlantic, Ga, 1984.

[2] MIL-HDBK 259, Life-Cycle Cost in Navy Acquisitions.

LCC analysis provides input in deciding between alternative designs, replacing aging equipment, making purchasing comparisons between equipment, vendors, or contractors, developing operations and support concepts and maintenance methods, and preparing logistics support plans. Because LCC deals to a large extent with future costs, it is subject to considerable uncertainty. The predicted service life is a critical parameter whose estimation is quite uncertain. Life can be unexpectedly shortened by the appearance of a superior competing product or the introduction of a new technology that suddenly makes the existing technology obsolete. Or an unrecognized design defect may produce fatigue failure at a life that is much less than the design life. The use of "constant dollars" is recommended in LCC analyses.

LCC focuses attention on the costs of ownership of an item. Maintenance costs, especially maintenance labor costs, usually dominate other support costs. Most analyses divide maintenance costs into scheduled or preventative maintenance and unscheduled or corrective maintenance. The *mean time between failure* and the *mean time to repair* are important parameters from reliability theory (see Sec. 12-3) that impact LCC. Other costs that must be projected for the operations and support phase are: maintenance of support equipment; maintenance facility costs; pay and fringe benefits for support personnel; fuel and other provisions; and utilities.

9-13   COST MODELS

The usefulness of modeling in the design process was illustrated in Chap. 4. Modeling can show which elements of a design contribute most to the cost; i.e., it can identify cost drivers. With a cost model it is possible to determine the conditions that minimize cost or maximize production (cost optimization). Another important use for cost models is to aid in the decision process in R&D studies. If competing processes to manufacture a product are being pursued, then cost models may be able to determine which processes have the best chance of achieving the goal at minimum cost.

Machining Cost Model

Extensive work has been done[1] on cost models for metal-removal processes. Broken down into its simplest cost elements, a machining process can be

---

[1] E. J. A. Armarego and R. H. Brown, "The Machining of Metals," chap. 9, Prentice-Hall, Inc., Englewood Cliffs, N.J., 1969; G. Boothroyd, "Fundamentals of Metal Machining and Machine Tools," pp. 161–164, McGraw-Hill Book Company, New York, 1975; M. Field, N. Zlatin, R. Williams, and M. Kronenberg, *Trans. ASME,* Ser. B, *J. Eng. Ind.,* vol. 90, pp. 455–466, 585–590, 1968; S. M. Wu and S. S. Ermer, ibid., vol. 88, pp. 435–442, 1966; M. Y. Friedman and V. A. Tipnis, ibid., vol. 98 pp. 481–496, 1976; "Mathematical Modeling of Material Removal prevesses," AFML-TR-154, September 1977.

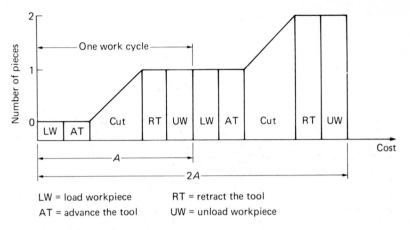

LW = load workpiece   RT = retract the tool
AT = advance the tool   UW = unload workpiece

FIGURE 9-7
Elements of the machining process.

described by Fig. 9-7. The time designated A is the machining plus work-handling costs per piece. If B is the tool cost, including the costs of tool changing and tool grinding, in dollars per tool, then

$$\text{Cost/piece} = \frac{n A + B}{n} = A + \frac{B}{n} \qquad (9\text{-}12)$$

where n is the number of pieces produced per tool.

We shall now consider a more detailed cost model for turning a bar on a lathe (Fig. 9-8). The machining time for one cut is

$$t_c = \frac{L}{V_{\text{feed}}} = \frac{L}{fN} = \frac{L}{f}\frac{D}{12v} \qquad (9\text{-}13)$$

where V_{feed} = feed velocity, in/min
    f = feed rate, in/rev
    N = rotational velocity, rev/min
    D = work diameter
    v = cutting velocity, ft/min

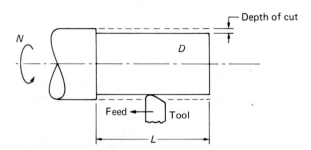

FIGURE 9-8
Details of lathe cutting.

Equation (9-13) holds in detail only for the process of turning a cylindrical bar. For other geometries or other processes such as milling or drilling, different expressions would be used for L or V_{feed}.

The total cost of a machined part is the sum of the machining cost C_{mc}, the cost of the tooling C_t, and the cost of the material C_m.

$$C_u = C_{\text{mc}} + C_t + C_m \qquad (9\text{-}14)$$

where C_u is the total unit (per piece) cost. The machining cost usually is the major part of the unit production cost. It depends on the machining time and the costs of the machine, labor, and overtime.

$$C_{\text{mc}} = C_1 t_{\text{unit}} \qquad (9\text{-}15)$$

$$C_1 = \frac{1}{60}\left[\frac{M(1 + \text{OH}_m)}{100} + \frac{W(1 + \text{OH}_{\text{op}})}{100}\right] \qquad (9\text{-}16)$$

where C_1 = cost rate, \$/min
M = machine cost, \$/h
OH_m = machine overhead rate, %
W = labor rate for operator, \$/h
OH_{op} = operator overhead rate, %

The machine cost includes the cost of interest, depreciation, and maintenance. It is found by the methods of Chap. 8 by determining these costs on an annual basis and converting them to per hour costs on the basis of the number of hours the machine is used in the year. The machine overhead cost includes the cost of power and other services and a proportional share of the building, taxes, insurance, etc.

The production time for a unit t_{unit} is the sum of the machining time t_m and the nonproduction or idle time t_i.

$$t_{\text{unit}} = t_m + t_i \qquad (9\text{-}17)$$

The machining time is the number of cuts times the time for a cut t_c. The idle time is

$$t_i = t_{\text{set}} + t_{\text{change}} + t_{\text{hand}} + t_{\text{down}} \qquad (9\text{-}18)$$

where t_{set} = total time for job setup divided by number of parts in the batch
t_{change} = prorated time for changing the cutting tool
= tool change time $\times \dfrac{t_m}{\text{tool life}}$
t_{hand} = time the machine operator spends loading and unloading the work on the machine
t_{down} = downtime lost because of machine or tool failure, waiting for material or tools, or maintenance operations. Downtime is prorated per units of production.

The cost of the tooling is the cost of the cutting tools and the prorated cost of any special jigs and fixtures used to hold the workpiece. The cost of the cutting tool per unit depends on the cost of the tool and the life of the tool.

$$C_t = C_{\text{tool}} \frac{t_m}{T} \tag{9-19}$$

where C_{tool} = cost of a cutting tool, \$
  t_m = machining time, min
  T = tool life, min

Tool life usually is expressed by the Taylor tool life equation, which relates surface velocity v and feed f to tool life:

$$vT^n f^m = K \tag{9-20}$$

For a cutting tool that is brazed to the toolholder, the tool cost is given by

$$C_{\text{tool}} = \frac{K_t + rK_s}{r + 1} \tag{9-21}$$

where K_t = cost of tool, \$
  r = number of resharpenings
  K_s = cost of resharpening, \$

For an insert (throwaway) tool

$$C_{\text{tool}} = \frac{K_i}{n_i} + \frac{K_h}{n_h} \tag{9-22}$$

where K_i = cost of tool insert, \$
  n_i = number of cutting edges
  K_h = cost of toolholder
  n_h = number of cutting edges in life of toolholder

Substituting for T in Eq. (9-19) results in

$$C_t = C_{\text{tool}} \frac{t_m v^{1/n} f^{m/n}}{K^{1/n}} \tag{9-23}$$

Substituting into the original equation, Eq. (9-14), yields for the total unit cost

$$C_u = C_1 \left[t_m \left(1 + \frac{t_{\text{tool}}}{T} \right) + t_0 \right] + C_t \frac{t_m}{T} + C_m \tag{9-24}$$

where t_{tool} = tool change time
  t_0 = the time elements in Eq. (9-18) that are independent of tool life

In Eq. (9-24) both t_m and T depend on cutting velocity and feed, and the former is the more important variable. If we plot unit cost vs. cutting velocity, for a constant feed (Fig. 9-9), there will be an optimum cutting velocity to minimize cost. That is so because machining time decreases with increasing velocity; but as velocity increases, tool wear and tool costs increase also. Thus,

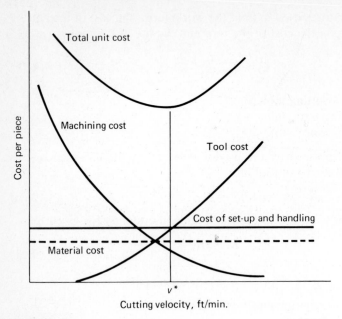

FIGURE 9-9
Variation of unit cost with cutting velocity.

there is an optimum cutting velocity. An alternative strategy would be to operate at the cutting speed that results in maximum production rate. Still another alternative is to operate at the speed that maximizes profit. The three criteria do not result in the same operating point.

Process Cost Model

The comparative cost of competitive processes can be determined directly from the process flowchart.[1] The steps for producing turbine blades from directionally solidified (DS) superalloys are shown in Fig. 9-10. In the directionally solidified structure the mold is slowly withdrawn from the furnace, so that the grains grow in a directional manner parallel to the length of the blade. This produces a structure with practically no grain boundaries perpendicular to the direction of bending stress, so that failure from stress rupture becomes less prevalent.

In this cost model the unit cost is expressed by a modification of Eq. (9-11) in which all overhead costs are lumped into a single factor LOH that is a

[1] C. F. Barth, D. E. Blake, and T. S. Stelson, "Cost Analysis of Advanced Turbine Blade Manufacturing Processes," NASA CR-135203, October 1977.

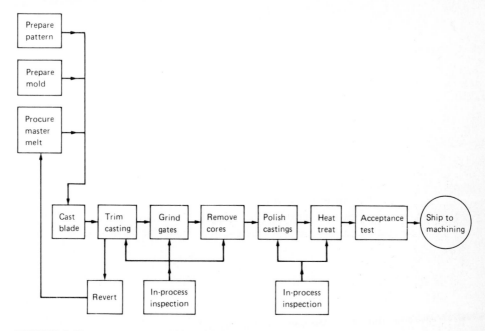

FIGURE 9-10
Manufacturing sequence for directionally solidified turbine blades.

composite factor of the hourly labor rate plus an overhead rate.

$$C_u = \frac{LOH}{nm} \Sigma t + C_m \tag{9-25}$$

where C_u = total unit (per piece) cost
$\quad$ C_m = material cost
$\quad$ n = number of parts per batch
$\quad$ m = number of batch processes operated simultaneously by one operator
$\quad$ t = time required to complete process step for the batch

A detailed cost breakdown is prepared for each process step. The first step in the DS casting process is the preparation of a cluster of wax patterns of turbine blades that will be invested in the next step by coating them with a ceramic slurry. The steps involved in producing the cluster are 1) wax patterns containing internal cores are prepared by injection molding; 2) the parts of the pattern are assembled into a cluster of blades; 3) any defects are smoothed out (dressing the cluster); and 4) the finished cluster is inspected. The cost model for this step in the process sequence is

$$C_{pp} = \frac{1}{n_{cl}} LOH(t_{amp} + t_{dc} + t_{ci}) + C_p + C_{cl} \tag{9-26}$$

where C_{pp} = cost of pattern preparation

n_{cl} = number of blades per cluster

t_{amp} = time to assemble the patterns in a cluster

t_{dc} = time required to dress the cluster

t_{ci} = time required to inspect the cluster

C_p = cost of a wax pattern

$$C_p = \text{LOH}(t_{dic} + t_{ip} + t_{ifp}) + C_w W_{wp} + C_k \tag{9-27}$$

where t_{dic} = time to dress and inspect cores

t_{ip} = time required to inject the pattern

t_{ifp} = time required to inspect final pattern

C_w = cost of wax, \$/lb

W_{wp} = wax weight in pattern

C_k = cost of ceramic cores

C_{cl} = cost of components in cluster other than the pattern

$$C_{cl} = \text{LOH}\, t_{inp} + C_w W_{wmp} \tag{9-28}$$

where t_{inp} = injection time

W_{wmp} = weight of wax

It is obvious that the costs are established with a high level of detail. An equation similar to Eq. (9-26) is developed for each process step shown in Fig. 9-10. Usually it is just a summation of the time required to carry out an operation in the process step, but occasionally some aspect of the process itself enters into the model.

The process model for the casting (solidification) step is:

$$C_c = \frac{1}{n_c m_f} \text{LOH}\left(t_{ct} + \frac{L}{v}\right) \tag{9-29}$$

where C_c = unit cost for the casting step

n_c = number of blades per casting

m_f = number of furnaces per operator

t_{ct} = constant time needed to cast a batch regardless of the withdrawal rate

L = withdrawal length

v = withdrawal velocity

Thus, since the withdrawal rate of the casting may be very slow, Eq. (9-29) shows that this is a significant cost driver.

An important aspect in process costs, and especially when the technology is new, is the yield obtained in the process. The simple (uncoupled) yield factor is

$$Y_i = \frac{\text{no. of acceptable parts from } i\text{th step}}{\text{no. of parts that enter } i\text{th step}} \tag{9-30}$$

where Y_i = the yield factor for the ith process step. Thus, the cost of producing one acceptable part from the casting step is

$$(C_c)_{ap} = \frac{C_c}{Y_i} \qquad (9\text{-}31)$$

The cost of manufacturing DS blades is shown below in terms of four main cost centers. This is important because it shows which aspect of the manufacturing sequence contributes the most to the product cost. Note the importance of including product yield in the cost analysis. Without considering yield the coating of the blade would be the chief cost driver, but since that process has a very high yield the chief contributor to the cost is the casting step.

Percentage contribution of cost

Cost center	Without yield consideration	Considering process yield
Fabrication (casting)	30.2	42.3
Machining and finishing	16.6	24.4
Coating	47.4	25.3
Final acceptance testing	5.8	8.0
	100.0	100.0

Early Prediction of Cost

The chief reason for choosing one material/process combination over another which would satisfy the same functional requirements is the total cost of the manufactured part. In order that this type of decision making can be done in the conceptual phase of design, before design details begin to become frozen, it is important to have an early cost estimating procedure. This methodology should be microcomputer based, so as to permit quick computation and easy change of ideas, as well as providing for input of a large library of material, processing, and cost data.

Dewhurst and Boothroyd[1] have shown that it is possible to devise such an early cost estimating method provided that a good cost model exists for the manufacturing process. They have done this for machining, forging,[2] and injection molding[3] and have developed it into a computer-based design tool.

BIBLIOGRAPHY

Clark, F. D., and A. A. Lorenzoni: "Applied Cost Engineering," Marcel Dekker, Inc., New York, 1978.

---

[1] P. Dewhurst and G. Boothroyd, *Jnl. of Manufacturing Systems*, vol. 7, no. 3, pp. 183–191, 1987.

[2] W. A. Knight and C. R. Poli, *Annals of CIRP*, vol. 31, no. 1, pp. 159–163, 1982.

[3] P. Dewhurst and D. Kupparajan, *Int. J. Prod. Res.*, vol. 27, no. 1, pp. 21–29, 1989.

Dudick, T. S.: "Manufacturing Cost Control," Prentice-Hall Inc., Englewood Cliffs, NJ, 1985.

Gutman, N.: "How to Keep Product Costs in Line," Marcel Dekker Inc., New York, 1985.

Malstrom, E. M.: "What Every Engineer Should Know About Manufacturing Cost Estimating," Marcel Dekker Inc., New York, 1981.

—— E. M. (ed.): "Manufacturing Cost Engineering Handbook," Marcel Dekker Inc., New York,

—— (ed.): "Manufacturing Cost Engineering Handbook," Marcel Dekker Inc., New York, 1984.

Michaels, J. V., and W. P. Wood: "Design to Cost," John Wiley & Sons, New York, 1989.

Ostwald, P. F.: "Cost Estimating for Engineering and Management," 2d ed. Prentice-Hall, Inc., Englewood Cliffs, NJ 1984.

—— (ed.): "Manufacturing Cost Estimating," Society of Manufacturing Engineers, Dearborn, MI, 1980.

Peters, M. S., and K. D. Timmerhaus: "Plant Design and Economics for Engineers," 3d ed., McGraw-Hill Book Company, New York, 1980.

Valle-Riestra, J. F.: "Project Evaluation in the Chemical Process Industries," McGraw-Hill Book Co., New York, 1983.

Winchell, W. (ed.): "Realistic Cost Estimating for Manufacturing," 2d ed., Society of Manufacturing Engineers, Dearborn, MI, 1968.

CHAPTER
10

PROJECT
PLANNING
AND SCHEDULING

10-1 IMPORTANCE OF PLANNING

It is an old business axiom that time is money. Therefore, planning future events and scheduling them so they are accomplished with a minimum of time delay is an important part of the engineering design process. For large construction and production projects, detailed planning and scheduling is a must. Computer-based methods for handling the large volume of information have become commonplace. However, engineering design projects of all magnitudes of scale can profit greatly by applying the simple planning and scheduling techniques discussed in this chapter.

One of the most common criticisms leveled at the young graduate engineer is an overemphasis on technical perfection of the design and not enough concern for completing the design on time and below the estimated cost. Therefore, the planning and scheduling tools presented in this chapter can profitably be applied at the personal level as well as to the more complex engineering project.

In the context of engineering design, *planning* consists of identifying the key activities in a project and ordering them in the sequence in which they should be performed. *Scheduling* consists of putting the plan into the time frame of the calendar. The design process generally is divided into the following phases.

- Feasibility study
- Preliminary design—the concept phase
- Detail design
- Production phase
- Operational phase

Usually, a detailed design review is conducted at the end of each phase to establish whether the results warrant advancing into the next phase. The alternatives may be to repeat the phase or abandon the project. Frequently, well-defined decision points or milestones are established partway through a phase in order to provide a target to strive for and a way to control the project.

A "campus comedian" once said that the real stages in an engineering design project are:

- Enthusiasm
- Disillusionment
- Panic
- Search for the guilty
- Punishment of the innocent
- Praise and honors for the nonparticipants

Certainly, these could be the stages of a project that was poorly planned and controlled.

The major decisions that are made over the life cycle of a project fall into four areas: performance, time, cost, and risk.

Performance. The design must possess an acceptable level of operational capability or the resources expended on it will be wasted. The design process must generate satisfactory specifications to test the performance of prototypes and production units.

Time. In the early phases of a project the emphasis is on accurately estimating the length of time required to accomplish the various tasks and scheduling to ensure that sufficient time is available to complete those tasks. In the production phase the time parameter becomes focused on setting and meeting production rates, and in the operational phase it focuses on reliability, maintenance, and resupply.

Cost. The importance of cost in determining what is feasible in an engineering design has been emphasized in earlier chapters. Keeping costs and resources within approved limits is one of the chief functions of the project manager.

Risk. Risks are inherent in anything new. Acceptable levels of risk must be established for the parameters of performance, time, and cost, and they must be monitored throughout the project. The subject of risk is considered in detail in Chap. 12.

The most crucial step for planning is the preliminary design phase. At its completion the design concept has been formulated and must be expressed in terms of performance standards, a time schedule, a cost estimate, and a risk assessment. At this stage, performance characteristics are usually the deciding factors in the tradeoff studies.

The first step in developing a plan is to identify the activities that need to be controlled. The usual way to do that is to start with the entire system and identify the 10 or 20 activities that are critical. Then the larger activities are broken down into subactivities, and these in turn are subdivided until you get to tasks performed by single persons. Generally the work breakdown proceeds in a hierarchical fashion from the system to the subassembly to the component to the individual part.

10-2 BAR CHART

The simplest scheduling tool is the bar, or Gantt, chart (Fig. 10-1). The activities are listed in the vertical direction, and elapsed time is recorded horizontally. This shows clearly the date by which each activity should start and finish, but it does not make clear how the ability to start one activity depends upon the successful completion of other activities.

The dependence of one activity on another can be shown by a network logic diagram like Fig. 10-2. This diagram clearly shows the precedence relations, but it loses the strong relation with time that the bar chart displays.

Note that a *critical path* through the network can be determined. In this case it is the 20 weeks required to traverse the path *a–b–c–d–e–f–g,* and it is shown on the modified bar chart, Fig. 10-3. The parts of the schedule that have slack time are shown dashed. *Slack* is the time by which an activity can exceed its estimated duration before failure to complete the activity becomes critical. For example, for the activities of installing heaters, there is a 7-week slack before the activities must be completed to proceed with the leak testing. Thus, the identification of the longest path focuses attention on the activities that must be given special management attention, for any delay in those activities would critically lengthen the project. Conversely, identification of activities with slack indicates the activities in which some natural slippage can occur without serious consequences. This, of course, is not license to ignore the activities with slack.

10-3 CRITICAL-PATH METHOD

Two computer-based scheduling systems based on networks were introduced in the late 1950s to aid in scheduling large engineering projects. The critical-path

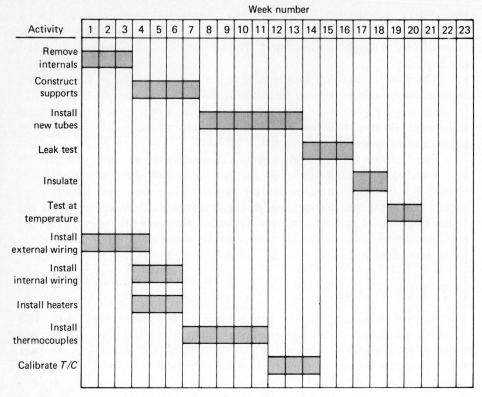

FIGURE 10-1
Bar chart for prototype testing a heat exchanger.

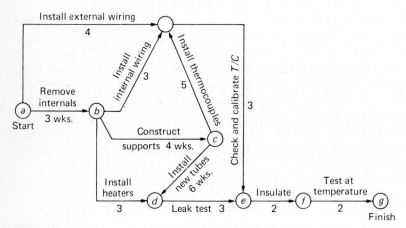

FIGURE 10-2
Network logic diagram for heat exchanger.

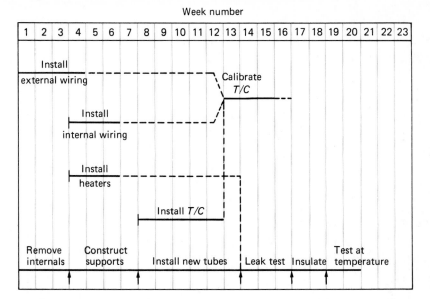

FIGURE 10-3
Modified bar chart for heat exchanger tests.

method (CPM), developed by Du Pont and Remington Rand, is a deterministic system that uses the best estimate of the time to complete a task. The program evaluation and review technique (PERT), developed for the U.S. Navy, uses probabilistic time estimates. The techniques have much in common. We shall start by considering CPM.

The basic tool of CPM is an arrow network diagram similar to Fig. 10-2. The chief elements of this diagram are:

1. An *activity*—a time-consuming effort that is required to perform part of a project. An activity is shown on an arrow diagram by a line with an arrowhead pointing in the direction of progress in completion of the project.
2. An *event*—the end of one activity and the beginning of another. An event is a point of accomplishment and/or decision. A circle is used to designate an event.

There are several logic restrictions to constructing the network diagram.

1. An activity cannot be started until its tail event is reached. Thus, if activity B cannot begin until activity A has been completed. Similarly, if activities D and E cannot begin until activity C has been completed.

2. An event cannot be reached until all activities leading to it are complete. If

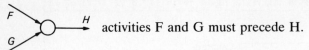 activities F and G must precede H.

3. Sometimes an event is dependent on another event preceding it, even though the two events are not linked together by an activity. In CPM we record that situation by introducing a *dummy activity*, denoted ┈┈→. A *dummy activity* requires zero time and has zero cost. Consider two examples:

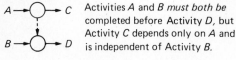

Activities A and B *must both be* completed before Activity D, but Activity C depends only on A and is independent of Activity B.

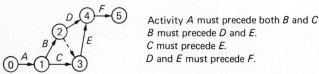

Activity A must precede both B and C
B must precede D and E.
C must precede E.
D and E must precede F.

The longest time through the network (the critical path) may be determined by inspection for a relatively simple network like the one in Fig. 10-2, but a methodology for the much more complex problems found in engineering project management must be established. To do so we establish the following parameters.

Earliest start time (ES). The earliest time an activity can begin when all preceding activities are completed as rapidly as possible.

Latest start time (LS). The latest time an activity can be initiated without delaying the minimum completion time for the project.

Earliest finish time (EF). $EF = ES + D$, where D is the duration of each activity.

Latest finish time (LF). $LF = LS + D$

Total float (TF). The slack between the earliest and latest start times. $TF = LS - ES$. An activity on the critical path has zero total float.

In CPM the estimate of each activity duration is based on the most likely estimate of time to complete the activity. All durations should be expressed in the same time units, such as days or weeks. The sources of time estimates are records of similar projects, calculations involving the manpower needs, legal restrictions, and technical considerations.

The network diagram in Fig. 10-2 has been redrawn as a CPM network in Fig. 10-4. To facilitate solution with computer methods, the events that occur at the nodes have been numbered serially. The node number at the tail of each activity must be less than that at the head. The ES times are determined by

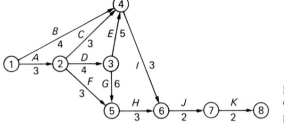

FIGURE 10-4
CPM diagram for heat exchanger project.

starting at the first node and making a forward pass through the network while adding each activity duration in turn to the ES of the preceding activity. The details are shown in Table 10-1.

The LS times are calculated by a reverse procedure. Starting with the last event, a backward pass is made through the network while subtracting the activity duration from the limiting LS at each event. The calculations are detailed in Table 10-2. We note that, for calculating LS, each activity arising from a common event can have a different late start time, whereas all activities starting from the same event had the same early start time.

The chief work is in establishing ES and LS times. Once that is accomplished, the remaining boundary time parameters can be determined by routine operations; see Table 10-3. The critical path is identified by the activities with zero total float.

Generally in a CPM problem we are interested in two classes of solutions:

1. The least-cost solution using costs associated with the normal time to complete the activities.

TABLE 10-1
Calculation of early start time based on Fig. 10-4

Event	Activity	ES	Comment
1	A, B	0	Conventional to use ES $= 0$ for the initial event
2	C, D, F	3	$ES_2 = ES_1 + D = 0 + 3 = 3$
3	E, G	7	$ES_3 = ES_2 + D = 7$
4	I	12	At a merge like 4 the largest $ES + D$ of the merging activities is used
5	H	13	$ES_5 = ES_3 + 6 = 13$
6	J	16	$ES_6 = ES_5 + 3 = 16$
7	K	18	
8	—	20	

TABLE 10-2
Calculation of late start times based on Fig. 10-4

Event	Activity	LS	Event	Activity	LS
8	—	20	5-2	F	10
8-7	K	18	4-3	E	8
7-6	J	16	4-2	C	10
6-5	H	13	4-1	B	9
6-4	I	13	3-2	D	3
5-3	G	7	2-1	A	0

TABLE 10-3
Summary of boundary timetable

Activity	Description	D, weeks	ES	LS	EF	LF	TF
A	Remove internals	3	0	0	3	3	0
B	Install external wiring	4	0	9	4	13	9
C	Install internal wiring	3	3	10	6	13	7
D	Construct supports	4	3	3	7	7	0
E	Install thermocouples	5	7	8	12	13	1
F	Install heaters	3	3	10	6	13	7
G	Install new tubes	6	7	7	13	13	0
H	Leak test	3	13	13	16	16	0
I	Check thermocouples	3	12	13	15	16	1
J	Insulate	2	16	16	18	18	0
K	Test prototype at temperature	2	18	18	20	20	0

ES is determined by forward pass through network.
LS is determined by backward pass through network.
$EF = ES + D$.
$LF = LS + D$.
$TF = LS - ES$.

2. The least-time solution in which crash costs are incurred to reduce the time, e.g., by employing overtime, extra workers, or bringing in extra production equipment.

10-4 PERT

The program evaluation and review technique (PERT) uses the same ideas as CPM; but instead of using the most likely time estimate, it uses a probabilistic estimate of time for completion of an activity. The designer is asked to make an optimistic time estimate o if everything goes smoothly and a pessimistic time estimate p if everything goes badly. The most likely time m is bracketed between those values. The time estimates are assumed to follow a beta frequency distribution that gives the expected time as

$$t_e = \frac{o + 4m + p}{6} \tag{10-1}$$

The expected time is a mean value that divides the area under the frequency distribution into two equal parts. In PERT the expected time is computed for each activity, and the expected times are used to determine the critical path and the boundary times as illustrated in Sec. 10-3 for the CPM technique.

The expected time for each activity also has a standard deviation (see Sec. 11-5), which describes its scatter, given by

$$\sigma = \frac{p - o}{6} \tag{10-2}$$

The standard deviation along a path in the PERT network is the square root of the sum of the individual variances for the separate activities along that path.

$$\sigma_{\text{path}} = \sqrt{\Sigma \sigma^2} \tag{10-3}$$

Knowing the variance for each activity permits the calculation of the probability that a certain scheduled event will be completed on schedule. If SS is the scheduled start of a particular event, called a milestone, and ES is the earliest start time for the event, then

$$z = \frac{SS - ES}{\sigma_{\text{path}}} \tag{10-4}$$

where z is the standard normal deviate and represents the area under the standardized normal frequency distribution (see Sec. 11-5). If, for example, $z = 0$, there is a 50 percent probability of completing the event on the scheduled date. If $z = -0.5$, there is a 30 percent probability.

PERT/COST is an attempt to include cost data in the CPM-PERT type of network scheduling program. The original concept involved costs at a very high level of detail, but that has proved very cumbersome because of the need for continual updating, reestimating, and cost changes due to design changes. In most cases, PERT/COST is operated with costs aggregated to a considerable degree.

Project managment software is common for the personal computer and workstation. The three scheduling techniques discussed in this chapter can be found in many software versions in a range of complexity and price.[1]

---

[1] H. Fersko-Weiss, "Project Management Software", *PC Magazine*, Sept. 29, 1987, pp. 153–209.

10-5 OTHER SCHEDULING METHODS

CPM and PERT systems are most applicable to scheduling nonrepetitive operations such as occur in the early stages of a design project. Scheduling and sequencing for repetitive jobs, such as occur on a production line, are standard problems in the field of operations research.[1] Considered are such problems as how to schedule to minimize the number of jobs produced beyond the time schedule, how to minimize the waiting time, and how to schedule when different jobs have different degrees of importance. When there are many different processing stations, the analysis can become very detailed.

CPM and PERT are deterministic network models in that they do not permit probabilistic occurrence of activities. For example, if a PERT network contains two branches, all calculations must flow down both branches even though we have decided on the basis of earlier results in the calculation not to pursue one of the paths. However, stochastic networking techniques have the capability of associating a probability at each event. The most widely used technique of this type is graphical evaluation and review technique[2] (GERT). An important advantage of a probabilistic network is the ability to simulate project outcomes. By repeated runs of the network, a distribution of likely outcomes can be built up to assist the project manager in assessing the probability of success and the risk of failure.

A network type of analysis is a useful tool for optimizing the manpower assigned to a project. In *manpower leveling* the objective is to smooth out the peaks and valleys of manpower requirements over the life of the project. This is an important consideration when limitations are imposed on manpower "slots" or when certain skills are in short supply. Reduction in the peak manpower needs can be accomplished by utilizing the slack in the schedule to utmost advantage. *Time compression* for the completion of the project may be achieved by transferring people from slack paths to critical paths in the network.

BIBLIOGRAPHY

Cleland, D. I., and W. R. King: "Systems Analysis and Project Management," 2d ed., McGraw-Hill Book Co., New York, 1975.

Hajek, V. G.: "Management of Engineering Projects", 3d ed., McGraw-Hill Book Co., New York, 1984.

Kerzner, H.: "Product Management: A Systems Approach to Planning, Scheduling, and Controlling," Van Nostrand Reinhold Company, New York, 1979.

---

[1] K. R. Baker, "Introduction to Sequencing and Scheduling," John Wiley & Sons, Inc., New York, 1974; L. A. Johnson and D. C. Montgomery, "Operations Research in Production Planning, Scheduling and Inventory Control," ibid.

[2] A. Pritsker and W. Happ, *J. Ind. Eng.*, vol. 17, no. 5, pp. 267–274, 1966.

Meredith, D. D., K. W. Wong, P. W. Woodhead, and R. H. Wortman: Design and Planning of Engineering Systems," 2d ed., Prentice-Hall, Englewood Cliffs, NJ, 1985.

Whitehouse, G. E.: "Systems Analysis and Design Using Network Techniques," Prentice-Hall, Englewood Cliffs, NJ, 1973.

Wiest, J. D., and F. K. Levy: "A Management Guide to PERT/CPM," 2d ed., Prentice-Hall, Englewood Cliffs, NJ, 1979.

CHAPTER
11

ENGINEERING
STATISTICS

11-1 STATISTICS AND DESIGN

Probability and statistics have become working tools of the engineer in many areas of engineering practice. Since in engineering design we typically deal with poorly defined situations or are forced to use data that have low precision, it is easy to appreciate how the proper application of statistical analysis can help greatly with engineering design. This fact is recognized in some engineering curricula by requiring extensive course work in statistics. Unfortunately, that is far from the usual situation, and many engineers are forced to acquire their statistical background in bits and pieces obtained from several courses or from independent reading.

Thus, in no other area within this book is there greater disparity in prior background. Those who have taken a complete course in engineering statistics will have received a far richer background than can be imparted by this chapter. You lucky ones should use the chapter as a review while paying particular attention to the examples. For those of you have not had previous instruction in engineering statistics, the chapter aims at providing a basic background. However, emphasis is on the application of statistical methods rather than the mathematical concepts underlying the methods. Many references to in-depth texts and technical papers are provided for those who wish to pursue the topics further. Also, the discussion of many topics is keyed to the excellent statistics handbook by Natrella.[1] This inexpensive reference volume can serve as a source of tried-and-true procedures and useful statistical tables. It is highly recommended for the library of every engineer.

Statistical techniques are important in engineering decision making. The underlying philosophy is that we use observed samples to estimate the statistical population. The known properties of the statistical distribution provide the basis for decision making.

In making physical measurements we are concerned with their precision and accuracy. The *precision* of an instrument indicates the instrument's capacity to reproduce a certain reading time after time. The fact that the same reading is not always obtained is due to the existence of *random error.* The reading also may contain systematic error or *bias,* such that the readings are consistently high or low. The presence of both random error and systematic error affects the accuracy of the reading. The *accuracy* is the deviation of the reading from the true value.

At least four major aspects of statistical analysis are important in engineering design. At the most basic level we are interested in the analysis of experimental data, so that the results can be unequivocally described by appropriate statistical parameters. Next we are concerned with *statistical inference,* which uses statistics to make reliable decisions utilizing tests of hypotheses and confidence limits. *Hypotheses tests* are used to determine whether there is a significant difference between the characteristics of an observed set of data and a proposed mathematical model of the data. *Confidence limits* allow us to determine a range in which the true characteristic of a population is likely to lie. A third important area, *analysis of variance,* is a test for the equality of means and/or variances of groups of observations, such as the means resulting from different experimental procedures. Finally, these concepts lead to consideration of the most efficient way to collect data through the *statistical design of experiments.* They also permit us to determine empirical relations between variables with regression analysis and response surface methodology.

11-2 PROBABILITY

The concept of probability is part of our general base of knowledge. When asked the probability that a tossed coin will come up heads, most people will answer "one-half," or when asked the probability that a six-sided die will come up a 4, they will answer "one-sixth." Thus, they recognize the concept of a

---

[1] M. G. Natrella, "Experimental Statistics," Nat. Bur. Standards Handbook 91, Government Printing Office, 1963; also available from John Wiley & Sons, New York, 1983.

probability scale in which a non-event has a probability of zero $[P(A) = 0]$ and an assured event has a probability of one $[P(A) = 1]$. If there are N possible outcomes, then the probability that an event A will occur is

$$P(A) = \frac{\text{number of ways in } N \text{ that produce A}}{N} = \frac{n}{N} \qquad (11\text{-}1)$$

The probability that an event will not occur is $P(\bar{A})$, and

$$P(\bar{A}) = 1 - P(A) = \frac{N - n}{N} \qquad (11\text{-}2)$$

Since the event A will either occur or not occur,

$$P(A) + P(\bar{A}) = 1.0 \qquad (11\text{-}3)$$

The odds of event A occurring are given by $P(A)/P(\bar{A})$.

A basic underlying assumption of probability theory is that it deals with random events. A *random event* is one in which the conditions are such that each member of the population N has an equal chance of being chosen.

Notation of Probability

A special and precise system of language and notation is used in probability theory. Two events A and B are said to be *independent* if the occurrence of either one has no effect on the occurrence of the other. Two events that have no elements in common are said to be mutually exclusive events.

1. If A and B are independent events, than the probability of *both* A and B occurring (joint probabilities) is

$$P(A \text{ and } B) = P(AB) = P(A)P(B) \qquad (11\text{-}4)$$

If A and B are *mutually exclusive* events,

$$P(AB) = 0 \qquad (11\text{-}5)$$

2. The probability of A *or* B occurring (for non-mutually exclusive events) is

$$P(A \text{ or } B) = P(A + B) = P(A) + P(B) - P(AB) \qquad (11\text{-}6)$$

If A and B are mutually exclusive events,

$$P(A + B) = P(A) + P(B) \qquad (11\text{-}7)$$

3. A conditional probability is one in which the probability of the event depends upon whether the other event has occurred.

$P(A \mid B) =$ probability that A will occur given B has occurred

$$P(A \mid B) = \frac{P(AB)}{P(B)} \qquad (11\text{-}8)$$

or

$$P(B \mid A) = \frac{P(AB)}{P(A)} \qquad (11\text{-}9)$$

4. Bayes' theorem (see Sec. 3-9) allows us to modify a probability estimate as additional information becomes available. From Eqs. (11-8) and (11-9), $P(AB) = P(A \mid B)P(B) = P(B \mid A)P(A)$, since $AB = BA$, and

$$P(A \mid B) = \frac{P(B \mid A)P(A)}{P(B)} \qquad (11\text{-}10)$$

Since $P(A) + P(\bar{A}) = 1$, it follows that event B must occur jointly with either A or $\bar{A}$

$$P(B) = P(AB) + P(\bar{A}B)$$

and from Eq. (11-9)

$$P(B) = P(A)P(B \mid A) + P(\bar{A})P(B \mid A) \qquad (11\text{-}11)$$

Substituting Eq. (11-11) into Eq. (11-10) gives

$$P(A \mid B) = \frac{P(A)P(B \mid A)}{P(A)P(B \mid A) + P(\bar{A})P(B \mid \bar{A})} \qquad (11\text{-}12)$$

If event A has more than two available alternatives, then Eq. (11-12) is expressed as

$$P(A_i \mid B) = \frac{P(A_i)P(B \mid A_i)}{\sum_i P(A_i)P(B \mid A_i)} \qquad (11\text{-}13)$$

Example 11-1. The number of defective and acceptable parts received from vendor 1 (V_1) and vendor 2 (V_2) is:

	V_1	V_2	Total
Defective	300	750	1,050
Acceptable	9,700	6,250	15,950
	10,000	7,000	17,000

Let A_1 = event "part from vendor 1"
A_2 = event "part from vendor 2"
B_1 = event "acceptable part"
B_2 = event "defective part"
$B_1 \mid A_1$ = event "acceptable part from vendor 1"

$$P(A_1) = \frac{10,000}{17,000} = 0.59 \qquad P(A_2) = \frac{7000}{17,000} = 0.41$$

$$P(B_1 \mid A_1) = 0.97 \qquad\qquad P(B_2 \mid A_1) = 0.03$$

$$P(B_1 \mid A_2) = 0.89 \qquad\qquad P(B_2 \mid A_1) = 0.11$$

We wish to know the probability of selecting a part that is made by vendor 1 and is also defective.

Solution. From Eq. (11-8)

$$P(A_1B_2) = P(B_2A_1) = P(A_1)P(B_2 \mid A_1)$$
$$= 0.59(0.03) = 0.018$$

In a similar way, the probability of selecting a defective part that is made by vendor 2 is

$$P(A_2B_2) = P(A_2)P(B_2 \mid A_2)$$
$$= 0.41(0.11) = 0.045$$

Because the selection of a defective part made by vendor 1 and vendor 2 are mutually exclusive events, the probability of selecting a defective part, irrespective of which vendor provided it, is given by

$$P(B_2) = P(A_1B_2) + P(A_2B_2)$$
$$= 0.018 + 0.045 = 0.063$$

Example 11-2. Referring to Example 11-1, if we find a part to be defective, we can then ask, what is the *posterior probability* that it came from vendor 1?

Solution. This uses Bayes' theorem.

$$P(A_1 \mid B_2) = \frac{P(A_1)P(B_2 \mid A_1)}{P(A_1)P(B_2 \mid A_1) + P(A_2)P(B_2 \mid A_2)}$$

$$= \frac{0.59(0.03)}{0.59(0.03) + 0.41(0.11)} = 0.28$$

Thus, using the new information that the part selected did fail, the posterior probability that it came from vendor 1 is much less than the prior probability 0.59.

11-3 ERRORS AND SAMPLES[1]

The act of making any type of experimental observation involves two types of errors: systematic errors (which exert a nonrandom bias) and experimental, or random, errors. Systematic errors arise because of faulty control of the experiment. Experimental errors are due to limitations of the measuring

---

[1] Sections 11-3 to 11-6 are from G. E. Dieter, "Mechanical Metallurgy," McGraw-Hill Book Company, New York, 1961.

equipment or to inherent variability in the material being tested. As an example, in the measurement of the reduction of area of a fractured tensile specimen, a systematic error could be introduced if an improperly zeroed micrometer were used for measuring the diameter, whereas random errors would result from slight differences in fitting together the two halves of the tensile specimen and from the inherent variability of reduction-of-area measurements on metals. By averaging a number of observations, the random error will tend to cancel out. The systematic error, however, will not cancel upon averaging. One of the major objectives of statistical analysis is to deal quantitatively with random error.

When a tensile specimen is cut from a steel forging and the reduction of area is determined for it, the observation represents a sample of the *population* from which it was drawn. The population, in this case, is the collection of all possible tensile specimens that could be cut from the forging or from all other forgings. As more and more tensile specimens are cut from the forging and reduction-of-area values are measured, the sample estimate of the population values becomes more accurate. However, it is obviously impractical to sample and test the entire forging. Therefore, one of the main purposes of statistical techniques is to determine the best *estimate* of the population parameters from a randomly selected sample. The approach that is taken is to postulate that, for each sample, the population has fixed and invariant parameters. However, the corresponding parameters calculated from samples contain random errors, and, therefore, the sample provides only an estimate of the population parameters. It is for this reason that statistical methods lead to conclusions having a given *probability* of being correct.

11-4 FREQUENCY DISTRIBUTION

When a large number of observations are made from a random sample, a method is needed to characterize the data. The most common method is to arrange the observations into a number of equal-valued *class intervals* and determine the frequency of the observations falling within each class interval. In Table 11-1, out of a total sample of 449 measurements of yield strength, 4 observations fell between 114,000 and 115,900 psi, 26 fell between 120,000 and 121,900 psi, etc. An estimate of the frequency distribution of the observations can be obtained by plotting the frequency of observations against the class intervals of the yield strength measurements (Fig. 11-1). This type of bar diagram is known as a *histogram*. As the number of observations increases, the size of the class interval can be reduced until we obtain the limiting curve, which represents the *frequency distribution* of the sample (Fig. 11-2). Note that most of the values of yield strength fall within the interval from 126,000 to 134,000 psi.

If the frequency of observations in each class interval is expressed as a percentage of the total number of observations, the area under a frequency-distribution curve, which is plotted on this basis, is equal to unity (Fig. 11-3).

TABLE 11-1
Frequency tabulation of yield strength of steel*

(1) Yield strength, 1000 psi, class interval	(2) Class midpoint x_i	(3) Frequency f_i	(4) f_ix_i	(5) Frequency, % of total	(6) Cumulative frequency	(7) Cumulative frequency, %
114–115.9	115	4	460	0.9	4	0.9
116–117.9	117	6	702	1.3	10	2.2
118–119.9	119	8	952	1.6	18	3.8
120–121.9	121	26	3146	5.8	44	9.6
122–123.9	123	29	3657	6.5	73	16.1
124–125.9	125	44	5500	9.8	117	25.9
126–127.9	127	47	5969	10.5	164	36.4
128–129.9	129	59	7611	13.1	223	49.5
130–131.9	131	67	8777	15.0	290	64.5
132–133.9	133	45	5985	10.0	335	74.5
134–135.9	135	49	6615	10.9	384	85.4
136–137.9	137	29	3973	6.5	413	91.9
138–139.9	139	17	2363	3.8	430	95.7
140–141.9	141	9	1269	2.0	439	97.7
142–143.9	143	6	858	1.3	445	99.0
144–145.9	145	4	580	0.9	449	99.9

$$\sum f_i = 449 \qquad \sum f_ix_i = 58{,}417$$
$$\bar{x} = 58{,}417/449 = 130{,}000 \text{ psi}$$

* Data from F. B. Stulen, W. C. Schulte, and H. N. Cummings, *in* D. E. Hardenbergh (ed.), "Statistical Methods in Materials Research," Pennsylvania State University, University Park, Pa., 1956.

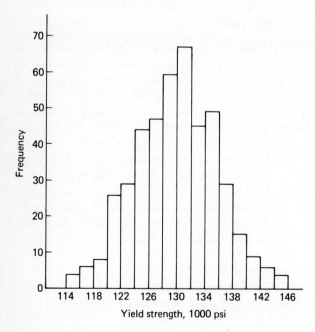

FIGURE 11-1
Frequency histogram of data in Table 11-1.

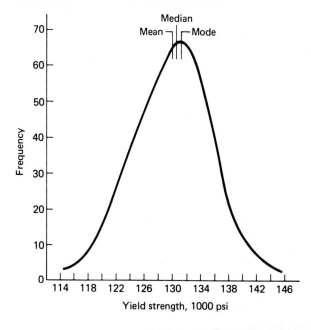

FIGURE 11-2
Frequency distribution of data in Table 11-1.

The probability that a single random measurement of yield strength will be between the value x_1 and a slightly higher value $x_1 + \Delta x$ is given by the area under the frequency-distribution curve bounded by those two limits. Similarly, the probability that a single observation will be greater than some value x_2 is given by the area under the curve to the right of x_2, while the probability that the single observation will be less than x_2 is given by the area to the left of x_2. Before inferring probabilities from the frequency distribution, we need to fit the experimental results to a standard statistical distribution such as the normal or log-normal distribution (Sec. 11-6).

Another way to present these data is to arrange the frequency in a cumulative manner. In Table 11-1, column 6, the frequency for each class

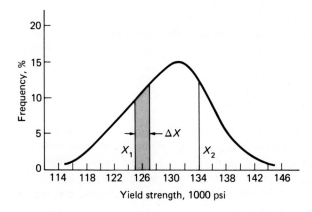

FIGURE 11-3
Frequency distribution based on relative frequency.

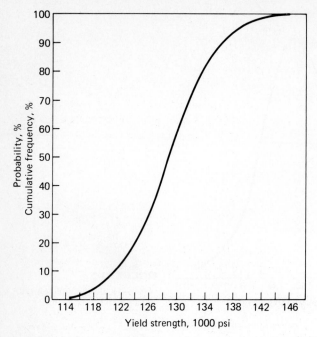

FIGURE 11-4
Cumulative frequency distribution.

interval is accumulated with the values of frequency for the class intervals below it to indicate the total number of observations with a value of yield strength less than or equal to the value of the upper limit of the class interval. As an example, for the sample described in Table 11-1, 164 out of 449 observations of yield strength have a value of 127,900 psi or less. If the cumulative frequency is expressed as the percentage of the total (Table 11-1, column 7), the values represent the probability that the yield strength will be less than or equal to the value of the observation. Figure 11-4 shows the cumulative frequency distribution plotted on that basis. The presentation of data as a cumulative distribution is sometimes preferred to a frequency distribution because it is much less sensitive than the frequency distribution to the choice of class intervals.

11-5 MEASURES OF CENTRAL TENDENCY AND DISPERSION

A frequency distribution such as that of Fig. 11-2 can be described with numbers that indicate the central location of the distribution and how the observations are spread out from the central region (dispersion). The most common and important measure of the central value of an array of data is the *arithmetic mean,* or *average.* The mean of $x_1, x_2, \ldots, x_n$ observations is denoted by $\bar{x}$ and is given by

$$\bar{x} = \frac{\sum\limits_{i=1}^{n} x_i}{n} \tag{11-14}$$

The arithmetic mean is equal to the summation of the individual observations divided by the total number of observations n. If the data are arranged in a frequency table, as in Table 11-1, the mean can be most conveniently determined from

$$\bar{x} = \frac{\sum\limits_{i=1}^{k} f_i x_i}{\sum f_i} \tag{11-15}$$

where f_i is the frequency of observations in a particular class interval with midpoint x_i. The summation is taken over all class intervals (see Table 11-1, column 4).

Two other common measures of central tendency are the mode and the median. The *mode* is the value of the observations that occurs most frequently. The *median* is the middle value of a group of observations. For a set of discrete data the median can be obtained by arranging the observations in numerical sequence and determining which value falls in the middle. For a frequency distribution the median is the value that divides the area under the curve into two equal parts. The mean and the median are frequently close together; but if there are extreme values in the observations (either high or low), the mean will be more influenced by these values than the median. As an extreme example, the situation sometimes arises in fatigue testing that out of a group of specimens tested at a certain stress a few do not fail in the time allotted for the test and therefore presumably have infinite lives. These extreme values could not be grouped with the failed specimens to calculate a mean fatigue life; yet they could be considered in determining the median. The positions of the mean, median, and mode are indicated in Fig. 11-2.

The most important measure of the dispersion of a sample is given by the unbiased estimate of the variance s^2.

$$s^2 = \frac{\sum\limits_{i=1}^{n} (x_i - \bar{x})^2}{n - 1} \tag{11-16}$$

The term $x_i - \bar{x}$ is the deviation of each observation x_i from the arithmetic mean $\bar{x}$ of the n observations. The quantity $n - 1$ in the denominator is called the number of degrees of freedom and is equal to the number of observations minus the number of linear relations between the observations. Since the mean represents one such relation, the number of degrees of freedom for the variance about the mean is $n - 1$. For computational purposes it is often convenient to calculate the variance from the following relation:

$$s^2 = \frac{n \sum_{i=1}^{n} x_i^2 - \left(\sum_{i=1}^{n} x_i\right)^2}{n(n-1)} \tag{11-17}$$

Equation (11-16) is preferable to Eq. (11-17) when the number of significant digits is important. When the data are arranged in a frequency table, the variance can be most readily computed from the following equation:

$$s^2 = \frac{\sum_{i=1}^{k} f_i x_i^2 - \frac{\left(\sum_{i=1}^{k} f_i x_i\right)^2}{n}}{n-1} \tag{11-18}$$

In dealing with the dispersion of data it is usual practice to work with the standard deviation s, which is defined as the positive square root of the variance.

$$s = \left[\frac{\sum_{i=1}^{n} (x_i - \bar{x})^2}{n-1}\right]^{1/2} \tag{11-19}$$

Sometimes it is desirable to describe the variability relative to the average. The coefficient of variation v is used for this purpose.

$$v = \frac{s}{\bar{x}} \tag{11-20}$$

A measure of dispersion that is sometimes used because of its extreme simplicity is the *range*. It is simply the difference between the largest and smallest observation. The range does not provide as precise estimates as does the standard deviation.

The need to distinguish between the sample and the population has been emphasized above. Standard notation is adopted in statistics to make this distinction between the characteristics of the sample and the population.

	Sample	Population
Arithmetic mean	$\bar{x}$	μ
Standard deviation	s	σ

11-6 THE NORMAL DISTRIBUTION

Many physical measurements follow the symmetrical, bell-shaped curve of the normal, or Gaussian, frequency distribution; repeated measurements of the length or diameter of a bar would closely approximate it. The distributions of yield strength, tensile strength, and reduction of area from the tension test have been found to follow the normal curve to a suitable degree of approximation. The equation of the normal curve is

$$f(x) = \frac{1}{\sigma\sqrt{2\pi}} \exp\left[-\frac{1}{2}\left(\frac{x-\mu}{\sigma}\right)^2 \right] \tag{11-21}$$

where $f(x)$ is the height of the frequency curve corresponding to an assigned value x, μ is the mean of the population, and σ is the standard deviation of the population.[1]

The properties of the normal distribution are well known and are readily available in tabular form. The central limit theorem of statistics states that the distribution of the mean of N independent observations from *any* frequency distribution with a finite mean and variance approaches the normal distribution as N approaches infinity. When a random variable represents the total effect of a number of independent small causes, the central limit theorem leads one to expect that the distribution of the measured variable will be normal. For example, the measurement of the yield of a chemical process contains errors in chemical analysis, sampling error, and process errors in factors like temperature, pressure, and flow rate. However, because of the prominence of the normal distribution, there is a tendency to assume that every experimental variable is normally distributed unless proved otherwise. This is wrong.

The normal distribution extends from $x = -\infty$ to $x = +\infty$ and is symmetrical about the population mean μ. The existence of negative values and long "tails" makes the normal distribution a poor model in certain engineering problems.

In order to place all normal distributions on a common basis in a standardized way, the normal curve frequently is expressed in terms of the standard normal variable z.

$$z = \frac{x-\mu}{\sigma} \tag{11-22}$$

Now, the equation of the standard normal curve becomes

$$f(z) = \frac{1}{\sqrt{2\pi}} \exp\left(-\frac{z^2}{2} \right) \tag{11-23}$$

Figure 11-5 shows this standardized normal curve, where $\mu = 0$ and $\sigma = 1$. The total area under the curve is unity. The relative frequency of a value of z falling between $z = -\infty$ and a specified value of z are given by the total area under the curve. Some typical values are listed in Table 11-2.[2]

---

[1] It should be noted that σ is used for normal stress in other chapters of this book.

[2] For complete values see M. G. Natrella, op. cit., tables A-1 and A-2, and Table C-1 in appendix, or any other statistics text.

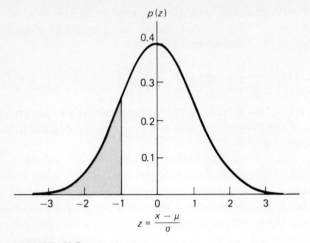

FIGURE 11-5
Standardized normal frequency distribution.

Example 11-3. A highly automated factory is producing ball bearings. The average ball diameter is 0.2152 in and the standard deviation is 0.0125 in. These dimensions are normally distributed.
(a) What percentage of the parts can be expected to have a diameter less than 0.2500 in.?
Determining the standard normal variable

$$z = \frac{x - \mu}{\sigma} \approx \frac{x - \bar{x}}{s} = \frac{0.2500 - 0.2512}{0.0125} = \frac{-0.0012}{0.0125} = -0.096$$

From Table C-1, $P(z < -0.09) = 0.4641$ and $P(z < 0.10) = 0.4602$. Interpolating, the area under the z distribution curve at $z = -0.096$ is 0.4618. Therefore, 46.18 percent of the ball bearings are below 0.2500 in diameter.

TABLE 11-2
Areas under standardized normal frequency curve

	$z = x - \mu/\sigma$	Area	z	Area
	−3.0	0.0013	−3.090	0.001
	−2.0	0.0228	−2.576	0.005
	−1.0	0.1587	−2.326	0.010
	−0.5	0.3085	−1.960	0.025
	0.0	0.5000	−1.645	0.050
	+0.5	0.6915	1.645	0.950
	+1.0	0.8413	1.960	0.975
	+2.0	0.9772	2.326	0.990
	+3.0	0.9987	2.576	0.995
			3.090	0.999

(*b*) What percentage of the balls are between 0.2574 and 0.2512 in.?

$$z = \frac{0.2512 - 0.2512}{0.0125} = 0.0 \quad \begin{array}{l} \text{Area under curve from } -\infty \text{ to } z = 0 \\ \text{is } 0.5000. \end{array}$$

$$z = \frac{0.2574 - 0.2512}{0.0125} = \frac{0.0062}{0.0125} = +0.50 \quad \begin{array}{l} \text{Area under curve from} \\ -\infty \text{ to } z = 0.5 \text{ is } 0.6915. \end{array}$$

Therefore, percentage of ball diameters in interval 0.2512 to 0.2574 is 0.6915 − 0.5000 = 0.1915 or 19.15 percent.

Tests for Normal Distribution

In a convenient method for determining whether a sample frequency distribution approximates a normal distribution a cumulative frequency plot on normal probability paper is used. First the sample mean and standard deviation are computed. Plot $\bar{x}$ at $P = 0.5$ and $\bar{x} + s$ at $P = 0.84$. A straight line is then drawn through those points. The individual points are then plotted on the normal probability paper. For small sample populations the probability is obtained by rank-ordering the data from smallest to largest value. The probability of an event occurring is

$$P_m = \frac{m}{n + 1} \tag{11-24}$$

where m = rank number of the value

n = total number of data points

The degree of fit to the normal dsitribution is determined by how well the data group around the straight line (Fig. 11-6).

The statistical advantages of the normal distribution are quite considerable; so that when the data do not fit the normal curve, it is worthwhile to search for a simple transformation that will normalize them. The most common transformations are $x' = \log x$ and $x' = x^{1/2}$. As an example, the distribution of fatigue life at constant stress is skewed with respect to the number of cycles to failure, but it is approximately normal with respect to the logarithm of the number of cycles to failure. This is called a log-normal distribution.[1]

---

[1] C. H. Lipson and N. J. Sheth, "Statistical Design and Analysis of Engineering Experiments," pp. 32–36, McGraw-Hill Book Company, New York, 1973; ASM Metals Handbook, 9th ed., vol. 8, pp. 630–632, ASM International, Metals Park, OH, 1985.

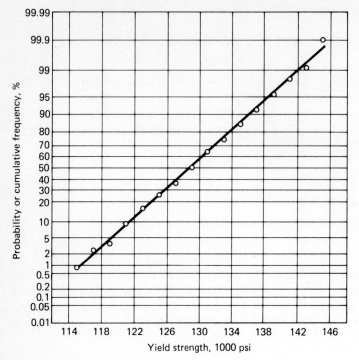

FIGURE 11-6
Normal probability plot of data in Table 11-1.

11-7 WEIBULL AND OTHER FREQUENCY DISTRIBUTIONS

The normal distribution is an unbounded symmetrical distribution with long tails extending to $-\infty$ and to $+\infty$. However, many random variables follow a bounded, nonsymmetrical distribution. A good example is a distribution describing the life of a component for which all values are positive (there are no negative lives) and for which there are occasional long-lived results.

Weibull Distribution

The Weibull distribution[1] is widely used for many engineering problems because of its versatility (see example in Sec. 1-7). Originally proposed for describing fatigue life, it is widely used to describe the life of parts or components, such as ball bearings, gears, and electronic components. The

---

[1] W. Weibull, *J. Appl. Mech.*, vol. 18, pp. 293–297, 1951; *Materials Research and Stds.*, pp. 405–411, May 1962; ASM Metals Handbook, op. cit., pp. 632–634.

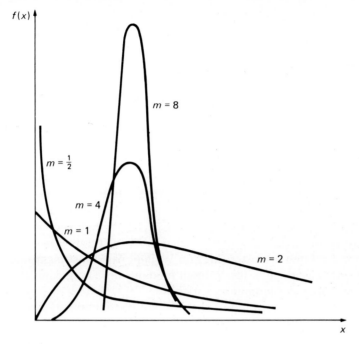

FIGURE 11-7
The Weibull distribution for $\theta = 1$ and different values of m.

two-parameter Weibull function is described by:

$$f(x) = \frac{m}{\theta} \left(\frac{x}{\theta}\right)^{m-1} \exp\left[-\left(\frac{x}{\theta}\right)^m\right], \qquad x > 0 \qquad (11\text{-}25)$$

where $f(x)$ = frequency distribution of the random variable x
      m = shape parameter, which is sometimes referred to as the Weibull modulus
      θ = scale parameter, sometimes called the characteristic value

The change in the Weibull distribution for various values of shape parameter is shown in Fig. 11.7. The mean of a Weibull population is given by

$$\bar{x} = \theta\Gamma\left(1 + \frac{1}{m}\right) \qquad (11\text{-}26)$$

where Γ is the standard gamma function (see next under gamma distribution). The variance of a Weibull population is given by

$$\text{Variance} = \theta^2\left[\Gamma\left(1 + \frac{2}{m}\right) - \Gamma^2\left(1 + \frac{1}{m}\right)\right] \qquad (11\text{-}27)$$

The cumulative frequency distribution is given by

$$F(x) = 1 - \exp\left[-\left(\frac{x}{\theta}\right)^m\right], \quad x > 0 \tag{11-28}$$

Rewriting Eq. (11-28) as

$$\frac{1}{1 - F(x)} = \exp\left(\frac{x}{\theta}\right)^m$$

$$\ln\frac{1}{1 - F(x)} = \left(\frac{x}{\theta}\right)^m$$

$$\ln\left(\ln\frac{1}{1 - F(x)}\right) = m\ln x - m\ln\theta = m(\ln x - \ln\theta) \tag{11-29}$$

which is a straight line of the form $Y = mx + C$. Special Weibull probability paper is available to assist in the analysis according to Eq. (11-29). When the probability of failure is plotted against x (life) on Weibull paper a straight line is obtained (Fig. 11-8). The slope is the Weibull modulus m. The greater the slope the smaller the scatter in the random variable x.

θ is called the characteristic value of the Weibull distribution. If $x = \theta$, then

$$F(x) = 1 - \exp\left[-\left(\frac{\theta}{\theta}\right)^m\right] = 1 - e^{-1} = 1 - \frac{1}{e}$$

Therefore,

$$F(x) = 1 - \frac{1}{2.718} = 0.632$$

Thus, for any Weibull distribution the probability of being less than or equal to the characteristic value is 0.632. Therefore, θ will divide the area under the probability distribution function into 0.632 and 0.368 for all values of m.

If the data do not plot as a straight line on Weibull graph paper, then either the sample was not taken from a population with a Weibull distribution or it may be that the Weibull distribution has a minimum value x_0 which is greater than $x_0 = 0$. This leads to the three-parameter Weibull distribution.

$$F(x) = 1 - \exp\left[-\left(\frac{x - x_0}{\theta - x_0}\right)^m\right] \tag{11-30}$$

For example, in the distribution of fatigue life at a constant stress, it is unrealistic to expect a minimum life of $x_0 = 0$. The easiest procedure for finding x_0 is to use the Weibull probability plot. First, plot the data as in the two-parameter case where $x_0 = 0$. Now, pick a value of x_0 between 0 and the lowest value of x_i and subtract it from each of the observed x. Continue adjusting x_0 and plotting $x - x_0$ until a straight line is obtained on the Weibull graph paper.

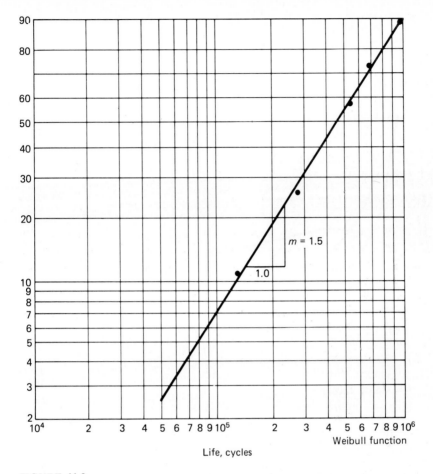

FIGURE 11-8
Weibull plot for life of ball bearings. (*C. Lipman and N. J. Sheth, "Statistical Design and Analysis of Engineering Experiments," p. 41. Copyright © 1973 by McGraw-Hill. Used with the permission of McGraw-Hill Book Company.*)

Gamma Distribution

The two-parameter gamma density function is used to describe random variables that are bounded at one end.[1] It has the form:

$$f(x) = \frac{\lambda^{\eta} x^{\eta-1} e^{-\lambda x}}{\Gamma(\eta)} \qquad \text{for } x > 0; \ \lambda > 0; \ \eta > 0 \qquad (11\text{-}31)$$

---

[1] G. J. Hahn and S. S. Shapiro, "Statistical Models in Engineering," John Wiley & Sons, Inc., New York, 1967.

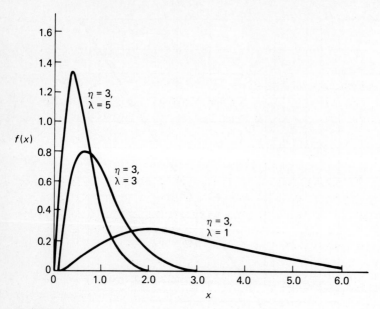

FIGURE 11-9
Gamma frequency distributions with $\eta = 3$ and various values of λ.

where $\Gamma(\eta) =$ gamma function $= \int_0^\infty x^{\eta-1} e^{-x}\, dx$
  $\eta =$ shape factor
  $\lambda =$ scale factor

Figure 11-9 illustrates the wide variety of shaped distributions that can be described by the gamma distribution. The gamma distribution models the time required for a total of η independent events to take place if the events occur at a constant rate λ. For example, in a complex engineering system, if failure occurs when η subfailures have taken place at a rate λ, the time for system failure is distributed according to the gamma distribution. The gamma distribution also is important in queuing theory (the analysis of waiting lines). It has been used as an empirical distribution to describe such factors as the distribution of family income and the time to failure for capacitors. The chi-square and exponential distributions are special cases of the gamma distribution.

Exponential Distribution

The exponential distribution is a special case of the gamma distribution for $\eta = 1$.

$$f(x) = \lambda e^{-\lambda x} \qquad \text{for } x > 0 \text{ and } \lambda > 0 \qquad (11\text{-}32)$$

We also can consider the exponential distribution to be a special case of the

Weibull distribution when $m = 1$, and $x_0 = 0$.

$$f(x) = \frac{1}{\theta} e^{-x/\theta} \tag{11-33}$$

The Weibull distribution models the probability of failure of a component like a bearing or a shaft, but for the complete system, such as a pump, the exponential distribution is a better model.

Distributions for Discrete Variables

The normal distribution and the other distributions discussed in this section deal with continuous random variables, which can take on any value over a considerable range. However, there are important engineering problems in which the random variable takes on only discrete values. Such a situation occurs at a quality-control station on a production line, where a part is either accepted or rejected.

A random variable follows a binomial distribution when the following four characteristics occur: 1) there are n trials; 2) the trials are independent; 3) each trial has only two possible outcomes; and 4) the probability of the event remains constant from trial to trial. Let the overall probability of an unsuccessful occurrence be p and the probability of a success be $q = 1 - p$. The probability of occurrence of any value of $r = 0, 1, 2, \ldots, n$ in n trials is given by

$$P(r) = \frac{n!}{r!\,(n-r)!} p^r q^{n-r} \tag{11-34}$$

The hypergeometric distribution is related to the binomial distribution. It applies when samples are randomly drawn from a finite population without replacement, so that the probability of the event cannot remain constant from trial to trial. As with the binomial distribution, each trial has only two possible outcomes.

$$P \text{ (number of defects } d) = \frac{\binom{N_p}{d}\binom{N_q}{n-d}}{\binom{N}{n}} \tag{11-35}$$

where N = lot size
$\quad n$ = sample size
$\quad N_p$ = number of failures in the lot
$\quad N_q$ = number of successes in the lot
$\quad d$ = number of failures in a sample

and

$$\binom{a}{b} = \frac{a!}{b!\,(a-b)!} \quad \text{for the generalized case}$$

Note that $0! = 1$.

Example 11-4. A parts supplier receives 12 transmissions and sells 9 of them. Later he is informed that 3 of the 12 transmissions were defective, but which three parts were defective was not specifically known. What is the probability that all three of the defective parts were sold to customers?

Solution.

$$N = 12$$
$$p = \text{fraction defective} = \tfrac{3}{12} = 0.25$$
$$q = 1 - p = 0.75$$
$$N_p = 12(0.25) = 3$$
$$N_q = 12(0.75) = 9$$
$$n = 9$$
$$d = 3$$

$$P(d = 3) = \frac{\binom{3}{3}\binom{9}{6}}{\binom{12}{9}} = \frac{3!}{3!\,(3-3)!} \times \frac{9!}{6!\,(9-6)!} \times \frac{9!\,(12-9)!}{12!}$$

$$= \frac{3!}{3!\,0!} \times \frac{9!}{6!\,3!} \times \frac{9!\,3!}{12!} = 0.382$$

With both the binomial and hypergeometric distributions, we know the number of failures and successes. There are situations in which we know the number of times the event occurred but it is not practical to observe the number of times an event did not occur. The distribution of isolated events in a continuum of events is described by the Poisson distribution. If y represents the average number of occurrences of an event, the probability of observing the occurrence of r events is given by

$$P(r) = \frac{e^{-y}y^r}{r!} \qquad \text{for } r = 0, 1, 2, \ldots \tag{11-36}$$

Example 11-5. The following data were obtained for the number of breakdowns of sheet-forming presses in a stamping shop. Records were taken for 350 consecutive days. Determine the probability of having zero, one, two, three, four or five breakdowns per day.

Number of breakdowns per day	Number of days breakdowns occurred	Number of breakdowns
0	80	$0 \times 80 = 0$
1	134	$1 \times 134 = 134$
2	62	$2 \times 62 = 124$
3	41	$3 \times 41 = 123$
4	23	$4 \times 23 = 92$
5	6	$5 \times 6 = 30$
6	3	$6 \times 3 = 18$
7	1	$7 \times 1 = 7$
	350	528

Solution. Average breakdown per day $= \frac{528}{350} = 1.51 = y$

$$e^{-y} = e^{-1.51} = 0.22$$

$$P(0) = 0.22 \frac{1.51^0}{0!} = 0.22$$

$$P(1) = 0.22 \frac{1.51^1}{1!} = 0.33$$

$$P(2) = 0.22 \frac{1.51^2}{2!} = 0.25$$

$$P(3) = 0.22 \frac{1.51^3}{3!} = 0.13$$

$$P(4) = 0.22 \frac{1.51^4}{4!} = 0.05$$

$$P(5) = 0.22 \frac{1.51^5}{5!} = 0.01$$

11-8 SAMPLING DISTRIBUTIONS

We have already emphasized that the central problem in statistics is relating the population and the samples that are drawn from it. This problem is viewed from two perspectives: 1) what does the population tell us about the behavior of the samples drawn from it and 2) what does a sample or series of samples tell us about the population from which the sample came? The first question is the subject of this section. Section 11-9 is concerned with the second question.

Suppose we have a population consisting of 300 items. We take all possible samples of $n = 10$ and determine the mean of each sample $\bar{x}$. This would give a population of $\bar{x}$'s, one for each sample of 10. The frequency distribution of this population is called the sampling distribution of $\bar{x}$ for samples of $n = 10$. Similarly, we could have sampling distributions for s^2 or for any other sample statistic. Such sampling distributions have been worked out by exact mathematical methods for random samples from any normal distribution, but comparable information from nonnormal populations still is quite incomplete.

Distribution of Sample Means

The mean of a sample provides an unbiased estimate of the mean of the population from which the sample was drawn. The sample values of the mean will be normally distributed about the population mean μ. The mean of the distribution of $\bar{x}$, $\mu_{\bar{x}}$, is equal to the mean of the population of x's if the population is very large. The standard deviation of the sampling distribution is called the standard error of the mean, and it is equal to

$$\sigma_{\bar{x}} = \frac{\sigma}{\sqrt{n}} \tag{11-37}$$

If, for example, $n = 36$, the scatter of $\bar{x}$ about $\mu_{\bar{x}}$ is only one-sixth what it would be for the distribution of x about μ.

We can use the above to establish the standardized variable z, which is normally distributed with a mean 0 and standard deviation 1.

$$z = \frac{\bar{x} - \mu}{\sigma_{\bar{x}}} = \frac{\bar{x} - \mu}{\sigma/\sqrt{n}} \tag{11-38}$$

t Distribution

It is a common situation not to know the standard deviation of the population. When we do not know it, we cannot substitute s for σ in Eq. (11-38) and assume that the statistic will be normally distributed, unless n is very large. However, if x is normally distributed, the sampling distribution for the mean is the *t distribution*:

$$t = \frac{\bar{x} - \mu}{s/\sqrt{n}} \quad \text{for } v = n - 1 \text{ degrees of freedom} \tag{11-39}$$

Values of the t distribution are given in Appendix C. The distribution is symmetrical, with a different curve for each value of v. As v increases, the distribution of t approaches the normal distribution.

Distribution of Sample Variances

The distribution of sample variances s^2 from a normal population with a variance σ^2 is given by the *chi-square distribution*.

$$\chi^2 = \frac{vs^2}{\sigma^2} = \frac{(n-1)s^2}{\sigma^2} \tag{11-40}$$

F Distribution

The F distribution, Appendix C, describes the sampling distribution from two normally distributed populations $N_1(\mu_1, \sigma_1^2)$ and $N_2(\mu_2, \sigma_2^2)$. If we take n_1 and n_2 independent samples, respectively, for which the variances are s_1^2 and s_2^2, then the F distributions tells us whether the two samples come from populations having equal variances.

$$F_{v_1, v_2} = \frac{s_1^2}{s_2^2} \tag{11-41}$$

where $v_1 = n_1 - 1$
$v_2 = n_2 - 1$

11-9 STATISTICAL TESTS OF HYPOTHESESES

The statistical decision-making process can be put on a rational, systematic basis by considering various statistically based hypotheses. The procedure is to establish the appropriate hypothesis and its alternative *before the experiment is conducted*. Then the hypothesis can be tested with the appropriate statistics determined from the sample data.

The usual approach is to set up a null hypothesis H_0, that is, that there is no real change or difference, and test this null hypothesis against an alternative hypothesis, of which there are several alternatives. For example, suppose the minimum acceptable yield strength for steel to be used in the construction of a building is μ_0. We test a sample of 10 specimens to get a mean μ. The null hypothesis is that the mean of the sample equals the required minimum value. The alternative hypothesis H_1 is that the mean of the sample is less than μ_0.

Null hypothesis H_0: $\mu = \mu_0$
Alternative hypothesis H_1: $\mu < \mu_0$

One cannot expect the sample mean to equal exactly the expected population mean. Therefore, we must allow for variation between the two values. The variation is reflected in two types of errors. One error would be if we concluded that the mean of the sample was not equal to the minimum standard when, in fact, $\mu = \mu_0$. We call this a type I error. The probability of making a type I error is α, which is also called the level of significance.

The other type of error is in accepting the null hypothesis when it is, in fact, false. This is called a type II error. The probability of making a type II error is β. The kinds of errors associated with hypothesis testing are summarized in the following decision table.

Decision	State of nature	
	H_0 is true	H_0 is false
Accept H_0	No error	Type II error
Reject H_0	Type I error	No error

We can visualize these errors in Fig. 11-10. For the decision point located as shown, the type I error is the part of the distribution of the population near μ_0 that lies below the decision point. These are observations that belong to the distribution with mean μ_0 but which we reject. Similarly, the type II error is that part of the sample distribution that lies above the decision point. We see that α and β change with the location of the decision point. For a fixed sample size, as β decreases, α increass.

We need to consider the implications in each problem of deciding whether to minimize type I or type II errors. For example, in the simple example described above, a type I error to reject H_0 when, in fact, H_0 is true

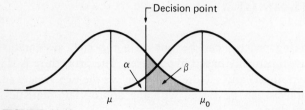

FIGURE 11-10
Graphical representation of type I and II errors.

would mean that steel would be shipped back to the manufacturer, with a resulting financial loss and penalty to the builder due to delay in construction. On the other hand, a type II error, accepting H_0 when it is false, would mean that steel of insufficient strength would be used to construct the building. A possible result would be failure of the building, with likely loss of life and inevitable litigation and associated costs. Thus, a type II error should be minimized in most engineering situations.

To gain an understanding of how hypothesis testing is used, consider the following example.[1] Five measurements were made of the sulfur content of a steel

> 0.0307wt%; 0.0324; 0.0314; 0.0311; 0.0307
> for which $\bar{x} = 0.03126$, $s = 0.000702$, $n = 5$

Because of difficulty in hot-working if the sulfur becomes too great, the process standard sets a limit on sulfur of 0.0300 wt percent. In this problem we want to know whether the sample exceeds this value or if a value of $\bar{x} = 0.03126$ could have been obtained by chance. Therefore, we establish the null hypothesis as

H_0 $$\mu = \mu_0 = 0.0300$$

whereas the alternative hypothesis is

H_1: $$\mu > 0.0300$$

We select $\alpha = 0.005$ because of the difficulty and expense associated with a process change. Since the population standard deviation is not known, we must use the t distribution. From Table C-2, in Appendix C, the critical value for t with $\alpha = 0.005$ in a single-tailed distribution is 4.604. Therefore, if the observed t exceeds 4.604, we shall reject the null hypothesis and conclude that the sample of five values exceeds the process standard for sulfur. However, if $t < 4.604$, we will conclude that there is no reliable evidence as a reason to reject the null hypothesis.

$$t = \frac{\bar{x} - \mu_0}{s/\sqrt{n}} = \frac{0.03126 - 0.0300}{0.000702/\sqrt{5}} = 4.01$$

---

[1] I. W. Burr, "Applied Statistical Methods", p. 200, Academic Press, New York, 1974.

Therefore, we do not reject the null hypothesis. We note, however, that if the type I error had been set at $\alpha = 0.05$, the critical value of t would be 2.01, and we would have rejected the null hypothesis. The type II error, β, is obtained from an operating characteristic curve (OC curve)[1] for the statistic. Some useful OC curves are shown in Appendix D. If we accept a value of $d = 5 = [(\mu - \mu_0)\sqrt{n}]/\sigma$, then for $\alpha = 0.005$ and $n = 5$ the value of $\beta = 0.35$. However, if we increase n to 25, then β decreases to 0.02. This illustrates the importance of sample size in establishing errors and the value of using OC curves to preselect n so as to minimize the error.

We have seen that the use of hypothesis testing consists of the following steps.

1. Identify the null hypothesis H_0 and the alternative hypothesis H_1.
2. Select the appropriate sample statistic.
3. Select the test criteria (α and β) and determine the necessary sample size.
4. Obtain the sample and compute the sample characteristics.
5. Determine the critical region for rejecting H_0.
6. Make your decision

The various statistics used to test the common hypotheses, and the critical region for rejecting the null hypothesis, are summarized in Table 11-3. Entries 1, 2, and 3 deal with a single sample that is compared with population values. These cases often arise in design or production when standards or design requirements establish the population values. Another common situation is when we may wish to compare two distinct samples, e.g., two materials or two processing conditions. The statistical problem is to test two samples and determine whether the observed difference between the results is large enough to be significant, i.e., whether it is a real difference and not one that could be produced by chance variation in the measurements. In entries 4, 6, and 7 this situation is considered. Entry 5 lists the hypothesis needed to establish whether both samples have the same standard deviation. This is important information for establishing how to apply the t test in this type of situation.

In entry 8 the situation in which pairing is achieved between two sets of observations is considered. Examples of test conditions in which pairing can be achieved are 1) the testing of two different bearing designs by putting two bearings on the same shaft in a single test engine and 2) the testing the corrosion resistance of two protective coatings when opposite ends of the test coupon have been given different coatings. In paired observations we are dealing with the difference between two observations made under conditions that are experimentally as similar as possible. In this case the t test is based on the difference $d = x_{1j} - x_{2j}$.

---

[1] See M. G. Natrella, op. cit., chap. 3.

TABLE 11-3
Summary of statistical tests of hypotheses

	Hypothesis	Conditions	Test statistic	Distribution of test statistic	Alternative hypothesis	Critical region
1.	$\mu = \mu_0$	σ known	$z = \dfrac{\bar{x} - \mu_0}{\sigma/\sqrt{n}}$	Normal (0, 1)	$\mu \neq \mu_0$	$z < -z_{\alpha/2}$ $z > z_{\alpha/2}$
					$\mu < \mu_0$	$z < -z_\alpha$
					$\mu > \mu_0$	$z > z_\alpha$
2.	$\mu = \mu_0$	σ unknown	$t = \dfrac{\bar{x} - \mu_0}{s/\sqrt{n}}$	t distribution $\nu = n - 1$	$\mu \neq \mu_0$	$t < -t_{\alpha/2}$ $t > t_{\alpha/2}$
					$\mu < \mu_0$	$t < -t_\alpha$
					$\mu > \mu_0$	$t > t_\alpha$
3.	$\sigma = \sigma_0$	μ unknown	$\chi^2 = (n - 1)s^2/\sigma_0^2$	chi-square $\nu = n - 1$	$\sigma \neq \sigma_0$	$\chi^2 < \chi^2_{\alpha/2}$ $\chi^2 > \chi^2_{1-\alpha/2}$
					$\sigma < \sigma_0$	$\chi^2 < \chi^2_\alpha$
					$\sigma > \sigma_0$	$\chi^2 > \chi^2_{1-\alpha}$
4.	$\mu_1 - \mu_2 = \delta$	σ_1 and σ_2 known	$z = \dfrac{(\bar{x}_1 - \bar{x}_2) - \delta}{\sigma_{\bar{x}_1 - \bar{x}_2}}$ $\sigma_{\bar{x}_1 - \bar{x}_2} = \left(\dfrac{\sigma_1^2}{n_1} + \dfrac{\sigma_2^2}{n_2}\right)^{1/2}$	Normal (0, 1)	$(\mu_1 - \mu_2) \neq \delta$	$z < -z_{1-\alpha/2}$ $z > z_{1-\alpha/2}$
					$(\mu_1 - \mu_2) < \delta$	$z < z_{1-\alpha}$
					$(\mu_1 - \mu_2) > \delta$	$z > z_{1-\alpha}$

	Assumptions	Test statistic	Distribution	Alternative	Rejection region
5. $\sigma_1 = \sigma_2$	x's normal and independent	$F = s_1^2/s_2^2$	F_{ν_1, ν_2}	$\sigma_1 \neq \sigma_1$ $\sigma_1 > \sigma_2$	$F > F_{1-\alpha/2}$ $F < F_{1-\alpha/2}$ $F > F_{1-\alpha}$
6. $\mu_1 - \mu_2 = \delta$	σ unknown but $\sigma_1 = \sigma_2$	$t = \dfrac{(\bar{x}_1 - \bar{x}_2) - \delta}{s_{\bar{x}_1 - \bar{x}_2}}$ $s_{\bar{x}_1 - \bar{x}_2} = \left[s_p^2\left(\dfrac{1}{n_1} + \dfrac{1}{n_2}\right)\right]^{1/2}$ $s_p^2 = \dfrac{(n_1 - 1)s_1^2 + (n_2 - 1)s_2^2}{n_1 + n_2 - 2}$	t distribution $\nu = n_1 + n_2 - 2$	$(\mu_1 - \mu_2) \neq \delta$ $(\mu_1 - \mu_2) > \delta$ $(\mu_1 - \mu_2) < \delta$	$t < -t_{\alpha/2}$ $t > t_{\alpha/2}$ $t > t_\alpha$ $t < -t_\alpha$
7. $\mu_1 - \mu_2 = \delta$	σ unknown and $\sigma_1 \neq \sigma_2$	$t = \dfrac{(\bar{x}_1 - \bar{x}_2) - \delta}{\left(\dfrac{s_1^2}{n_1} + \dfrac{s_2^2}{n_2}\right)^{1/2}}$	t distribution $\nu = \dfrac{1}{\left(\dfrac{k^2}{\nu_1}\right) + \dfrac{(1 - k)^2}{\nu_2}}$ $k = \dfrac{s_1^2/n_1}{s_1^2/n_1 + s_2^2/n_2}$	$(\mu - \mu_2) \neq \delta$ $(\mu_1 - \mu_2) < \delta$ $(\mu_1 - \mu_2) > \delta$	$t < -t_{\alpha/2}$ $t > t_{\alpha/2}$ $t < -t_\alpha$ $t > t_\alpha$
8. $\mu_{1j} - \mu_{2j} = \mu_{dj} = 0$	Matched pairs; equal variability	$t = \dfrac{\bar{d}}{s/\sqrt{n}}$	t distribution $\nu = n - 1$, where n is the number of pairs	$\mu_{dj} = \mu_d \neq 0$ $\mu_{dj} = \mu_d > 0$	$t > t_{\alpha/2}$ $t < -t_{\alpha/2}$ $t > t_\alpha$

Example 11-6. (*a*) The lifetime of a mechanical switch produced by a company has been determined to have a population mean of $\mu = 2000h$ and $\sigma = 200h$. The temper of a phosphor bronze leaf spring in the switch is changed slightly by the supplier. To determine whether this has changed the product, a sample of 100 switches is tested to give the sample values $\bar{x} = 1960$ and $s = 180$. Has there been a change in the product?

Solution. To decide whether there has been a change in the product, we establish the null hypothesis

H_0: $\qquad\qquad\qquad\qquad\qquad \mu = \mu_0 = 2000$

H_1: $\qquad\qquad\qquad\qquad\qquad \mu \neq \mu_0$

We let $\alpha = 0.05$; for $n = 100$ this establishes $\beta = 0.02$. The critical region is $z < -z_{0.025}$ and $z > z_{0.025}$; or, from Table 11-2, if z lies below -1.96 or above 1.96, the null hypothesis is rejected. For this example,

$$z = \frac{1960 - 2000}{180/\sqrt{100}} = \frac{-40}{18} = -2.22$$

Since z falls in the critical region, the null hypothesis is rejected and we conclude that the change in the temper of the leaf spring has significantly altered the life of the switch.

There are 5 chances in 100 that we have rejected the null hypothesis when it is true. We can minimize this error by selecting $\alpha = 0.01$. Now the critical region is $z < -2.58$ and $z > 2.58$. Since the z of -2.22 falls outside of the critical region, we accept H_0 at the 0.01 level of significance.

(*b*) In reality, we are concerned only with whether the change in spring material reduced the life of the switch; it is not a serious consequence if the life was increased. Thus, we need a different alternative hypothesis

H_0: $\qquad\qquad\qquad \mu = \mu_0 = 2000 \qquad \alpha = 0.01$

H_1: $\qquad\qquad\qquad \mu < \mu_0 = 2000$

Now we are dealing with a one-tailed test and the critical region for rejecting H_0 is $z < -z_{0.01}$, or $z < -2.326$. Since $z = -2.22$, we do not reject H_0.

Example 11-7. A vendor of steel wire advertises a mean breaking load of 10,000 lb. A sample of eight tests shows a mean breaking load of 9250 lb and a standard deviation of 110 lb. Do our tests support the vendor's claim?

Solution. Since the population standard deviation is not known, we use entry 2 in Table 11-3.

H_0: $\qquad\qquad\qquad \mu = \mu_0 = 10,000 \qquad \alpha = 0.05$

H_1: $\qquad\qquad\qquad \mu < \mu_0 = 10,000$

The critical region for rejecting H_0 is $t < -t_{0.05}$, or from Appendix C, if $t < 1.90$ for $8 - 1 = 7$ degrees of freedom,

$$t = \frac{\bar{x} - \mu_0}{s/\sqrt{n}} = \frac{9250 - 10,000}{110/\sqrt{8}} = -19.3$$

Therefore, we reject H_0 and conclude that the sample did not come from a population with $\mu = 10,000\,\text{lb}$.

Example 11-6. The manager of a quality-control lab is concerned with whether two stress rupture machines are giving reliable test results. Six specimens of the same material are run on each machine at the same temperature and stress level. The logarithm of the rupture time is used as the test response. The results for the two machines are:

Machine 1	Machine 2
$\bar{x}_1 = 0.91$	$\bar{x}_2 = 1.78$
$s_1 = 0.538$	$s_2 = 0.40$
$n_1 = 6$	$n_2 = 6$

Does the manager have cause for concern?

Solution. Since σ is not known, we must first establish whether there is a significant difference in the variances of the two machines

$$F = \frac{s_1^2}{s_2^2} = \frac{0.538^2}{0.40^2} = \frac{0.289}{0.160} = 1.88$$

From a table of the F distribution for $v_1 = v_2 = 5$, we find the critical region is $F > 11.0$ for $\alpha = 0.01$. Therefore, $\sigma_1 = \sigma_2$ and we can follow entry 6 in Table 11-3.

H_0: $\qquad\qquad\qquad\qquad \mu_1 - \mu_2 = \delta = 0$

H_1: $\qquad\qquad\qquad\qquad \mu_1 - \mu_2 \neq 0 \qquad \alpha = 0.05$

The critical region is $t < -t_{\alpha/2}$ and $t > t_{\alpha/2}$, which for $v = n_1 + n_2 - 2$ is $t < -2.23$ and $t > 2.23$.

$$t = \frac{(\bar{x}_1 - \bar{x}_2) - \delta}{s_{\bar{x}_1 - \bar{x}_2}} \qquad s_{\bar{x}_1 - \bar{x}_2} = \left[s_p^2 \left(\frac{1}{n_1} + \frac{1}{n_2} \right) \right]^{1/2}$$

$$s_p^2 = \frac{(n_1 - 1)s_1^2 + (n_2 - 1)s_2^2}{n_1 + n_2 - 2} = \frac{5(0.289) + 5(0.16)}{10} = 0.225$$

$$s_{\bar{x}_1 - \bar{x}_2} = \left[0.225 \left(\frac{1}{6} + \frac{1}{6} \right) \right]^{1/2} = [0.225(0.333)]^{1/2} = 0.274$$

$$t = \frac{(0.91 - 1.78) - 0}{0.274} = \frac{-0.87}{0.274} = -3.17$$

Since the calculated value of t is well into the critical region, we reject the null hypothesis and conclude that the two stress rupture machines do not give comparable results.

11-10 STATISTICAL INTERVALS

Interval estimation is commonly used to make probability statements about the population from which a sample has been drawn or to predict the results of a future sample from the same population. A common method is to determine the confidence limits for a parameter so that we can have a specified degree of confidence that the parameter lies within the interval. For example, if we have determined the 90 percent confidence limits of the mean, then, in the long run, the true mean of the population will lie within the limits 90 percent of the time. A number of statistical intervals have been established.

Confidence Interval

The confidence interval on the mean gives limits that can be claimed with a $(1 - \alpha)$ 100 percent degree of confidence to contain the unknown value of the population mean. When σ is unknown, the two-sided confidence interval is

$$\bar{x} - z_{\alpha/2} \frac{\sigma}{\sqrt{n}} < \mu < \bar{x} + z_{\alpha/2} \frac{\sigma}{\sqrt{n}} \tag{11-42}$$

For the usual case in which the population standard deviation is unknown, we must use the t distribution with $n - 1$ degrees of freedom.

$$\bar{x} - t_{\alpha/2} \frac{s}{\sqrt{n}} < \mu < \bar{x} + t_{\alpha/2} \frac{s}{\sqrt{n}} \tag{11-43}$$

The confidence interval on the variance, from Sec. 11-8, is given by

$$(n - 1) \frac{s^2}{\chi^2_{1-\alpha/2}} < \sigma^2 < (n - 1) \frac{s^2}{\chi^2_{\alpha/2}} \tag{11-44}$$

Tolerance Interval

An interval that can be claimed to contain at least a specified proportion p of the population with a confidence $(1 - \alpha)$ 100 percent is known as a statistical tolerance interval. This type of interval is especially useful in quality-control situations in mass production. For example, if we have obtained data on a random sample of 30 ball bearings, a tolerance interval will provide limits to specify a given proportion, for example, $p = 0.8$, of the population of bearings from which the sample was selected.

Prediction Interval

A *prediction interval* will contain all of k future observations with a confidence level $(1 - \alpha)$ 100 percent. Alternatively, the prediction interval could be selected to contain the mean of k future observations. Prediction intervals often are of interest to manufacturers of equipment that is produced in small lots. Thus, a prediction interval that contains all future k items of production would be required to establish limits on performance for a small shipment of equipment when the satisfactory performance of all units is to be guaranteed.

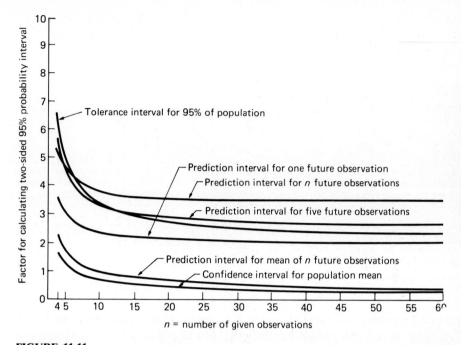

FIGURE 11-11
Comparison of c factor for calculating some two-sided 95 percent probability intervals. (*G. J. Hahn, J. Qual. Tech., vol. 2, no. 3, p. 117, 1970.*)

The statistical interval can be determined from

$$\bar{x} \pm cs \tag{11.45}$$

Details on how to compute the appropriate value of c are given by Hahn.[1] The appropriate factors for two-sided 95 percent intervals are given in Fig. 11-11.

Rejection of Outliers

It is not uncommon to find a bad data point in a set of observations, i.e., a point that "looks out of place" when compared with the bulk of the data. It is incorrect just to discard the outlier, but statistics can lend assistance on this point.[2]

If we take n measurements, and n is large enough that we can expect a Gaussian error distribution, then we can use the distribution to compute the probability that a given point will deviate a certain amount from the mean. It

---

[1] G. J. Hahn, *J. Qual. Tech.*, part I, vol. 2, no. 3, pp. 115–125, 1970; part II, vol. 2, no. 4, pp. 195–206, 1970.

[2] See M. G. Natrella, op. cit., chap. 17.

TABLE 11-4
Rejection of outliers by Chauvenet's criterion

Number of observations n	Ratio of max. acceptable deviation to standard deviation d_{max}/σ
2	1.15
3	1.38
4	1.54
5	1.65
6	1.73
7	1.80
10	1.96
15	2.13
25	2.33
50	2.57
100	2.81

would be unlikely to expect the outlier to occur with a probability less than $1/n$. Thus, if the probability for the observed deviation of the outlier is less than $1/n$, it confirms our suspicions that this an extreme observation. There are a number of statistical criteria for rejecting outliers. The *Chauvenet criterion* specifies that a reading may be rejected if the probability of obtaining the particular deviation from the mean is less than $1/2n$. Table 11-4 gives the value of the deviation from the point to the mean that must be exceeded in order to reject the point according to the Chauvenet criterion. Once all the outliers are rejected, a new mean and standard deviation are calculated for the sample.

11-11 ANALYSIS OF VARIANCE

Section 11-9 provided methods for making statistically based decisions between two samples or treatments. Now we shall consider the situation in which we have three or more treatments. The statistical procedure is called the analysis of variance (ANOVA). An understanding of ANOVA will be important later when we consider the important topic of design of experiments. With ANOVA we determine:

1. The total spread of results *between* the different treatments
2. The spread of results *within* each treatment

The variability between treatments is compared with the variability within each treatment. This approach is based on the idea that if all treatments are merely samples from the same distribution, then the mean standard deviation within treatments should be approximately the same as the overall standard deviation between samples.

One-way Classification

Consider k independent random samples, each of size n, from k different populations. Let the jth observations in the ith sample be y_{ij}.

Mean

Sample 1:	y_{11} $y_{12} \cdots y_{1j} \cdots y_{1n}$		$\bar{y}_1$
Sample 2:	y_{21} $y_{22} \cdots y_{2j} \cdots y_{2n}$		$\bar{y}_2$
Sample i:	y_{i1} $y_{i2} \cdots y_{ij} \cdots y_{in}$		$\bar{y}_i$
Sample k:	y_{ki} $y_{k2} \cdots y_{kj} \cdots y_{kn}$		$\bar{y}_k$

$$\text{Grand mean} = \bar{y}_{..} = \frac{\Sigma\, y_{ij}}{kn}$$

To test the null hypothesis that the k sample means are equal, we compare two estimates of variance:

1. Variance based on variation *between* sample means σ_b^2
2. Variance based on variation *within* the samples σ_w^2

The estimate of the variance of one sample is given by

$$s_i^2 = \sum_{j=1}^{n} \frac{(y_{ij} - \bar{y}_i)^2}{n-1}$$

The variance of k sample *means* is given by

$$s_{\bar{x}}^2 = \sum_{i=1}^{k} \frac{(\bar{y}_i - \bar{y}_.)^2}{k-1}$$

This is an estimate of σ^2/n; therefore,

$$\sigma_b^2 = n s_{\bar{x}}^2 = n \sum_{i=1}^{k} \frac{(\bar{y}_i - \bar{y}_.)^2}{k-1}$$

For k samples, the variance within samples is given by

$$\sigma_w^2 = \sum_{i=1}^{k} \frac{s_i^2}{k} \quad \text{i.e., the mean of the sample variances}$$

or

$$\sigma_w^2 = \sum_{i=1}^{k} \sum_{j=1}^{n} \frac{(y_{ij} - \bar{y}_i)^2}{k(n-1)}$$

We test the null hypothesis (equal sample means) with the F statistic $F = \sigma_b^2/\sigma_w^2$ and reject the null hypothesis if $F > F_{1-\alpha}$ with $k-1$ and $k(n-1)$ degrees of freedom.

The above equations contain terms consisting of the sum of the squares of the difference between the individual observation and a mean value. Thus, the equations of the analysis of variances often are expressed in terms of "sum of the squares."

$$
\begin{array}{ccc}
\text{Total sum} & \text{error sum} & \text{treatment sum} \\
\text{of squares (SST)} = & \text{of squares (SSE)} + & \text{of squares [SS(Tr)]}
\end{array}
$$

$$
\sum_{i=1}^{k}\sum_{j=1}^{n}(y_{ij}-\bar{y}.)^2 = \sum_{i=1}^{k}\sum_{j=1}^{n}(y_{ij}-\bar{y}_i)^2 + n\sum_{i=1}^{k}(\bar{y}_i-\bar{y}.)^2
$$

The error sum of squares SSE is an estimate of the random or chance error and is related to σ_w^2. The treatment sum of squares SS(Tr) is related to σ_b^2. Thus, we can write the F statistic as

$$
F = \frac{\dfrac{\text{SS(Tr)}}{k-1}}{\dfrac{\text{SSE}}{k(n-1)}}
$$

For computation, the following shortcut formulas often are used

$$
\text{SST} = \sum_{i=1}^{k}\sum_{j=1}^{n}y_{ij}^2 - C
$$

$$
\text{SS(Tr)} = \frac{\sum_{i=1}^{k}T_i^2}{n} - C
$$

$$
\text{SSE} = \text{SST} - \text{SS(Tr)}
$$

where $C = \dfrac{T_.^2}{kn}$

$T_.$ = grand total of kn observations

T_i = total of n observations in the ith sample

Two-way Classification

Consider $1, 2, \ldots, a$ treatments divided between $1, 2, \ldots, b$ blocks. A block is a portion of an experimental design that is expected to be more homogeneous than the total design space. Greater precision is obtained by confining treatment comparisons to within blocks.

	B_1	B_2	$\cdots$	B_j	$\cdots$	B_b	Mean
Treatment 1:	y_{11}	y_{12}	$\cdots$	y_{1j}	$\cdots$	y_{1b}	$\bar{y}_{1.}$
Treatment 2:	y_{21}	y_{22}	$\cdots$	y_{2j}	$\cdots$	y_{2b}	$\bar{y}_{2.}$
Treatment i:	y_{i1}	y_{i2}	$\cdots$	y_{ij}	$\cdots$	y_{ib}	$\bar{y}_{i.}$
Treatment a:	y_{a1}	y_{a2}	$\cdots$	y_{aj}	$\cdots$	y_{ab}	$\bar{y}_{a.}$
Mean:	$\bar{y}_{.1}$	$\bar{y}_{.2}$		$\bar{y}_{.j}$		$\bar{y}_{.b}$	$\bar{y}_{..}$

We note that a dot . used instead of a subscript indicates that the mean was obtained by summing over that subscript.

The sum of the squares is given by:

$$\overset{\text{SST}}{\sum_{i=1}^{a}\sum_{j=1}^{b}(y_{ij}-\bar{y}..)^2} = \overset{\text{SSE}}{\sum_{i=1}^{a}\sum_{j=1}^{b}(y_{ij}-\bar{y}_{i.}-\bar{y}_{.j}+\bar{y}..)^2} + \overset{\text{SS(Tr)}}{b\sum_{i=1}^{a}(y_{i.}-\bar{y}..)^2}$$

$$+ a\overset{\text{SS}(Bl)}{\sum_{j=1}^{b}(y_{.j}-\bar{y}..)^2}$$

For computational purposes we can use

$$\text{SST} = \sum_{i=1}^{a}\sum_{j=1}^{b}y_{ij}^2 - C$$

where

$$C = \frac{T_{..}^2}{ab}$$

$$\text{SS(Tr)} = \frac{\sum_{i=1}^{a}T_{i.}^2}{b} - C$$

$$\text{SS}(Bl) = \frac{\sum_{j=1}^{b}T_{.j}^2}{a} - C$$

$$\text{SSE} = \text{SST} - \text{SS(Tr)} - \text{SS}(Bl)$$

Example 11-9. A sensitive measure of the quality of a steel forging is the reduction of area of a tensile specimen that is cut from a direction normal to the forging fiber, i.e., the transverse reduction of area, abbreviated RAT. Table 11-5 gives the values of RAT that were determined from forgings with 10:1 and 30:1 reduction from the ingot, with and without a homogenization heat treatment. We are interested in answering the following questions.

1. Is there a significant difference in RAT for 10 to 1 and 30 to 1 forging reductions?

TABLE 11-5

Effect of forging reduction and homogenization treatment on transverse reduction of area (RAT)

Forging reduction	Homogenization		
	None	2400°F, 50 h	
	25.8	26.4	
	30.4	35.6	
	28.6	30.2	
10:1	35.6	21.4	
	20.3	27.2	
	$T = 140.7$	$T = 140.8$	$T_{i.} = 281.5$
	22.8	30.6	
	19.5	35.3	
	30.6	27.9	
30:1	28.5	28.2	
	25.5	31.5	
	$T = 126.9$	$T = 153.5$	$T_{i.} = 280.4$
	$T_{.j} = 267.8$	$T_{.j} = 294.3$	$T.. = 561.9$

2. Is there a significant difference in RAT between two forgings one of which was given homogenization treatment and the other was not?

3. Is there a significant interaction[1] between forging reduction and homogenization that affects the level of RAT!

The experiment outlined in Table 11-5 is a two-way classification with two columns and two rows and with five replicate measurements of RAT for each combination of forging reduction and homogenization condition. Therefore, $a = 2$, $b = 2$, $r = 5$, and $n = 2 \times 2 \times 5 = 20$.

$$\text{SST} = \sum y_{ij}^2 - \frac{T_{..}^2}{abr} = 16,208.67 - \frac{561.9^2}{2(2)(5)} = 442.09$$

The treatment sum of squares (between rows) is

$$\text{SS(Tr)} = \sum \frac{T_{i.}^2}{br} - \frac{T_{..}^2}{abr} = \frac{281.5^2 + 280.4^2}{2(5)} - \frac{561.9^2}{2(2)(5)} = 0.06$$

The blocks sum of squares (between columns) is

$$\text{SS}(Bl) = \sum \frac{T_{.j}^2}{ar} - \frac{T_{..}^2}{abr} = \frac{267.8^2 + 294.3^2}{2(5)} - \frac{561.9^2}{2(2)(5)} = 35.65$$

---

[1] An interaction is present if the observed values for different levels of one factor are altered by the presence of another factor.

TABLE 11-6
Analysis of variance table

Source of variability	Sum of squares	Degrees of freedom	Mean square
Forging reduction (between rows)	0.06	1	0.06
Homogenization (between columns)	35.65	1	35.65
Interaction	35.11	1	35.15
Within cells (error)	351.27	16	21.92
Total		19	

By introducing replication, we are able to separate out the variation due to interaction effects from that due to random error. The sum of the squares due to random error within each test condition (within cells) is determined by first determining the subtotal sum of squares (SSS).

$$SSS = \sum \frac{T^2}{r} - \frac{T_{..}^2}{abr} = \frac{140.7^2 + 140.8^2 + 126.9^2 + 153.5^2}{5} - \frac{561.9^2}{2(2)(5)} = 70.82$$

Then, the within-cells sum of squares is

$$SSWC = SST - SSS = 422.09 - 70.82 = 351.27$$

Finally, the interaction sum of squares is given by

$$SSI = SSS - SS(Tr) - SS(Bl) = 70.82 - 0.06 - 35.65 = 35.11$$

Once these computations are made, the results are entered into an analysis of variance table (ANOVA table), Table 11-6.

The values of mean square are obtained by dividing the values of sum of the squares by their respective degrees of freedom. We use the mean squares in the F test to determine whether there are significant differences. For example, to determine whether there is a significant effect of homogenization on the level of RAT, the following F ratio is determined.

$$F = \frac{\text{between columns mean square}}{\text{within cells (error) mean square}} = \frac{35.65}{21.92} = 1.63$$

$$F_{0.05}(1, 16) = 4.45$$

Since the calculated F ratio does not exceed 4.45, the effect of homogenization is not significant at the 95 percent level.

To determine whether there is a significant interaction between forging reduction and homogenization treatment, the following F ratio is determined.

$$F = \frac{\text{interaction mean square}}{\text{error mean square}} = \frac{35.11}{21.92} = 1.61$$

Once again the effect is not significant at the 5% level.

11-12 STATISTICAL DESIGN OF EXPERIMENTS

The greatest benefit can be gained from statistical analysis when the experiments are planned in advance so that data are taken in a way that will provide the most unbiased and precise results commensurate with the desired expenditure of time and money. This can best be done through the combined efforts of a statistician and the engineer during the planning stage of the project.

Probably the most important benefit from statistically designed experiments is that more information per experiment will be obtained that way than with unplanned experimentation. A second benefit is that statistical design results in an organized approach to the collection and analysis of information. Conclusions from a statistically designed experiment very often are evident without extensive statistical analysis, whereas with a haphazard approach the results often are difficult to extract from the experiment even after detailed statistical analysis. Still another advantage of statistical planning is the credibility that is given to the conclusions of an experimental program when the variability and sources of experimental error are made clear by statistical analysis. Finally, an important benefit of statistical design is the ability to discover interactions between experimental variables.

A major impetus to the use of experimental design has been given by the strong emphasis being shown in design for quality. The techniques developed in Japan by Genichi Taguchi employ experimental design methods in a unique way to produce *robust designs*. This chapter sets the statistical basis for considering this and other concepts of quality engineering in Chap. 13.

In any experimental program involving a large number of tests, it is important to randomize the order in which the specimens are selected for testing. By randomization we permit any one of the many specimens involved in the experiment to have an equal chance of being selected for a given test. In this way, bias due to uncontrolled second-order variables is minimized. For example, in any extended testing program errors can arise over a period of time owing to subtle changes in the characteristics of the testing equipment or in the proficiency of the operator of the test. In taking metal specimens from large forgings or ingots the possibility of the variation of properties with position in the forging must be considered. If the objective of the test is to measure the average properties of the entire forging, randomization of the test specimens will minimize variability due to position in the forging.

One way to randomize a batch of specimens is to assign a number to each specimen, put a set of numbered tags corresponding to the specimen numbers in a jar, mix the tags thoroughly, and then withdraw the tags from the jar. Each tag should be placed back in the jar after it is withdrawn to allow an equal probability of selecting that number. The considerable labor involved in this procedure can be minimized by using a table of random numbers, as discussed in Sec. 4-4.

We have previously considered the various stages in the design process. In the same way, any experimental investigation starts with preliminary familiarization experiments. These may take the form of trying to duplicate earlier results or learning to operate new equipment. Much of the work at this stage is mainly intuitive and aimed at defining the problem; generally, statistical design is not very applicable. However, after the goals of the investigation are better defined, the next step is to reduce the large number of possible variables to the few most important ones. Here statistical design can be very useful. As the experimentation moves into the optimization stage, the number of variables has been reduced to only a few. Once again statistical design can be effective in finding the optimum values for the variables. Finally, statistical analysis can assist in choosing between possible models for the process or phenomenon under study.

The response variables are the data we obtain from an experimental run. Responses can be classified as quantitative, qualitative, or quantal. A quantitative response, which is measured by a continuous scale, is the most common and easiest to work with in statistical analysis. Qualitative responses, like luster or odor, can be ranked on an ordinal scale, for example, 0 (worst) to 10 (best). Quantal or binary responses produce one of two values, go or no-go, pass or fail. An example is fatigue specimens tested near the fatigue limit that either fail or run out. The case of the quantal response is treated by special statistical techniques.[1]

Factors are experimental variables that are controlled by the investigator. An important part of planning an experimental program is identifying the important variables that affect the response and deciding how to exploit them in the experiment. This is best when jointly done by the engineer and the statistician. However, identifying the variables, and assuring that no important factors are forgotten, is the responsibility of the engineer. Frequently the scientific model of the problem is examined for important variables. Previous experience should always be called upon. We often take advantage of dimensional analysis in establishing the factors.

Factors may be *independent* in the sense that the level of one factor is independent of the level of the other factors. However, two or more factors may *interact* with one another, i.e., the effect on the response of one variable depends upon the levels of the other variables. Figure 11-12 illustrates different types of behavior between the factors and the response. In this case the response y is the yield strength of an alloy as it is influenced by two factors, temperature x_1 and aging time x_2. Interactions between the factors are determined by varying factors simultaneously under statistical control rather than one at a time.

---

[1] D. J. Finney, "Probit Analysis," 3d ed., Cambridge University Press, New York, 1971; Metals Handbook, 9th ed., vol. 8, pp. 695–720, ASM International, Metals Park, OH, 1985.

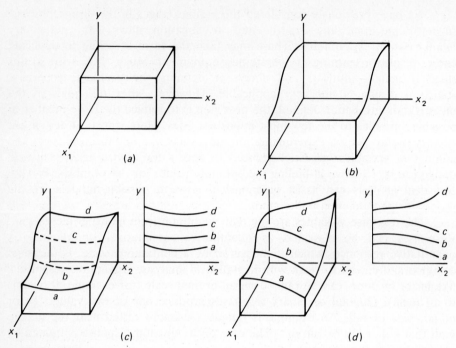

FIGURE 11-12

Different behavior of response y as a function of the factors x_1 and x_2, (a) No effect of x_1 and x_2 on y. (b) Main effect of x_1 on y. No effect of x_2 on y. (c) Effect of x_1 and x_2 on y but no $x_1 - x_2$ interaction; (d) Main effects of x_1 and x_2. Interaction between x_1 and x_2.

In addition to the primary variables that are under the control of the experimenter, there are other variables, which may not be under strict experimental control. An example would be subtle differences in the way different machine operators run an experiment or carry out a test or slight differences in humidity or other environmental factors. The use of randomization of the test runs so as to remove unconscious bias in results has been discussed above. Also, the use of an experimental block to produce relatively homogeneous test conditions has been mentioned in Sec. 11-9. When an experiment is run with blocking, the effect of a background variable is removed from the experimental error, but randomization alone does not usually produce this effect.

It is important to realize that not all primary factors may be capable of variation with equal facility. Thus, completely randomizing the sequence of testing may be impractical. Often the final experimental plan is a compromise between the information that can be obtained and the cost of the information. In developing the experimental design, physical reality should take precedence over slavish adherence to statistics. Whenever faced with this decision, the experimental design should be modified to accommodate the real-world situation, not the reverse.

It is important to make some initial estimates of the overall repeatability before embarking on a major experimental test program. Often this information is available from previous experiments; but if it is not, it would be wise to conduct some preliminary experiments under identical conditions over a reasonable time period. If these experiments show large variability in response, then the primary variables may not have been properly identified.

It is not necessary to conduct a statistically designed experiment all the way through to completion. In fact, there are advantages to conducting the experimental program in stages. That permits changes to be made in later tests based on the information gained from early results. Conducting the experiment in stages is attractive when the experimenter is searching for an optimum response, since it takes investigation closer to the optimum at each stage. On the other hand, if there are large start-up costs for each stage or if there is a long time delay between preparing the samples and measuring their performance (as in agricultural research), then carrying the designed experiment straight through to completion may be preferred.

In general, there are three classes of statistically designed experiments.

1. Blocking designs use blocking techniques to remove the effect of background variables from the experimental error. The most common designs are the randomized block plan and the balanced incomplete block,[1] which remove the effect of a single extraneous variable. The Latin square and the Youden square designs[2] remove the effects of two extraneous variables. Graeco-Latin square and hyper-Latin square designs remove the effects of three or more extraneous variables.

2. Factorial designs are experiments in which all levels of each factor in an experiment are combined with all levels of every other factor. This important class of experimental design is discussed in Sec. 11-13.

3. Response surface designs are used to determine the empirical functional relation between the factors (independent variables) and the response (performance variable). The central composite design and rotatable designs are frequently used for this purpose. This is discussed in Sec. 11-15.

We emphasize that a working partnership between the engineer and the statistician leads to the best results. Frequently the best experimental design is specially made for the problem and deviates significantly from the standard textbook designs. By getting involved in the project at an early stage, the

---

[1] M. G. Natrella, op. cit., pp. 13-1 to 13-29.

[2] Ibid., pp. 13–30 to 13–46.

statistician can help evolve an experimental design to minimize testing effort and maximize information content.

11-13 FACTORIAL DESIGN

A factorial experiment is one in which we control several factors and investigate their effects at each of two or more levels. The experimental design consists of making an observation at each of all possible combinations that can be formed for the different levels of the factors. Each different combination is called a treatment combination. The approach in a factorial experiment is much different than in the traditional experiment, in which all factors but one are held constant.

The simplest, and most common type of factorial design is one that uses two levels, i.e., a 2^n factorial design. Not only does it reduce the number of experimental conditions but convenient computational methods exist for the 2^n case. A disadvantage is that with only two levels it is not possible to distinguish between linear and higher-order effects.

In a 2^n experimental design, factors that are set at the low level are indicated $(-)$ and those at the high level $(+)$. The particular combination of treatments can be easily determined by using the special Yates notation. Each factor is represented by a lowercase letter arranged in a standard order. When a lowercase letter representing a factor is missing it indicates that the factor is at a lower level. Otherwise the factor is at the higher level. Table 11-7 illustrates this for a three-factor, 2^3, design.

The main effect of a factor is the change in response produced by a change in level of the factor.

$$\text{Main effect of } x_i = \frac{\Sigma \text{ responses at high } x_i - \Sigma \text{ responses at low } x_i}{\text{half the number of runs in expt.}}$$

TABLE 11-7
Three-factor, 2^3, a factorial design

Yeates std. order	Run no.	Level of factor		
		x_A	x_B	x_C
1	1	−	−	−
a	2	+	−	−
b	3	−	+	−
ab*	4	+	+	−
c	5	−	−	+
ac*	6	+	−	+
bc*	7	−	+	+
abc†	8	+	+	+

* First-order interaction.
† Second-order interaction.

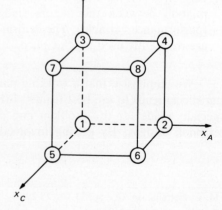

TABLE 11-8
Interactions for 2^3 factorial design

Yates std. order	Level of factor						
	x_A	x_B	x_C	$x_A x_B$	$x_A x_C$	$x_B x_C$	x_{ABC}
(1)	−	−	−	+	+	+	−
a	+	−	−	−	−	+	+
b	−	+	−	−	+	−	+
ab	+	+	−	+	−	−	−
c	−	−	+	+	−	−	+
ac	+	−	+	−	+	−	−
bc	−	+	+	−	−	+	−
abc	+	+	+	+	+	+	+

Thus, the main effect of factor x_B is

$$\text{Main effect } x_B = \frac{(y_b + y_{ab} + y_{bc} + y_{abc}) - (y_1 + y_a + y_c + y_{ac})}{4}$$

To find the interaction effects, extend Table 11-7 as shown in Table 11-8. Then, the effect for the BC interaction is written as:

$$BC \text{ interaction} = \frac{(y_1 + y_a + y_{bc} + y_{abc}) - (y_b + y_{ab} + y_c + y_{bc})}{4}$$

The main effects and interactions for a 2^3 factorial design are given by the following equations, where $a = y_a$, $b = y_b$, etc.

$$x_A = \tfrac{1}{4}(a-1)(b+1)(c+1)$$
$$x_B = \tfrac{1}{4}(a+1)(b-1)(c+1)$$
$$x_C = \tfrac{1}{4}(a+1)(b+1)(c-1)$$
$$x_{AB} = \tfrac{1}{4}(a-1)(b-1)(c+1) \qquad (11\text{-}46)$$
$$x_{AC} = \tfrac{1}{4}(a-1)(b+1)(c-1)$$
$$x_{BC} = \tfrac{1}{4}(a+1)(b-1)(c-1)$$
$$x_{ABC} = \tfrac{1}{4}(a-1)(b-1)(c-1)$$

These equations can be generalized for a 2^n factorial design with $A, B, C, \ldots, Q$ factors as,

$$x_A = \tfrac{1}{2}^{n-1}(a-1)(b+1)(c+1)\cdots(q+1)$$
$$x_{AB} = \tfrac{1}{2}^{n-1}(a-1)(b-1)(c+1)\cdots(q+1)$$
$$x_{ABC\cdots Q} = \tfrac{1}{2}^{n-1}(a-1)(b-1)(c-1)\cdots(q-1)$$

Example 11-10. The strength of steel oil well drill pipe is controlled by the austenitization temperature A, the temperature at which it enters the tube mill B, and the amount of deformation produced in the mill C. A 2^3 factorial design is developed. The tensile strength of the steel tube, in 1000 psi, is the measured response y.

Treatment	Response UTS (y)	x_A	x_B	x_C	x_{AB}	x_{AC}	x_{BC}	x_{ABC}
(1)	151	−	−	−	+	+	+	−
a	147	+	−	−	−	−	+	+
b	147	−	+	−	−	+	−	+
ab	101	+	+	−	+	−	−	−
c	182	−	−	+	+	−	−	+
ac	159	+	−	+	−	+	−	−
bc	173	−	+	+	−	−	+	−
abc	158	+	+	+	+	+	+	+

The main effect of factor A

$$A = \frac{(147 + 101 + 59 + 158) - (151 + 147 + 182 + 173)}{4}$$

$$= 141 - 163 = -22$$

The minus sign signifies that the lower level of austenization temperature (factor A) resulted in higher values of the response.

We can compare the mean value of factor A at its high level with the mean at its low level by using the paired t test, Table 11-3, Entry 8.

Low A	High A	Difference	Deviation from $\bar{d}$
151	147	4	−18
147	101	46	24
182	159	23	1
173	158	15	−7
		$\bar{d} = \dfrac{88}{4} = 22$	

$$s^2 = \frac{\sum \text{deviation}^2}{n-1} = \frac{-18^2 + 24^2 + 1^2 - 7^2}{3} = \frac{950}{3} = 316.67$$

$$s = 17.79$$

$$H_0: \mu = 0 \quad \text{for } \bar{d} = \bar{x}_L - \bar{x}_H \quad t = \frac{\bar{d}}{s/\sqrt{n}} = \frac{22}{17.79/\sqrt{4}} = 2.44$$

$$H_1: \mu \neq 0 \qquad\qquad\qquad v = 4 - 1 = 3$$

For the two-sided t distribution (Appendix C) $t_{0.10} = 2.35$ and $t_{0.05} = 3.18$. Thus, the null hypothesis is rejected at a 10 percent level of significance.

To increase the precision of the experiment, we perform duplicate observations for each treatment combination. This is called replicating the experiment. For each of the eight treatments there are duplicate observations of the response, y_1 and y_2. The standard deviation of these duplicates about their own mean is $s = [(y_1 - y_2)/2]^{1/2}$. These individual deviations are combined into a pooled standard deviation s_p.

$$s_p = \left[\frac{\dfrac{(y_1 - y_2)_1^2}{2} + \dfrac{(y_1 - y_2)_2^2}{2} + \dfrac{(y_1 - y_2)_3^2}{2} + \cdots + \dfrac{(y_1 - y_2)_8^2}{2}}{2 + 2 + 2 + \cdots + 2} \right]^{1/2}$$

and in general, for n duplicates

$$s_p = \left[\frac{\Sigma (y_1 - y_2)^2}{2n} \right]^{1/2} \tag{11-47}$$

When duplicate observations are made for the 2^3 experiment (data not given), $x_p = 4.8$. The mean difference between the low and high levels of factor A has changed slightly from $\bar{x}_L - \bar{x}_H = 22$ to 20. We observe a mean difference of 20 ksi based on two groups of tests each with eight observations. The standard error of the mean difference is given by

$$s_{\bar{x} - \bar{x}} = \left(\frac{4.8^2}{8} + \frac{4.8^2}{8} \right)^{1/2} = 2.4$$

Now, applying the null hypothesis for paired observations,

$$t = \frac{\bar{d}}{s_{\bar{x} - \bar{x}}} = \frac{20}{2.4} = 8.35 \qquad v = 8 - 1 = 7$$

$t_{0.01} = 3.50$, so the null hypothesis is rejected with a very high (0.999) level of confidence. The main effect of austenitization temperature is shown to be highly significant. This illustrates how replication greatly increases the precision of the statistical analysis.

Let us now look at the interaction between austenitization temperature A and the temperature of working in the tube mill B. The AB interaction is given by

$$AB \text{ interaction} = \frac{(151 + 101 + 182 + 158) - (147 + 147 + 159 + 173)}{4}$$

$$= \frac{592 - 626}{4} = \frac{-34}{4} = -8.5$$

Applying the null hypothesis with the paired t test

$$t = \frac{\bar{d}}{s_{\bar{x} - \bar{x}}} = \frac{-8.5 - 0}{2.4} = 3.54 \qquad v = 7$$

$$t_{0.01} = 3.50$$

so the AB interaction is highly statistically significant.

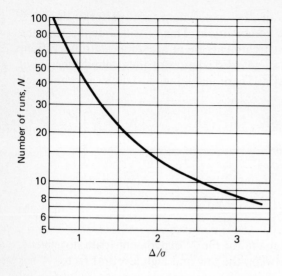

FIGURE 11-13
Number of experimental runs required
to achieve a given level of precision.

Size of the Experiment

In a 2^n design all of the treatment combinations are used in each estimation of
main effect and interaction. However, it is important to have some way to
estimate the total number of runs N required to achieve a desired level of
precision.[1]

We first need to decide on the minimum change in the response that will
be of practical interest Δ. Then σ is our best estimate of the standard
deviation. The relation between N and Δ/σ for Δ to be declared significant
with a high degree of assurance is given in Fig. 11-13. We see that for a given
Δ, if σ is large, the number of runs is correspondingly large. If N is larger than
the number of runs that will be provided by an experimental design,
replication of the design is required. This was illustrated by Example 11-10.
Although its primary purpose is to increase the precision of the estimate,
replication works in another way to give a more precise estimate of σ.

Fractional Factorial Designs

The number of treatment combinations in a factorial design increases rapidly
with an increase in the number of factors. Thus, if $n = 4$, $2^n = 16$, but if $n = 8$,
$2^n = 256$. Since engineering experimentation can easily involve 6 to 10 factors,
the number of experiments required can rapidly become prohibitive in cost.

---

[1] T. D. Murphy, Jr., *Chem. Eng.*, pp. 168–182, June 6, 1977.

TABLE 11-9
Fractional factorial design for 2^3 experiment

	Treatment combinations	Effects						
		x_A	x_B	x_C	x_{AB}	x_{AC}	x_{BC}	x_{ABC}
Block I	(1)	−	−	−	+	+	+	−
	ab	+	+	−	+	−	−	−
	ac	+	−	+	−	+	−	−
	bc	−	+	+	−	−	+	−
Block II	a	+	−	−	−	−	+	+
	b	−	+	−	−	+	−	+
	c	−	−	+	+	−	−	+
	abc	+	+	+	+	+	+	+

Consider a 2^6 factorial experiment. There are:

6 main effects	n
15 first-order interactions	$n(n-1)/2$
20 second-order interactions	$n(n-1)(n-2)/2 \times 3$
15 third-order interactions	
6 fourth-order interactions	
1 fifth-order interaction	

Since we usually are not interested in higher-order interactions, we are collecting a great deal of extraneous information if we perform all 64 treatment combinations. This permits us to design an experiment that is some fraction of the total factorial design.

As a simple example, we consider a one-half replicate of a 2^3 factorial design.

$$N = r(2^n) = \tfrac{1}{2}(2^3) = \tfrac{8}{2} = 4$$

We split the treatment combinations into two blocks, Table 11-9. The main effect of factor A in block I is:

$$x_A = (ab + ac) - [(1) + bc]$$

Also, in block I the BC interaction is:

$$x_{BC} = [(1) + bc] - (ab + ac)$$

Therefore, we see that $x_A = -x_{BC}$, and we can also show that $x_B = -x_{AC}$ and $x_C = -x_{AB}$. We say that BC, AC, and AB are *aliased* with A, B, and C, respectively. Therefore, by working with just one-half of the total factorial experiment, i.e., block I, we can determine the main effects for A, B, C if we assume there are no interactions.

The half replicate of the 2^3 factorial is the smallest practical fractional design. Fractional factorial designs usually are used when four or more factors must be considered. For example, in the half replicate of a 2^5 factorial design no first-order interactions are aliased with each other. Sometimes one-quarter replicates are used. A great many fractional factorial designs are given in Natrella.[1] The use of a fractional factorial design will be illustrated in Sec. 11-15.

11-14 REGRESSION ANALYSIS

It is a frequent experience in engineering design and experimentation to be interested in determining whether there is a relation between two or more variables. Regression analysis is the statistical technique for establishing such relations. There are two different but related situations. The first determines a functional relation between variables when one or more variables are set at fixed values (independent variables) and the response (dependent variable) is established; here the emphasis is on *prediction*. In the second situation two or more variables are assumed to be *associated* with each other, i.e., there is a correlation between the variables, which vary jointly. Although the mathematical relations for regression and correlation analysis are identical, the concepts are related but distinct.

Method of Least Squares

The simplest case of regression analysis is a linear relation between an independent variable x and the dependent variable y. The mathematical model for the population is

$$y_i = \alpha + \beta x_i + \varepsilon_i \tag{11-48}$$

in which y_i and x_i are the ith observations of the dependent and independent variables, respectively. α and β are the population values of the intercept and slope regression coefficients. ε_i is the error.

The sample equation is

$$\hat{y} = a + bx \tag{11-49}$$

in which $\hat{y}$ is the predicted value of the dependent variable and a and b are the sample estimates of the regression coefficients. The assumptions that underlie the method of least squares are:

1. The errors are normally distributed with a mean of zero and a constant variance (σ_e^2).
2. The errors are independent of each other.

---

[1] M. G. Natrella, op. cit., pp. 12–14 to 12–21.

The method of least squares establishes the line through the data such that the sum of the squares of the vertical deviations of observations from the line is smaller than for any other line that could be drawn through the data.

$$\varepsilon_i = a + bx_i - y_i$$

and

$$\sum \varepsilon_i^2 = \sum (a + bx_i - y_i)^2 \tag{11-50}$$

We want to establish the values of the parameters a and b that result in the least value of $\sum \varepsilon_i^2$. This is done by taking the partial derivatives of Eq. (11.50) with respect to a and b and setting the resulting equations equal to zero.

$$\frac{\partial \varepsilon_i^2}{\partial a} = 2 \sum_{i=1}^{n} (a + bx_i - y_i) = 0$$

$$\frac{\partial \varepsilon_i^2}{\partial b} = 2 \sum_{i=1}^{n} (a + bx_i - y_i)x_i = 0 \tag{11-51}$$

If the terms within the patentheses are separated, the two derivatives can be rearranged to provide two simultaneous equations, which are called the normal equations. Note that all summations extend over the observations in the sample of size n.

$$\sum y_i = an + b \sum x_i$$
$$\sum x_i y_i = a \sum x_i + b \sum x_i^2 \tag{11-52}$$

When the normal equations are solved simultaneously, we get the equations for the parameters a and b.

$$b = \frac{n \sum xy - \sum x \sum y}{n \sum x^2 - (\sum x)^2} = \frac{\sum (x - \bar{x})(y - \bar{y})}{\sum (x - \bar{x})^2} \tag{11-53}$$

$$a = \frac{\sum y - b \sum x}{n} = \bar{y} - b\bar{x} \tag{11-54}$$

The sample *correlation coefficient* is given by

$$r = \frac{n \sum xy - \sum x \sum y}{\sqrt{[n \sum x^2 - (\sum x)^2][n \sum y^2 - (\sum y)^2]}} \tag{11-55}$$

The value of r vary between -1 and $+1$. When $r = \pm 1$, the experimental points fit the linear model $y = a + bx$ perfectly. When r approaches zero, there is very low correlation between the data, and the linear model does not fit the data. The procedure for establishing the confidence interval for the regression line is given by Natrella.[1]

---

[1] M. G. Natrella, op. cit., pp. 5-36 to 5-46.

Linear Multiple Regression Analysis

The engineering design situation usually involves more than one independent variable. A frequently used model has the form:

$$y = b_0 + b_1x_1 + b_2x_2 + \cdots + b_mx_m \qquad (11\text{-}56)$$

The bivariate model of Eq. (11-48) is a special case of Eq. (11-56) in which $m = 1$. To solve Eq. (11-56), the same procedure as for when $m = 1$ is used; but matrix methods and digital computer techniques are used to solve the $m + 1$ linear equations.[1]

An example of a linear multiple regression equation without interaction effects is

$$YS(1000\,\text{psi}) = 13.29 + 5.90\,\text{Mn} + 10.21\,\text{Si} + 0.220\,(\%\ \text{pearlite}) + 0.476d^{-1/2}$$

This equation predicts the yield strength of hot rolled ferritic–pearlitic steel based on chemical composition, volume fraction of pearlite, and the grain size. The composition range was 0 to 0.3 wt% carbon, 0 to 1.60% manganese, and 0 to 0.8% silicon. The term d is the mean linear ferrite intercept grain size, in inches. The correlation coefficient is 0.89 and the 95 percent confidence limits for prediction of yield strength is ± 3800 psi.

Nonlinear Regression Analysis

We are well aware that nonlinearity is encountered in many engineering problems. A second-order model with two independent variables is given by:

$$y = b_0 + b_1x_1 + b_2x_2 + b_{11}x_1^2 + b_{12}x_1x_2 + b_{22}x_2^2 \qquad (11\text{-}57)$$

Such a model is needed to detect nonlinearity and second-order effects. Standardized computer programs are usually used to determine the parameters of this type of equation.[2]

The reader is cautioned against indiscriminate use of "canned" statistical programs in computer-based data reduction. There is no question that the availability of these programs in almost every computer center, together with the proliferation of home computers and programmable hand calculators, has increased the use of regression analysis, but it is important to understand what the statistical package really does. The user should take the time and trouble to read the backup documentation for the program and, in the event of failing to understand everything, seek the advice of a statistical consultant.

---

[1] C. Daniel and F. S. Wood, "Computer Analysis of Multifactor Data," 2d ed., John Wiley & Sons, Inc., New York, 1980; N. R. Draper and H. Smith, "Applied Regression Analysis," 2d ed., ibid., 1981.

[2] C. Daniel and F. S. Wood, op. cit.

TABLE 11-10
Some common linearization transformation, $W = c + dV$

Nonlinear equation	Linearized equation	Linearized variables	
		W	**V**
1. $y = a + bx$ (linear)	$y = a + bx$	y	x
2. $y = ax^b$ (logarithmic)	$\ln y = \ln a + b \ln x$	$\ln y$	$\ln x$
3. $y = ae^{bx}$ (exponential)	$\ln y = \ln a + bx$	$\ln y$	x
4. $y = 1 - e^{-bx}$ (exponential)	$\ln \dfrac{1}{1-y} = bx$	$\ln \dfrac{1}{1-y}$	x
5. $y = a + b\sqrt{x}$ (square root)	$y = a + bx$	y	$\sqrt{x}$
6. $y = a + \dfrac{b}{x}$ (inverse)	$y = a + bx$	y	$\dfrac{1}{x}$
7. $y = 1 - \exp - \left(\dfrac{x - x_0}{\theta - x_0}\right)^m$ (Weibull)	$\ln \ln \dfrac{1}{1-y} =$ $-m \ln (\theta - x_0)$ $+ m(x - x_0)$	$\ln \ln y$	$\ln(x - x_0)$

Linearization Transformations

The computational difficulties associated with nonlinear regression analysis sometimes can be avoided by using simple transformations that convert a problem that is nonlinear into one that can be handled by simple linear regression analysis. The most common transformations are given in Table 11-10. The reader needs to be aware that logarithmic transformation can introduce bias in the prediction of the response variable.[1]

11-15 RESPONSE SURFACE METHODOLOGY

In the preceding section we saw that complex engineering problems can be analyzed to give a regression equation that describes either a linear or a nonlinear model. The equations give the response in terms of the several independent variables of the problem. If the response is plotted as a function of x_1, x_2, etc., we obtain a response surface. A powerful statistical procedure

---

[1] R. W. McCuen, R. B. Leahy, and P. A. Johnson, *J. Hydraulic Engr*, vol. 11b, March 1990, pp. 414–428.

that employs factorial analysis and regression analysis has been developed for the determination of the optimum operating condition on a response surface.[1]

Response surface methodology (RSM) has two objectives:

1. To determine with one experiment where to move in the next experiment so as to continually seek out the optimal point on the response surface.

2. To determine the equation of the response surface near the optimal point.

Response surface methodology (RSM) uses a two step procedure aimed at rapid movement from the current operating position into the central region of the optimum. This is followed by the characterization of the response surface in the vicinity of the optimum by a mathematical model. The basic tools used in RSM are two level factorial designs and the method of least squares model and its simpler polynomial forms.

Initially, a small factorial experiment is run over a small area of the response surface where the surface may be considered to be planar, point A, Fig. 11-14. This is modeled by Eq. (11-56). To move most efficiently toward the optimum we follow a path of steepest ascent, along the gradient vector of the surface. Experiments are conducted along this path until a maximum is reached. Then another factorial experiment is centered around that point and the new gradient vector is followed. When a maximum is reached on that path another small designed set of experiments is laid out. Since the surface is definitely not planar in this region it is modeled by Eq. (11-57). The factorial design is changed to a *central composite design* in which a point is added at the center of the cube (Table 11-7) and a "star point" is added along each of the axes. This will provide information for fitting a model for the response surface, determining the shape of the surface, and establishing the optimum values of the dependent variables (x).

Example 11-11. An experimental investigation was undertaken[2] to explore the possibility of developing a precipitation-hardened cobalt-nickel high-temperature alloy. Cobalt-base superalloys have traditionally been strengthened by solid-solution alloy additions, such as tungsten and tantalum, and by a dispersion of carbides. Little use has been made of the finely dispersed coherent γ', $Ni_3(Al, Ti)$, precipitates that serve to provide the high-temperature strength in nickel-base superalloys. The base starting composition was Co, 25 W, 0.5 Zr,

[1] G. E. P. Box and K. B. Wilson, *J. Roy. Stat. Soc.,* sec. B., vol. 13, pp. 1–45, 1951; J. S. Hunter, *Ind. Qual. Control,* December 1950 and February 1959; R. H. Meyers, "Response Surface Methodology," Allyn & Bacon, Inc., Boston, 1971; O. L. Davies and P. L. Goldsmith (eds.), "Statistical Methods in Research and Production," 4th ed., Longman Group Ltd., London, 1976; G. E. P. Box and N. R. Draper, "Empirical Model-Building and Response Surfaces," John Wiley and Sons, New York, 1987.

[2] G. D. Sandrock and A. G. Holms, NASA TN D-5587, December 1969.

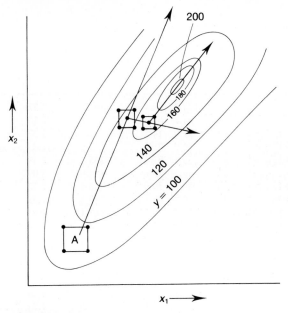

FIGURE 11-14
Concept of response surface methodology.

3.12 Cr, 0.6 C. The objective was to add Ni and enough Al and Ti to produce γ' strengthening.[1]

Solution. Before a statistically designed series of experiments was initiated, preliminary survey experiments were conducted to see whether significant strengthening due to γ' dispersion could be produced in an alloy with the above base composition. Several preliminary experiments showed that it could be, and the base composition was changed to 38 Co, 38 Ni, 14 W, 0.5 Ti, 0.25 Zr, 1.75 Cr, 0.35 C, 7 Al.

The high-temperature performance of a material depends on properties such as stress rupture life, high-temperature tensile strength, elongation to fracture, and oxidation resistance.[2] The stress rupture life as measured at 1850°F and a stress of 15,000 psi was chosen as the response variable. The independent variables selected were:

1. Ti content, because this element plays an active role in the formation of γ' and MC carbides.

---

[1] This project was conducted in a period of time when cobalt was in reasonably plentiful supply. Subsequent politico-economic changes have periodically increased the cost and scarcity of cobalt. Today, it is unlikely these experiments would have been conducted, because cobalt is a strategic material.

[2] G. E. Dieter, "Mechanical Metallurgy," 3d ed., chap. 13, McGraw-Hill Book Company, New York, 1986; C. T. Simms, N. S. Stoloff and W. C. Hagel (eds.), "Superalloys II," John Wiley & Sons, Inc., New York, 1987.

TABLE 11-11
Treatment combinations and responses

Alloy	x_{Ti}	x_{Cr}	x_C	x_{Al}	x_T	Stress rupture* life, t, h	t_{avg}	$y = \log t_{avg}$
						Levels		
1	−1	−1	−1	−1	−1	175.1, 199.4	187.3	2.27
2	+1	−1	−1	−1	+1	83.2, 166.5	124.8	2.10
3	−1	+1	−1	−1	+1	22.9, 24.5	23.7	1.37
4	+1	+1	−1	−1	−1	14.7, 21.1	17.9	1.25
5	−1	−1	+1	−1	+1	153.5, 237.6	195.5	2.29
6	+1	−1	+1	−1	−1	119.5, 129.6	124.6	2.09
7	−1	+1	+1	−1	−1	28.2, 39.0	33.6	1.52
8	+1	+1	+1	−1	+1	30.0, 38.1	34.1	1.53
9	−1	−1	−1	+1	+1	55.1, 79.2	67.1	1.83
10	+1	−1	−1	+1	−1	29.2, 47.0	38.1	1.58
11	−1	+1	−1	+1	−1	3.5, 11.1	7.3	0.86
12	+1	+1	−1	+1	+1	17.7, 19.6	18.6	1.27
13	−1	−1	+1	+1	−1	132.1, 190.7	161.4	2.21
14	+1	−1	+1	+1	+1	94.1, 95.1	94.6	1.97
15	−1	+1	+1	+1	+1	12.7, 19.1	15.9	1.20
16	+1	+1	+1	+1	−1	16.7, 16.8	16.8	1.22
								26.56

2. Cr content, generally detrimental to high-temperature strength yet essential for oxidation resistance.
3. C content, important in producing carbide strengthening.
4. Al content, the chief element influencing γ', $Ni_3(Al, Ti)$, formation.
5. Pour temperature. Test bars were prepared by casting, and this variable controls the cooling rate and resulting grain size.

All changes in composition were made at the expense of the cobalt content. The composition of the remaining elements was held constant at 38 Ni, 14 W, and 0.25 Zr.

With five factors, there are $2^5 = 32$ treatment combinations. Because we are willing to ignore any interactions at this stage, a one-half replicate of the 2^5 factorial experiment was used.[1] Table 11-11 shows the treatment combinations for the one-half replicate of the 2^5 factorial design. Duplicate values for the response (stress rupture life) are given. The design center for this experimental design is given in Table 11-12. We note that the experimental conditions were changed slightly from the base composition in the hope of getting more γ' strengthening from a higher Ti content and because it was recognized that more Cr was needed for oxidation resistance. These are examples of the types of engineering judgments that must be allowed for in a statistically designed plan. The level of factors

---

[1] M. G. Natrella, op. cit., table 12-4, pp. 12–16.

TABLE 11-12

Levels of factors in experimental design

Contents of Ni, W, and Zr held constant at 38, 14, and 0.25 wt percent, respectively; balance, Co

| Variable | Level, wt percent | | | |
	Design center	Lower level	Upper level	Scale factor
x_{Ti}	1.0	0.5	1.5	0.5
x_{Cr}	4.0	2.0	6.0	2.0
x_C	0.4	0.3	0.5	0.1
x_{Al}	7.0	6.75	7.25	0.25
x_T				
°F	2900	2850	2950	50
(°C)	(1593)	(1566)	(1621)	(28)

is converted from the natural units shown in Table 11-12 to the design units given in Table 11-11 by the following relations:

$$x_{Ti} = \frac{Ti - 1.0}{0.5} \qquad x_{Cr} = \frac{Cr - 4.0}{2.0} \qquad x_T = \frac{T - 2900}{50}$$

$$x_C = \frac{C - 0.4}{0.1} \qquad x_{Al} = \frac{Al - 7.0}{0.25}$$

We should emphasize that the 16 experimental heats listed in Table 11-11 should be made in random order and the 32 measurements of stress rupture life should also be made in random order.

The response model that these experiments represent is

$$y = b_0 + b_1 x_1 + b_2 x_2 + b_3 x_3 + b_4 x_4 + b_5 x_5 + b_{12} x_1 x_2 + b_{13} x_1 x_3$$
$$+ b_{14} x_1 x_4 + b_{15} x_1 x_5 + b_{23} x_2 x_3 + b_{24} x_2 x_4 + b_{25} x_2 x_5 + b_{34} x_3 x_4$$
$$+ b_{35} x_3 x_5 + b_{45} x_4 x_5$$

Since stress rupture life is known to be lognormally distributed, the response will be taken as $\log t_f$. The mean value of each of the duplicate observations will be used. The value of each coefficient in the regression equation is given by

$$b = \frac{\sum yx}{\sum x^2} \tag{11-58}$$

In this equation values of the independent variables are expressed in design units. For b_0, all values of x are at $+1$, so that, from Table 11-11,

$$b_0 = \frac{26.56}{16} = 1.66$$

and

$$b_1 = b_{Ti} = \tfrac{1}{16}(-2.27 + 2.10 - 1.37 + 1.25 - 2.29 + 2.09 - 1.52$$
$$+ 1.53 - 1.83 + 1.58 - 0.86 + 1.27 - 2.21 + 1.97 - 1.20 + 1.22)$$

$$b_1 = \frac{13.01 - 13.55}{16} = \frac{-0.54}{16} = -0.034$$

The resulting regression equation is

$$y = 1.66 - 0.034x_{Ti} - 0.383x_{Cr} + 0.100x_C - 0.146x_{Al} + 0.037x_T$$
$$+ 0.078x_{Ti}x_{Cr} - 0.016x_{Ti}x_C + 0.033x_{Ti}x_{Al} + 0.052x_{Ti}x_T$$
$$- 0.003x_{Cr}x_C - 0.002x_{Cr}x_{Al} + 0.036x_{Cr}x_T$$
$$+ 0.042x_C x_{Al} - 0.045x_C x_T + 0.022x_{Al}x_T$$

By ranking the b coefficients, we see that only the factors Cr, Al, and possibly C are really significant. This can be confirmed with the "half-normal probability" plot suggested by Daniels.[1] Thus, we can ignore interactions, and the initial experiment can be described by a linear model of the form

$$y = \log t = 1.66 - 0.383x_{Cr} - 0.146x_{Al} + 0.100x_C + 0.037x_T - 0.034x_{Ti}$$

We are now in a position to determine the direction of steepest ascent on the response surface. The b coefficients in the regression equation are the direction numbers of the steepest ascent path.

$$\frac{\Delta x_{Cr}}{b_{Cr}} = \frac{\Delta x_{Al}}{b_{Al}} = \text{etc.}$$

where the coefficients are expressed in design units. To convert to natural units multiply, by the scale factor.

$$\frac{\Delta x_{Ti}}{-0.034(0.50)} = \frac{\Delta x_{Cr}}{-0.383(2.0)} = \frac{\Delta x_C}{0.100(0.1)} = \frac{\Delta x_{Al}}{-0.146(0.25)} = \frac{\Delta x_T}{0.037(50)}$$

We start at the center of the design and decide to decrease the Cr content, letting the other four factors be determined by the coefficients of the steepest ascent vector. For example, in run 2, if Cr = 2.0, the carbon content is determined by

$$\frac{\Delta x_{Cr}}{-0.766} = \frac{\Delta x_C}{0.010} \quad \text{and} \quad \Delta x_C = \frac{0.10(-2.00)}{-0.766} = +0.026$$

The various treatment combinations are given in Table 11-13.

We note that the procedure does not tell us how far along the steepest ascent vector we should proceed. This is determined by experience or "feel for the problem." In general, you should proceed far enough to approach a maxima (or minima) in the response surface.

In this case, we decide to carry out a second factorial experiment around

---

[1] C. Daniel, *Technometrics*, vol. 1, no. 4, pp. 311–341, 1959.

TABLE 11-13
Treatment combinations and responses for steepest ascent

Run	Ti	Cr	C	Al	T, °F	t_f	t_{avg}
1 (design center)	1.0	4.0	0.4	7.0	2900	60.3, 55.6	57.9
2	0.96	2.0	0.426	6.90	2905	101.6, 92.1	96.6
3	0.92	0.0	0.452	6.80	2910	141.9, 171.1	156.5
4	0.88	−2.0*	0.478	6.70	2915	116.0, 133.7	124.9

* Cr content kept at 0 percent.

the conditions of run 3. Because the variations in pour temperature really are within the scatter of our control over this temperature, we decide to fix T at 2900°F and drop it as an independent variable. Also, realistically we conclude that we cannot go with an alloy without chromium; for such an alloy would have very poor oxidation resistance. As a result, the design center and lower and upper values for the four factors are as given in Table 11-14. Because we are approaching the optimum of the response surface, the curvature is becoming more significant relative to the gradient and a second-order response surface equation most likely would apply. To evaluate the full set of coefficients, we add star points at ±2 from the design center for each independent variable axis. This adds $2 \times 4 = 8$ experiments to the $2^4 = 16$ experiments for the basic factorial design. Also, in order to estimate the run-to-run variability, we add 4 experiments at the design center. This new design is called a rotatable composite design.

Using the techniques already described, the 28 runs, each of which is duplicated, can be analyzed to give the following regression equation.

$$y = \log t = 2.354 - 0.011x_{Ti} + 0.013x_{Cr} + 0.030x_C - 0.037x_{Al}$$
$$- 0.085x_{Ti}^2 - 0.026x_{Cr}^2 - 0.026x_C^2 - 0.060x_{Al}^2 - 0.004x_{Ti}x_{Cr}$$
$$+ 0.025x_{Ti}x_C - 0.054x_{Ti}x_{Al} - 0.011x_{Cr}x_C - 0.021x_{Cr}x_{Al}$$
$$+ 0.077x_C x_{Al}$$

TABLE 11-14
Levels of factors (in natual units) for 2^4 design

Factor	Level, wt percent			
	Design center	Lower level	Upper level	Scale factor
Ti	1.0	0.50	1.5	0.5
Cr	2.0	1.50	2.5	0.5
C	0.5	0.4	0.6	0.1
Al	6.75	6.5	7.0	0.25

where t is stress rupture life, in hours, and the variables x_{Ti}, x_{Cr}, x_C, and x_{Al} are in *design units*. This equation predicts the log of rupture life so long as the independent variables are somewhere within the region of two design units of the design center.

Next, we need to find the stationary point, i.e., the point of zero slope, on the response surface. By taking partial derivatives of the regression equation, that is, $\partial y / \partial x_{Ti} = 0$, etc., we get four simultaneous equations that are eventually solved to yield:

Ti = 1.01%   C = 0.54%

Cr = 2.11%   Al = 6.72%

We observe that the stationary point is very close to the design center.

Finally, we need to determine whether the stationary point represents a true maximum or is a saddle point or even a minimum. This is handled by the canonical reduction procedure, which involves transformation to new axes x_1, x_2, x_3, and x_4 that are centered at the stationary point. Carrying out the canonical transformation results in:

$$y - 2.363 = -0.1148x_1^2 - 0.0293x_2^2 + 0.0031x_3^2 - 0.0561x_4^2$$

This shows that the stationary point is located on a ridge, with y decreasing in the x_1, x_2, and x_4 directions. The surface does not decrease in the x_3 direction, but the coefficient is so small that for all practical purposes it can be considered constant. Thus, within practical limits the observed stationary point is the optimum condition.

Evolutionary Operation (EVOP)

Evolutionary operation (EVOP) is a technique similar to the steepest ascent procedure that is aimed at improving the operating conditions of processes and machines while production is underway.[1] EVOP is designed to be run by process operators on a full-size manufacturing process while it continues to produce a satisfactory product. Thus, the technique is relatively simple and highly structured. It uses 2^2 or 2^3 factorial experiments to make small changes in operating factors around the initial reference conditions. As these changes are made with time the process is moved to a more optimum set of operating conditions.

Since the changes in the factors are small, so as to not upset the process too much, the variation in the response may be small and difficult to detect above the background scatter. Therefore, each cycle of operating conditions is repeated a number of times so any real differences will become statistically significant.

The EVOP technique is illustrated in Fig. 11-15. In (*a*) a 2^2 factorial is

---

[1] G. E. P. Box and N. R. Draper, "Evolutionary Operations," John Wiley & Sons, Inc., New York, 1969.

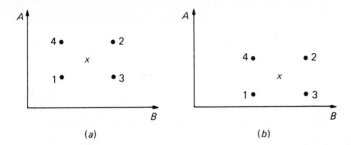

FIGURE 11-15
Test plan for EVOP.

grouped around the reference condition x. If condition 3 gives the best response, then in the next phase of the EVOP procedure condition 3 becomes the new reference condition (Fig. 11-15b). However, the old reference condition is included in the new design as condition 4 to ensure that we have not moved the operating conditions prematurely. By repeated trials, varying the process factors by small amounts from the reference conditions, and then choosing the optimum to become the new reference, the process conditions will gradually evolve in the direction of the final optimum condition.

Computer Tools for Experimental Design

The ubiquitous personal computer and workstation has placed statistical analysis, and especially statistical design of experiments, at the engineer's easy grasp. Many programs are available.[1] This will certainly make it easier to incorporate statistical methods in design. However, the admonition given previously, to consult with a competent statistician early in the planning stages of a project, requires repeating.

BIBLIOGRAPHY

Blank, L.: "Statistical Procedures for Engineering, Management, and Science," McGraw-Hill Book Company, New York, 1980.
Bowker, A. H., and G. J. Lieberman: "Engineering Statistics," 2d ed., Prentice-Hall, Inc., Englewood Cliffs, N.J., 1972.
Box, G. P., W. G. Hunter, and J. S. Hunter: "Statistics for Experimenters," John Wiley & Sons, Inc., New York, 1978.
Davies, O. L., and P. L. Goldsmith (eds.): "Statistical Methods in Research and Production," 4th ed., Longman Group Ltd., London, 1976.

---

[1] C. J. Nachtsheim, *Jnl. of Quality Tech.*, vol. 19, no. 3, pp. 132–160, 1987.

Duckworth, W. E.: "Statistical Techniques in Technological Research," Methuen & Co., Ltd., London, 1968.

Lipson, C., and N. J. Sheth: "Statistical Design and Analysis of Engineering Experiments," McGraw-Hill Book Company, New York, 1973.

McCuen, R. H.: "Statistical Methods for Engineers", Prentice-Hall, Inc., Englewood Cliffs, N.J., 1985.

Mason, R. J., R. F. Gunst, and J. L. Hess, "Statistical Design and Analysis of Experiments", John Wiley & Sons, Inc., New York, 1989.

Miller, I., and J. E. Freund: "Probability and Statistics for Engineers," 3d ed., Prentice-Hall, Inc., Englewood Cliffs, N.J., 1985.

Wadsworth, H. M. (ed.): "Handbook of Statistical Methods for Engineers and Scientists," McGraw-Hill Book Co., New York, 1990.

CHAPTER
12

RISK AND RELIABILITY

12-1 RISK AND SOCIETY

Risk is the potential for realizing some unwanted and negative consequence of an event. Risk is part of our individual existence and that of society as a whole. As young children we were taught about risks. "Don't touch the stove." Doon't chase the ball into the street." As adults we are made aware of the risks of society in our everyday newspaper and newscast. Thus, depending upon the particular week, the news makes us concerned about the risk of all-out nuclear war or of infestation of the California fruit crop by the Mediterranean fruit fly or of midair collision because of the strike by air traffic controllers. The list of risks in our highly complex technological society is endless.

Table 12-1 lists the six classes of hazards to which society is subject. We can see that categories 3 and 4 are directly within the realm of responsibility of the engineer and categories 2, 5, and possibly 6 provide design constraints in many situations. We must realize that risk exists only when a hazard exists and something of value is exposed to that hazard.

$$\text{Risk}\left(\frac{\text{consequence}}{\text{unit time}}\right) = \text{frequency}\left(\frac{\text{events}}{\text{unit time}}\right) \times \text{magnitude}\left(\frac{\text{consequence}}{\text{event}}\right)$$

TABLE 12-1
Classification of societal hazards

Category of hazard	Examples
Infections and degenerative diseases	Influenza, heart disease, aids
Natural disasters	Earthquakes, floods, hurricanes
Failure of large technological systems	Failure of dams, power plants, aircraft, ships, buildings
Discrete small-scale accidents	Automotive accidents, power tools, consumer and sport goods
Low-level, delayed-effect hazards	Asbestos, PCB, microwave radiation, noise
Sociopolitical disruption	Terrorism, nuclear weapons proliferation, oil embargo

From W. W. Lawrence, *in* R. C. Schwing and W. A. Albus (eds.), "Social Risk Assessment," Plenum Press, New York, 1980.

Engineering risk can be considered the link between technological growth and social values as they are reflected in public policy.[1] The engineer has a special responsibility to assess the impact of the technology that his efforts are turning into reality. Ethical issues relating to technological growth and risk assessment are concerned with:

1. Using specialized engineering knowledge and skill in making accurate estimates of risk.
2. Providing an estimate of the precision of the estimated risk as well as an estimate of the risk itself.
3. Communicating with laymen and other professionals about the role of risk assessment in decision making.
4. Understanding the roles of other professionals in risk assessment and settlement of value conflicts.
5. Maintaining competence so that risks can be properly evaluated.

Risk assessment has become increasingly important in engineering design as the complexity of engineering systems has increased. The risks associated with engineering systems do not arise because risk avoidance was ignored in the design. One category of risks arises from external factors that were considered acceptable at the time of design but which subsequent research has revealed to be a health or safety hazard. A good example is the extensive use of sprayed asbestos coating as an insulation and sound barrier before the toxicity of asbestos fibers was known. A second category of risks comes from abnormal conditions that are not a part of the basic design concept in its

[1] R. H. McCuen, ASME Paper 81-WA/TS-5.

normal mode of operation. Usually these abnormal events stop the operation of the system without harm to the general public, although there may be danger to the operators. Other systems, such as passenger aircraft or Three-Mile Island, pose a potential risk and cost to the larger public. Risks in engineering systems are often associated with operator error. Although these should be designed for, it is a difficult task to anticipate all possible future events. This topic is discussed in Secs. 12-5 and 12-6. Finally, there are the risks associated with design errors and accidents. Clearly, these should be eliminated, but since design is a human activity, errors and accidents will occur.

Most reasonable people will agree that society is not risk-free and cannot be made so.[1] However, an individual's reaction to risk depends upon three main factors: 1) whether the person feels in control of the risk or whether the risk is imposed by some outside group, 2) whether the risk involves one big event (like an airplane crash) or many small, separate occurrences, and 3) whether the hazard is familiar or is some strange, puzzling risk like a nuclear reactor. Through the medium of mass communications the general public have become increasingly better informed about the existence of risks in society, but they have not been educated concerning the need to accept some level of risk and to balance risk avoidance against cost. It is inevitable that there will be conflict between various special-interest groups when trying to decide on what constitutes an acceptable risk.

Regulation as a Result of Risk

In a democracy when the public perception of a risk reaches sufficient intensity, the result is legislation to control the risk. That usually means the formation of a regulatory commission that is charged with overseeing the regulatory act. In the United States the first regulatory commission was the Interstate Commerce Commission (ICC). The following federal organizations have a major role to play in some aspects of technical risk.

Consumer Product Safety Commission (CPSC)
Environmental Protection Agency (EPA)
Federal Aviation Agency (FAA)
Federal Highway Administration (FHA)
Federal Railway Administration (FRA)
Nuclear Regulatory Commission (NRC)
Occupational Safety and Health Administration (OSHA)

---

[1] E. Wenk, "Tradeoffs: Imperatives of Choice in a High-Tech World," Johns Hopkins Univ. Press, Baltimore, 1986.

TABLE 12-2
Federal laws concerning product safety

Year	Legislation
1938	Food,. Drug and Cosmetic Act
1953	Flammable Fabrics Act
1960	Federal Hazardous Substance Act
1966	National Traffic and Motor Vehicle Safety Act
1968	Fire Research and Safety Act
1969	Child Protection and Toy Safety Act
1970	Poison Prevention Packaging Act
1970	Lead-Based Paint Poison Prevention Act
1970	Occupational Safety and Health Act
1971	Federal Boat Safety Act
1972	Consumer Product Safety Act

Some of the chief federal laws concerning product safety are listed in Table 12-2. The rapid acceleration of interest in consumer safety legislation is shown by the dates of enactment of these regulatory laws.

Legislation has the important result that it charges all producers of a product with the cost of complying with the product safety regulations. Thus, we are not faced with the situation in which the majority of producers spend money to make their product safe but an unscrupulous minority cuts corners on safety to save on cost. However, in complex engineering systems it may be very difficult to write regulations that do not conflict with each other and work at cross purposes. The automobile is a good example.[1] Here, separate agencies have promulgated regulations to influence fuel economy, exhaust emissions, and crash safety. The law to control emissions also reduces fuel efficiency by 7.5 percent, but the fuel efficiency law has forced the building of smaller cars that have increased crash fatalities by an additional 1400 per year. The need for a strong technical input into the regulatory process should be apparent from this example.

A common criticism of the regulatory approach is that decisions are often made arbitrarily. That is understandable when we consider that a regulatory agency often has a congressional mandate to protect the public from "unreasonable risk." Since there usually are no widely agreed-on definitions of unreasonable risk, the regulators are accused of being hostile to or soft on the regulated industry, depending upon the individual's point of view. Sometimes the regulating agency specifies the technology for meeting the target level of risk. This removes the incentive for innovation in developing more effective methods of controlling the risk.

---

[1] L. B. Lave, *Science*, vol. 212, pp. 893–899, May 22, 1981.

Standards

generally public

busines

Standards are one of the most important ways in which the engineering profession makes sure that society receives a minimum level of safety and performance. They are a set of rules that tell what must be done in particular situations. For example, EPA Standard AP-50 sets the maximum annual average concentration of sulfur dioxide at $80 \mu g/m^3$ and the maximum 24-h average at $365 \mu g/m^3$. Standards may be voluntary or mandatory. Mandatory standards are issued by governmental agencies, and violations are treated like criminal acts for which fines and/or imprisonment may be imposed. Voluntary standards are prepared by a committee of interested parties (industry suppliers and users, government, and the general public), usually under the sponsorship of a technical society or a trade association. Approval of a new standard generally requires agreement by nearly all participants in the committee. Therefore, voluntary standards are consensus standards. They usually specify only the lowest performance level acceptable to all members of the standards committee. Thus, a voluntary standard indicates the lowest safety level that an industry intends to provide in the product it manufactures, whereas a mandatory standard indicates the lowest safety level the government will accept. Because mandatory standards frequently set more stringent requirements than voluntary standards do, they force manufacturers to innovate and advance the state-of-the-art, but often at increased cost to the consumer. That, of course, requires that the mandatory standards be based on realistically obtainable levels.

Another classification of standards is design (specification) standards versus performance standards.[1] Design standards specify the acceptable levels of technical details, such as minimum flow rate and minimum yield strength. Performance standards specify the minimum performance characteristics without specifying the individual technical details. Thus, the supplier has more freedom to innovate and arrive at the same end point. Performance standards can include design specifications as examples of current state-of-the-art methods of meeting the performance criteria.

Although the concept of performance standards has gained considerable attention recently, the vast majority of standards are design standards. It is much more difficult and expensive to write a good performance standard because it must be more general and all-encompassing. Performance standards do not eliminate the problems of determining proper quality levels, and reliable laboratory tests of actual product performance may not be available. Therefore, one often is forced to fall back on specifying a series of well-known and accepted test standards, which are related in some not too well defined way to the product performance.

---

[1] D. Hemenway, "Performance vs. Design Standards," NBS GCR-80-287, October 1980 (PB81-120362).

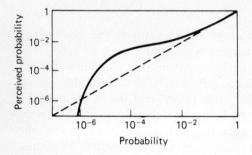

FIGURE 12-1

Perception of probability of risk. (After Starr.)

Risk Assessment

The assessment of risk is an imprecise process involving much judgment and intuition and a poorly defined data base. However, a large effort has gone into this subject in recent years, and a considerable literature is developing.[1] A major effort to develop analytical methods of risk assessment is underway. These efforts are in contrast to the intuitive approach based on individual and collective psychology of risk perception and evaluation. When the results of intuitive risk assessments differ significantly from those of analytical methods, conflict results.

It appears that an individual's perception of the probability of a risk deviates from the true (statistical) probabilities as shown in Fig. 12-1. At some very low level of risk, the perceived interpretation drops to a negligible probability even though in actuality the risk still remains. This may explain the public attitude toward use of seat belts, wherein the probability of an accident is very low per trip but much higher over a lifetime of driving. Another important characteristic of Fig. 12-1 is that the average individual overestimates the probability of occurrence of high-consequence but low-probability risks like nuclear power plant explosions or aircraft crashes. The nonlinearity in Fig. 12-1 suggests that the unit of exposure used to evaluate a risk can be important. Starr[2] has suggested that the proper unit of exposure to risk should be the hour, because that is more consistent with an individual's intuitive risk perception than a year.

Data on risk are subject to considerable uncertainty and variability. In general, three classes of statistics are available: 1) financial losses (chiefly from the insurance industry), 2) health information, and 3) accident statistics. Usually the data are differentiated between fatalities and injuries. Risk is usually expressed as the probability of the risk of a fatality per person per year

---

[1] W. D. Rowe, "An Anatomy of Risk," John Wiley & Sons, Inc., New York, 1977; C. Starr and C. Whipple, *Science*, vol. 208, pp. 1114–1119, June 6, 1980; P. Slovic, *Science*, vol. 236, pp. 280–285, April 17, 1987.

[2] C. Starr, *Science*, vol. 165, pp. 1232–1238, September 19, 1969.

TABLE 12-3
Probability of individual risk

Type of accident	Number in U.S. in 1969	Probability of fatality per person per year
Motor vehicle	55,791	3×10^{-4}
Falls	17,827	9×10^{-5}
Fires	7,451	4×10^{-5}
Drowning	6,181	3×10^{-5}
Poison	4,516	2×10^{-5}
Machinery	2,054	1×10^{-5}
Water transport	1,743	9×10^{-6}
Air travel	1,778	9×10^{-6}
Railway accidents	884	4×10^{-6}
Lightning	160	5×10^{-7}

or per hour of exposure. Table 12-3, taken from the famous study of the safety of nuclear reactors,[1] gives some typical values of accidental fatalities. One point to note is that no general category has a risk rate in the range of 10^{-3} per person per year. A rate of 10^{-6} is not of concern to the average person, but categories with fatality rates between 10^{-6} and 10^{-3} are those in which social concern and expenditures for accident avoidance are practiced. An important point is that the risk values in Table 12-3 are based on the entire population of the United States. Thus, if only a fraction of the population were exposed to the hazard, the effective probability for a particular individual would be increased.

The most usual analytical approach in risk assessment is a benefit-cost analysis. Given a specific set of values and criteria, a benefit-cost analysis could pinpoint the decisions that would best balance technological risk and benefit. However, we have seen that it is difficult to determine true group values. One particularly sticky issue is how to place a monetary value on a human life. This is important because the benefits usually are in lives saved and the costs usually are in dollars. Table 12-4 gives some data[2] on this issue.

12-2 PROBABILISTIC APPROACH TO DESIGN

Conventional engineering design uses a deterministic approach. It disregards the fact that material properties, the dimensions of the components, and the

---

[1] N. Rasmussen, *et al.*, "Reactor Safety Study," WASH-1400-D, U.S. Atomic Energy Commission, August 1974.

[2] B. L. Cohen, *Health Physics*, vol. 38, pp. 35–51, 1980.

TABLE 12-4
Cost per fatality averted by various societal actions

Societal action	$ per fatality averted (1975 $)
Collapsible auto steering column	100,000
Highway guardrail improvements	34,000
Breakaway sign and lighting posts	116,000
Sulfur scrubbers in coal-fired power plants	500,000
Coke oven emission standards	4,500,000

externally applied loads are statistical in nature. In conventional design these uncertainties are handled by applying a factor of safety. In critical design situations, such as aircraft, space, and nuclear applications, however, there is a growing trend toward using a probabilistic approach to better quantify uncertainty and thereby increase reliability.[1]

There are three typical approaches for allowing for probabilistic effects in design. In order of increasing sophistication they are: the use of a factor of safety, the use of the absolute worst case design, and the use of probability in design. We will present the probabilistic approach first, and then show how the other approaches can be folded in.

Consider a structural member subjected to a static load that develops a stress σ. The variation in load or sectional area results in the distribution of stress shown in Fig. 12-2, where the mean is $\bar{\sigma}$ and the standard deviation[2] of

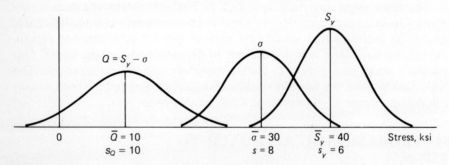

FIGURE 12-2.
Distributions of yield strength S_y and stress.

---

[1] E. B. Haugen, "Probabilistic Mechanical Design," Wiley-Intersience, New York, 1980; J. N. Siddall," Probabilistic Engineering Design," Marcel Dekker Inc., New York, 1983.

[2] Note that probabilistic design is at the intersection of two engineering disciplines: mechanical design and engineering statistics. Thus, confusion in notation is a problem.

TABLE 12-5
Mean and standard deviation of independent random variables x and y

Algebraic functions	Mean, $\bar{Q}$	Std. deviation
$Q = C$	C	0
$Q = Cx$	$C\bar{x}$	$C\sigma_x$
$Q = x + C$	$\bar{x} + C$	σ_x
$Q = x \pm y$	$\bar{x} \pm \bar{y}$	$\sqrt{\sigma_x^2 + \sigma_y^2}$
$Q = xy$	$\bar{x}\bar{y}$	$\sqrt{\bar{x}^2\sigma_y^2 + \bar{y}^2\sigma_x^2}$
$Q = x/y$	$\bar{x}/\bar{y}$	$(\bar{x}^2\sigma_y^2 + \bar{y}^2\sigma_x^2)^{1/2}/\bar{y}^2$
$Q = 1/x$	$1/\bar{x}$	$\sigma_x/\bar{x}^2$

the sample of stress values is s. The yield strength of the material S_y has a distribution of values given by $\bar{S}_y$ and s_y. However, the two frequency distributions overlap and it is possible for $\sigma > S_y$, which is the condition for failure. The probability of failure is given by

$$P_f = P(\sigma > S_y) \tag{12-1}$$

The reliability R is defined as

$$R = 1 - P_f \tag{12-2}$$

If we subtract the stress distribution from the strength distribution, we get the distribution $Q = S_y - \sigma$ shown at the left in Fig. 12-2.

We now need to be able to determine the mean and standard deviation of the distribution Q constructed by performing algebraic operations on two independent random variables x and y, that is, $Q = x \pm y$. Without going into statistical details,[1] the results are as given in Table 12-5. Referring now to Fig. 12-2, and using the results in Table 12-5, we see that the distribution $Q = S_y - \sigma$ has a mean value $\bar{Q} = 40 - 30 = 10$ and $s_Q = \sqrt{6^2 + 8^2} = 10$. The part of the distribution to the left of $Q = 0$ represents the area for which $S_y - \sigma$ is a negative number, that is, $\sigma > S_y$, and failure occurs. If we transform to the standard normal variable (see Sec. 11-6), $z = (x - \mu)/\sigma$, we get, at $Q = 0$,

$$z = \frac{0 - \bar{Q}}{\sigma_Q} = -\frac{10}{10} = -1.0$$

From a table of cumulative normal distribution[2] we find that 0.16 of the area falls between $-\infty$ and $z = -1.0$. Thus, the probability of failure is $P_f = 0.16$,

---

[1] E. B. Haugen, op. cit., pp. 26–56.
[2] M. G. Natrella, op. cit, table A-2, p. T-3.

TABLE 12-6
**Value of z to give different levels
of probability of failure**

Probability of failure P_f	$z = x - \mu/\sigma$
10^{-1}	-1.28
10^{-2}	-2.33
10^{-3}	-3.09
10^{-4}	-3.72
10^{-5}	-4.26
10^{-6}	-4.75

and the reliability is $R = 1 - 0.16 = 0.84$. Clearly, this is not a particularly satisfactory situation. If we select a stronger material with $\bar{S}_y = 50$ ksi, $\bar{Q} = 20$ and $z = 2.0$. The probability of failure now is about 0.02 Values of z corresponding to various values of failure probabilities are given in Table 12-6.

Variability in Material Properties

The mechanical properties of engineering materials exhibit variability. Fracture and fatigue properties show greater variability than do the static tensile properties of yield strength and tensile strength. Most published mechanical property data do not give mean values and standard deviations. MIL Handbook 5C presents statistical values, and Haugen[1] has presented most of the published statistical data. Much other statistical data resides in the files of companies and government agencies.

Minimum design values for material properties are required in the design of minimum weight, highly-reliable structural members. Often the minimum mechanical property value given in the specification is used as the design value. However, usually there is no statistical assurance associated with that number. A procedure for determining statistically valid design allowables for inclusion in MIL HDBK-5, Metallic Materials and Elements for Aerospace Vehicle Structures, has evolved since 1950.[2] Data are recorded as one-sided tolerance limits of a normal distribution (see Sec. 11-10). Three bases for analysis exist, A, B, and S. For the A basis, at least 0.99 of the population values are expected to equal or exceed the mechanical property allowable, with a confidence of 0.95. This is expressed as

$$\text{Design allowable } (A) = \bar{x} - k_A s \tag{12-3}$$

---

[1] E. B. Haugen, op. cit., chap. 8 and app. 10A and 10B.

[2] Metals Handbook, 9th ed., Vol. 8, pp. 662–677, ASM International, Metals Park, OH, 1985.

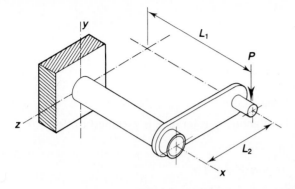

FIGURE 12-3
Cantilever supported crank attached to steel tube. (*From Steffen, Hayes, Wiedemeir, and Tennyson, ASEE Proceedings, 1988.*)

where $\bar{x}$ is the sample mean based on n observations, s is the standard deviation and k_A is the one-sided tolerance limit factor.

For the B basis at least 0.90 of the population values are expected to exceed the mechanical property allowable, with a confidence of 0.95.

$$\text{Design allowable } (B) = \bar{x} - k_B s \qquad (12\text{-}4)$$

The S bases is the minimum value given by the specification. The statistical assurance of this value is not known.

Although certainly not all mechanical properties are normally distributed, a normal distribution is a good first approximation that usually results in a conservative design.[1] When statistical data are not available, a rough estimate of the standard deviation still is possible. If we are given two values x_a and x_b and we have reason to believe they are the end points of a considerable sample ($n \approx 500$), then

$$s = \frac{x_a - x_b}{6}$$

If the sample size is about 25, then the range is divided by 4; if it is only 4 observations, then the range is divided by 2.

Probabilistic Design

We illustrate the probabilistic approach to design with the example[2] of a crank that must support a single static load P, Fig. 12-3. The shaft is considered to be

---

[1] E. B. Haugen and P. H. Wirsching, *Machine Design*, pp. 80–85, May 1, 1975.

[2] This example is taken from a paper "Introducing Reliability Concepts in Machine Design" by J. R. Steffen, L. T. Hayes, D. W. Wiedemeir, and S. Tennyson, *1988 ASEE Annual Conference Proceedings*, pp. 902–909.

a cantilever supported beam and is to be made from AISI 4140 steel tubing. Each design variable in the problem is a random variable which is assumed to be normally distributed. The mean and standard deviation $(\bar{x}, \sigma)$ for the parameters are estimated as follows:

Load $\qquad\qquad P = (150, 10) \text{ lb}$

Yield Stress $\qquad S_y = (129.3, 3.2) \text{ ksi}$

Shaft dia. $\qquad d = (\bar{x}_d, \sigma_d)$ Manufacturing tolerance is assumed to be

$$\sigma_d = 0.015\bar{x}_d$$

$$L_1 = 6 \text{ in} \pm 1/4 \text{ in}$$

$$L_2 = 4 \text{ in} \pm 3/32 \text{ in}$$

In order to prevent buckling of the tubular shaft, the allowable tube thickness is determined by $d/t \leq 100$. We choose the minimum value, $t = d/100$. The moment of inertia of this cross-section is $I = 0.003811d^4$. The polar moment of inertia of the shaft then is given by $J = 2I = 2kd^4$.

The critical stress in the tabular shaft, which will determine the outside diameter, occurs at both top and bottom on the outside surface where it connects to the wall. The torsional stress and the bending stress are both maximum at these locations. If the yield strength S_y is exceeded, failure will initiate here. Because this is a combined stress situation, the von Mises yield stress, S_0, will be the controlling stress. For the geometry of this problem this is given by

$$S_0 = (P/4kd^3)(4L_1^2 + 3L_2^2)^{1/2}$$

The mean value for the von Mises stress is given by

$$\bar{x}_{S_0} = (\bar{x}_P/4k\bar{x}_d^3)(4\bar{x}_{L_1}^2 + 3\bar{x}_{L_2}^2)^{1/2} = 0.1364(10^6)/\bar{x}_d^3$$

The standard deviation of the von Mises stress can be determined from the relationship $\sigma_f = [\Sigma (\partial f/\partial x_i)^2(\sigma_{x_i})^2]^{1/2}$, which holds when the dispersion of each random variable, $c = \sigma/\bar{x}$, is less than 0.20. For the von Mises stress,

$$\sigma_{S_0} \simeq \left[\left(\frac{\partial S_0}{\partial P}\right)^2 \sigma_P^2 + \left(\frac{\partial S_0}{\partial d}\right)^2 \sigma_d^2 + \left(\frac{\partial S_0}{\partial L_1}\right)^2 \sigma_{L_1}^2 + \left(\frac{\partial S_0}{\partial L_2}\right)^2 \sigma_{L_2}^2\right]^{1/2}$$

The partial derivatives are computed from the above equation for S_0.

$$\frac{\partial S_0}{\partial P} = (4L_1^2 + 3L_2^2)^{1/2}/4kd^3 = 0.9090(10^3)/\bar{x}_d^3$$

$$\frac{\partial S_0}{\partial d} = -3P(4L_1^2 + 3L_2^2)^{1/2}/4kd^4 = -0.4090(10^6)/\bar{x}_d^4$$

$$\frac{\partial S_0}{\partial L_1} = \frac{PL_1}{kd^3}(4L_1^2 + 3L_2^2)^{-1/2} = 17.04(10^3)/\bar{x}_d^3$$

$$\frac{\partial S_0}{\partial L_2} = \frac{3PL_2}{4kd^3}(4L_1^2 + 3L_2^2)^{-1/2} = 8.522(10^3)/\bar{x}_d^3$$

Substituting the above four equations and the values for the σ_{xi}^2 terms results in:

$$\sigma_{S_0} = [82.62(10^6)/\bar{x}_d^6 + 37.64(10^6)/\bar{x}_d^6 + 2.017(10^6)/\bar{x}_d^6$$
$$+ 0.0709(10^6)/\bar{x}_d^6]^{1/2} = 11.06(10^3)/\bar{x}_d^3$$

We can use Fig. 12-2 to visualize the solution. We know the distribution of strength, $\bar{s}_y = 129.3$ and $s_y = 3.2$ and we have just calculated the distribution of von Mises combined stress, $\bar{x}_{S_0} = 136.4/\bar{x}_d^3$ and $\sigma_{S_0} = 11.06/\bar{x}_d^3$. All values are in ksi. If we set the allowable failure probability at 0.001 the corresponding value of z from Table 12-6 is -3.09. Thus, following the earlier analysis of subtracting the stress distribution from the strength distribution, we get

$$z = -3.09 = \frac{0 - (129.3 - 136.4/\bar{x}_d^3)}{\left[(3.2)^2 + \left(\dfrac{11.06}{\bar{x}_d^3} \right)^2 \right]^{1/2}}$$

Solving for $\bar{x}_d$ gives $\bar{x}_d = 1.102$ in. The tube thickness is then $t = \bar{x}_d/100 = 0.011$ in.

Safety Factor

The use of a safety factor is far simpler than the above, but with much less information content. Using a safety factor is a form of "derating", but the extent of reduction from the true capacity is not known. In the usual case for static loading the safety factor N_0 is the yield strength divided by the working stress, in this example the von Mises stress.

$$N_0 = S_y/S_0 = \bar{S}_y/\bar{x}_{S_0}$$

We can see that the safety factor is the ratio of the mean capacity to the mean load or demand. If we choose $N_0 = 1.25$, then we can calculate the shaft diameter.

$$1.25 = 129.3/(136.4/\bar{x}_d^3)$$
$$\bar{x}_d = 1.096 \text{ in.}$$

Since we already know the pertinent statistics from the probabilistic design calculation we can readily determine the probability of failure associated with the choice of safety factor.

$$z = \frac{136.4/\bar{x}_d^3 - 129.3}{[(3.2)^2 + (11.06/\bar{x}_d^3)^2]^{1/2}}$$

and on substituting $\bar{x}_d = 1.096$ we compute $z = -2.858$, which corresponds to a failure probability of about 0.002.

A conservative approach is to determine the minimum safety factor. If we view the safety factor as the ratio of capacity C to demand D, then the minimum safety factor would be the lowest expected material property value divided by the greatest expected demand.

$$\text{Minimum safety factor} = C/D = \frac{\left[1 + \frac{\Delta D}{D}\right]}{\left[1 - \frac{\Delta C}{C}\right]}$$

Some very interesting examples of the calculation of factors of safety have been given by Mischke.[1]

Absolute Worse Case Design

In absolute worse case (AWC) design the variables are set at either the lowest or largest expected values. AWC design, like the use of the safety factor, is an approach that accounts for the statistical nature of the design environment in a deterministic way.

Using the previous analysis for the tubular shaft under probabilistic design, failure occurs when the von Mises stress equals the material yield strength, S_y. Thus, the shaft diameter can be expressed by

$$d = [(P/4kS_y)(4L_1^2 + 3L_2^2)^{1/2}]^{1/3}$$

To determine whether each design variable should be at its high or low end for AWC design take the partial derivative of d with respect to that variable. Thus,

$$\frac{\partial d}{\partial P} = \left[\left(\frac{1}{4kS_y}\right)(4L_1^2 + 3L_2^2)^{1/2}\right]^{1/3} \frac{1}{3P^{2/3}} = \text{Positive value}$$

Similarly, we find that maximum values should be used for L_1 and L_2 and a minimum value for S_y. The bounds on the design variables are given as either a tolerance limit, e.g., $L_1 = 6 \pm 1/4$ in or using the 3σ limit as the tolerance, e.g. $P = 150 \pm 3(10)$. Putting the appropriate values into the equation for shaft diameter yields d_{min}.

$$d = \left[\frac{180\{4(6.25)^2 + 3(4.094)^2\}^{1/2}}{4(0.003811)(119,700)}\right]^{1/3} = 1.123 \text{ In.}$$

Note, that while the analysis is easy for AWC design, this approach leads to the use of excess material.

12-3 RELIABILITY THEORY

Reliability is the probability that a system, component, or device will perform without failure for a specified period of time under specified operating

---

[1] C. R. Mischke, Some Guidance of Relating Factor of Safety to Risk of Failure, ASME Paper, 86-WA/DE-22.

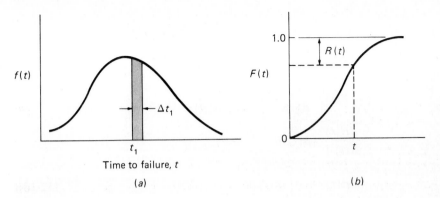

FIGURE 12-4
(a) Distribution of time to failure; (b) cumulative distribution of time to failure.

conditions. The discipline of reliability engineering basically is a study of the causes, distribution, and prediction of failure.[1] If $R(t)$ is the reliability with respect to time t, then $F(t)$ is the unreliability (probability of failure) in the same time t. Since failure and nonfailure are mutually exclusive events,

$$R(t) + F(t) = 1 \qquad (12\text{-}5)$$

If N_0 components are put on test, the number surviving to or at time t is $N_s(t)$ and the number that failed between $t = 0$ and $t = t$ is $N_f(t)$.

$$N_s(t) + N_f(t) = N_0 \qquad (12\text{-}6)$$

From the definition of reliability

$$R(t) = \frac{N_s(t)}{N_0} = 1 - \frac{N_f(t)}{N_0} \qquad (12\text{-}7)$$

The hazard rate, or instantaneous failure rate, is the number of failures per unit time per the number of items exposed for the same time.

$$h(t) = \frac{dN_f(t)}{dt} \frac{1}{N_s(t)} \qquad (12\text{-}8)$$

In more statistical terms we also can define the hazard rate $h(t)$ as the probability that a given test item will fail between t_1 and $t_1 + dt_1$ when it already has survived to t_1. Referring to Fig. 12-4,

$$h(t) = \frac{f(t)}{1 - F(t)} = \frac{f(t)}{R(t)} = P(t_1 \leq t \leq t_1 + dt \mid t \geq t_1) \qquad (12\text{-}9)$$

---

[1] J. H. Bompas-Smith, "Mechanical Survival," McGraw-Hill Book Company, London, 1973; C. O. Smith, "Introduction to Reliability in Design," ibid., New York, 1976.

From the nature of statistical frequency distributions,

$$f(t) = \frac{dF(t)}{dt} = \frac{d[1 - R(t)]}{dt} = \frac{-dR(t)}{dt}$$

and

$$h(t) = \frac{f(t)}{R(t)} = \frac{-dR(t)}{dt} \cdot \frac{1}{R(t)}$$

$$\frac{-dR(t)}{R(t)} = h(t)\, dt$$

$$\ln R(t) = -\int_0^t h(t)\, dt \tag{12-10}$$

$$R(t) = \exp\left[-\int_0^t h(t)\, dt \right] \tag{12-11}$$

The hazard or failure rate is given in terms like 1 percent per 1000 h or 10^{-5} per hour. Components in the range of failure rates of 10^{-5} to 10^{-7} per hour exhibit a good commercial level of reliability. Another way to define the occurrence of failures is to state the mean time between failures (MTBF).

The general form of the failure curve is shown in Fig. 12-5. The three-stage curve shown in Fig. 12-5a is typical of electronic components. At short time there is a high failure rate due to "infant mortality" arising from design errors, manufacturing defects, or installation defects. This is a period of shakedown, or debugging, of failures. These early failures can be minimized by improving production quality control, subjecting the parts to a proof test before service, or "running in" the equipment before sending it out of the plant. As these early failures leave the system, failure will occur less and less frequently until eventually the failure rate will reach a constant value. The time period of constant failure rate is a period in which failures can be considered to occur at random from random overloads or random flaws. These

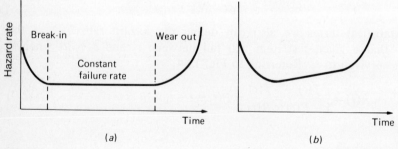

FIGURE 12-5
Forms of the failure curve: (a) three-stage (bath tube) curve typical of electronic equipment; (b) failure curve more typical of mechanical equipment.

failures follow no predictable pattern. Finally, after what is hopefully a long time, materials and components begin to age and wear rapidly and the wearout period of accelerating failure rate begins. Mechanical components, Fig. 12-5*b,* do not exhibit a region of constant failure rate. After an initial break-in period, wear mechanisms operate continuously until failure occurs.

Constant Failure Rate

For the special case of a constant failure rate, $h(t) = \lambda$, and Eq. (12-11) can be written

$$R(t) = \exp\left[-\int_0^t \lambda \, dt \right] = e^{-\lambda t} \tag{12-2}$$

The probability distribution of reliability, for this case, is a negative exponential distribution.

$$\lambda = \frac{\text{number of failures}}{\text{number of time units during which all items were exposed to failure}}$$

The reciprocal of λ, $\bar{T} = 1/\lambda$, is the mean time between failures (MTBF).

$$\bar{T} = \frac{1}{\lambda} = \frac{\text{number of time units during which all items were exposed to failure}}{\text{number of failures}}$$

so

$$R(t) = e^{-t/\bar{T}} \tag{12-13}$$

Note that if a component is operated for a period equal to MTBF, the probability of survival is $1/e = 0.37$.

Although an individual component may not have an exponential reliability distribution, in a complex system with many components the overall reliability may appear as a series of random events and the system will follow an exponential reliability distribution.

Variable Failure Rate

Mechanical failures and some electronic components, e.g., relays and thermionic devices, do not exhibit a period of constant failure rate such as shown in Fig. 12-5*a,* but instead have a curve like Fig. 12-5*b.* Since the failure rate is a function of time, the simple exponential relation for reliability does not apply. The most common practice is to consider that failure is distributed according to the Weibull function (see Sec. 11-7), so that the reliability is given by

$$R(t) = \exp\left[-\left(\frac{t - t_0}{\theta - t_0} \right)^m \right] \tag{12-14}$$

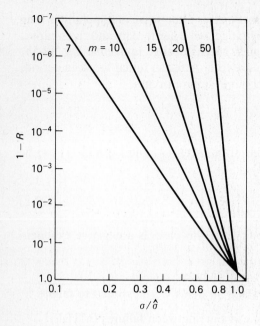

FIGURE 12-6
Relation between probability of failure $(1 - R)$ and strength and Weibull shape parameter. (*From A. A. Bondi.*)

where t = time to failure
t_0 = time at which $F(t) = 0$
θ = characteristic life (scale parameter)
m = slope of the Weibull plot (shape parameter)

Substituting in Eq. (12-9) gives us

$$h(t) = \frac{f(t)}{R(t)} = m\frac{(t - t_0)^{m-1}}{(\theta - t_0)^m} \tag{12-15}$$

For the special case $t_0 = 0$ and $m = 1$, Eq. (12-14) reduces to the exponential distribution with $\theta =$ MTBF. When $m < 1$, $h(t)$ decreases as t increases, as in the break-in period of a three-stage failure curve. When $1 < m < 2$, $h(t)$ increases with time. When $m = 3.2$, the Weibull distribution becomes a good approximation to the normal distribution. The method of plotting data as a Weibull function and determining the parameters θ, m, and t_0 was discussed in Sec. 11-7.

Figure 12-6 shows the probability of failure plotted in a two-parameter Weibull plot for the tensile strength of various materials.[1] The applied stress is σ, and the characteristic parameter is $\hat\sigma \approx \theta$, where $\bar\sigma = \hat\sigma\Gamma(1 + 1/m)$ and $\Gamma(\ )$ is the gamma function. Thus $\hat\sigma/\sigma$ is the safety factor required to obtain a desired level of probability of failure. Each line is for a different value of the

---

[1] A. A. Bondi, *Trans. ASME*, Ser. H, *J. Eng. Materials Tech.*, vol. 101, pp. 27–33, 1979.

shape parameter (Weibull slope). The larger the value of m the less the variability in the strength property of the material. Figure 12-6 shows the strong influence of strength variability on the reciprocal of the safety factor.

Example 12-1. If a device has a failure rate of 2×10^{-6} failures/h what it its reliability for an operating period of 500 h? If there are 2000 items in the test, how many failures are expected in 500 h? Assume that strict quality control has eliminated premature failures so we can assume a constant failure rate.

$$R(500) = \exp(-2 \times 10^{-6} \times 500) = e^{-0.001} = 0.999$$
$$N_s = N_0 R(t) = 2000(0.999) = 1998$$
$$N_f = N_0 - N_s = 2 \text{ failures expected}$$

If the MTBF for the device is 100,000 h, what is the reliability if the operating time equals 100,000 h?

$$t = \bar{T} = 1/\lambda$$
$$R(t) = e^{-t/\bar{T}} = e^{-100,000/100,000} = e^{-1} = 0.37$$

We note that a device has only a 37 percent chance of surviving as long as the MTBF.

If the length of the constant failure rate period is 50,000 h, what is the reliability for operating for that length of time.

$$R(50,000) = \exp(-2 \times 10^{-6} \times 5 \times 10^4) = e^{-0.1} = 0.905$$

If the part has just entered the useful life period, what is the probability it will survive 100 h?

$$R(100) = \exp(-2 \times 10^{-6} \times 10^2) = e^{-0.0002} = 0.9998$$

If the part has survived for 49,900 h, what is the probability it will survive for the next 100 h?

$$R(100) = \exp(-2 \times 10^{-6} \times 10^2) = e^{-0.002} = 0.9998$$

We note that the reliability of the device is the same for an equal period of operating time so long as it is in the constant-failure-rate (useful-life) region.

System Reliability

Most mechanical and electronic systems comprise a collection of components. The overall reliability of the system depends on how the individual components with their individual failure rates are arranged.

If the components are so arranged that the failure of any component causes the system failure, it is said to be arranged in series.

$$R_{\text{system}} = R_A \times R_B \times \cdots \times R_n \tag{12-16}$$

It is obvious that if there are many components exhibiting series reliability, the system reliability quickly becomes very low. For example, if there are 20 components each with $R = 0.99$, the system reliability is $0.99^{20} = 0.818$. Most consumer products exhibit series reliability.

If we are dealing with a constant-failure-rate system

$$R_{system} = R_A \times R_B = e^{-\lambda_A t} \times e^{-\lambda_B t} = e^{-(\lambda_A + \lambda_B)t}$$

and the value of λ for the system is the sum of the values of λ for each component.

A much better arrangement of components is one in which it is necessary for all components in the system to fail in order for the system to fail. This is called parallel reliability.

$$R_{system} = 1 - (1 - R_A)(1 - R_B) \cdots (1 - R_n) \qquad (12\text{-}17)$$

If we have a constant-failure-rate system

$$R_{system} = 1 - (1 - R_A)(1 - R_B) = 1 - (1 - e^{-\lambda_A t})(1 - e^{-\lambda_B t})$$
$$= e^{-\lambda_A t} + e^{-\lambda_B t} - e^{-(\lambda_A + \lambda_B)t}$$

Since this is not in the form $e^{-\text{const}}$, the parallel system has a variable failure rate.

A system in which the components are arranged to give parallel reliability is said to be redundant; there is more than one mechanism for the system functions to be carried out. In a system with full active redundancy all but one component may fail before the system fails.

Other systems have partial active redundancy, in which certain components can fail without causing system failure but more than one component must remain operating to keep the system operating. A simple example would be a four-engine aircraft that can fly on two engines but would lose stability and control if only one engine were operating. This type of situation is known as an n-out-of-m unit network. At least n units must function normally for the system to succeed rather than only one unit in the parallel case and all units in the series case. The reliability of an n-out-of-m system is given by a binomial distribution, on the assumption that each of the m units is independent and identical.

$$R_{n \mid m} = \sum_{i=n}^{m} \binom{m}{i} R^i (1 - R)^{m-i} \qquad (12\text{-}18)$$

where

$$\binom{m}{i} = \frac{m!}{i!\,(m - i)!}$$

Example 12-2. A complex engineering design can be described by a reliability block diagram as shown in Fig. 12-7. In subsystem A two components must operate for the subsystem to function successfully. Subsystem C has true parallel reliability. Calculate the relability of each subsystem and the overall system reliability.

Subsystem A is an n-out of-m model for which $n = 2$ and $m = 4$. Using Eq. (12-18).

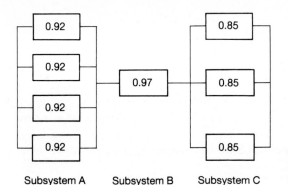

FIGURE 12-7
Reliability block diagram depicting complex design network.

Subsystem A   Subsystem B   Subsystem C

$$R_A = \sum_{i=2}^{4} \binom{4}{i} R^i (1 - R)^{4-i}$$

$$= \binom{4}{2} R^2 (1 - R)^2 + \binom{4}{3} R^3 (1 - R) + \binom{4}{4} R$$

$$= 6R^2(1 - 2R + R^2) + 4R^3(1 - R) + (1)R^4$$

$$= 3R^4 - 8R^3 + 6R^2 = 3(0.92)^4 - 8(0.92)^3 + 6(0.92)^2 = 0.998$$

Since subsystem B is a single component, $R_B = 0.97$.

Subsystem C is a parallel system. Using Eq. (12-17)

$$R_C = 1 - (1 - R_1)(1 - R_2)(1 - R_3) = 1 - (1 - R)^3$$

$$= 1 - (1 - 0.85)^3 = 1 - (0.15)^3 = 1 - 3.375 \times 10^{-3} = 0.9966$$

The total system reliability can be calculated by visualizing the system reduced to three subsystems in series, of value $R_A = 0.998$, $R_B = 0.970$ and $R_C = 0.997$. From Eq. (12–16)

$$R_{Syst.} = R_A \times R_B \times R_C = (0.998)(0.970)(0.997) = 0.965$$

Another approach to redundancy is to employ a standby system, which is activated only when it is needed. An emergency diesel generating unit in a hospital is a common example. In the analysis of the standby redundant system,[1] the Poisson distribution is used. The reliability of a system of two components, one of which is on standby, is

$$R(t) = e^{-\lambda t}(1 + \lambda t) \tag{12-19}$$

If the units are not identical, but have failure rates λ_1 and λ_2, the reliability of

---

[1] C. O. Smith, op. cit., pp. 50–59.

the systems is given by

$$R(t) = \frac{\lambda_1}{\lambda_2 - \lambda_1}(e^{-\lambda_1 t} - e^{-\lambda_2 t}) + e^{-\lambda_1 t} \tag{12-20}$$

On a theoretical basis the use of standby redundancy results in higher reliability than active redundancy. However, the feasibility of standby redundancy depends completely on the reliability of the sensing and switching unit that activates the standby unit. When this key factor is considered, the reliability of a standby system is little better than that of an active redundant system.

Maintenance and Repair

An important category of reliability problem deals with maintenance and repair of systems. If a failed component can be repaired while a redundant component has replaced it in service, then the overall reliability of the system is improved. If components subject to wear can be replaced before they have failed, then the system reliability will be improved.

Preventive maintenance is aimed at minimizing system failure. Routine maintenance, such as lubricating, cleaning, and adjusting, generally does not have a major positive effect on reliability, although the absence of routine maintenance can lead to premature failure. Replacement before wear-out is based on knowledge of the statistical distribution of failure time; components are replaced sooner than they would normally fail. Here a small part of the useful life is traded off for increased reliability. This approach is greatly facilitated if it is possible to monitor some property of the component that indicates degradation toward an unacceptable performance.

Repairing a failed component in a series system will not improve the reliability, since the system is not operating. However, decreasing the repair time will shorten the period during which the system is out of service, and thus the maintainability and availability will be improved.

A redundant system continues to operate when a component has failed, but it may become vulnerable to shutdown unless the component is repaired and placed back in service. To consider this fact we define some additional terms.

$$\text{MTBF} = \text{MTTF} + \text{MTTR} \tag{12-21}$$

where MTBF = mean time between failures
    MTTF = mean time to fail
    MTTR = mean time to repair

If the repair rate $r = 1/\text{MTTR}$, then for an active redundant system

$$\text{MTTF} = \frac{3\lambda + r}{2\lambda^2} \tag{12-22}$$

As an example of the importance of repair, let $r = 6\,h^{-1}$ and $\lambda = 10^{-5}$ per h. With repair, the MTTF $= 3 \times 10^{10}$ h, but without repair it is 1.5×10^5 h.

Maintainability is the probability that a component or system that has failed will be restored to service within a given time. The MTTF and failure rate are measures of reliability, but the MTTR and repair rate are measures of maintainability.

$$M(t) = 1 - e^{-rt} = 1 - e^{-t/\text{MTTR}} \qquad (12\text{-}23)$$

where $M(t) =$ maintainability
$\qquad r =$ repair rate
$\qquad t =$ permissible time to carry out the required repair

It is important to try to predict maintainability during the design of an engineering system.[1] The components of maintainability include 1) the time required to determine that a failure has occurred and to diagnose the necessary repair action, 2) the time to carry out the necessary repair action, and 3) the time required to check out the unit to establish that the repair has been effective and the system is operational. An important design decision is to establish what constitutes the least replaceable assembly, i.e., the unit of the equipment beyond which diagnosis is not continued but the assembly simply is replaced. An important design trade-off is between MTTR and cost. If MTTR is set too short for the man-hours to carry out the repair, then a large maintenance crew will be required at an increased cost.

Availability is the concept that combines both reliability and maintainability; it is the proportion of time the system is working "on line" to the total time, when that is determined over a long working period.

$$\text{Availability} = \frac{\text{total on-line time}}{\text{total on-line time} + \text{total downtime}}$$

$$= \frac{\text{total on-line time}}{\text{total on-line time} + (\text{no. of failures} \times \text{MTTR})}$$

$$= \frac{\text{total on-line time}}{\text{total on-line time} + (\lambda \times \text{total on-line time} \times \text{MTTR})}$$

$$= \frac{1}{1 + \lambda\,\text{MTTR}} \qquad (12\text{-}24)$$

---

[1] B. S. Blanchard, "Logistics Engineering and Management," 2d ed., Prentice-Hall, Inc., Englewood Cliffs, N.J., 1981; C. E. Cunningham and W. Cox, "Applied Maintainability Engineering," John Wiley & Sons, Inc., New York, 1972; A. K. S. Jardine, "Maintenance, Replacement and Reliability," ibid., 1973.

If MTTF $= 1/\lambda$, then

$$\text{Availability} = \frac{\text{MTTF}}{\text{MTTF} + \text{MTTR}} \tag{12-25}$$

Further Topics

We have just begun to scratch the surface of such a dynamic and fertile subject as reliability engineering. Models for the realistic situation where the reliability varies with time are important. Another extension is models for the case where each element in the network fails by two mutually exclusive failure modes, and its further refinement where failure is by a common cause. A *common cause failure* is one where there is loss of redundancy due to a single event like a fire. Additional complexity yet realism arises when the repairability condition is added to the above models. These topics are beyond the scope of this book. The interested reader is referred to the references in the Bibliography at the end of this chapter.

12-4 DESIGN FOR RELIABILITY

The design strategy used to ensure reliability can fall between two broad extremes. The *fail-safe approach* is to identify the weak spot in the system or component and provide some way to monitor that weakness. When the weak link fails, it is replaced, just as the fuse in a household electrical system is replaced. At the other extreme is what can be termed "the one-horse shay" approach. The objective is to design all components to have equal life so the system will fall apart at the end of its useful lifetime just as the legendary one-horse shay did. Frequently a *worst-case approach* is used; in it the worst combination of parameters is identified and the design is based on the premise that all can go wrong at the same time. This is a very conservative approach, and it often leads to overdesign.

Two major areas of engineering activity determine the reliability of an engineering system. First, provision for reliability must be established during the earliest design concept stage, carried through the detailed design development, and maintained during the many steps in manufacture. Once the system becomes operational, it is imperative that provision be made for its continued maintenance during its service.

The various aspects of building reliability into the design process are illustrated in Fig. 12-8. The preliminary design establishes the design concept based on the customer's requirements. Both an empirical experience base and theoretical base are used to arrive at the design criteria and a preliminary design concept. Frequently a proof-of-concept machine, pilot plant, or other laboratory-scale experimental vehicle is used to validate and search out the weak points of the design concept. In the major undertaking labeled "design development," the concept is further refined and translated into a complete,

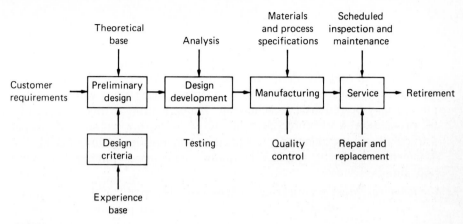

FIGURE 12-8
Elements of integrating reliability into design process. (*Adapted from E. K. Walker, J. C. Ekvall, and J. E. Rhodes, Trans. ASME, vol. 102, pp. 32–39, 1980.*)

detailed design ready for manufacture. Extensive amounts of testing and analysis are conducted to set allowable stress levels, make the final selection of materials, and draw up the design details. Verification tests are conducted on full-size prototypes. Theses tests involve simulated service conditions and are aimed at uncovering design and material deficiencies before the production stage begins.

The information obtained from the testing and analysis is used to establish material and process specifications and quality-control procedures that are used in manufacturing. These specifications are modified as a result of manufacturing experience and become the basis for inspection and maintenance practices that should be applied while the system is in service. As operational experience accumulates, the inspection and maintenance practices are modified. Finally, when the system becomes obsolete or no longer cost-competitive, it is ready for retirement.

Causes of Unreliability

The malfunctions that an engineering system can experience can be classified into five general categories.[1]

1. *Design mistakes.* Among the common design errors are failure to include all important operating factors, incomplete information on loads and environmental conditions, erroneous calculations, and poor selection of materials.

---

[1] W. Hammer, "Product Safety Management and Engineering." chap. 8, Prentice-Hall, Inc., Englewood Cliffs, NJ., 1980.

2. *Manufacturing defects.* Although the design may be free from error, defects introduced at some stage in manufacturing may degrade it. Some common examples are 1) poor surface finish or sharp edges (burrs) that lead to fatigue cracks and 2) decarburization or quench cracks in heat-treated steel. Elimination of defects in manufacturing is a key responsibility of the manufacturing engineering staff, but a strong relationship with the R&D function may be required to achieve it. Manufacturing errors produced by the production work force are due to such factors as lack of proper instructions or specifications, insufficient supervision, poor working environment, unrealistic production quota, inadequate training, and poor motivation.

3. *Maintenance.* Most engineering systems are designed on the assumption they will receive adequate maintenance at specified periods. When maintenance is neglected or is improperly performed, service life will suffer. Since many consumer products do not receive proper maintenance by their owners, a good design strategy is to make the products maintenance-free.

4. *Exceeding design limits.* If the operator exceeds the limits of temperature, speed, etc., for which it was designed, the equipment is likely to fail.

5. *Environmental factors.* Subjecting equipment to environmental conditions for which it was not designed, e.g., rain, high humidity, and ice, usually greatly shortens its service life.

Minimizing Failure

A variety of methods are used in engineering design practice to improve reliability. We generally aim at a probability of failure of $P_f < 10^{-6}$ for structural applications and $10^{-4} < P_f < 10^{-3}$ for unstressed applications.

Margin of safety. We saw in Sec. 12-2 that variability in the strength properties of materials and in loading conditions (stress) leads to a situation in which the overlapping statistical distributions can result in failures. In Fig. 12-6 we saw that the variability in strength has a major impact on the probability of failure, so that failure can be reduced with no change in the mean value if the variability of the strength can be reduced.

Derating. The analogy to using a factor of safety in structural design is derating electrical, electronic, and mechanical equipment. The reliability of such equipment is increased if the maximum operating conditions (power, temperature, etc.) are derated below their nameplate values. As the load factor of equipment is reduced, so is the failure rate. Conversely, when equipment is operated in excess of rated conditions, failure will ensue rapidly.

Redundancy. One of the most effective ways to increase reliability is with redundancy (see Sec. 12-3). In parallel redundant designs the same system

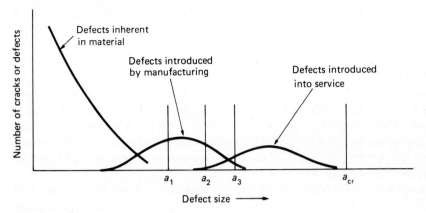

FIGURE 12-9
Distribution of defects in engineering components.

functions are performed at the same time by two or more components even though the combined outputs are not required. The existence of parallel paths may result in load sharing so that each component is derated and has its life increased by a longer than normal time.

Another method of increasing redundancy is to have inoperative or idling standby units that cut in and take over when an operating unit fails. The standby unit wears out much more slowly than the operating unit does. Therefore, the operating strategy often is to alternate units between full-load and standby service. The standby unit must be provided with sensors to detect the failure and switching gear to place it in service. The sensor and/or switching units frequently are the weak link in a standby redundant system.

Durability. The material selection and design details should be performed with the objective of producing a system that is resistant to degradation from such factors as corrosion, erosion, foreign object damage, fatigue, and wear.[1] This usually requires the decision to spend more money on high-performance materials so as to increase service life and reduce maintenance costs. Life cycle costing is the technique used to justify this type of decision.

Damage tolerance. Crack detection and propagation have taken on great importance since the development of the fracture mechanics approach to design (Sec. 7-14). A damage-tolerant material or structure is one in which a crack, when it occurs, will be detected soon enough after its occurrence so that the probability of encountering loads in excess of the residual strength is very remote. Figure 12-9 illustrates some of the concepts of damage tolerance. The

---

[1] A. L. Smith (ed.), "Reliability of Engineering Materials," Butterworth, London, 1984.

initial population of very small flaws inherent in the material is shown at the far left. These are small cracks, inclusions, porosity, surface pits, and scratches. If they are less than a_1, they will not grow appreciably in service. Additional defects will be introduced by manufacturing processes. Those larger than a_2 will be detected by inspection and eliminated as scrapped parts. However, some cracks will be present in the components put into service, and they will grow to a size a_3 that can be detected by the nondestructive evaluation (NDE) techniques that can be used in serice. The allowable design stresses must be so selected that the number of flaws of size a_3 or greater will be small. Moreover, the material should be damage-tolerant so that propagation to the critical crack size a_{cr} is slow (see Sec. 7-15).

In conventional fracture mechanics analysis (Sec. 7-14) the critical crack size is set at the largest crack size that might be undetected by the NDE technique used in service. The value of fracture toughness of the material is taken as the minimum reasonable value. This is a safe but overly conservative approach. These worst-case assumptions can be relaxed and the analysis based on more realistic conditions by using probabilistic fracture mechanics[1] (PFM).

Ease of inspection. The importance of detecting cracks should be apparent from Fig. 12-9. Ideally it should be possible to employ visual methods of crack detection, but special design features may have to be provided in order to do so. In critically stressed structures special features to permit reliable NDE by ultrasonics or eddy current techniques may be required. If the structure is not capable of ready inspection, then the stress level must be lowered until the initial crack cannot grow to a critical size during the life of the structure. For that situation the inspection costs will be low but the structure will carry a weight penalty because of the low stress level.

Simplicity. Simplification of components and assemblies reduces the chance for error and increases the reliability. The components that can be adjusted by operation or maintenance personnel should be restricted to the absolute minimum. The simpler the equipment needed to meet the performance requirements the better the design.

Specificity. The greater the degree of specification that is provided the greater the inherent reliability of design. Whenever possible, be specific with regard to material characteristics, sources of supply, tolerances and characteristics of the manufacturing process, tests required for qualification of materials and components, procedures for installation, maintenance, and use. Specifying

[1] C. A. Rau and P. M. Besuner, *Trans. ASME*, ser. H, *J. Eng. Materials Tech.*, vol. 102, pp. 56–63, 1980; C. A. Rau, P. M. Besuner, and K. G. Sorenson, *Metal Science*, vol. 14, pp. 463–472, 1980.

standard items increases reliability. It usually means that the materials and components have a history of use so that their reliability is known. Also, replacement items will be readily available. When it is necessary to use a component with a high failure rate, the design should especially provide for the easy replacement of that component.

Sources of Reliability Data

Data on the reliability of a product clearly is highly proprietary to its manufacturer. However, the U.S. defense and space programs have created a strong interest in reliability, and this has resulted in the compilation of a large amount of data on failure rate and failure mode.

The Reliability Analysis Center, Rome Air Development Center, Griffiss Air Force Base, NY. 13441 is the chief repository of defense-related reliability data. Two important publications from RAC are:

Digital Evaluation and Generic Failure Analysis Data, MDR-10, 1978. Covers data on microelectronic and semiconductor devices.

Non-electronic Parts Reliability Data, NPRD-1, 1978. Covers data on a wide variety of mechanical, electrical, pneumatic, hydraulic, and rotating parts.

The Government-Industry Data Exchange Program (GIDEP), Corona, CA 91720, has been operating since the late 1950s to build a comprehensive bank of failure rate data.

Cost of Reliability

Reliability costs money, but the cost nearly always is less than the cost of unreliability. The cost of reliability comes from the extra costs associated with designing and producing more reliable components, testing reliability, and training and maintaining a reliability organization. Figure 12-10 shows the cost to a manufacturer of increasing the reliability of a product. The costs of design and manufacture increase with product reliability. Moreover, the slope of the curve increases, and each incremental increase in reliability becomes harder to achieve. The costs of the product after delivery to the customer, chiefly warranty or replacement costs, reputation of the supplier, etc., decrease with increasing reliability. The summation of the curves produces an optimum level of reliability.[1] Other types of analyses establish the optimum schedule for part replacement to minimize cost.[2]

---

[1] R. G. Fenton, *Trans. ASME,* Ser. B, *J. Eng. Ind.,* vol. 98, pp. 1066–1068, 1976.

[2] N. W. Nelson and K. Hayashi, ibid., vol. 96, pp. 311–316, 1974.

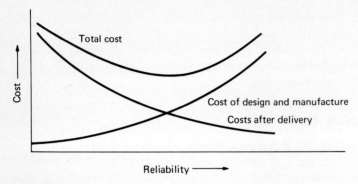

FIGURE 12-10
Influence of reliability on cost.

12-5 PROCEDURE FOR ASSESSING DESIGN RELIABILITY

Here we briefly discuss the procedure for assessing the reliability of a product during the design process.[1]

The first step is to have a well-established problem statement or system definition. This should lay out the criteria for success or failure of the design, the performances expected, the environmental factors, duty cycles, and all boundary conditions and physical constraints on the design.

Next, draw a reliability block diagram of the system. This will be similar to the functional block diagram, which may be part of the problem definition, but it will stress those areas which influence reliability.

Prepare a list of parts in each block of the reliability block diagram.

Collect data on failure and performance for each part or component. The data will come from company records, user data banks, military sources, and the general technical literature.

Calculate the hazard rate or failure rate for each component with the aid of the data that has been collected. Often extensive plotting of data and analysis is required.[2]

Knowing the failure rate for each component in each block of the reliability block diagram, combine the failure rates to calculate the failure rate of each block.

With the above information, compute the system reliability, its failure rate, and the mean time between failures. This information identifies the weak aspects of the system and the extent of improvement that is required. These facts should be fed back into the design process to take corrective action.

---

[1] B. S. Dhillon, "Quality Control, Reliability, and Engineering Design," pp. 257–259, Marcel Dekker Inc., New York, 1985.

[2] W. B. Nelson, "Applied Life Data Analysis," John Wiley & Sons, New York, 1982.

12-6 HAZARD ANALYSIS

A number of techniques have been developed to identify potential causes of failure, rate them in terms of criticality, and establish the conditions under which the failure has the greatest likelihood of occurrence and/or the gravest consequence. These techniques are generally called hazard analysis. Although hazard analysis methods frequently are employed in post-mortem examination of failures, obviously the best time to apply them is in the initial decision-making stages of the design process.

Preliminary Hazard Analysis

The preliminary hazard analysis is a broad study made in the early stages of design. It consists of breaking the engineering system down into subsystems or assemblies, or even to individual components, and for each item answering the following questions.[1]

1. Subsystem or item under investigation
2. Mode of operation
3. Hazardous element
4. Event that triggers the hazardous condition
5. Hazardous condition
6. Event that triggers the potential accident
7. Potential accident
8. Possible effects of the accident
9. Measures taken to contain or prevent occurrences
10. Classification of the severity of the hazard

The classification of the criticality of the hazard can be done by the qualitative descriptors given in Table 12-7.

Information to complete the preliminary hazard analysis comes from such sources as personal experience, interviews with operating personnel and operating supervisors, failure report forms, and published literature. The review of carefully prepared hazards checklists[2] is an important technique to ensure that no obvious points are overlooked.

Failure Modes and Effects Analysis

Failure modes and effects analysis (FMEA) is a detailed analysis of the malfunctions that can be produced in the components of an engineering

---

[1] C. R. James (ed.), "Systems Safety Analytical Technology—Preliminary Hazard Analysis," Report DZ-113073-1, Boeing Co., Seattle, 1969.

[2] W. Hammer, op. cit., app. A.

TABLE 12-7
Criticality of hazards

Rating	Injury or death	Property loss and system performance
I Safe	None; warranty service only	No property loss
II Minor	Minor injury	Minor property loss; degradation of performance does not go beyond acceptable limits
III Major	Major injury	Large property loss; system performance degraded beyond acceptable limits but can be controlled
IV Critical	Multiple major injuries or death	Critical loss; failure degrades system beyond acceptable limits and requires immediate corrective action for personnel or equipment survival
V Catastrophic	Multiple deaths	Critical loss; significant system damage as to preclude mission accomplishment; may threaten corporate survival

system.[1] The emphasis is less on identifying hazards and potential safety problems and more on how to redesign the components to increase system reliability.

To conduct an FMEA, the system is broken into assemblies and the engineering design data are reviewed to determine the interrelations of assemblies and the interrelations of the components of each subassembly. This is best done by making block diagrams like Fig. 4-4. A complete list of the components of each assembly and the function of each component is prepared. From an analysis of the operating and environmental conditions the failure mechanisms that could affect each component are determined. Then the failure modes of all components are identified. Some components may have more than one failure mode. Each failure mode is analyzed as to whether it has an effect on the next higher item in the assembly and whether it has an effect on the entire system or product. The preventive measures or corrective actions that have been taken to control or eliminate the hazard are listed. The probability of failure of each component, based on published data or company experience, is listed, and the probabilities of failure of the subassemblies, assemblies, and the complete system are calculated from reliability theory. Often FMEA is used in conjunction with fault tree analysis (Sec. 12-7), which pinpoints the areas in a complex system where FMEA is needed.

In an extension of FMEA the critically of each assembly is examined and

---

[1] W. Hammer, opt. cit., pp. 175–179; H. E. Lambert, "Systems Safety Analysis and Fault Tree Analysis," UCID-16238, Univ. of California, Lawrence, 1973; Procedures for Performing a Failure Mode and Effect Analysis, MIL-STD-1629A, U.S. Dept. of Defense.

the components and assemblies to which special attention should be given are identified. A component that can give rise to a single-point failure is one example. A *single-point failure* is one in which an accident could result from the failure of a single component, a single human error, or any other single undesirable event. This extended version of FMEA is known as failure modes effects and criticality analysis[1] (FMECA).

Fault Hazard Analysis

The FMEA method concentrates on the malfunction of components, but it ignores the three other major categories of hazards: hazardous characteristics, environmental effects, and human error. Fault hazard analysis[2] was developed by safety professionals to be sure that all categories of hazards are included in the analysis. It is a qualitative method and does not require the estimation of probability of failure. It can be considered more as an extension of the preliminary hazard analysis in that it considers the downstream consequences (damage to higher assemblies in the system) of the failure.

The questions involved in a typical fault hazard analysis are:

1. What is the item?
2. What is the event or condition?
3. What is the potential problem?
4. Why can the potential problem be one?
5. Will failure cause downstream damage?
6. What upstream input or component can "command" the undesirable event?
7. What method of compensation or control is available?

12-7 FAULT TREE ANALYSIS

Fault trees are diagrams that show how the data developed by FMEA should be interrelated to lead to a specific event. FMEA is very effective when applied to a single unit or single failure. When it is applied to a complex system, however, the number of failure modes can become very voluminous, so that economics and/or human frailty make it difficult to identify all potential failures.

Fault tree analysis (FTA) is a technique that provides a systematic description of the combinations of possible occurrences in a system that can result in failure or severe accidents.[3] In FTA the emphasis is on "how things

---

[1] W. Hammer, op. cit., pp. 179–182; R. A. Collacott, *Engineering,* pp. 791–796, June 1979.

[2] W. Hammer, op., cit., p. 183.

[3] J. B. Fassell, *Nuc. Sci. Eng.,* vol. 2, pp. 433–438, 1973; G. J. Powers and F. C. Tompkins, Jr., *AIChE J.,* vol. 20, pp. 376–387, 1974.

can fail to work" rather than the more traditional emphasis on "design performance." Basically, a fault tree is a logic diagram in which logic gates are used to determine the relations between input events and output events. A full quantitative FTA uses boolean algebra in the logical analysis, and probabilities of failure are computed for each event. However, considerable insight can be gained from the graphical relations portrayed by the fault tree. Our discussion of FTA will be restricted to this qualitative level.

Each fault tree deals with a specific event, e.g., failure of a lawn mower engine to start (see Fig. 12-11). FTA is a "top-down approach" that starts with the top event and then determines the contributory events that would lead to the top event. Most top events would be suggested from a preliminary hazard analysis. They can be either hardware failures or human errors.

Different kinds of events are identified with specific symbols on the fault tree diagram.

An output event ▭ is an event that should be developed or analyzed further to determine how it can occur. It is the only symbol on the fault tree diagram that will have a logic gate and input events below it. Except when it represents the undesired top event, the rectangle also serves as an input event to another output event.

An independent event ◯ is an event that does not depend on other components within the system for its occurrence. A common example is a failure of a system component.

A normal event ⌂ is an event that is expected to occur during system operation. It always occurs unless a failure occurs.

An undeveloped event ◇ is an event that has not been developed further because of the lack of information or because it is not of sufficient consequence.

A transfer symbol △ is a connection to another part of the fault tree within the same branch.

The AND gate ⌂ is used to represent a logic condition in which all inputs below the gate must be present for the output (at the top of the gate) to occur.

The OR gate ⌂ is used to represent the situation in which any of the input events will lead to the output.

These symbols are employed in Fig. 12-11, which is a fault tree for the failure of a lawn mower engine. The bottom branch of the tree consists of failures or initiating events that are errors. They point out events where FEMA should be done. Starting with the top event and moving down the

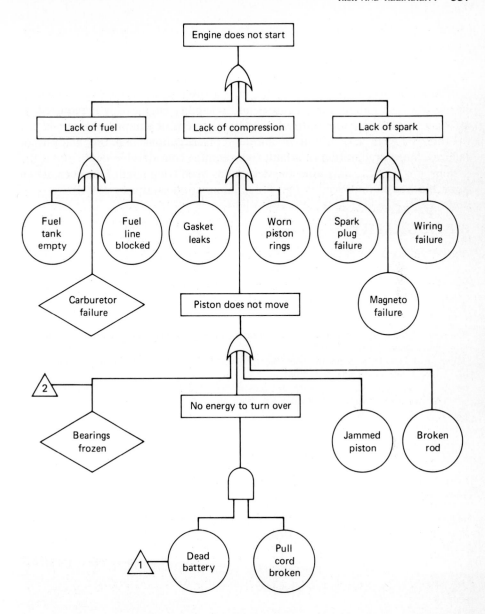

FIGURE 12-11
Fault tree for the failure of a lawn mower engine to start.

branches, we enter a hierarchy of causes. Each event branches into other events. In constructing the tree, it is important to proceed slowly and deliberately down from the top event and list every direct immediate cause before going on to consider the next level of causes. The description used for each event must be carefully chosen to indicate the cause.

The fault tree clearly indicates corrective action that should be taken. For example, the need for diassembling the piston and/or bearings is indicated. Also, the need for preventive maintenance and inspection of the battery or pull cord is called for.

In summary, FTA points out the critical areas in a complex system where further study in failure mode analysis and reliability engineering is required. It also can be used as a tool in troubleshooting a piece of equipment that will not operate. FTA is very useful in accident investigations and the analysis of failures. Since preparation of a fault tree requires considerable detail about the system, it is possible only after the design has been completed. However, it can point the way to improved design for reliability and safety.

BIBLIOGRAPHY

Bloch, H. P. and F. K. Geitner: "Introduction to Machinery Reliability and Assessment," Van Nostrand Reinhold, New York, 1989.

Carter, A. D. S.: "Mechanical Reliability," 2d ed., John Wiley & Sons, New York 1986.

Conrad, J. (ed.): "Society, Technology and Risk Assessment," Academic Press, New York, 1980.

Dhillon, B. S.: "Quality Control, Reliability, and Engineering Design," Marcel Dekker Inc., New York, 1985.

Hammer, W.: "Occupational Safety, Management and Engineering," 3d ed., Prentice-Hall Inc., Englewood Cliffs, N.J., 1985.

Henley, E. J., and H. Kumamoto: "Reliability Engineering and Risk Assessment," Prentice-Hall, Inc., Englewood Cliffs, N.J., 1981.

Ireson, W. G., and C. F. Coombs Jr. (eds.): "Handbook of Reliability Engineering and Management," McGraw-Hill Book Co., New York, 1988.

Kapur, K. C., and L. R. Lamberson: "Reliability in Engineering Design," John Wiley & Sons, New York, 1977.

Lowrance, W. W.: "Of Acceptable Risk," Wm. Kaufman Co., Los Altos, Calif., 1976.

O'Connor, P. D. T.: "Practical Reliability Engineering," 2d ed., John Wiley & Sons, New York, 1985.

Rowe, W. D.: "An Anatomy of Risk, John Wiley & Sons, New York, 1977.

Schwing, R. C., and W. A. Alpers, Jr. (eds.): "Societal Risk Assessment: How Safe is Enough?", Plenum Publishing Co., New York, 1980.

Smith, C. O.: "Introduction to Reliability in Design," McGraw-Hill Book Co, New York, 1976.

Wells, G. L.: "Safety in Process Plant Design," John Wiley & Sons, New York, 1980.

CHAPTER 13

QUALITY ENGINEERING

13-1 TOTAL QUALITY CONCEPT

There is growing awareness of the lack of quality of American-made products when compared with some foreign-made products. U.S. companies have begun to realize that it is not enough to manufacture high-tech products at a competitive price, but also the products must be able to meet and withstand the customer's expectations. High among these is that the product must be durable, and when there is a problem the manufacturer must provide prompt and reliable service. But total quality goes beyond these important market related issues. There is an important savings realized by doing the job right the first time and avoiding the costs of reworking and repairing defective products. It is becoming clear that to remain cost competitive requires high quality.

An important lesson, learned principally from Japan, is that the best way to achieve quality is to design it into the product from the beginning and then assure that it is maintained throughout the manufacturing stage. The key concept of robust design, which was advanced by Dr. Genichi Taguchi, is discussed in Sec. 13-4. The challenge is for the product development team to produce a product design that has been systematically optimized to meet the customer's needs as early as possible in the development process. The design must be robust enough to ensure that the product will provide customer satisfaction even when subject to the conditions of the factory floor and real-world service.

The phrase *total quality* used in the heading of this section denotes a broader concept of quality[1] than simply checking the parts for defects as they come off the production line. The idea of preventing defects by improved design, manufacturing, and process control plays a big role in total quality. We refer to the first aspect as off-line quality control, while the latter is on-line quality control. In order for total quality to be achieved it must be made the number one priority of the organization. This is rooted in the belief that quality is the best way to assure long term profitability. In a study in which companies were ranked by an index of perceived quality, the firms in the top third showed an average return on assets of 30 percent while the firms in the bottom third showed an average return of 5 percent.

For total quality to be achieved requires a customer focus. Quality is meeting customer requirements consistently. To do this we must know who our customers are and what they require. This attitude should not be limited to external customers. Within our organization we should consider those we interact with as our customers. This means that a manufacturing unit providing parts to another unit for further processing should be just as concerned about defects as if the parts were shipped directly to the customer.

Total quality is achieved by the use of facts and data to guide in decision making. Thus, data should be used to identify problems and to help determine when and if action should be taken. Because of the complex nature of the work environment this requires a considerable skill in data acquisition and analysis with statistical methods.

Finally, the quest for quality must be continuous and require total involvement. Total quality will not be achieved by spurts or campaigns for quality. Also, it must involve all employees, especially those engaged in production.

Deming's 14 Points

Work by Walter Shewart, Edwards Deming, and Joseph Juran in the 1920s and 1930s pioneered the use of statistics for the control of quality in production. These quality control methods were mandated by the War Department in World War II for all ordnance production and were found to be very effective. After the war, with a pent up demand for civilian goods and relatively cheap labor and materials costs, these statistical quality control (SQC) methods were largely abandoned as unnecessary and an added expense. However, in Japan, whose industry had been largely destroyed by aerial bombing, it was a different story. The Japanese Union of Scientists and

---

[1] A. V. Feigenbaum, "Total Quality Control," 3d ed., McGraw-Hill Book Co., New York, 1983.

Engineers invited Dr. W. Edwards Deming to Japan in 1950 to teach them SQC. His message was enthusiastically received and SQC became an integral part of the rebuilding of Japanese industry. An important difference between how Americans and Japanese were introduced to SQC is that in Japan the first people converted were top management while in America it was largely engineers who adopted it. The Japanese have continued to be strong advocates of SQC methods and have extended it and developed new adaptations. Today the world looks to Japanese products as a standard of quality and is rapidly adopting statistical methods in design and production. In Japan, the national award for industrial quality, a very prestigious award, is called the Deming Prize.

Dr. Deming views quality in a broader philosophy of management.[1] This is expressed by his fourteen points.

1. Create a constancy and consistency of purpose toward improvement of product and service. Aim to become competitive and to stay in business and to provide jobs.

2. Adopt the philosophy that we are in a new economic age. Western management must awaken to the challenge, must learn their responsibilities, and take on the leadership of change.

3. Stop depending on inspection to achieve quality. Eliminate the need for production line inspection by building quality into the product in design.

4. Stop the practice of awarding business only on the basis of price. The goal should be to minimize total cost, not just acquisition cost. Move toward a single supplier for any one item. Create a relationship of loyalty and trust with your suppliers.

5. Search continually for problems in the system and seek ways to improve it.

6. Institute modern methods of training on the job. Management and workers alike should know statistics.

7. The aim of supervision should be to help people and machines to do a better job. Provide the tools and techniques for people to have pride of workmanship.

8. Eliminate fear, so that everyone may work effectively for the company. Encourage two-way communication.

9. Break down barriers between departments. Research, design, sales, and production must work as a team.

10. Eliminate the use of numerical goals, slogans, posters for the workforce. 80 to 85 percent of the causes of low quality and low productivity are the fault of the system, 15 to 20 percent are because of the workers.

---

[1] W. E. Deming, "Out of Crisis", MIT Center for Advanced Engineering Study, Cambridge, MA, 1986. M. Tribus, Mechanical Engineering, Jan. 1988, pp. 26–30.

11. Eliminate work standards (quotas) on the factory floor and substitute leadership. Eliminate management by objective, management by numbers, and substitute leadership.

12. Remove barriers to the pride of workmanship.

13. Institute a vigorous program of education and training to keep people abreast of new developments in materials, methods, and technology.

14. Put everyone in the company working to accomplish this transformation. This is not just a management responsibility—it is everybody's job.

13-2 QUALITY CONTROL AND ASSURANCE

"Quality control"[1] refers to the actions taken throughout the engineering and manufacturing of a product to prevent and detect product deficiencies and product safety hazards. The American Society for Quality Control (ASQC) defines *quality* as the totality of features and characteristics of a product or service that bear on ability to satisfy a given need. In a narrower sense, "quality control" refers to the statistical techniques employed in sampling production and monitoring the variability of the product. *Quality assurance* refers to those systematic actions vital to provide satisfactory confidence that an item or service will fulfil defined requirements.

Quality control received its initial impetus in the United States in World War II when war production was facilitated and controlled with QC methods. The traditional role of quality control has been to control the quality of raw materials, control the dimensions of parts during production, eliminate imperfect parts from the production line, and assure functional performance of the product. With increased emphasis on tighter tolerance levels, slimmer profit margins, and stricter interpretation of liability laws by the courts (see Sec. 13-11), there has been even greater emphasis on quality control. More recently the heavy competition for U.S. markets from overseas producers who have emphasized quality in the extreme has placed even more emphasis on QC by U.S. producers.

Fitness for Use

An appropriate engineering viewpoint of quality is to consider that it is *fitness for use*. The consumer may confuse quality with luxury, but in an engineering context quality has to do with how well a product meets its design and

---

[1] J. M. Juran and F. M. Gryna (eds.), "Juran's Quality Control Handbook," 4th ed., McGraw-Hill Book Co., New York, 1988; J. M. Duran and F. M. Gryna, "Quality Planning and Analysis," 2d ed., McGraw-Hill Book Co., New York, 1980.

performance specifications. With that concept in mind, the various components for achieving quality can be listed as follows.

1. Quality in design
   Market research
   Design concept
   Design specifications
2. Quality in conformance
   Manufacturing technology
   Production personnel
   Management
3. Availability
   Reliability
   Maintainability
   Logistic support
4. Field service
   Promptness
   Competence
   Integrity

Most of these components of quality have been discussed previously in this text. However, it is not redundant to emphasize that quality is established in the design phase. As we shall see in the discussion of product liability, the courts have greatly increased the responsibility of the design engineer by insisting that all of the improper uses a product can be put to be anticipated and that the misuse of the product be prevented by design.

The particular technology used in manufacturing has an important influence on quality. We saw in Chap. 7 that each manufacturing process has an inherent capability for maintaining tolerances. The same is true for defect generation. As the computer pervades manufacturing, there is a growing trend toward automated inspection, which permits higher volume of part inspection and removes human variability from the inspection process. An important but often overlooked aspect of QC is the design of inspection fixtures and gaging.[1]

The skill and attitude of production workers can have a great deal to do with quality. Where there is pride in the quality of the product there is greater concern for quality on the production floor. A technique used successfully in Japan and meeting with growing acceptance in the United States is the quality circle, in which small groups of production workers meet regularly to suggest quality improvements in the production process.

---

[1] C. W. Kennedy and D. E. Andrews, "Inspection and Gaging," 6th ed., Industrial Press Inc., New York, 1987.

Management must be solidly behind it, or quality control will not be achieved. There is an inherent conflict between achieving quality and wanting to meet production schedules at minimum cost. This is another manifestation of the perennial conflict between short- and long-term goals. There is general agreement that the greater the autonomy of the quality function in the management structure the higher the level of quality in the product. Most often the quality control and manufacturing departments are separate and both the QC manager and the production manager report to the plant manager.

The concept of availability was discussed in Chap. 12. Reliability should be designed into the product; maintainability is a function of design plus manufacturing quality. Logistic support includes the cost and ease of obtaining the parts needed to keep the equipment operational.

Field service comprises all the services provided by the manufacturer after the product has been delivered to the customer: equipment installation, operator training, repair service, warranty service, and claim adjustment. The level of field service is an important factor in establishing the value of the product to the customer, so that it is a real part of the fitness-for-use concept of quality control. Customer contact by field service engineers is one of the major sources of input about the quality level of the product. Information from the field "closes the loop" of quality assurance and provides needed information for redesign of the product.

Quality-control Concepts

A basic tenet of quality control is that variability is inherent in any manufactured product. Someplace there is an economic balance between reducing the variability and the cost of manufacture.[1] Statistical quality control considers that part of the variability is inherent in the materials and process and can be changed only by changing those factors. The remainder of the variability is due to assignable causes that can be reduced or eliminated if they can be identified.

The basic questions in establishing a QC policy for a part are four in number: 1) What do we inspect? 2) How do we inspect? 3) When do we inspect? 4) Where do we inspect?

What to inspect. The objective should be to focus on a few critical characteristics of the product that are good indicators of performance. This is chiefly a technically based decision. Another decision is whether to emphasize non-destructive or destructive inspection. Obviously, the chief value of an NDI technique is that it allows the manufacturer to inspect a part that will actually

---

[1] J. L. Plunkett and B. G. Dale, *Int. J. Prod. Res.*, vol. 26, pp. 1713–1726, 1988.

be sold. Also, the customer can inspect the same part before it is used. Destructive tests, like tensile tests, are done with the assumption that the results derived from the test are typical of the population from which the test samples were taken. Often it is necessary to use destructive tests to verify that the nondestructive test is measuring the desired characteristic.

How to inspect. The basic decision is whether the characteristic of the product to be monitored will be measured on a continuous scale (inspection by variables) or by whether the part passes or fails some go no-go test. The latter situation is known as measurement by attribute. Inspection by variables uses the normal, lognormal, or some similar frequency distribution. Inspection by attributes uses the binomial and Poisson distributions.

When to inspect. The decision on when to inspect determines the QC method that will be employed. Inspection can occur either while the process is going on (process control) or after it has been completed (acceptance sampling). A process control approach is used when the inspection can be done nondestructively at low unit cost. An important feature of process control is that the manufacturing conditions can be continuously adjusted on the basis of the inspection data to reduce the percent defectives. Acceptance sampling often involves destructive inspection at a high unit cost. Since not all parts are inspected, it must be expected that a small percentage of defective parts will be passed by the inspection process. The development of sampling plans[1] for various acceptance sampling schemes is an important aspect of statistical quality control.

Where to inspect. This decision has to do with the number and location of the inspection steps in the manufacturing process. There is an economic balance between the cost of inspection and the cost of passing defective parts to the later stages of the production sequence or to the customer. The number of inspection stations will be optimal when the marginal cost of another inspection exceeds the marginal cost of passing on some defective parts. Inspection operations should be conducted before production operations that are irreversible, i.e., such that rework is impossible, or that are very costly. Inspection of incoming raw material to a production process is one such place. Steps in the process that are most likely to generate flaws should be followed by an inspection. In a new process, inspection operations might take place after every process step; but as experience is gathered, the inspection would be maintained only after steps that have been shown to be critical.

---

[2] See MIL-STD-105D and MIL-STD-414.

New ideas

The success of the Japanese in designing and producing quality products has led to new ideas about quality control. Rather than flooding the receiving dock with inspectors who establish the quality of incoming raw material and parts, it is cheaper and faster to require the supplier to provide statistical documentation that the incoming material meets quality standards. This can only work where the buyer and seller work in an environment of cooperation and trust.

In traditional QC an inspector makes the rounds every hour, picks up a few parts, takes them back to the inspection area, and checks them out. By the time the results of the inspection are available it is possible that bad parts have been manufactured and it is likely that these parts have either made their way into the production stream or have been placed in a bin along with good parts. If the latter happens, the QC staff will have to perform a 100 percent inspection to separate good parts from bad. We end up with four grades of product—first quality, second quality, rework, and scrap. To achieve close to real-time control inspection must be an integral part of the manufacturing process. Ideally, those responsible for making the parts should also be responsible for acquiring the process performance data so that they make appropriate adjustments. There is a trend to using electronic data collectors to eliminate human error and to speed up analysis of data.

Quality Assurance

Quality assurance is concerned with all corporate activities that affect customer satisfaction with the quality of the product. There must be a quality assurance department with sufficient independence from manufacturing to act to maintain quality. This group is responsible for interpreting national codes and standards in terms of each purchase order and for developing written rules of operating practice. Emphasis should be on clear and concise written procedures. A purchase order will generate a great amount of in-plant documentation, which must be accurate and be delivered promptly to each work station. Much of this paper flow has been computerized, but there must be a system by which it gets on time to the people who need it. There must also be procedures for maintaining the identity and traceability of materials and semifinished parts while in the various stages of processing. Definite policies and procedures for dealing with defective material and parts must be in place. There must be a way to decide when parts should be scrapped, reworked, or downgraded to a lower quality level. A quality assurance system must identify which records should be kept, and establish procedures for accessing those records as required.

Quality control is not something that can be put in place and then forgotten. There must be procedures for training, qualifying, and certifying inspectors and other QC personnel. Funds must be available for updating inspection and laboratory equipment and for the frequent calibration of instruments and gages.

An important aspect of quality assurance is the periodic audit of the QC system against written standards. The audit should be performed by persons who have no responsibility in the area under audit. It should be reviewed by top management.

13-3 QUALITY IMPROVEMENT

Four basic costs are associated with quality.

Prevention—those costs incurred in planning, implementing, and maintaining a quality system. Included are the extra expense in design and manufacturing to ensure the highest quality product.

Appraisal—costs incurred in determining the degree of conformance to the quality requirements. The cost of inspection is the major contributor.

Internal failure—costs incurred when materials, parts, and components fail to meet the quality requirements for shipping to the customer. These parts are either scrapped or reworked.

External failure—costs incurred when products fail to meet customer expectation. These result in warranty claims, ill will, or product liability suits. The techniques of failure analysis are considered at the end of this chapter.

To simply collect statistics on defective parts and weed them out of the assembly line is not sufficient for quality improvement and cost reduction. A proactive effort must be made to determine the root causes of the problem so that permanent corrections can be made. Two commonly used techniques in this area of problem solving are the *Pareto diagram* and *cause-and-effect analysis*.

Pareto diagram

In 1897 an Italian economist, Vilfred Pareto, studied the distribution of wealth in Italy and found that a large percentage of the wealth was concentrated in about 10 percent of the population. This was published and became known as Pareto's law. Shortly after World War II, inventory control analysts observed that about 20 percent of the items in the inventory accounted for about 80 percent of the dollar value. In 1954 Joseph Juran generalized Pareto's law as the "80/20 rule", e.g., 80 percent of sales are generated by 20 percent of the customers, 80 percent of the product defects are caused by 20 percent of the parts, etc. While there is no widespread validity of the 80/20 rule it is widely quoted. Certainly Juran's admonition "to concentrate on the vital few and not on the trivial many" is excellent advice in quality improvement, as in other aspects of life.

Cause-and-Effect Analysis

Cause-and-effect analysis uses the "fishbone diagram" or Ishikawa diagram,[1] Fig. 13-3. The poor quality is associated with four categories of causes: operator(man), machine, method, and material. The likely causes of the problem are listed on the diagram under these four main categories. Suggested causes of the problem are generated by the manufacturing engineers, technicians and production workers meeting to discuss the problem. The use of the cause and effect diagram provides an orderly, step-by-step approach to improving manufacturing quality. Manufacturing management should develop a cause-and-effect diagram for each manufacturing process and update it as more knowledge about the process becomes available.

> **Example 13-1.** A manufacturing plant is producing injection molded automobile grilles.[2] The process was newly installed and the parts produced had a number of defects. Therefore, a quality improvement team consisting of operators, setup people, manufacturing engineers, production supervisors, quality control staff, and statisticians was assembled to improve the situation. The first task was to agree on what the defects were and how to specify them. Then a sampling of 25 grilles was examined for defects. Figure 13-1a shows the control chart (see Sec. 13-4 for more details) for the grilles produced by the process. It shows a mean of 4.5 defects per part. The pattern is typical of a process out of control.
>
> A Pareto diagram was prepared to portray the relative frequency of the various types of defects, Fig. 13-2. This was based on the data in Fig. 13-1a. It shows that black spots (degraded polymer patches on the surface) are the most prevalent type of defect. Therefore, it was decided to focus attention initially on this defect.
>
> Focusing on the causes of the black spots resulted in the "fishbone" diagram shown in Fig. 13-3. The causes are grouped under the four-m of manufacturing. Note that for some items, like the injector screw, the level of detail is greater. The group decided that the screw had been worn through too much use and needed to be replaced.
>
> When the screw was changed the black spots completely disappeared, see control chart in Fig. 13-1b. Then after a few days the black spots reappeared to about the same level of intensity as before. Thus, it must be concluded that the root cause of black spots had not been identified. The quality team continued to meet to discuss the black spot problem. It was noted that the design of the vent tube on the barrel of the injection molding machine was subject to clogging and was difficult to clean. It was hypothesized that either polymer accumulated in the vent tube port, became overheated and periodically broke free and continued down the barrel, or was pushed back into the barrel during cleaning. A new vent tube design which minimized these possibilities was designed, constructed, and when installed the black spots disappeared, Fig. 13-1c.

---

[1] K. Ishikawa, "Guide to Quality Control," 2d ed., UNIPUB, New York, 1982.

[2] This example is based on "Tool and Manufacturing Engineer's Handbook," 4th ed., vol. 4, pp. 2–20 to 2–24, Soc. Manufacturing Engineers, Dearborn, MI, 1987.

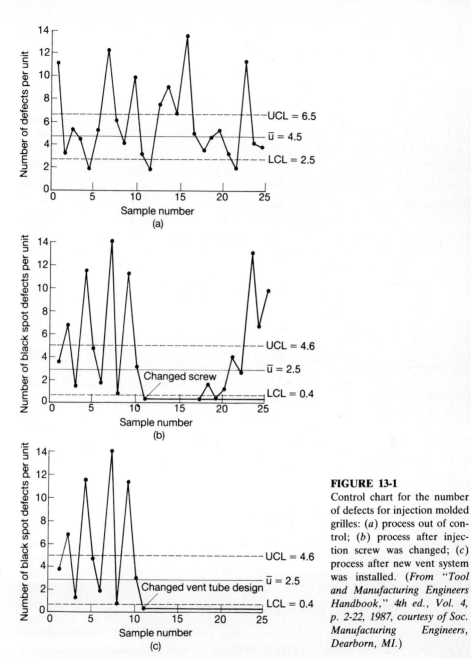

FIGURE 13-1
Control chart for the number of defects for injection molded grilles: (*a*) process out of control; (*b*) process after injection screw was changed; (*c*) process after new vent system was installed. (*From "Tool and Manufacturing Engineers Handbook," 4th ed., Vol. 4, p. 2-22, 1987, courtesy of Soc. Manufacturing Engineers, Dearborn, MI.*)

Having solved the most prevalent defect problem the team turned their attention to scratches, the defect with second highest frequency. A press operator proposed that the scratches were caused by the hot plastic parts falling on the metal lacings of the conveyor belt. He proposed using a continuous belt without metal lacings. However, this type of belt cost twice as much. Therefore, an

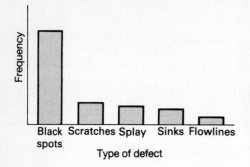

FIGURE 13-2
Pareto diagram for defects in automotive grille. (*From "Tool and Manufacturing Engineers Handbook," 4th ed., Vol. 4, p. 2–22, 1987, courtesy of Soc. Manufacturing Engineers, Dearborn, MI.*)

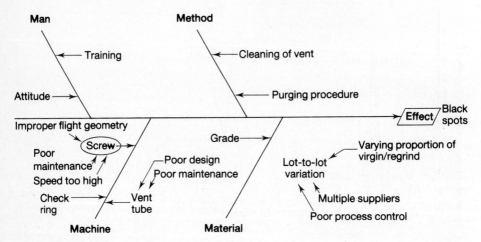

FIGURE 13-3
Cause-and-effect (Ishikawa) diagram for black spot defects on automobile grille. (*From "Tool and Manufacturing Engineers Handbook, 4th ed., Vol. 4, p. 2–23, 1987, courtesy of Soc. Manufacturing Engineers, Dearborn, MI.*)

experiment was proposed in which the metal lacings were covered with a soft latex coating. When this was done the scratches disappeared, but after time they reappeared as the latex coating wore away. With the evidence from this experiment the belt with metal lacings was replaced by a continuous vulcanized belt, not only on the machine under study but for all the machines in the shop.

13-4 STATISTICAL PROCESS CONTROL

Collecting manufacturing performance data and keeping charts on this data is common practice in industrial plants. William Shewhart[1] showed that such data

---

[1] W. A. Shewhart, "Economic Control of Quality in Manufactured Product," Van Nostrand Reinhold Co, New York, 1931.

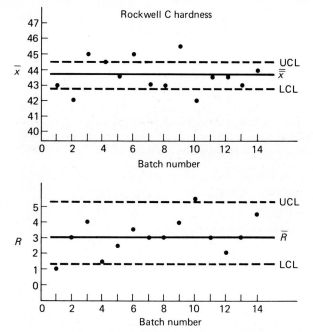

FIGURE 13-4
Control charts for $\bar{x}$ and R.

could be interpreted and made useful through a simple but statistically sound method called a control chart. The use of the control chart is based on the viewpoint that every manufacturing process is subject to two sources of variation: 1) chance variation, also called common causes of variation, and 2) assignable variation, or that due to special causes. Chance variation arises from numerous factors which are individually of small importance. Generally it is not feasible to detect or identify them individually. An assignable variation is due to a cause like poorly trained operators or worn production machines, which it is possible and important to detect and identify. The control chart is one of the chief methods used in the important branch of applied statistics known as quality control.[1]

Example 13-2. Consider a commercial heat-treating operation in which bearing races are being quenched and tempered in a conveyor-type furnace on a continuous 24-h basis. Every 2 h the Rockwell hardness is measured in 10 bearing races to determine whether the product conforms to the specifications. The mean of each sample $\bar{x}$ is computed, and the dispersion is determined by computing the range R. Separate control charts are kept for the mean and the range (Fig. 13-4).

If the population parameters are known, then the $\pm 3\sigma$ limits represent the

[1] E. L. Grant and R. S. Leavenworth, "Statistical Quality Control," 6th ed., McGraw-Hill Book Co., New York, 1988; W. S. Messina, "Statistical Quality Control for Manufacturing Managers," John Wiley & Sons, New York, 1987.

upper control limit (UCL) and the lower control limit (LCL). In the more usual case, in which the population parameters are unknown, it is necessary to estimate the parameters from preliminary samples. If k samples, each of size n, are used, then we compile the mean for each sample $\bar{x}_i$ and determine the grand mean from

$$\bar{\bar{x}} = \frac{1}{k} \sum_{i=1}^{k} \bar{x}_i \tag{13-1}$$

The variability of the process usually is determined from the range, $R = x_{max} - x_{min}$. For samples with small n there is little loss in efficiency in estimating σ from the sample range. The mean range $\bar{R}$ is determined from

$$\bar{R} = \frac{1}{k} \sum_{i=1}^{k} R_i \tag{13-2}$$

The upper and lower control limits on the mean are determined by multiplying $\bar{R}$ by a constant A_2, which depends on the sample size n.

$$UCL = \bar{\bar{x}} + A_2 \bar{R}$$
$$LCL = \bar{\bar{x}} - A_2 \bar{R} \tag{13-3}$$

Also, the limits on $\bar{R}$ are obtained from

$$UCL = D_4 \bar{R}$$
$$LCL = D_3 \bar{R} \tag{13-4}$$

where typical values of the statistical parameters are given in Table 13-1.

In interpreting the control chart it is important to start with the R chart because the limits on the $\bar{x}$ chart depend on the magnitude of the chance variation of the process as measured by R. If some points on the R chart are initially out of control the limits on the $\bar{x}$ chart will be inflated.

On examining the R chart we see two points outside of the control limits. We go back into the records to see if we can establish reasonable cause for these events. We discover that the first point was for the initial heat treat batch on Monday morning when the furnace had been started up and the temperature strip chart record showed that the furnace was not up to the proper temperature when

TABLE 13-1
Factors for use in determining control limits of control charts

No. of observations in sample n	A_2	D_3	D_4
2	1.88	0	3.27
4	0.73	0	2.28
6	0.48	0	2.00
8	0.37	0.14	1.86
10	0.27	0.22	1.78
12	0.22	0.28	1.71

the parts were quenched. No assignable cause can be given to the second point, but with the warning given here the production supervisor decides to give closer scrutiny to the operators on this furnace. Examining the $\bar{x}$ chart we conclude that the frequent out of control values, even after the $\bar{x}$ have been recalculated to omit the two points from the R chart, indicate that hardenability variation in the steel is too great to hold such a tight tolerance. This is an example of the use of the control chart to establish whether a process has the capability of being controlled within the specification limits.

The control chart is a model that describes the way the process variability is expected to appear when only chance causes of variability are present. Once this pattern is established it sends a signal when variation occurs that does not fit the model of common cause variation. If a process is operating within the control limits then any improvement of the process by simple machine adjustment is not likely to improve the mean but it is likely to induce additional variation in the process. The occurrence of points outside the control limits points to the existence of special causes of variation. However, there are other patterns that may also indicate that the process is out of statistical control. These patterns may be trends, cycles, stratification, etc.[1] The control chart can be helpful in tracking down these causes, as in Sec. 13-3. Some powerful statistical techniques to improve the process and product performance are discussed in Sec. 13-5.

The above discussion of control charts was based on a parameter measured on a quantitative scale. Often in inspection it is quicker and cheaper to check the product on a go/no-go basis; the part either passes or fails some predetermined specification. In this type of *attribute testing* different control charts are used. The p chart, based on the binomial distribution, deals with the fraction of defective parts in a sample over a succession of samples. The c chart, based on the Poisson distribution, monitors the number of defects per sample. Other important issues in statistical quality control are the design of sampling plans and the intricacies of sampling parts on the production line.

13-5 TAGUCHI METHOD

A systematized statistical approach to product and process improvement has developed in Japan under the leadership of Dr. Genichi Taguchi.[2] This took a total quality emphasis but developed quite unique approaches and terminology. It emphasizes moving the quality issue upstream to the design stage and

---

[1] A. J. Duncan, "Quality Control and Industrial Statistics," 4th ed., pp. 386–393, Irwin, Inc., 1974.

[2] G. Taguchi, "Introduction to Quality Engineering," Asian Productivity Organization, Tokyo, 1986, available from Kraus Int. Publ., White Plains, N.Y.; G. Taguchi, E. A. Elsayed, and T. Hsiang, "Quality Engineering in Production Systems," McGraw-Hill Book Co., New York, 1989.

focusing on prevention of defects by process improvement. Taguchi has placed great emphasis on the importance of minimizing variation as the primary means of improving quality. Special attention is given to the idea of designing products so that their performance is insensitive to environmental conditions. The process of achieving this through the use of statistically designed experiments has been called *robust design* (see Sec. 13-6).

Loss Function

Taguchi defines the quality level of a product to be the total loss incurred by society due to the failure of the product to deliver the expected performance and due to harmful side effects of the product, including its operating cost. This may seem a backwards definition of quality because the word quality usually connotes desirability, while the word loss conveys the impression of undesirability. In the Taguchi concept some loss is inevitable from the time a product is shipped to the customer so that the smaller the loss the more desirable the product. It is important to be able to quantify this loss so that alternative product designs and manufacturing processes can be compared. This is done with a quadratic loss function, Fig. 13-5a,

$$L(y) = k(y - m)^2 \qquad (13\text{-}5)$$

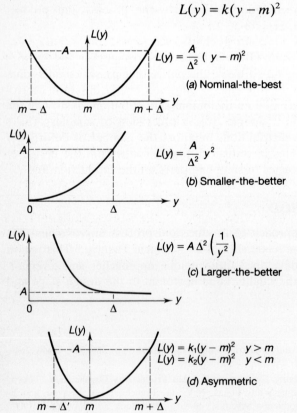

$$L(y) = \frac{A}{\Delta^2}(y - m)^2$$

(a) Nominal-the-best

$$L(y) = \frac{A}{\Delta^2} y^2$$

(b) Smaller-the-better

$$L(y) = A\Delta^2 \left(\frac{1}{y^2}\right)$$

(c) Larger-the-better

$$L(y) = k_1(y - m)^2 \quad y > m$$
$$L(y) = k_2(y - m)^2 \quad y < m$$

(d) Asymmetric

FIGURE 13-5
Variations of the quadratic loss function.

where $L(y)$ is the quality loss when the quality characteristic is y, m is the target value for y and k is a constant, the quality loss coefficient.

We note that when $y = m$ the loss is zero and so is the slope of the loss function. The loss increases slowly when y is near m, but as y deviates further from m the loss increases more rapidly. If $m + \Delta$ and $m - \Delta$ are the customer's tolerance limits, i.e., the product performance is unsatisfactory when y is outside this interval, and if the cost to the customer for repairing or replacing the product is A dollars, then

$$k = A/\Delta^2 \tag{13-6}$$

and

$$L(y) = \frac{A}{\Delta^2}(y - m)^2 \tag{13-7}$$

There is a very important concept of quality engineering inherent in the loss function. In the usual practice of manufacturing quality control the producer specifies a mean (target) value of the performance characteristic and the tolerance interval around that value. *Any value* of the performance characteristic which falls within the interval is defined to be quality product, even if it is barely inside the -3σ limit. With the loss function as a definition of quality the emphasis is on achieving the target value of the performance characteristic and deviations from that value are penalized. The greater the deviation from the target value the greater the quality loss.

The type of loss function described by Eq. (13-7) is called nominal-the-best. Other situations require a modification of this equation. The smaller-the-better type of function, Fig. 13-5b, describes the case where the ideal value is equal to zero. An example would be if y described pollution from an automobile exhaust. The situation is produced by letting $m = 0$ in Eq. (13-7). Another situation can be called larger-the-better, Fig. 13-5c. This describes the situation where $y = 0$ is the worst case and as y increases the quality loss becomes progressively smaller. A situation where strength is the performance characteristic is a good example. Finally, there is the asymmetric loss function, which has a different k on each side of the target value, Fig. 13-5d.

The average quality loss to customers is obtained by averaging the quadratic loss function. Because of noise factors the performance characteristic y of a product varies from piece to piece and over time during the use of the product. If $y_1, y_2, \ldots, y_n$ are measurements of y taken from n units, then the average quality loss, Q, is given by

$$Q = \frac{1}{n}[L(y_1) + L(y_2) + \cdots + L(y_n)]$$

$$= \frac{k}{n}[(y_1 - m)^2 + (y_2 - m)^2 + \cdots + (y_n - m)^2]$$

$$= k\left[(\mu - m)^2 + \frac{n-1}{n}\sigma^2\right] \tag{13-8}$$

In Eq. (13-8) μ is the mean of y and σ^2 is the variance

$$\mu = \frac{1}{n}\sum_{i=1}^{n} y_i \quad \text{and} \quad \sigma^2 = \frac{1}{n-1}\sum_{i=1}^{n}(y_i - \mu)^2$$

When n is large we can write Eq. (13-8) as

$$Q = k[(\mu - m)^2 + \sigma^2] \tag{13-9}$$

Equation (13-9) shows that the average quality loss consists of two components. The first, $k(\mu - m)^2$ results from the deviation of the average value of y from the target m. The second, $k\sigma^2$, results from the *mean squared error* of y around m. Of the two components of quality loss, it is usually easier to minimize the first by appropriate parameter design. Reducing the second component is more difficult, because it requires reducing the variance. This can be achieved by screening out bad products and searching for and fixing causes of poor process performance. Usually, both of these approaches add cost to the product. The preferred approach under the Taguchi philosophy is to apply the concepts of robust design (Sec. 13-6) to search for design parameters that minimize the product's sensitivity to noise.

> **Example 13-3.** A diesel truck engine has a target horsepower of 300 hp. Customer acceptance falls off below that value and at 260 the agitation with underpower becomes strong enough to require some action. On the high side of 300 hp concern arises because an overpowered engine may cause transmission failure. The average cost to correct the underpower by replacing injectors is $300. What is the loss function?
>
> $$A = \$300 \text{ and } \Delta = 40$$
>
> $$L(y) = k(y - m)^2 = \frac{A}{\Delta^2}(y - m)^2 = \frac{300}{(40)^2}(y - 300)^2 = 0.19(y - 300)^2$$
>
> To inspect and replace injectors in the plant during manufacture costs $200. At what tolerance from the target would it be economical to do this, and in the process increase quality to the customer?
>
> This reduces to the question, when the loss is $200 per unit, what is the value of y?
>
> $$200 = 0.19(y - 300)^2; \quad y = 300 \pm 32.4 \text{ and the rework tolerance is 32 hp.}$$

Tolerance Selection

The proper selection of tolerances is an important quality issue as well as a vital economic issue. Taguchi[1] shows how the quality loss function can deal with this problem.

---

[1] G. Taguchi, E. A. Elsayed, and T. Hsiang, op. cit., pp. 45–59.

Example 13-4. From Example 13-3, the loss function for the diesel truck horsepower was

$$L(y) = 0.19(y - m)^2$$

The average total quality is given by Eq. (13-9) and if we assume the engine horsepower will be centered around the target value, then $Q = k\sigma^2$. If the present variance of engine horsepower is $\sigma^2 = (30)^2 = 900$, then the average quality loss per engine is $Q = \$0.19(900) = \$171$. One of the variables controlling horsepower is the valve that controls fuel input. A higher performance control valve would increase cost by $12 and would reduce the standard deviation to 20 hp. Would this be a wise investment?

$$Q = 0.19(20) = \$76.$$

Thus, the quality loss is decreased by $95 at a cost of $12. This, obviously, is a wise investment.

Noise

The input parameters that affect the quality of the product or process may be classified as *design factors* and *disturbance factors*. The former are parameters that can be specified freely by the designer. It is the designer's responsibility to select the optimum levels of the design factors. Disturbance factors are the parameters that are either inherently uncontrollable or impractical to control.

The variability of the input and output parameters plays an important role in the Taguchi methodology. These will be classified into four categories.

Output variability

Variational noise is the short term unit-to-unit variation due to the manufacturing process.

Inner noise is the long-term change in product characteristics over time due to deterioration and wear.

Input variability

Tolerances (design factor variability) is the normal variability in design factors.

Outer noise represents the variability of the disturbance factors that contribute to output variability. Examples are temperature, humidity, dust, vibration, and human error.

Taguchi has borrowed the concept of *signal-to-noise ratio* from electrical engineering and applied it in a specific way to quality engineering. Since we need in control the performance characteristic with respect both to the mean and the variation around the mean it is important to have a performance measure that combines both of these parameters. Taguchi uses the signal-to-noise ratio, *S/N,* as the objective function to be optimized in many situations.

For the nominal-the-best type problem

$$S/N = 10 \log(\mu^2/\sigma^2) \qquad (13\text{-}10)$$

where

$$\mu = \frac{1}{n}\sum_{i=1}^{n} y_i \quad \text{and} \quad \sigma^2 = \frac{1}{n-1}\sum_{i=1}^{n} (y_i - \mu)^2$$

We note that S/N is the inverse of the coefficient of variation introduced in Sec. 11-5.

For a smaller-the-better type of problem

$$S/N = -10 \log\left(\frac{1}{n}\sum y_i^2\right) \qquad (13\text{-}11)$$

In this case the signal is a constant value aimed at making $y = 0$.

For a larger-the-better type problem the quality performance characteristic is continuous and nonnegative; we would like y to be as large as possible. To find the S/N we turn this into a smaller-the-better problem by considering the reciprocal of the performance characteristic

$$S/N = -10 \log\left(\frac{1}{n}\sum \frac{1}{y_i^2}\right) \qquad (13\text{-}12)$$

Of the many design factors involved in an experimental design, we look for two particular types.

Control factors affect primarily the S/N ratio, but not the mean. These factors are first set at the appropriate levels so as to minimize the output variability.

Signal factors affect primarily the mean response of the performance characteristic.

The strategy in setting design factors is to first use the control factors to minimize output variability and then employ the signal factors to move the mean to the desired target. The design factors that prove to be neither control nor signal factors are set at their low cost settings, since they do not affect the performance. Fundamental to the Taguchi method is the approach of economical maximization of the output performance characteristic while minimizing the effect of output variability.

13-6   ROBUST DESIGN

Robust design is the systematic approach to finding optimum values of design factors which result in economical designs with low variability. Taguchi achieves this goal by first performing *parameter design,* and then, if the conditions still are not optimum, by performing *tolerance design.*

Parameter design is the process of identifying the settings of the design parameters or process variables that reduce the sensitivity of the design to sources of variation. In parameter design an accurate modeling of the mean response is not as important as finding the factor levels that optimize robustness. Thus, once the variance has been reduced the mean response should be easily adjusted by using a suitable design factor, known as the signal factor. An important tenet of robust design is that a design found optimum in laboratory experiments should also be optimum under manufacturing and service conditions. Also, since product designs are often broken down into subsystems for design purposes, it is vital that the robustness of a subsystem not be affected by changes in other subsystems. Therefore, interactions among control factors are highly undesirable.

Parameter Design

Parameter design makes heavy use of statistically planned experiments. Two- and three-level orthogonal arrays are most often used.[1] Orthogonal arrays are generalized Graeco–Latin squares. All common fractional factorial designs are orthogonal arrays. These arrays have the pairwise balancing property that every setting of a design parameter occurs with every setting of all other design parameters the same number of times. They keep this balancing property while minimizing the number of test runs.

A Taguchi-type parameter design of experiments consists of two parts: (1) a design parameter matrix and (2) a noise matrix, Fig. 13-6. The design parameter matrix specifies the test settings of the design parameters. In Fig. 13-6 there are four parameters θ_1, θ_2, θ_3, θ_4, each tested at three levels for nine test runs in the design parameter matrix. The noise matrix consists of three noise factors, w_1, w_2, w_3, each at two levels. The complete experiment consists of a combination of the design parameter matrix and the noise matrix. Each test run of the design parameter matrix is crossed with all the rows of the noise matrix. Thus, for test run 1, there are four trials, one for each combination of the factors in the noise matrix, like humidity, operator experience, etc. For test run 2, there are another four experiments, etc., so that all told $9 \times 4 = 36$ test conditions will be run. The performance characteristic is evaluated for each of the four trials in the first test run and performance statistics like the mean and the signal-to-noise ratio are determined. This is done for each of the nine trials of the design performance matrix.

---

[1] G. Taguchi, "System of Experimental Design:Engineering Methods to Optimize Quality and Minimize Cost," two vols, Quality Resources, White Plains, N.Y., 1987; M. S. Phadke, "Quality Engineering Using Robust Design," Prentice-Hall Inc., Englewood Cliffs, N.J., 1989; T. B. Barker, "Engineering Quality by Design: Interpreting the Taguchi Approach," Marcel Dekker, Inc., New York, 1990.

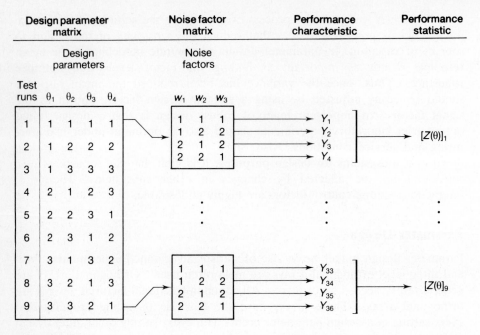

FIGURE 13-6
A typical Taguchi design of experiments for parameter (robust) design. (*After R. N. Kackar, Quality Progress, p. 27, Dec. 1986.*)

The ability to carry out an experimental design of this type will depend on the cost of the experiments and the time required. If experiments are very expensive to run then it will not be possible to employ a complete outer array and only the noise factor thought to be the most important will be used. In cases where product and process performance can be modeled accurately then an extensive statistical design can be done rather inexpensively. However, Japanese experience has found that time and money spent in establishing a robust design or process conditions pays off handsomely in improved quality and reduced costs.

Example 13-5 (A case of "larger-is-better"). To reinforce the above, and continue the understanding of robust design, we will use an example.[1] This concerns the process optimization of a twin-screw polymer extruder that is used to make a small plastic part where strength is important. There are six design parameters that are felt to be important. These parameters, the ranges of study (three levels), and the low cost tolerances are listed below. The latter represent tolerances that can be easily attained by low cost components or readily available facilities. Later on, if necessary, we will tighten the critical tolerances.

---

[1] T. B. Barker, *Quality Progress,* vol. 19, December 1986, pp. 32–42.

Parameter	Range of study			Low cost tolerance
Feed rate, gms/min	1000	1200	1400	20%
First screw, rpm	400	440	480	10%
Second screw, rpm	850	900	950	10%
Gate size, in.	−0.030	0	+0.030	10%
First temp., °F	280	320	380	15%
Second temp., °F	320	360	400	15%

We have used three levels for each design factor since we anticipate the possibility of nonlinear interactions between the factors. With 6 factors this would require $3^6 = 729$ runs. Because that is impractical, and since it has been shown[1] that good results can be obtained with far fewer tests in fractional experiments, we will use a $L_{27}(3^{13})$ orthogonal array.[2] Table 13-2 shows the assignment of levels among the six design parameters. The particular columns used were chosen on the basis of statistical advice[3] about aliases (see p. 509). Table 13-3 shows the specific values of the six design parameters. These constitute the design matrix or the inner array. Each design factor is set at its lower tolerance value, the nominal factor, and the upper tolerance factor, see Table 13-3. For the noise matrix or outer array we take a different approach from that shown in Fig. 13-6. Here the outer array represents the process noise factors. To save resources an orthogonal design with 18 runs is used. Table 13-3 shows the value for the first run of the inner array. For the next run a different outer array is determined. There will be 27 outer arrays in all, and with 18 runs in the outer array this means that $27 \times 18 = 486$ runs will be carried out in the experiment. This is possible to achieve in this instance since it takes 30 seconds to produce a part at a material cost of less than one dollar. The entire experiment can be done in less than a week at a cost under $5000.

Table 13-4 shows the breaking strengths (performance characteristic) obtained for the first run of the outer array. The average strength and standard deviation are determined, as are the signal-to-noise ratios for nominal (N), smaller (S), and larger (B) is best conditions. Since we are interested in the largest value of strength we shall be interested in maximizing the type B S/N. We note that the average strength for this first run of the design parameters is rather low. However, the objectives of these experiments is not to make good parts every time, but rather to make changes in the process conditions so that we learn what conditions make good and bad parts.

---

[1] G. Taguchi, op. cit.; J. S. Hunter, *Jnl. of Quality Tech.*, vol. 17, no. 4, pp. 210–221, 1985.

[2] M. S. Phadke, op. cit., pp. 293–295.

[3] Taguchi developed a theory of linear graphs which aids in assigning factors and interactions to columns in the orthogonal array; see M. S. Phadke, op. cit., pp. 159–166.

TABLE 13-2
Assignment of test levels among six design parameters based on $L_{27}(3^{13})$ orthogonal design. Column headings refer to headings in full design

Column No.	1	2	5	9	10	12
1	1	1	1	1	1	1
2	1	1	2	2	2	2
3	1	1	3	3	3	3
4	1	2	1	2	2	3
5	1	2	2	3	3	1
6	1	2	3	1	1	2
7	1	3	1	3	3	2
8	1	3	2	1	1	3
9	1	3	3	2	2	1
10	2	1	1	2	3	2
11	2	1	2	3	1	3
12	2	1	3	1	2	1
13	2	2	1	3	1	1
14	2	2	2	1	2	2
15	2	2	3	2	3	3
16	2	3	1	1	2	3
17	2	3	2	2	3	1
18	2	3	3	3	1	2
19	3	1	1	3	2	3
20	3	1	2	1	3	1
21	3	1	3	2	1	2
22	3	2	1	1	3	2
23	3	2	2	2	1	3
24	3	2	3	3	2	1
25	3	3	1	2	1	1
26	3	3	2	3	2	2
27	3	3	3	1	3	3

The results of all 27 test runs of the design parameters are given in Table 13-5. Because we have used orthogonal designs we can use the analysis of variance (ANOVA) technique[1] (see Sec. 11-11) to determine which factors contribute to the variation in strength and which factors move the strength on target. Table 13-6 shows that the polymer feed rate, the speed of the first screw

---

[1] M. S. Phadke, op. cit., pp. 51–59; P. J. Ross, "Taguchi Techniques for Quality Engineering," pp. 23–61, McGraw-Hill Book Co., New York, 1988.

TABLE 13-3

Specific values chosen for the design parameters (Inner Array) and values of Outer Array for Run no. 1

"Inner Array"

FEED RATE	FIRST RPM	SECOND RPM	GATE	FIRST TEMP	SECOND TEMP
1000	400	850	-30	280	320
1000	400	900	0	320	360
1000	440	950	+30	360	400
1000	440	900	0	360	400
1000	440	950	+30	360	320
1000	480	850	-30	280	360
1000	480	850	+30	360	360
1000	480	900	-30	280	400
1000	400	950	0	320	320
1200	400	850	+30	360	360
1200	400	900	-30	280	400
1200	400	950	-30	320	320
1200	440	850	+30	280	360
1200	440	900	-30	320	400
1200	480	950	0	360	320
1200	480	900	0	360	360
1400	480	850	+30	280	400
1400	400	900	-30	360	360
1400	400	950	0	320	320
1400	400	850	-30	280	360
1400	440	900	0	360	360
1400	440	950	+30	280	280
1400	440	850	0	320	320
1400	480	900	+30	280	360
1400	480	950	-30	360	400

"Outer Array"

800	360	765	-27	238	272
800	400	850	-30	280	320
800	440	935	-33	322	368
1000	360	765	-30	280	368
1000	400	850	-33	322	272
1000	440	935	-27	238	320
1200	360	850	-27	238	320
1200	400	935	-30	280	368
1200	440	765	-33	322	272
800	360	935	-33	280	320
800	400	765	-27	322	368
800	440	850	-30	238	272
1000	360	850	-33	238	368
1000	400	935	-27	280	272
1000	440	765	-30	322	320
1200	360	935	-30	322	320
1200	400	765	-33	238	272
1200	440	850	-27	280	368

Reprinted with permission from T. B. Barker, *Quality Progress*, Dec. 1986.

TABLE 13-4

Values of breaking strength (performance characteristic) obtained for the first run of the outer array

```
800   360   765 -27   238   272 BREAK STRENGTH=   27.7541
800   400   850 -30   280   320 BREAK STRENGTH=   72.1158
800   440   935 -33   322   368 BREAK STRENGTH=  102.184
1000  360   765 -30   280   368 BREAK STRENGTH=   96.6834
1000  400   850 -33   322   272 BREAK STRENGTH=   61.0035
1000  440   935 -27   238   320 BREAK STRENGTH=  118.554
1200  360   850 -27   322   320 BREAK STRENGTH=  106.238
1200  400   935 -30   238   368 BREAK STRENGTH=  143.226
1200  440   765 -33   280   272 BREAK STRENGTH=   59.0063
800   360   935 -33   280   320 BREAK STRENGTH=   84.8466
800   400   765 -27   322   368 BREAK STRENGTH=   65.4833
800   440   850 -30   238   272 BREAK STRENGTH=   29.2238
1000  360   850 -33   238   368 BREAK STRENGTH=  120.354
1000  400   935 -27   280   272 BREAK STRENGTH=   83.7825
1000  440   765 -30   322   320 BREAK STRENGTH=   94.6045
1200  360   935 -30   322   272 BREAK STRENGTH=   81.4962
1200  400   765 -33   238   320 BREAK STRENGTH=   98.8986
1200  440   850 -27   280   368 BREAK STRENGTH=  128.075
BREAK STRENGTH=  104.345

NOM FOR TC#  1  104.345   STD. DEV.=  31.5154
AVERAGE=  87.4183
TYPE N S/N=  10.399
TYPE S S/N= -39.3349
TYPE B S/N=  35.8897
```

Reprinted with permission from T. B. Barker, *Quality Progress*, Dec. 1986

and the second screw, and the barrel temperature of the second screw are statistically significant in affecting the S/N, i.e. these are control variables, while all of the design parameters are significant to some degree in affecting the mean, i.e., they are signal factors. The strategy is to first use the control variables to set the variability at a minimum level and then use the signal variables to move the mean to the target value. If we plot the effect of each design parameter on the mean and the S/N ratio over its range of three levels we will get a good feel of what to do, provided there are no significant interactions. This results in the following set points:

Polymer feed: 1200 gms/min

First screw, rpm: 480

Second screw, rpm: 950

Gate size: zero deviation from nominal

First temperature, °F: 360—note S/N is higher at 280, but this is not a significant factor with respect to variation so we set to maximize mean

Second temperature, °F: 360

The next step is to validate the prediction of the experimental design and ANOVA. We note that the set of optimum conditions listed above did not

TABLE 13-5
Results of all 27 runs of design parameters

FEED RATE	FIRST RPM	SECOND RPM	GATE	FIRST TEMP	SECOND TEMP	$\bar{X}$	s	S/N_b
1000	400	850	0	280	320	87.4	31.5	35.9
1000	400	900	+30	320	360	115.6	24.6	40.5
1000	400	950	0	360	400	106.2	36.3	36.8
1000	440	850	0	320	400	101.5	34.6	37.3
1000	440	900	+30	360	320	117.6	35.4	39.1
1000	440	950	-30	280	360	115.2	24.3	40.5
1000	480	850	+30	360	360	131.1	30.7	41.4
1000	480	900	-30	280	400	93.9	35.4	36.8
1000	480	950	0	320	320	134.3	35.5	40.8
1200	400	850	0	360	360	111.6	21.9	40.5
1200	400	900	+30	280	400	108.6	30.5	39.2
1200	400	950	-30	320	320	111.9	29.3	40
1200	440	850	+30	280	320	105.7	28	39.3
1200	440	900	-30	320	360	118.3	20.1	41.1
1200	440	950	0	360	400	133.1	34	41.5
1200	480	850	-30	320	400	104.1	33.2	38.8
1200	480	900	0	360	320	144.5	35.4	42.3
1200	480	950	+30	280	360	133.5	21.4	42.2
1400	400	850	+30	320	400	82.5	41.6	33.2
1400	400	900	0	360	320	85.8	42	29.5
1400	400	950	-30	280	360	120.4	33.5	40.4
1400	440	850	-30	360	360	99.3	36.7	38
1400	440	900	0	280	400	99.1	41.5	36.2
1400	440	950	+30	320	320	115.3	41.1	38.8
1400	480	850	0	280	320	96.2	40.9	34.1
1400	480	900	+30	320	360	121.6	36.7	40.4
1400	480	950	-30	360	400	120.8	46.4	39.2

Reprinted with permission from T. B. Barker, *Quality Progress*, Dec. 1986.

TABLE 13-6
Results of Analysis of Variance (ANOVA)

	For S/N			For $\bar{X}$		
Source	**MS**	**F**	**Source**		**MS**	**F**
Feed rate	33.0	10.2*	Feed rate		471	10.7*
First RPM	12.0	3.8*	First RPM		625	14.3*
Second RPM	13.0	4.0*	Second RPM		815	18.6*
Gate	5.8	1.8	Gate		421	9.6*
First temperature	1.1	0.3	First temperature		225	5.1*
Second temperature	24.0	7.4*	Second temperature		381	8.7*

* Significant at the 0.05 level.

occur in any of the 486 tests that have been run. Once again we test each factor at the lower limit of the tolerance, at the nominal value, and at the upper limit of the tolerance range. These results are given in Table 13-7. We see that the mean is much increased and the S/N is at a new high. The loss function for the part is $L(y) = \$0.01389(y - m)^2$. From Eq. (13-9) we can write $Q = k\sigma^2 = \$0.01389(28.0234)^2 = \$10.90$ per finished component, which is still too high.

Tolerance Design

A great feature of the Taguchi method is that often the parameter design gives acceptable results. However, in this example we need to reduce the variance by tightening up on the tolerances. To decide where to do this, we perform ANOVA on the data in Table 13-7. This shows that the significant parameters affecting variation are, in order of decreasing importance: first screw rpm; first temperature, second screw rpm, feedrate, and second temperature. We decide that the standard deviation should be reduced from 28 to 12, or the variance must be reduced to 0.1836 of its original value. To do this requires drastic surgery on the tolerances, as follows.

Parameter	Range of study	Tolerance % of mean
Feed rate, gms/min	1200 ± 126	±10.5
First screw, rpm	480 ± 15.84	±3.3
Second screw, rpm	950 ± 31.35	±3.3
First, temp., °F	360 ± 21.6	±6
Second temp., °F	360 ± 21.6	±6

Using these settings and the reduced tolerances yields the results shown in Table 13-8. The standard deviation is below the new target and the S/N is at 45 db, well above its original value. The calculated quality loss per component

TABLE 13-7
Results of validation experiments with process conditions around new set points

FEED   1200

1ST RPM   480

2ND RPM   950

GATE   0

1ST TEMP   360

2ND TEMP   360

```
960   432   855   0   306   306 BREAK STRENGTH=  108.777
960   480   950   0   360   360 BREAK STRENGTH=  168.682
960   528   1045  0   414   414 BREAK STRENGTH=  214.871
1200  432   855   0   360   414 BREAK STRENGTH=  129.069
1200  480   950   0   414   306 BREAK STRENGTH=  182.115
1200  528   1045  0   306   360 BREAK STRENGTH=  192.842
1440  432   950   0   414   360 BREAK STRENGTH=  154.933
1440  480   1045  0   306   414 BREAK STRENGTH=  151.943
1440  528   855   0   360   306 BREAK STRENGTH=  150.739
960   432   1045  0   360   360 BREAK STRENGTH=  168.195
960   480   855   0   414   414 BREAK STRENGTH=  148.562
960   528   950   0   306   306 BREAK STRENGTH=  140.859
1200  432   950   0   306   414 BREAK STRENGTH=  142.331
1200  480   1045  0   360   306 BREAK STRENGTH=  183.806
1200  528   855   0   414   360 BREAK STRENGTH=  212.603
1440  432   1045  0   414   306 BREAK STRENGTH=  154.586
1440  480   855   0   306   360 BREAK STRENGTH=  134.206
1440  528   950   0   360   414 BREAK STRENGTH=  168.823
BREAK STRENGTH=  184.152
```

NOM FOR TC#   184.152   STD. DEV.=   28.0234
AVERAGE=   161.552
TYPE B S/N=   43.7907

Reprinted with permission from T. B. Barker, *Quality Progress*, Dec. 1986.

is $2, compared with an original value of $39. Based on the projected annual production of the part the loss avoidance of the new design is about $3 million.

Taguchi's methods of quality engineering have created great interest in the United States as certain major manufacturing corporations have embraced the approach. While his approach is new in focusing on elimination of variability and in emphasizing off-line quality control through robust design many of the statistical techniques have been in existence for over fifty years. Statisticians point out[1] that less complicated and more efficient methods exist

---

[1] R. N. Kackar, *Jnl of Quality Tech.*, vol. 17, no. 4, pp. 176–209, 1985.

TABLE 13-8
Results from experiments based on tolerance design.

FEED   1200

1ST RPM   480

2ND RPM   950

GATE   0

1ST TEMP   360

2ND TEMP   360

```
1074   464.16   918.65   0   338.4   338.4 BREAK STRENGTH=   159.814
1074   480   950   0   360   360 BREAK STRENGTH=   179.888
1074   495.84   981.35   0   381.6   381.6 BREAK STRENGTH=   197.12
1200   464.16   918.65   0   360   381.6 BREAK STRENGTH=   169.108
1200   480   950   0   381.6   338.4 BREAK STRENGTH=   187.761
1200   495.84   981.35   0   338.4   360 BREAK STRENGTH=   188.163
1326   464.16   950   0   381.6   360 BREAK STRENGTH=   178.789
1326   480   981.35   0   338.4   381.6 BREAK STRENGTH=   176.35
1326   495.84   918.65   0   360   338.4 BREAK STRENGTH=   177.1
1074   464.16   981.35   0   360   360 BREAK STRENGTH=   179.728
1074   480   918.65   0   381.6   381.6 BREAK STRENGTH=   177.529
1074   495.84   950   0   338.4   338.4 BREAK STRENGTH=   174.983
1200   464.16   950   0   338.4   381.6 BREAK STRENGTH=   170.046
1200   480   981.35   0   360   338.4 BREAK STRENGTH=   187.171
1200   495.84   918.65   0   381.6   360 BREAK STRENGTH=   192.397
1326   464.16   981.35   0   381.6   338.4 BREAK STRENGTH=   181.808
1326   480   918.65   0   338.4   360 BREAK STRENGTH=   167.363
1326   495.84   950   0   360   381.6 BREAK STRENGTH=   183.068
BREAK STRENGTH=   184.152
```

NOM FOR TC# 1 184.152 STD. DEV.= 9.29678
AVERAGE= 179.344
TYPE B S/N= 45.0401

Reprinted with permission from T. B. Barker, *Quality Progress* Dec. 1986.

to do what the Taguchi methods accomplish. However, the fact is that Taguchi's methods have catalyzed a major focus on quality engineering, and this is very important for society.

13-7 CAUSES OF DEFECTS AND FAILURES

Lack of quality results in defects, and defects in the material can result in failure. At the lowest level of severity in the classification of defects is a lack of conformance to a stated specification. This was the type of defect that was generally assumed in the discussion of quality in the earlier sections of this chapter. An example would be dimension "out of spec" or a strength level below specification. Next in severity is a lack of satisfaction by the user or customer. This may be caused by a critical performance characteristic set at an improper value, or it may be a system type problem caused by rapid deterioration. Finally, the ultimate defect is one which causes failure of the product. Failure may denote an actual fracture or disruption of physical continuity of the part, but failure can occur in less dramatic ways by simply causing the component to fail to function.

In Chap. 12 we considered the broad subject of risk and some of the techniques that can be used to anticipate failure and prevent them from occurring in the design process. In the rest of this chapter we consider the techniques for analyzing failures in components and translating failure analysis into increased design knowledge. Related to this is a consideration of nondestructive evaluation techniques for detecting cracks and other flaws. Finally, we discuss product liability, the legal implications of failure. Probably in no other area has the law relevant to engineers changed so rapidly and with such major impact on design.

Failures are caused by design errors or deficiencies in one or more of the following categories:

1. Design deficiencies
   Failure to adequately consider the effect of notches
   Inadequate knowledge of service loads and environment
   Difficulty of stress analysis in complex parts and loadings
2. Deficiency in selection material
   Poor match between service conditions and selection criteria
   Inadequate data on material
   Too much emphasis given to cost and not enough to quality
3. Imperfection in material due to manufacturing
4. Overload and other abuses in service
5. Inadequate maintenance and repair
6. Environmental factors
   Conditions beyond those allowed for in design
   Deterioration of properties with time of exposure to environment

Considering the importance of engineering failure to society, there is a small amount of literature on the subject. In addition to the volumes listed in the Bibliography, the report series of the British Engine, Boiler and Electrical Insurance Co., Ltd., Manchester, should be singled out for special mention. There is a natural understanding why organizations are reluctant to publish information about their failures; but until such information is available on a regular, routine basis, each generation will continue to make many of the mistakes of the preceding generation. There is a vital need for an industrywide program to collect information on failures and build up a national data base of failure experiences and solutions.

13-8 FAILURE MODES

Engineering components fail in service in the following general ways:

1. Excessive elastic deformation
2. Excessive plastic deformation
3. Fracture
4. Loss of required part geometry through corrosion or wear

The most common failure modes are listed in Table 13-9. For each failure mode the type of loading, type of stress, and operating condition generally associated with the failure mode are given. Also, the material property that usually is used in selecting a material to resist the failure mode is listed. This table should be combined with Fig. 6-3 for a more complete picture of failure mode versus material property.

It obviously is impossible in the space available to discuss in detail the complex interactions between metallurgical structure, stress state, and environment that produce these different modes of failure. For an understanding of material response the reader is referred to basic texts.[1] An introduction to mechanical failures is provided in the small book by Wulpi[2] and the books listed in the Bibliography. The most authoritative reference, which describes each failure mode in detail as well as the failure of specific components such as bearings, forgings, shafts, weldments, etc. is the ASM "Metals Handbook," 9th ed., vol. 11.

Failure-Experience Matrix

The failure-experience matrix[3] is an important attempt to place failure analysis on a firm analytical basis by classifying each failure with respect to failure mode, the elemental function that the component provided, and the corrective action that was taken to prevent recurrence of the failure. Thus, the failure-experience matrix is a three-dimensional assemblage of information cells (Fig. 13-7). The data for the failure-experience matrix was information on 500 failed parts from U.S. Army helicopters. Although the statistics on failure

---

[1] G. E. Dieter, "Mechanical Metallurgy," 3d ed., McGraw-Hill Book Company, New York, 1986; I. LeMay, "Principles of Mechanical Metallurgy," Elsevier North Holland, Inc., New York, 1981; M. G. Fontana and N. D. Greene, "Corrosion Engineering," 2d ed., McGraw-Hill Book Company, New York, 1978; W. Brostow and R. Corneliussen (eds.), "Failure of Plastics," Hanser Publ, 1986.

[2] D. J. Wulpi, "Understanding How Components Fail," ASM International, Metals Park, OH, 1985.

[3] J. A. Collins, B. T. Hagan, and H. M. Bratt, *Trans. ASME,* Ser. B, *J. Eng. Ind.,* vol. 98, pp. 1074–1079, 1976.

TABLE 13-9
Relation between common failure modes and conditions that produce the failure

Failure modes	Types of loading			Types of stress			Operating temperatures			Criteria generally useful for selection of material
	Static	Repeated	Impact	Tension	Compression	Shear	Low	Room	High	
Brittle fracture	×	×	×	×	—	—	×	×	—	Charpy V-notch transition temperature; notch toughness; K_{Ic} toughness measurements
Ductile fracture*	×	—	—	×	—	×	—	×	×	Tensile strength; shearing yield strength
High-cycle fatigue†	—	×	—	×	—	×	×	×	×	Fatigue strength for expected life, with typical stress raisers present
Low-cycle fatigue	—	×	—	×	—	×	×	×	×	Static ductility available and the peak cyclic plastic strain expected at stress raisers during prescribed life
Corrision fatigue	—	×	—	×	—	×	—	×	×	Corrosion-fatigue strength for the metal and contaminant and for similar time‡
Buckling	×	—	×	—	×	—	×	×	×	Modulus of elasticity and compressive yield strength
Gross yielding*	×	—	×	×	×	×	×	×	×	Yield strength
Creep	×	—	—	×	×	×	—	—	×	Creep rate or sustained stress rupture strength for the temperature and expected life‡
Caustic or hydrogen embrittlement	×	—	—	×	—	—	—	×	×	Stability under simultaneous stress and hydrogen or other chemical environment‡
Stress-corrosion cracking	×	—	—	×	—	×	—	×	×	Residual or imposed stress and corrosion resistance to the environment; K_{Iscc} measurements‡

Adapted from T. J. Dolan, *Experimental Mechanics*, pp. 1–14, January 1970.

From "Metals Handbook," 8th ed., vol. 10, p. 4, American Society for Metals, Metals Park, Ohio. Copyright American Society for Metals, 1975.

* Applies to ductile metals only.

† Millions of cycles.

‡ Items strongly dependent on elapsed time.

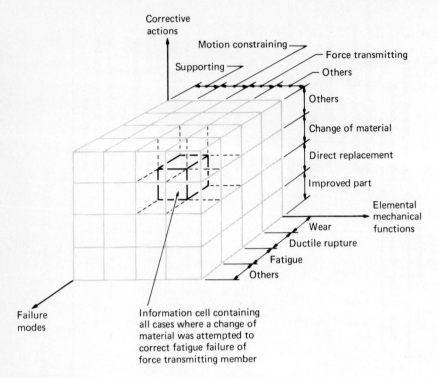

FIGURE 13-7
A portion of failure experience matrix for mechanical failures in helicopters.

modes most likely would be different for other types of equipment subjected to different service conditions, the method of approach should be applicable to all types of mechanical failures.

In Table 13-10 are listed the failure modes considered for this study. In the original paper they were ranked according to the frequency of occurrence in helicopter failures, but in this table they are grouped into more general categories following the scheme of Collins.[1] Note that not all failure modes listed in Table 13-10 are mutually exclusive. Several failure modes are combinations of two or more modes.

An elemental mechanical function is defined as a distinctive generic function of a machine part without reference to the specific application. That concept was applied by identifying 46 key words of mechanical functions and 40 antecedent adjectives. In many cases it was possible to describe the elemental mechanical function with a single key word, but in other cases an adjective was required. All told, 105 unique combinations of elemental

---

[1] J. A. Collins, "Failure of Materials in Mechanical Design," Wiley-Interscience, New York, 1981.

TABLE 13-10
Failure modes for mechanical components

1. Elastic deformation	9. Impact
2. Yielding	*a.* Impact fracture
3. Brinelling	*b.* Impact deformation
4. Ductile failure	*c.* Impact wear
5. Brittle fracture	*d.* Impact fretting
6. Fatigue	*e.* Impact fatigue
a. High-cycle fatigue	10. Fretting
b. Low-cycle fatigue	*a.* Fretting fatigue
c. Thermal fatigue	*b.* Fretting wear
d. Surface fatigue	*c.* Fretting corrosion
e. Impact fatigue	11. Galling and seizure
f. Corrosion fatigue	12. Scoring
g. Fretting fatigue	13. Creep
7. Corrosion	14. Stress rupture
a. Direct chemical attack	15. Thermal shock
b. Galvanic corrosion	16. Thermal relaxation
c. Crevice corrosion	17. Combined creep and fatigue
d. Pitting corrosion	18. Buckling
e. Intergranular corrosion	19. Creep buckling
f. Selective leaching	20. Oxidation
g. Erosion-corrosion	21. Radiation damage
h. Cavitation	22. Bonding failure
i. Hydrogen damage	23. Delamination
j. Biological corrosion	24. Erosion
k. Stress corrosion	
8. Wear	
a. Adhesive wear	
b. Abrasive wear	
c. Corrosive wear	
d. Surface fatigue wear	
e. Deformation wear	
f. Impact wear	
g. Fretting wear	

mechanical functions were discussed. They are listed in Table 13-11 in order of their occurrence in the helicopter failure study.

In the context of the failure-experience matrix a *corrective action* was defined as any measure or combination of steps taken to return a failed component or system to satisfactory performance. The 35 corrective actions are listed in Table 13-12 in descending order of frequency in the helicopter study.

If there were a computerized data base that encompassed a national inventory of failures, it would have great use in engineering design. An engineer who needed to design a critical component would enter the matrix with the classification of the elemental mechanical function and learn about the failure modes most likely to occur as well as the corrective actions most likely to avert failure. Such a system would also aid in correcting failures. When both

TABLE 13-11

Elemental mechanical functions

1. Supporting	54. Clutching
2. Attaching	55. Fastening
3. Motion-constraining	56. Information indicating
4. Force transmitting	57. Position indicating
5. Sealing	58. Movable lighting
6. Friction reducing	59. Partitioning
7. Protective covering	60. Position restoring
8. Liquid constraining	61. Flexible spacing
9. Pivoting	62. Electrical amplifying
10. Torque transmitting	63. Adjustable attaching
11. Pressure supporting	64. Shape constraining
12. Oscillatory sliding	65. Deflecting
13. Shielding	66. Disconnecting
14. Sliding	67. Electrical limiting
15. Energy transforming	68. Motion limiting
16. Removable fastening	69. Pressure limiting
17. Limiting	70. Sensing
18. Electrical conducting	71. Force sensing
19. Contaminant constraining	72. Spacing
20. Linking	73. Temporary supporting
21. Continuous rolling	74. Gas switching
22. Liquid transferring	75. Electrical transforming
23. Force amplifying	76. Power absorbing
24. Power transmitting	77. Information attaching
25. Covering	78. Sound absorbing
26. Oscillatory rolling	79. Constraining
27. Energy absorbing	80. Flexible coupling
28. Light transmitting	81. Removable coupling
29. Viewing	82. Damping
30. Energy dissipating	83. Electrical distributing
31. Guiding	84. Load distributing
32. Latching	85. Gas guiding
33. Electrical switching	86. Pressure indicating
34. Stabilizing	87. Electrical insulating
35. Gas constraining	88. Sound insulating
36. Permanent fastening	89. Temporary latching
37. Pressure increasing	90. Force limiting
38. Streamlining	91. Force maintaining
39. Motion reducing	92. Variable position maintenance
40. Filtering	93. Liquid pumping
41. Lighting	94. Electrical reducing
42. Pumping	95. Rolling
43. Gas transferring	96. Position sensing
44. Aero, force transmitting	97. Energy storing
45. Motion transmitting	98. Liquid storing
46. Signal transmitting	99. Flexible supporting
47. Motion damping	100. Switching
48. Force distributing	101. Pressure-to-torque transmitting
49. Reinforcing	102. Electrical transmitting
50. Pressure sensing	103. Flexible motion transmitting
51. Information transmitting	104. Flexible torque transmitting
52. Coupling	105. Torque limiting
53. Displacement indicating	

TABLE 13-12
Corrective actions for failure-experience matrix

Design change to improve part	Changed vendor
Direct replacement	Added adhesive
Change of material	Improved quality control
Supplement part	Changed lubricant type
Improved instructions to user	Improved run-in procedure
Changed dimensions	Applied surface treatment
Changed loading on part	Added sealant
Applied surface coating	Added or changed locking feature
Changed mechanism of operation	Adjusted part
Repositioned part	Provided drain
Repaired part	More easily replaceable part
Changed mode of attachment	Changed to correct part
Changed manufacturing procedure	Improved lubrication
Reinforced part	Made part interchangeable
Eliminated part	Relaxed replacement criteria
Strengthened part	Revised procurement specifications
Changed method or frequency	Provided for proper inspection
of lubrication	Changed electrical characteristics

the failure mode and the mechanical function were identified, the matrix would provide the corrective action most likely to solve the problem.

13-9 TECHNIQUES OF FAILURE ANALYSIS

When the problem of determining the cause of a failure and proposing corrective action must be faced, there is a definite procedure for conducting the failure analysis.[1] Frequently a failure analysis requires the efforts of a team of people, including experts in materials behavior, stress analysis, vibrations, and sophisticated structural and analysis techniques.

Field Inspection of the Failure

The most useful first approach is to inspect the failure at the site of the accident *as soon as possible* after the failure occurs. This site visit should be lavishly documented with photographs; for very soon the accident will be cleared away and repair begun. It is best to take photographs in color. Start taking pictures at a distance and move up to the site of the failure. Shoot pictures from several angles. Careful sketches and detailed notes help to orient

---

[1] G. F. Vander Voort, *Metals Eng. Quart.*, pp. 31–36, May 1975; R. Roberts and A. W. Pense, *Civil Eng.*, pp. 64–67, May 1980; pp. 60–62, July 1980.

the photographs and allow you to completely reconstruct the scene months or years later when you are in a design review or a courtroom.

The following critical pieces of information should be obtained during the field inspection.

1. Location of all broken pieces relative to each other
2. Identification of the origin of failure
3. Orientation and magnitude of the stresses
4. Direction of crack propagation and sequence of failure
5. Presence of obvious material defects, stress concentrations, etc.
6. Presence of oxidation, temper colors, or corrosion products
7. Presence of secondary damage not related to the main failure

It is important to interview operating and maintenance personnel to get their version of what happened and learn about any unusual operating history, such as unusual vibration or noise prior to failure. Whenever possible, the failure should be brought back to the laboratory for more detailed analysis. Any cutting that is required should be done well away from the fracture surface so as not to alter that surface. Whenever possible, samples should be obtained from identical material or components that did not fail. Samples of process fluids, lubricants, etc. should be obtained for corrosion-related failures. Be sure to label all pieces and key their identification to your notes.

Great care should be exercised in preserving the fracture surface. Never touch the fractured surfaces, and do not attempt to fit them back together. Avoid washing a fracture surface with water unless it has been contaminated with seawater or fire-extinguisher fluids. To prevent corrosion of a fracture surface, dry the surface with a jet of water-free compressed air and place the part in a desiccator or pack it with a suitable desiccant.

When the failure surface cannot be removed from the field for investigation in the laboratory, it is necessary to take the laboratory into the field. A portable metallographic laboratory has been developed for such a situation.[1]

Background History and Information

A complete case history on the component that failed should be developed as soon as possible. Ideally, most of this information should be obtained before making the site visit, since more intelligent questions and observations will result. The following is a list of data that need to be assembled.

1. Name of item, identifying numbers, owner, user, manufacturer or fabricator

---

[1] H. Crowder, "Metals Handbook," 8th ed., vol. 10, pp. 26–29.

2. Function of item
3. Data on service history, including inspection of operating logs and records
4. Discussion with operating personnel and witnesses concerning any unusual conditions or events prior to failure
5. Documentation on materials used in the item
6. Information on manufacturing and fabrication methods used, including any codes or standards
7. Documentation on inspection standards and techniques that were applied
8. Date and time of failure; temperature and environmental conditions
9. Documentation on design standards and calculations performed in the design
10. A set of shop drawings, including any modifications made to the design during manufacturing or installation

Macroscopic Examination

A macroscopic examination is made at magnifications ranging from $1 \times$ to about $100 \times$. Certainly this type of examination should occur at the site of the failure, but it is better repeated back in the laboratory where the lighting and other conditions are more favorable. The purpose of macroscopic examination is to observe the gross features of the fracture, the presence or absence of cracks, the presence of any gross defects, and the presence of corrosion or oxidation products. Most examination is done in the $1\times$ to $10\times$ range. An illuminated $10\times$ magnifier is a good tool for this type of study. Working in that magnification range, you should try to make an initial assessment of the origin of fracture and thus narrow down the region of the fracture for further study at higher magnification.

Often it is possible to identify the type of fracture from macroscopic examination. If there is gross deformation near the break and a dark fibrous texture to the fracture surface, you most likely have a ductile rupture. If the fracture surface is flat, with shiny flat grain facets visible in the surface, a brittle fracture is indicated. Often the surface of a brittle fracture has "chevron markings" pointing back to the origin of the fracture (Fig. 13-8*a*). Sometimes a

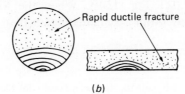

Rapid ductile fracture

(*a*)   (*b*)

FIGURE 13-8
Schematic drawings of fracture surfaces. (*a*) Brittle fracture of shaft and a plate; (*b*) fatigue fracture of a shaft and a plate.

brittle fracture has shear lips, regions of local deformation. In macroexamination, fatigue failures often show concentric rings or beach marks emanating from the origin of the fracture (Fig. 13-8b). The surface of a fatigue fracture generally is flat, with no shear lips, and is oriented normal to the largest tensile stress. A fracture surface that is heavily corroded or oxidized is usually evidence that the crack existed for a long time before finally propagating to failure. A large collection of macrophotographs of failures can be found in "Metals Handbook," 8th ed., vols. 9 and 10.

Microscopic Examination

A microscopic examination is made at magnifications greater than 100×. The term covers the use of such instruments as the metallurgical (reflected light) microscope, the scanning electron microscope (SEM), the transmission electron microscope (TEM), and the x-ray microprobe analyzer. All these instruments usually will be available in a modern metallurgical laboratory. When used properly, they provide highly valuable diagnostic information.

Metallographic examination with the metallurgical microscope requires a small section to be cut out, mounted, polished, and etched.[1] This type of examination is used to determine the microstructure of the material. The presence, size, and arrangement of phases is important documentation of the thermal and mechanical history of the metal.[2] Microstructural analysis will identify such features as grain size, inclusion size, and distribution of second phases. The technique also can be used to follow crack growth through the microstructure. For example, it can be used to determine whether a crack propagates in a transgranular or intergranular manner, whether a hard brittle phase cracks to initiate the fracture, or whether some microconstituent serves to impede crack propagation.

The scanning electron microscope (SEM) examines the actual surface of the fracture with a beam of electrons in an evacuated chamber (typically 1 by 2 by 5 in). A back-scattered image is recorded on a CRT display. Magnifications from 1000× to 40,000× are available. The image has great depth of field and a three-dimensional character. This makes SEM outstandingly useful for the examination of fractures. The crack propagation associated with a particular fracture mode leaves a characteristic appearance on the fracture surface. These *fractographs* are directly revealed by the SEM and provide an identification of the fracture mode[3] (see Fig. 13-9). When the SEM is equipped with an energy-dispersive x-ray probe, elements higher than atomic number 10 can be

---

[1] "Metals Handbook," 9th ed., vol. 10, American Society for Metals, Metals Park, Ohio, 1986.

[2] See "Metals Handbook," 9th ed, vol. 9, "Metallography and Microstructures", for a large collection of microstructures with extensive explanation and interpretation.

[3] See "Metals Handbook," 9th ed., vol. 12, "Fractography" and other references in Bibliography.

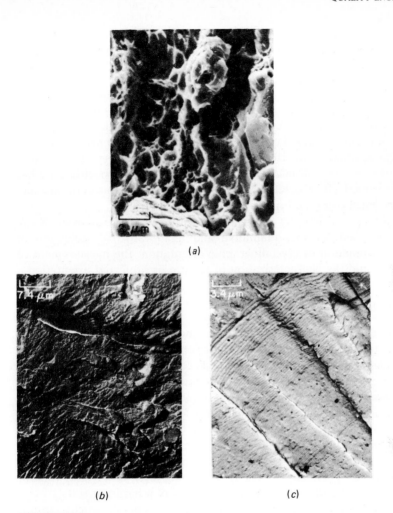

(a)

(b)   (c)

FIGURE 13-9
Examples of fractographs made with SEM. (a) Ductile rupture in 4340 steel; (b) cleavage type
features (river pattern) in steel; (c) striations from fatigue crack growth in aluminum alloy. Note:
(b) and (c) were made from plastic carbon replicas of the fracture surface. (*From "Metals
Handbook," 8th ed., vol. 9, American Society for Metals, Metals Park, Ohio. Copyright American
Society for Metals, 1974.*)

detected and analyzed quantitatively. This microanalysis capability is useful for
identifying inclusions, second phases, and corrosion products.

The transmission electron microscope (TEM) is used to examine the
microstructure and defect structure at magnifications up to 1,000,000× (on a
few special instruments). Since the electron beam must pass through it, the
specimen must either be a very thin foil or a plastic replica of the fracture
surface. The TEM is an important research instrument for microstructure
analysis; with selected area electron diffraction accessories, it is possible to

determine crystallographic relations between phases and to study solid-state transformations and the interplay of dislocations with structural features. However, because of its great depth of field and the ease of specimen preparation, the SEM has much greater application in fracture analysis.

Other sophisticated analytical instrumentation often is helpful in fracture analysis. In electron microprobe analysis the chemical composition of surface layers, precipitates, segregated regions, etc., is quantitatively determined. It is the same function that is carried out with the energy-dispersive probe on many SEM units. The scanning Auger electron spectrometer is a very sensitive tool for analyzing the first few atomic layers of a surface. Auger spectroscopy has been used very effectively in detecting trace elements such as bismuth, arsenic, and antimony, which cause embrittlement in steels.

X-ray diffraction techniques also may be useful in fracture analysis. X-ray methods can be used for the qualitative and quantitative identification of phases, the determination of crystallographic orientation, the measurement of residual stresses, and the characterization of texture or preferred orientation.

Additional Tests

It usually is necessary to obtain a number of other types of experimental data in order to put together the pieces of the puzzle that lead to the identification of the cause of failure. Determining the bulk composition of the material is part of the process of identifying the material and finding whether it meets the specification. Some failures are caused simply by a mixup of materials during manufacturing or maintenance. Quick spot tests for identifying materials in the field are available.[1]

It also is important to measure the mechanical properties of an unused specimen of the material that failed. Sometimes not enough unused material is available to machine specimens for tensile or impact tests, and it may then be necessary to "infer" the properties from the results of a hardness test.

It is important to probe the failed part with various nondestructive testing techniques (see Sec. 13-10) in order to search for flaws, seams, hidden cracks, etc. Sometimes a section of the material is deep-etched with acid to reveal defects such as segregation, hydrogen flakes, decarburized layer, or soft, spongy spots.

Analysis of the Data

The critical step is assembling the facts and pieces of data into a coherent picture of the cause of the failure. Early in the process it is common to develop

---

[1] "Identification of Metals and Alloys," bulletin published by International Nickel Co., New York; "Symposium on Rapid Methods for the Identification of Metals," ASTM Spec. Tech. Publ. No. 98, ASTM, Philadelphia, 1950; "Metals Handbook," 8th ed., vol. 11, pp. 270–286.

a hypothesis of the cause. All data should be cross-checked against the hypothesis, and any contradictions should be run down and either confirmed or discarded as spurious. When firm contradictions exist, they usually result in refining the hypothesis until gradually all pieces fit together. An experienced failure analyst not only considers the available data but will take note of the absence of features that experience suggests should be present. It is common for a failure to be caused by more than one factor. Therefore, developing a plausible hypothesis usually is not just a straightforward procedure.

Often when it becomes apparent that a critical piece of data is missing it will be necessary to reexamine the site of the failure and/or to start over again in analyzing the failure. At this stage it may be very helpful to assemble a team of experts from different disciplines so that as many varied viewpoints as possible can be brought to bear on the analysis.

Sometimes the failure resting of a model, or even a full-size duplicate of the failed unit, can be very illuminating. Computer models of the stress or temperature distribution, fatigue crack propagation rate, etc., may be needed in complex fracture problems.

Most failure analyses are failure-mechanism analyses which describe *what* failed and *how* the failure occurred. The question of *why* the failure occurred is often left to speculation or skipped over. To answer this important issue requires a *root-cause* failure analysis.[1] In a root-cause failure analysis every fact and item of evidence relating to the failed component and system is examined. Operating and maintenance people are interviewed in detail. The Kepner–Tregoe methodology for decision making[2] and fault tree analysis are useful tools for handling the large amount of information and its interrelationships.

Report of Failure

The report of the analysis of a failure is one of the most difficult written technical communications because a failure is often a matter of great sensitivity that may be fraught with legal implications. The best procedure is to stick to the hard facts, refrain from conjecture, and keep the technical jargon to a minimum. Additional tips on written technical communication will be found in Chap. 15.

13-10 NONDESTRUCTIVE EVALUATION

Nondestructive evaluation[3] is a method of engineering analysis in which the detection of material flaws and defects is combined with a prediction of the

[1] C. M. Jackson and R. D. Buchheit, *Mechanical Engineering*, July 1984, pp. 32–37.

[2] C. H. Kepner and B. B. Tregoe, "The New Rational Manager," Princeton Research Press, 1981.

[3] O. Buck and S. M. Wolf (eds.), "Nondestructive Evaluation: Microstructural Characterization and Reliability Strategies," TMS-AIME, Warrendale, Pa., 1981.

remaining life of the component with due consideration of the flaws. The detection of flaws by using a variety of physical probing technique is known as nondestructive testing (NDT) and nondestructive inspection[1] (NDI). The six principal methods used in NDI are listed and described briefly in Table 13-13. Great progress in NDI methods has been made in recent years by the application of microelectronics, signal processing methods, computer processing of data, and automation of NDI methods. Even with this progress, the inability of NDI methods to provide *quantitative* information on flaws is the main weak link in broadly implementing the nondestructive evaluation approach. Research continues[2] on improved NDI techniques.

Nondestructive Inspection Techniques

Radiography is based on the differential absorption of penetrating radiation, usually x-rays or gamma rays. Unabsorbed radiation passing through the part produces an image that is due to variations in thickness or density and is recorded on a photographic film, a fluorescent screen, or an electronic image intensifier. In general, radiography can detect only features that have an appreciable thickness in a direction parallel to the radiation beam. Thus, the ability of radiography to detect planar discontinuities such as cracks depends on proper orientation of the operator to obtain the optimum view and detect defects that might not show up on a single radiograph. The great advantage of radiography is that it is able to detect internal flaws that are located well below the surface of the part. Features that exhibit a 2 percent or greater difference in absorption compared to the surrounding material can be detected. Radiographic inspection is used extensively with castings and welded parts.

In *ultrasonic inspection* beams of high-frequency sound waves (1 to 25 MHz) are used to detect surface and subsurface flaws. Ultrasonics and radiography are the two main techniques for detecting internal defects. Flaws are detected by monitoring one or more of the following signals: 1) the reflection of energy from the surfaces of internal discontinuities, metal-gas interfaces, or metal-liquid interfaces, 2) the time of transit of a sound wave through the part, or 3) attenuation of a beam of sound waves by absorption and scattering within the metal. Ultrasonic inspection has superior penetrating power; high sensitivity, which permits detection of very small flaws; and greater accuracy than other NDI methods in determining the position of flaws and estimating their size and shape. However, ultrasonic signal interpretation can be very tricky. No other NDI method requires greater operator experience and knowledge.

---

[1] "Metals Handbook," 9th ed., vol. 17, "Nondestructive Evaluation and Quality Control," American Society for Metals, Metals Park, Ohio, 1989.

[2] R. B. Thompson and D. O. Thompson, *J. Metals*, pp. 29–34, July 1981.

TABLE 13-13

Comparison of basic nondestructive inspection methods

Method	Characteristics detected	Advantages	Limitations	Examples of use
Radiography	Changes in density from voids, inclusions, material variations; placement of internal parts	Can be used to inspect wide range of materials and thicknesses; versatile; film provides record of inspection	Radiation safety requires precautions; expensive; detection of cracks can be difficult	Pipeline welds for penetration, inclusions, voids
Ultrasonics	Changes in acoustic impedance caused by cracks, nonbonds, inclusions, or interfaces	Can penetrate thick materials; excellent for crack detection; can be automated	Normally requires coupling to material either by contact to surface or immersion in a fluid such as water	Adhesive assemblies for bond integrity
Visual-optical	Surface characteristics such as finish, scratches, cracks, or color; stain in transparent materials	Often convenient; can be automated	Can be applied only to surfaces, through surface openings, or to transparent material	Paper, wood, or metal for surface finish and uniformity
Eddy currents	Changes in electrical conductivity caused by material variations, cracks, voids, or inclusions	Readily automated; moderate cost	Limited to electrically conducting materials; limited penetration depth	Heat exchanger tubes for wall thinning and cracks
Liquid penetrant	Surface openings due to cracks, porosity, seams, or folds	Inexpensive; easy to use; readily portable; sensitive to small surface flaws	Flaw must be open to surface; not useful on porous materials	Turbine blades for surface cracks or porosity
Magnetic particles	Leakage magnetic flux caused by surface or near-surface cracks, voids, inclusions, or material or geometry changes	Inexpensive; sensitive both to surface and near-surface flaws	Limited to ferromagnetic materials; surface preparation and post-inspection demagnetization may be required	Railroad wheels or tracks

603

Visual-optical methods are limited to surface defects. The techniques are well established.

In *eddy current inspection* electromagnetic induction is used to detect surface defects like seams, laps, or cracks or to sort dissimilar metals to detect differences in composition or microstructure. It can also be used to measure the thickness of a nonconductive coating on a conductive metal. Because it is an electromagnetic induction technique, it does not require direct electrical contact with the part being inspected. Thus, the technique is adaptable to high-speed inspection in production. It is very versatile and can be applied to many inspection problems with electrically conducting materials. In using it one must be on guard for spurious signals due to material properties or characteristics unrelated to the problem of concern.

Liquid penetrant inspection depends on a liquid migrating into cavities and cracks that are open to the surface. The liquid should have high wetting ability, and the surface must be clean and dry before the penetrant is applied. Extremely rough surfaces are likely to give false indications of defects. After the liquid is allowed to penetrate any openings, the excess penetrant is removed and a developing agent is applied over the surface. The developer acts as a blotter to help the natural seepage of the penetrant out of cracks and to spread the penetrant at the edges so as to magnify the apparent width of the flaw. After the penetrant has been developed, it is examined visually for flaw indications in a good white light. Penetrants that contain fluorescent dyes are examined in ultraviolet (black) light. Liquid penetrant inspection is the simplest and least costly of NDI methods, but it has the major limitation that it can detect only flaws that are open to the surface.

Magnetic particle inspection is a technique for detecting surface and subsurface flaws in ferromagnetic materials. It utilizes the fact that surface flaws produce leakage magnetic fields that can be detected by the use of finely divided ferromagnetic particles applied over the surface. The collection of magnetized particles forms an outline of the surface flaw. Equipment is not overly complex, and there is little or no limitation because of the size or shape of the part being inspected. The basic limitation is that the method is applicable only to ferromagnetic materials. Demagnetization often is necessary after inspection, as is the cleaning of magnetic particles from the surface.

Other NDI techniques that are less well standardized but are growing in importance are acoustic emission, optical and acoustical holography, and thermography. Acoustic emission is the technique developing the most rapidly. It uses piezoelectric transducers to detect the transient elastic wave generated in a material by a rapid release of energy due to crack growth. The chief advantage of AE is the ability to continuously monitor a structure or machine and detect the degradation as it proceeds. Acoustic emission has been used in conjunction with ultrasonics and vibration signature analysis to detect incipient failure in rotating machinery.

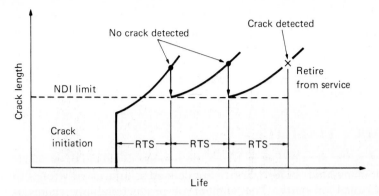

FIGURE 13-10
Schematic of the retirement-for-cause concept.

Retirement for Cause

For engineering systems of such nature that the consequences of failure are very great, e.g., aircraft engines and nuclear reactors, very conservative inspection procedures have been developed. For example, gas turbine engine rotors are limited by low-cycle fatigue life. Present performance standards require that engines be so designed that no more than one engine in 1000 would fail in the designed service life. After a turbine has served for one design life, all its parts are removed from service. However, because of the statistical distribution of fatigue life for nominally identical components, many of the components that are retired from service still have a large part of their useful life remaining.

A retirement-for-cause (RFC) approach[1] would allow each component to be used to the full extent of its total safe fatigue life, since it would be retired from service only when an observable defect reached a certain size. The defect size at which the component is no longer considered safe is established by combining NDI with fracture mechanics analysis. Figure 13-10 shows the retirement-for-cause concept. It is assumed that the crack initiation state is complete when the component has a crack just too small to be detected by the NDI method. A suitable fracture mechanics model and crack propagation data must be available to provide a reliable estimate of the crack length vs. life curve. This establishes the return-to-service (RTS) interval. At each RTS interval the component is subjected to NDI; and if no crack is detected, it is returned to service for an additional interval. To implement this maintenance

---

[1] C. G. Annis, J. S. Cargill, J. A. Harris, and M. C. Van Wanderham, *J. Metals*, pp. 24–27, July 1981.

philosophy, it is imperative to have an NDI capability with acceptable resolution and reliability. Because there is a probability of finding a given size flaw with a given NDI capability, it is necessary to use probabilistic fracture mechanics. Computer simulation methods have been developed for selection of the RTS interval to obtain an acceptably low failure probability with realistic NDI reliability.

13-11 PRODUCT LIABILITY

Product liability[1] is the area of law in which the liability of sellers of products falls. It is one of many kinds of *torts,* which are defined as injuries or wrongs to a person or personal property. The number of product liability claims is increasing very rapidly because of recent changes in the interpretation of the law. Not only has the number of legal actions increased by a factor of 20 over a 15-year period but the cost of product liability insurance also has increased five-fold in ten years.

For over 100 years the legal concept of privity restricted the widespread use of product liability actions. *Privity* is the relationship which exists between the buyer and the seller or two or more contracting parties. The courts held that the injured party could sue only the party in privity. Thus, if a consumer was injured by a power tool, he could sue only the retailer who sold him the tool; the retailer, in turn, could sue only the wholesaler, who in turn could sue the manufacturer. Now that the courts have abandoned the concept of privity in product liability, the injured consumer can sue all members in the manufacturing chain. From the viewpoint of recovering damages it obviously is an advantage to be able to directly sue the manufacturer, whose resources are likely to be much greater than those of the owner of the neighborhood hardware store.

A second major change in the law of product liability is the almost universal adoption by the courts of the standard of strict liability. Previously manufacturers or sellers were liable only when they could be proved negligent or unreasonably careless in what they made or how they made it. It had to be proved that a reasonable manufacturer using prudence would have exercised a higher standard of care. However, today in most states a standard of strict

---

[1] A. A. Weinstein, A. D. Twerski, H. R. Piehler, and W. A. Donaher, "Products Liability and the Reasonably Safe Product," John Wiley & Sons, Inc., New York, 1978; J. Kolb and S. S. Ross, "Product Safety and Liability—A Desk Reference," McGraw-Hill Book Company, New York, 1980; V. J. Colangelo and P. A. Thornton, "Engineering Aspects of Product Liability," American Society for Metals, Metals Park, Ohio, 1981; J. F. Thorpe and W. H. Middendorf, "What Every Engineer should Know about Product Liability," Marcel Dekker Inc., New York, 1979; C. O. Smith, "Products Liability: Are You Vulnerable?", Prentice-Hall Inc., Englewood Cliffs, N.J., 1981; C. E. Witherell, "How to Avoid Products Liability Lawsuits and Damages," Noyes Publications, 1985.

liability is applied. Under this theory of law the plaintiff must prove that: 1) the product was defective and unreasonably dangerous, 2) the defect existed at the time the product left the defendant's control, 3) the defect caused the harm, and 4) the harm is appropriately assignable to the identified defect. Thus, the emphasis on responsibility for product safety has shifted from the consumer to the manufacturer of products.

A related issue is the use for which the product is intended. A product intended to be used by children will be held to a stricter standard than one intended to be operated by a trained professional. Under strict liability a manufacturer may be held liable even if a well-designed and well-manufactured product injured a consumer who misused or outright abused it.

Goals of Product Liability Law

Only 100 years ago it was the practice in American and British law not to respond to accidental losses. It was generally held that the accident victim, not the manufacturer, should bear the economic burdens of injury. Starting in the mid-twentieth century the law began to assume a more active role. Product liability law evolved to serve four basic societal goals: loss spreading, punishment, deterrence, and symbolic affirmation of social values.[1] Loss spreading seeks to shift the accidental loss from the victim to other parties better able to absorb or distribute it. In a product liability suit the loss is typically shifted to the manufacturer, who theoretically passes this cost on to the consumer in the form of higher prices. Often the manufacturer has liability insurance, so the cost is spread further, but at the price of greatly increased insurance rates.

Another goal of product liability law is to punish persons or organizations responsible for causing needless loss. It is important to recognize that under liability law the designer, not just the company, may be held responsible for a design defect. In extreme cases, the punishment may take the form of criminal penalties, although this is rare. More common is the assessment of punitive damages for malicious or willful acts. A third function is to prevent similar accidents from happening in the future, i.e., deterrence. Substantial damage awards against manufacturers constitute strong incentives to produce safer products. Finally, product liability laws act as a kind of symbolic reaffirmation that society values human safety and quality in products.

Negligence

A high percentage of product litigation alleges engineering negligence. Negligence is the failure to do something that a reasonable man, guided by the

---

[1] D. G. Owen, The Bridge, Summer, 1987, pp. 8–12.

considerations that ordinarily regulate human affairs, would do. In product liability law the seller is liable for negligence in the manufacture or sale of any product that may *reasonably be expected* to be capable of inflicting substantial harm if it is defective. Negligence in design is usually based one of three factors.

1. That the manufacturer's design has created a concealed danger.
2. That the manufacturer has failed to provide needed safety devices as part of the design of the product.
3. That the design called for materials of inadequate strength or failed to comply with accepted standards.

   Another common area of negligence is failure to warn the user of the product concerning possible dangers involved in the product use. This should take the form of warning labels firmly affixed to the product and more detailed warnings of restrictions of use and maintenance procedures in the brochure that comes with the product.

Strict Liability

Under the theory of strict liability it is not necessary to prove negligence on the part of the manufacturer of the product. The plaintiff need only prove that 1) the product contained an unreasonably dangerous defect, 2) that the defect existed at the time the product left the defendant's hands, and 3) the defect was the cause of the injury. The fact that the injured party acted carelessly or in bad faith is not a defense under strict liability standards. More recently the courts have acted so as to the require the manufacturer to design his product in such a way as to anticipate foreseeable use and abuse by the user.

   The criteria by which the defective and unreasonably dangerous nature of any product[1] may be tested in litigation are:

1. The usefulness and desirability of the product
2. The availability of other and safer products to meet the same need
3. The likelihood of injury and its probable seriousness
4. The obviousness of the danger
5. Common knowledge and normal public expectation of the danger
6. The avoidability of injury by care in use of the warnings
7. The ability to eliminate the danger without seriously impairing the usefulness of the product or making the product unduly expensive

---

[1] H. R. Piehler, A. D. Twerski, A. S. Weinstein, and W. A. Donaher, *Science,* vol. 186, p. 1093, 1974.

Design Aspect of Product Liability

Court decisions on product liability coupled with consumer safety legislation have placed greater responsibility on the designer for product safety. The following aspects of the design process should be emphasized to minimize potential problems from product liability.

1. Take every precaution that there is strict adherence to industry and government standards. Conformance to standards does not relieve or protect the manufacturer from liability, but it certainly lessens the possibility of product defects.

2. All products should be thoroughly tested before being released for sale. An attempt should be made to identify the possible ways a product can become unsafe (see Sec. 12-6), and tests should be devised to evaluate those aspects of the design. When failure modes are discovered, the design should be modified to remove the potential cause of failure.

3. The finest quality-control techniques available will not absolve the manufacturer of a product liability if, in fact, the product being marketed is defective. However, the strong emphasis on product liability has placed renewed emphasis on quality engineering as a way to limit the incidence of product liability.

4. Make a careful study of the system relations between your product and upstream and downstream components. You are required to know how malfunctions upstream and downstream of your product may cause failure to your product. You should warn users of any hazards of foreseeable misuses based on these system relationships.

5. Documentation of the design, testing, and quality activities can be very important. If there is a product recall, it is necessary to be able to pinpoint products by serial or lot number. If there is a product liability suit, the existence of good, complete records will help establish an atmosphere of competent behavior. Documentation is the single most important factor in winning or losing a product liability lawsuit.

6. The design of warning labels and user instruction manuals should be an integral part of the design process. The appropriate symbols, color, and size and the precise wording of the label must be developed after joint meetings of the engineering, legal, marketing, and manufacturing staffs. Use international warning symbols.

7. Create a means of incorporating legal developments in product liability into the design decision process. It is particularly important to get legal advice from the product liability angle on new innovative and unfamiliar designs.

Business Procedures to Minimize Risk

In addition to careful consideration of the above design factors, there are a number of business procedures that can minimize product liability risk.

1. There should be an active product liability and safety committee charged with seeing to it that the corporation has an effective product liability loss control and product safety program. This committee should have representatives from the advertising, engineering, insurance, legal, manufacturing, marketing, materials, purchasing, and quality-control departments of the corporation.
2. Insurance protection for product liability suits and product recall expenses should be obtained.
3. Develop a product usage and incident-reporting system just as soon as a new product moves into the marketplace. It will enable the manufacturer to establish whether the product has good customer acceptance and detect early signs of previously unsuspected product hazards or other quality deficiencies.

Problems with Product Liability Law

As product liability has grown so rapidly certain problems have developed in the implementation of the law.[1] There has been a dramatic shift in the doctrine of product liability law from negligence to strict liability but the law has proved incapable of defining the meaning of strict liability in a useful fashion. The rules of law are vague, which gives juries little guidance, and as a result verdicts appear capricious and without any definitive pattern. Another problem concerns the computation of damages once liability is established. There is great uncertainty and diversity in awarding damages for pain and suffering. Our adversarial legal system and the unfamiliarity of juries with even the rudiments of technical knowledge lead to high costs and much frustration.

The great increases in the number of product liability claims and the dollars awarded by the courts to consumers, other companies, and government have brought a clamor to bring some restraint to the situation before we become a no-fault economy in which producers and sellers will be held responsible for all product-related injuries. Advocates of reform point to product liability insurance costs and damage awards as a significant factor in reducing American competitiveness. National product liability legislation has been introduced in the U.S. Congress to ease the situation. It aims at making tort law on product liability uniform in all the states and on speeding up product liability disputes. It proposes a limit on joint and several liability, a doctrine by which a defendant responsible for only a small portion of harm may be liable for an entire judgement award. It also calls for a limit on a product seller's liability to cases in which the harm was proximately caused by the seller's own lack of reasonable care or a breach of the seller's warranty.

Probably the best thing that could happen in product liability would be

---

[1] D. G. Owen, op. cit.

the adoption of standardized liability laws on a nationwide basis. Such standardization would mean more predictability, less litigation, and lower premiums for liability insurance.

BIBLIOGRAPHY

Quality engineering

Burgess, J. A.: "Design Assurance for Engineers and Managers," Marcel Dekker Inc., New York, 1984.
Crosby, P. B.: "Quality is Free," McGraw-Hill Book Co., New York, 1979.
Dhillon, B. S.: "Quality Control, Reliability, and Engineering Design," Marcel Dekker Inc., New York, 1985.
Groocock, J. M.: "The Chain of Quality," John Wiley & Sons, New York, 1986.
Ishikawa, K.: "What is Total Quality Control? The Japanese Way," Prentice-Hall Inc., Englewood Cliffs, NJ, 1985.
Juran, J. M., and F. M. Gryna (eds.): "Juran's Quality Control Handbook," 4th ed., McGraw-Hill Book Co., New York, 1988.
Stebbing, L.: "Quality Assurance," 2d ed., John Wiley & Sons, New York, 1989.

Failure analysis

Burke, J. J., and V. Weiss (eds.): "Nondestructive Evaluation of Materials," Plenum Press, New York, 1979.
"Case Histories in Failure Analysis," American Society for Metals, Metals Park, Ohio, 1979.
Colangelo, V. J., and F. A. Heiser: "Analysis of Metallurgical Failures," 2d ed., John Wiley & Sons, Inc., New York, 1987.
Fisher, J. W.: "Fatigue and Fracture in Steel Bridges (Case Studies)," John Wiley & Sons, New York, 1984.
Hutchings, F. R., and P. M. Unterweiser: "Failure Analysis: The British Engine Technical Reports," American Society for Metals, Metals Park, Ohio, 1981.
"Metals Handbook," 9th ed., vol. 11, "Failure Analysis and Prevention," American Society for Metals, Metals Park, Ohio, 1986.
Naumann, F. K.: "Failure Analysis: Case Histories and Methodology," translated from German, American Society for Metals, Metals Park, Ohio, 1980.
Petroski, H.: "To Engineer is Human," The Role of Failure in Successful Design," St. Martin Press, New York, 1985.
Rossmanith, H. P. (ed.): "Structural Failure, Product Liability and Technical Insurance," North-Holland, New York, 1984.
Shives, T. R. and W. A. Willard (eds.): "Mechanical Failure—Definition of the Problem," NBS Spec. Publ. 423, U.S. Government Printing Office, 1976.
Whyte, R. R. (ed.): "Engineering Progress Through Trouble," The Institution of Mechanical Engineers, London, 1975.

Fractography

Bhattacharya, S. (ed.): "ITTRI Fracture Handbook: Fracture Analysis of Metallic Materials by Scanning Electron Microscopy," ITTRI, Chicago, 1978.
"Metals Handbook," 9th ed., vol. 12, "Fractography," American Society for Metals, Metals Park, Ohio, 1987.
"SEM/TEM Electron Fractography Handbook," Metals and Ceramics Information Center, Battele Columbus Laboratories, Ohio, 1975.
Strauss, B. M., and W. H. Cullen Jr.: "Fractography in Failure Analysis," ASTM STP 645, AST, Philadelphia, 1978.

CHAPTER
14

SOURCES OF
INFORMATION

14-1 THE INFORMATION PROBLEM

We have already seen that the need for information can be crucial in a design project. You need information, and you need it quickly. Moreover, the information required for an engineering design may be more diverse and less readily available than that needed for conducting a research project, for which the published technical literature is the main source of information.

This chapter gives some suggestions for coping with the information problem. It is not intended to be encyclopedic or contain all the information on how and where to look. The first step you should take is to become familiar with your local information sources. Visit your university or company library and make friends with the librarian. Find out what is available and what your organization is prepared to do to help you with your information needs.

The next thing you should do is develop a personal plan for coping with information. The world technical literature is doubling every 10 to 15 years. That amounts to about 2 million technical papers a year, or a daily output that would fill seven sets of the Encyclopedia Britannica. This tremendous flood of information aids greatly in the development of new knowledge, but in the process it makes obsolete part of what you already know. To develop a personal plan for information processing is one of the most effective things you can do to combat your own technological obsolescence. Such a plan begins with the recognition that you cannot leave it entirely to your employer to finance your needs in this area. As a professional, you should be willing to allocate a small portion of your resources, e.g., 1 percent of your net annual salary, for adding to your technical library and your professional growth. This includes the purchase of new textbooks in fields of current or potential interest, specialized monographs, membership in professional societies, and subscriptions to technical journals and magazines. In one way or another, you should attend conference and technical meetings where new ideas on subjects related to your interest are discussed.

You should develop your own working files of technical information that is important to your work. A good way to do so is to put articles and information you want to have as ready reference into large three-ring binders with page dividers marking off different areas of interest. Material of less current interest may be stored in folders in filing cabinets. A common difficulty, once you start this activity, is compulsive saving. In order that your files will not grow without bound, be selective. Adopt the policy of discarding outdated material when you replace it with newer information. If you are concerned with losing track of possibly useful articles by discarding older material, you might compromise by keeping only the title page, which contains the abstract, and also possibly the last page, which contains the conclusions and references.

To have current awareness of your technical field, you should take a three-pronged approach:[1] 1) read the core journals in your chief area of interest, 2) utilize current awareness services, 3) participate in selective dissemination programs.

Every professional must read enough journals and technical magazines to keep up with the technology in the field and be able to apply the new concepts that have been developed. These journals, which should be read on a monthly basis, should come from three categories:

1. General scientific, technical, and economic (business) news. The monthly magazine of your main professional society would fit here.
2. Trade magazines in your area of interest or business responsibility.
3. Research-oriented journals in your area of interest.

Reading regularly in the above three categories is a major aspect of keeping current in your field. However, for many people this will not cover as wide a spectrum of the published literature as is required. Therefore, secondary current awareness services have been developed. Some of them provide abstracts of articles, others just the titles of articles. The abstract

---

[1] B. E. Holm, "How to Manage Your Information," Reinhold Book Corp., New York, 1968.

services available in each discipline, e.g., *Engineering Index* and *Metals Abstracts,* are scanned by some serious professionals each month to see what new information has been published. This usually requires spending several hours in the library. *Current Contents: Engineering and Technology* is a weekly publication of the Institute for Scientific Information, Philadelphia, which reproduces the title pages of a large number of engineering publications. A computerized index provides the addresses of authors if you want to write for reprints. The advantage of *Current Contents* is that it is less expensive than a full abstract service, so you can have an individual or shared subscription. Your technical library should have a copy.

Selective dissemination is concerned with sending specific information to the individual who has a need for and interest in it. Many company librarians provide such a service. Researchers in a common field will often develop a "community of interest" and keep each other informed by sharing their papers and ideas. As more and more technical information is put into computer data base, it becomes easier to provide selective dissemination.

There is general recognition that engineers as professionals make much less use of the published technical literature than do scientists and other technical professionals. Some of this difference may be due to differences in education, but mostly it is due to a basic difference in how technical information is organized and stored.[1] Almost all major indexing systems index by subjects, i.e., the names of things. This is ideal for the expert concerned with research or analysis, who is very familiar with the vocabulary in a narrow technical field. However, indexing by name is a severe handicap for the design engineer, who is involved in synthesis. The designer, concerned with finding the best way to solve a problem, wants to start the search with the function of what is needed. For example, Table 13-11 lists 105 unique combinations of mechanical functions. There is a vast array of devices for attaining those functions, but there is no organized way to help the designer identify the functions by name so they can be uncovered in a literature search. Also, the synthesis-oriented design engineer is concerned with attributes, e.g., higher energy efficiency and lower material cost. Although they frequently are described in technical articles, the attributes very seldom are included in indexing terms.

Thus, it is not surprising that a recent study revealed that engineers collect most of their advanced technical information from discussions with vendors, salesmen, consultants, and other engineers. Partly that is because information on new products, the buildings blocks of technology, is poorly handled in the technical literature. A 12-month study[2] of news releases on new

---

[1] E. J. Breton, *Mech. Eng.,* pp. 54–57, March 1981.

[2] ibid.

products in trade magazines showed that over 65 percent were actually old products or were old products with cosmetic alterations.

14-2 COPYRIGHT AND COPYING

A copyright is the exclusive legal right to publish a tangible expression of literary or artistic work, and it is therefore the right to prevent the unauthorized copying by another of that work. In the United States a copyright is awarded for a period of the life of the copyright holder.

A major revision of the copyright law of 1909 went into effect on January 1, 1978. The present copyright law covers original works of authorship that are literary works as well as pictorial, graphic, and sculptural works. Important for engineering design is the fact that the new law is broad enough to cover for the first time written engineering specifications, sketches, drawings, and models.[1] However, there are two important limitations to this coverage. Although plans, drawings, and models are covered under the copyright law, their mechanical or utilitarian aspects are expressly excluded. Thus, the graphic portrayal of a useful object may be copyrighted, but the copyright would not prevent the construction from the portrayal of the useful article that is illustrated.

The other limitation pertains to the fundamental concept of copyright law that one can copyright not an idea, but only its tangible expression. The protection offered the engineer under the new law lies in the ability to restrict the distribution of plans and specifications by restricting physical copying. An engineer who retains ownership of plans and specifications through copyrighting can prevent a client from using them for other than the original, intended use and can require that they be returned after the job is finished.

A major impetus for revising the copyright law was to make the law compatible with the technology of fast, cheap copying machines. The new law retains for the individual student or engineer the right to make a single copy of copyrighted material for personal use. However, fees must be paid for multiple copies of journal articles for classroom use or circulation to colleagues. The new law could have major impact in large industrial libraries that have standing orders for copies in selected areas and supply copies on request to large numbers of people. It is not likely to have much impact on university libraries, which rarely have offered such services.

14-3 HOW AND WHERE TO FIND IT

The search for information can be performed more efficiently if a little thought and planning are used at the outset. First, be sure you understand the purposes

---

[1] H. K. Schwentz and C. J. Hardy, *Professional Engineer,* pp. 32–33, July 1977.

TABLE 14-1
Sources of information for engineering design

I. Public sources
  A. Federal departments and agencies (Defense, Commerce, Energy, NASA, etc.)
  B. State and local government (highway department, departments dealing with land use, consumer safety, building codes, etc.)
  C. Libraries—community, university, special
  D. Universities, research institutions, museums
  E. Foreign governments—embassies, commercial attaches
II. Private sources
  A. Nonprofit organizations and services
    1. Professional societies
    2. Trade and labor associations
    3. Membership organizations (motorists, consumers, veterans, etc.)
  B. Profit-oriented organizations
    1. Vendors (include manufactures, suppliers, financiers). Catalogs, samples, test data, cost data and information on operation, maintenance, servicing and delivery
    2. Other business contacts with manufacturers and competitors
    3. Consultants
  C. Individuals
    1. Direct conversation or correspondence
    2. Personal friends, associates, "friends of friends"
    3. Faculty

for which the information is being sought. If you are looking for a specific piece of information, e.g., the yield strength of a new alloy or the cost of a miniature ball bearing, you should pursue one set of information sources. However, if your purpose is to become familiar with the state of the art in an area that is new to you, you should follow a different course. Table 14-1, based on a listing by Woodson,[1] shows the many sources of information that are open to you.

In reviewing this list, you can divide the sources of information, into 1) people who are paid to assist you, e.g., the company librarian or consultant, 2) people who have a financial interest in helping you, e.g., a potential supplier of equipment for your project, and 3) people who help you out of professional responsibility or friendship.

All suppliers of materials and equipment provide sales brochures, catalogs, technical manuals, etc., that describe features and operation of their products. Usually this information can be obtained at no cost by checking the reader service card that is enclosed in most technical magazines. Practicing engineers commonly build up a file of such information. Generally if a supplier

---

[1] T. T. Woodson, "Introduction to Engineering Design," chap. 5, McGraw-Hill Book Company, New York, 1966.

has reason to expect a significant order based on your design, he will most likely provide any technical information about his product that is needed for you to complete your design.

Large technology-based companies generate internally a great deal of technical information, most of which never is published in the technical literature. However, some of this information is made available to the general public for a subscription fee. General Electric Company markets an *Engineering Materials and Processes Information Systems* that provides detailed information on properties, specifications, and data for ordering on more than 11,000 materials. General Electric[1] also publishes the very detailed *Fluid Flow Data Book* and *Heat Transfer Book* on a subscription basis.

Important sources of critically evaluated design data and procedures are the data packages available from Engineering Sciences Data Unit (ESDU)[2] on a subscription basis. Originally developed by the Royal Aeronautical Society (Great Britain), the scope has been expanded to include heat transfer, thermodynamic and transport properties, strength of machine components, mechanisms, cams and gears, two-phase flow, and structural steel analysis.

It is only natural to concentrate on searching the published technical literature for the information you need, but don't overlook the resources available among your colleagues. The professional files or notebooks of engineers more experienced than you can be a gold mine of information if you take the trouble to communicate your problem in a proper way. Remember, however, that the flow of information should be a two-way street. Be willing to share what you know, and above all, return the information promptly to the person who lent it to you. The surest way to get shut off is to gain a reputation as a moucher.

In seeking information from sources other than libraries (see Sec. 14-4), a direct approach is best. Whenever possible, use a phone call rather than a letter. A direct dialogue is vastly superior to the written word. However, you may want to follow up your conversation with a letter. Open your conversation by identifying yourself, your organization, the nature of your project, and what it is you need to know. Preplan your questions as much as possible, and stick to the subject of your inquiry. Don't worry about whether the information you seek is confidential information. If it really is confidential, you won't get an answer, but you may get peripheral information that is helpful. Above all, be courteous in your manner and be considerate of the time you are taking from the other person.

It may take some detective work to find the person to call for the information. You may find the name of a source in the published literature or in the program from a recent conference you attended. The Yellow Pages in

---

[1] Available from Genium Publ. Co., 1145 Catalyn St., Schenectady, NY 12303–1836.

[2] ESDU International Ltd., 1495 Chain Bridge Road, McLean VA 22101.

the telephone directory are a good place to start. Most libraries contain telephone directories for major U.S. cities. For product information, you can start with the general information number that is listed for almost every major corporation. To locate federal officials, it is helpful to use one of the directory services that maintain up-to-date listings and phone numbers.

It is important to remember that information costs time and money. It is actually possible to acquire too much information in a particular area, far more than is needed to make an intelligent decision. One can consider that each decision in the design process is a balance between the risk of proceeding with what you have versus the cost of gaining more information to minimize the risk.

14-4 LIBRARY SOURCES OF INFORMATION

In the preceding section we considered the broad spectrum of information sources and focused mostly on the information that can be obtained in the business world. In this section we shall deal with the type of information that can be obtained from library sources. The library is the most important resource for students and young engineers who wish to develop professional expertise quickly.

A library is a repository of information that is published in the open or unclassified literature. Although the scope of the collection will vary with the size and nature of the library,[1] all technical libraries will have the capability of borrowing books and journals for you or providing, for a fee, copies of needed pages from journals and books. Many technical libraries also carry selected government publications and patents, and compnay libraries will undoubtedly contain a collection of company technical reports (which ordinarily are not available outside the company).

When you are looking for information in the library you will find a hierarchy of information sources, as shown in Table 14-2. These sources are arranged in increasing order of specificity. Where you enter the hierarchy depends on your own state of knowledge about the subject and the nature of the information you want to obtain. If you are a complete neophyte, it may be necessary to use a technical dictionary and read an encyclopeida article to get a good overview of the subject. If you are quite familiar with the subject, then you may simply want to use an index or abstract service to find pertinent technical articles. Most sources of information will be found in the reference section of the library.

---

[1] If you do not have a good technical library at your disposal, you can avail yourself via mail of the fine collection at the Engineering Societies Library, United Engineering Center, 345 East 47th Street, New york, NY 10017: (212) 705–7828.

TABLE 14-2
Hierarchy of library information sources

Technical dictionaries
Encyclopedias
Handbooks
Textbooks and monographs
Bibliographies
Indexing and abstract services
Technical and professional journals
Translations
Technical reports
Catalogs and manufacturers' brochures

The search for information can be visualized along the paths shown in Fig. 14-1. Starting with a limited information base, you should consult technical encyclopedias and library's public access catalog, today automated in most libraries, to search out broad introductory texts. As you become expert in the subject, you should move to more detailed monographs and/or use abstracts and indexes to find pertinent articles in the technical literature. Reading these articles will suggest other articles (cross references) that should be consulted. Another route to important design information is the patent literature (Sec. 14-6).

The task of translating your own search needs into the terminology that appears in the library card catalog is often difficult. As mentioned previously, library catalogs, whether in card form or on-line, are developed for more traditional scholarly and research activities. The kinds of questions raised in the context of engineering design may cut through the card catalog at an "oblique section." When trying to convey the needs and objectives of your search to the librarian, the best tactic is to tell the librarian *what it is you do not know* rather than describe what you already know. Two parameters that describe the efficiency of your information search are:

$$\text{Precision} = \frac{\text{number of relevant documents retrieved}}{\text{total number retrieved}}$$

$$\text{Recall} = \frac{\text{number of relevant documents retrieved}}{\text{number of relevant documents in the collection}}$$

Dictionaries and Encyclopedias

At the outset of a project dealing with a new technical area there may be a need to acquire a broad overview of the subject. English language technical dictionaries usually give very detailed definitions. Also, they often are very

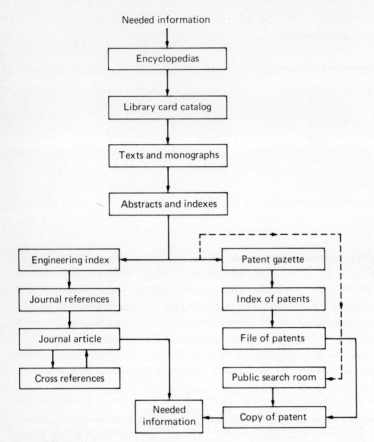

FIGURE 14-1
Flow diagram for an information search.

well illustrated. Some useful references are:

Ballentyne, D. W. G., and D. R. Lovett (eds.): "A Dictionary of Named Effects and Laws in Chemistry, Physics and Mathematics," 3d ed., Chapman and Hall, London, 1970, 335 pp.

Classon, W. E.: "Elsevier's Dictionary of Metallurgy and Metal Working," in English, French, Spanish, Italian, Dutch, and German, Elsevier Sci. Pub., New York, 1978, 848 pp.

Horner, J. G.: "Dictionary of Mechanical Engineering Terms," 9th ed., Heinemann, New York, 1967, 422 pp.

"McGraw-Hill Dictionary of Science and Engineering," McGraw-Hill Book Co., New York, 1984, 942 pp.

O'Bannon, L. S.: "Dictionary of Ceramic Science and Engineering," Plenum, New York, 1984, 303 pp.

Veilleux, R. (ed.): "Dictionary of Manufacturing Terms," Soc. of Manufacturing Engineers, Dearborn, MI, 1987, 133 pp.

Technical encyclopedias are written for the technically trained person who is just beginning to learn about a new subject. In using an encyclopedia, spend some time checking the index for the entire set of volumes to discover subjects that you would not have looked up by instinct. Some useful technical encyclopedias are:

Besancon, R. M. (ed.): "Encyclopedia of Physics," 2d ed., Van Nostrand Reinhold, New York, 1974, 848 pp.

Bever, M. B. (ed.): "Encyclopedia of Materials Science and Engineering," 8 vols., The MIT Press, Cambridge Mass., 1986, 5551 pp.

Considine, D. (ed.): "Chemical and Process Technology Encyclopedia," McGraw-Hill Book Co., New York, 1974, 1261 pp.

Grayson, M. (ed.): "Encyclopedia of Glass, Ceramics and Cement," John Wiley & Sons, New York, 1985.

"McGraw-Hill Encyclopedia of Science and Technology," 6th ed., McGraw-Hill Book Co., New York, 1987, 20 vols.

A vital reference book that is neither a dictionary or an encyclopedia is "Books in Print." This reference book is available in nearly every library. It will let you find information on books published under specific subject headings or by title or author.

Handbooks

Undoubtedly, some place in your engineering education a professor has admonished you to reason out a problem from "first principles" and not be a "handbook engineer." That is sound advice, but it may put handbooks in a poor perspective that is undeserved. Handbooks are compendia of useful technical data. Many handbooks also provide ample technical description of theory and its application, so they are good refereshers of material once studied in greater detail. You will find that an appropriately selected collection of handbooks will be a vital part of your professional library.

An extensive list of handbook dealing with the material properties of metals, ceramics, and polymers was given in Table 6-3. A compendium of handbooks and reference books concerned with various manufacturing processes was given in Sec. 7-2. We list below some basic engineering handbooks that cover most of the other engineering disciplines.

Baumeister, T. (ed.): "Marks' Standard Handbook for Mechanical Engineering," 8th ed., McGraw-Hill Book Co., New York, 1978.

R. C. King (ed.): "Piping Handbook," 5th ed., McGraw-Hill Book Co., New York, 1967.

Cummins, A. C., and I. A. Given (eds.): "SME Mining Engineering Handbook,"Society of Mining Engineers, AIME, New York, 1973, 2 vols.

Fink, D. G., and H. Beaty: "Standard Handbook for Electrical Engineers," 11th ed., McGraw-Hill Book Company, New York, 1978.

Fink, D. G., and D. Christiansen (eds.): "Electronics Engineers' Handbook," 2d ed., McGraw-Hill Book Co., New York, 1982.

Hanlon, J. F.: "Handbook of Package Engineering," McGraw-Hill Book Co., New York, 1984.

Higgins, L. R., and L. C. Morrow: "Maintenance Engineering Handbook," McGraw-Hill Book Co., New York, 1977.

Karassik, I. J., et al.: "Pump Handbook," McGraw-Hill Book Co., New York, 1976.

Kulwiec, R. A.: "Materials Handling Handbook," 2d. ed., John Wiley & Sons, New York, 1985.

Kutz, M. (ed.): "Mechanical Engineers' Handbook," John Wiley & Sons, New York, 1986.

Maynard, H. B. (ed.): "Industrial Engineering Handbook," 3d ed., McGraw-Hill Book Co., New York, 1971.

Merritt, F. S. (ed.): "Standard Handbook for Civil Engineers," McGraw-Hill Book Co., New York, 1968.

Parmley, R. O.: "Standard Handbook of Fastening and Joining," McGraw-Hill Book Co., New York, 1977.

Perry, J. H., and C. H. Chilton (eds.): "Chemical Engineers' Handbook," 5th ed., McGraw-Hill Book Co., New York, 1973.

Salvendy, G. (ed.): "Handbook of Industrial Engineering," John Wiley & Sons, New York, 1982.

Souders, M., and O. W. Eshbach: "Handbook of Engineering Fundamentals," 3d ed., John Wiley & Sons, New York, 1982.

White, John A.: "Production Handbook", 4th ed., John Wiley & Sons, New York, 1987.

Parmely, R. O. (ed.): "Mechanical Components Handbook," McGraw-Hill Book Co., New York.

Indexing and Abstract Services

In 1960 the world scientific community published 18,800 journals, but in 1980 the number exceeded 62,000. In scientific and technical fields a natural response to the explosion of knowledge is the division into subfields, and this brings with it the development of new journals. Not every library can afford to carry every journal. If you need a journal not covered by your library, you should consult the *Union List of Serials and New Serials Titles,* a set of volumes published by the Library of Congress. These volumes provide a worldwide, comprehensive listing of journals, with location codes showing many holding libraries. An even quicker way to determine which libraries may hold a particular journal title is by searching the national cataloging data base, called OCLC. Most libraries will routinely provide that service. There also exist local union serial lists, which show the periodicals held by a group or union of libraries that have banded together in a geographic region to provide a consolidated collection of serials. All of the above location devices are used by the Interlibrary Loan (ILL) service in all libraries. Consult the ILL office in your library, and they will acquire copies of journal articles for you.

Indexing and abstracting services provide current information on periodical literature, and they also provide a way to retrieve published literature. An *indexing service* cites the article by title, author, and bibliographic data. An *abstracting service* also provides a summary of the content in the article. Although indexing and abstracting services primarily are concerned with articles from periodicals, many often include books and conference proceedings, and some list technical reports and patents. The following is a list of indexes and abstracts that cover most of the engineering disciplines.

Applied Mechanics Reviews

Applied Science and Technology Index (formerly *Industrial Arts Index*)

Building Science Abstracts

Ceramic Abstracts

Chemical Abstracts

Computing Reviews

Corrosion Control Abstracts

Energy Index

Engineered Materials Abstracts

Engineering Index

Environment Index

Fuel Abstracts

Highway Research Abstracts

Instrument Abstracts

International Aerospace Abstracts

Metals Abstracts. Combine ASM *Review of Metals Literature* (U.S.) and *Metallurgical Abstracts* (British)

Nuclear Science Abstracts

Public Health Engineering Abstracts

Science Abstracts A: Physics

Science Abstracts B: Electrical and Electronics Abstracts

Science Abstracts C: *Computer and Control Abstract*
Science Abstracts D: *Information Technology Abstracts*
Solid State Abstracts

A useful abstract to much detailed information is *Dissertation Abstracts*, which gives abstracts of most doctoral dissertations completed in the United State and Canada. A copy of a dissertation can be ordered at modest cost.

Conducting a search of the published literature is like putting together a complex puzzle. One has to select a starting place, but some starts are better than others. A good strategy is to start with the most recent abstracts and work backwards. It is helpful to find a recent review article or a general technical paper; the references cited will be useful in finding other sources of relevant material. This approach to searching the literature is facilitated by the *Science Citation Index*. Once you have a reference of interest, you can go to the *Citation Index* to find all other references published in a given year that cited the key reference. Thus, there is a high probability that the additional references will be relevant to your original source reference.

An important tool in indexing, storing, and retrieving information is the thesaurus of technical terms. A *thesaurus* serves as a road map through the technical vocabulary, showing each term in relation to other terms. Perhaps you are familiar with Roget's "Thesaurus of English Words and Phrases," which relates the nontechnical terms in the English language. Two important technical thesauri are:

"Thesaurus of Engineering and Scientific Terms," Engineers Joint Council (AAES), New York, 1967.
"Thesaurus of Metallurgical Terms," 9th ed., American Society for Metals, Metals Park, Ohio, 1990.

A listing in the thesaurus will indicate which terminology is preferred, and which are narrower terms, broader terms, or related terms. For example, the listing *Induction melting* is given as:

Induction melting
  UF Air induction melting
        Inductovac process
  NT Vacuum induction melting
  RT Electric induction furnaces
        Levitation melting
        Multistage melting
        Oxygen melting
        Zone melting

This shows that induction melting should be used for (UF) the older terms air induction melting and inductovac process. A narrower term than induction

TABLE 14-3
Some computerized databases for indexing and abstracting

Data base	Supplier	Number of abstracts
Metadex	American Society for Metals in cooperation with The Metals Society (London)	700,000
CA Search	American Chemical Society	4,000,000
Compendex	Engineering Index, Inc.	2,000,000
NTIS	National Technical Information Service, US Department of Commerce	725,000

melting is indicated by NT and a broader term by BT. Five related terms (RT) are listed. This thesaurus listing would suggest that we broaden the search to include the related terms.

Many of the larger indexing and abstracting services have computerized their data bases. Table 14-3 gives information on some of the more prominent sources. Computerizing not only greatly speeds up searching the literature but also makes "pinpoint searching" for information a practical reality because it is possible to determine the intersection of several data bases. For example, by using a computerized search, we could determine whether there was any technical literature published in Chinese that deals with the vacuum induction melting of nickel-based superalloys in composition similar to alloy 713.

CD-ROM Databases

Libraries around the country are supplementing, and in some cases replacing, their printed indexes and abstracts and their online computer databases with CD-ROM (Compact Disk-Read Only Memory). For example, the U.S. patent and Trademakr Office is now providing its database (CASSIS) to Patent Depository Libraries on CD-ROM instead of online. Similarly, technical libraries are purchasing indexes, e.g. Applied Science and Technology Index and Science Citation Index, on CD-ROM instead of or in addition to their print subscriptions.

The advantage of CD-ROM over online searching is primarily cost, and that saving is passed on to the researcher. Most libraries charge researchers for some portion of the cost of searching commercial databases through online computer hookup. However, CD-ROM databases are normally purchased outright by the library and they are provided free of charge to researchers. Another important advantage of CD-ROM over print indexes is their more sophisticated search capability, including the possibility of Boolean searching and keyword searching.

Translations

Although English is the predominant world language for scientific literature, about one-third of the world's scientific and technical literature is produced in the Soviet Union, China, and Japan in languages that are unfamiliar to over 95 percent of scientists and engineers in the United States. Therefore, to avail yourself of information published in those languages, as well as in more common scientific languages like German and French, you should be familiar with sources of translations.

Unfortunately, machine translation by using the computer has not proved to be practical, so most technical translation must be done by an experienced translator. This is expensive, and therefore the need exists to make translations more widely available. Some of the most important indexes to translations are:

- Translations Register Index (1967–present)
  National Translations Center
  John Crerar Library
  Chicago, IL 60616
- World Index of Scientific Translations (1967–1971)
  World Transindex (1971–present)
  International Translations Centre
  101 Doelenstraat
  Delft, Netherlands
- British Industrial and Scientific International Translations Service (BISITS) available in the United States from American Society for Metals, Metals Park, OH 44073. Also, see ASM Translations Index published quarterly.

A number of commercial publishers regularly provide cover-to-cover translation of important foreign technical publications. In addition, U.S. federal agencies often commission translations of articles and books, and these are available through the channels described in Sec. 14-5. Finally, when all other sources of translations are exhausted, it may be necessary to commission a custom translation. Since this can be expensive, it is important to determine the qualifications of the translator, especially with respect to familiarity with the technical vocabulary of the specific field.

Catalogs, Brochures, and Business Information

An important source of information consists of directories that list manufacturers and suppliers of equipment, components, and materials. Because of the difficulty of keeping a collection of catalogs and brochures complete and up to date, some libraries are subscribing to commercial services that provide this information on microfilm. Some of the most common sources of information of this type are:

Chemical Engineering Catalog: The Process Industries' Catalog, Van Nostrand Reinhold; annual.

MacRae's Blue Book (Hinsdale, Ill.), a multivolume collection of manufacturers' catalogs; annual.

Thomas Register of American Manufacturers (New York), a comprehensive directory of addresses of manufacturers arranged by product.

Sweet's Catalog File (McGraw-Hill, New York), a compilation of manufacturers' catalogs with emphasis on machine tools, plant engineering, and building materials.

VSMF Design Engineering System, Information Handling Services (Englewood, Colo.), a compilation of vendors' catalogs and product specifications for mechanical and electronic parts; on microfilm, updated monthly.

Most technical libraries also contain certain types of business or commercial information that is important in many design problems. One is statistical information on consumption and sales of various raw materials, commodities, and manufactured goods by state and year. Such information is useful in the development of the needs analysis, although more detailed information most likely will be needed in the marketing phase. Some important sources of statistical business information are:

- "Census of Manufacturers"
U.S. Department of Commerce
- "Statistical Abstract of the United States,"
U.S. Bureau of the Census
- Metal Statistics," American Metals Market, New York
- "Minerals Yearbook," Vol. 1, "Metals and Minerals,"
U.S. Bureau of Mines

Three additional important sources of information for engineering design are technical reports generated with federal research funds, patents, and engineering codes and standards. Each of these subjects is treated in a succeeding section of this chapter.

14-5 GOVERNMENT SOURCES OF INFORMATION

The federal government either conducts or pays for about half of the research and development in this country. That generates an enormous amount of information. The federal government's activities are concentrated in defense, space, environmental, medical, and energy-related areas. It is such an important source of information that you should become familiar with it.

The Government Printing Office (GPO) is the largest publisher in the world. The *Monthly Catalog of U.S. Government Publications* lists all publications available from federal agencies. Reports prepared under contract by industrial and university RD&E organizations ordinarily are not available from the GPO. These reports may be obtained from the National Technical Information Service (NTIS), a branch of the Deparment of Commerce. NTIS provides custom computerized literature searches at nominal cost, provides bibliographies in many areas, and publishes the semimonthly *Government Reports Announcements,* which abstracts about 70,000 reports each year. The *Government Reports Index* indexes the GRA abstracts by subject, author, and government contract number.

NTIS provides a host of other services. Online computer searching of its databases is provided. By agreement with the Japan Information Center of Science and Technology, their computer database (in English) is available for searches. NTIS has over 400,000 reports on foreign technology on file and maintains formal agreements with over 90 foreign sources of technical reports. The Center for the Utilization of Federal Technology alerts industry and government to technology resulting from federal government R&D that is considered to have potential for commercial use. A directory also is maintained of federal laboratories and engineering centers willing to share expertise, equipment, and facilities with the private sector.

The National Aeronautics and Space Administration (NASA) conducts its own extensive information service. Both *Technology Utilization Reports* and *Technology Surveys* are aimed at transferring space technology to the more consumer-oriented industry. *Scientific and Technology Aerospace Reports* (STAR) is a semimonthly abstract service of aerospace-oriented literature. A brief listing of the major sources of government technical literature is given in Table 14-4.

In order to cope with the burgeoning technical information, the federal government has established a network of Information Centers. These centers are charged with acquiring and storing the world's literature in their designated fields, including unpublished reports. The output of these centers includes bibliographies on specialized topics and state-of-the-art reports. An information center differs from a documentation center or library in that it collects, reviews, analyzes, appraises, and summarizes the literature, rather than just storing and disseminating documents. Full information on available information centers can be obtained from the National Referral Center, Table 14-4. Some of the more active information centers are:

- Metals and Ceramics Information Center
  Battelle Columbus Laboratories
  Columbus, Ohio 43201
- Mechanical Properties Data Center
  Battelle Columbus Laboratories
  Columbus, Ohio 43201
- Plastics Technical Evaluation Center
  Picatinny Arsenal, N.J.
- Shock and Vibration Information Center
  Washington, D.C.

TABLE 14-4
Brief guide to government technical literature

"Directory of Information Resources in U.S.,"
  Vol. 1, "Physical Sciences Engineering," 1971
  National Referral Center, Sciences & Technology Division
  Library of Congress, Washington, DC 20540
  Telephone, (202) 426-5687

"Monthly Catalog of U.S. Government Publications,"
  Government Printing Office, Washington, DC 20401
  Telephone, (202) 541-3000

"Government Reports Announcements and Index" (semi-monthly)
  (Prior to 1971 was U.S. Government R&D Reports)
  National Technical Information Service, U.S. Dept. of Commerce
  Springfield, VA 22161
  Telephone, (703) 487-4650
  References R&D reports for government-sponsored reports plus translations of foreign material.
  Copies of reports may be purchased from NTIS (formerly DDIC and ASTIA).

"Scientific and Technical Aerospace Reports," (STAR)
  Abstracts compiled by NASA. Purchased from GPO

"Nuclear Science Abstracts,"
  Abstracts compiled by Department of Energy
  Purchased from GPO

- Machinability Data Center
  Metcut Research
  Cincinnati, Ohio
- Thermophysical and Electronic Properties
  Information Analysis Center
  Purdue University
  West Lafayette, Ind.
- Thermodynamics Information Center
  Bloomington, Ind.
- Atomic and Molecular Processes Center
  Oak Ridge, Tenn.
- Metal Matrix Composites Information Analysis Center
  Kaman Tempo
  Santa Barbara, Calif. 93102
- Manufacturing Technology Information Analysis Center,
  10 W. 35 Street
  Chicago, Illinois 60616
- Non-Destructive Testing Information Analysis Center
  Southwest Research Institute
  San Antonio, TX 78284

The Smithsonian Science Information Exchange (SSIE) is the only source of information on all federally funded research in progress. The data are in machine-readable form and can be searched by computer.

14-6 ONLINE COMPUTER DATABASES

The fastest growing segment of engineering information is online computer services. There are a number of vendors providing computerized access to technical databases. These are of two main types: bibliographic and source. Bibliographic databases are indexes or abstracts that refer the user to other sources, usually full text documents like papers in technical journals. We already have mentioned COMPENDEX (COMPuterized ENgineering inDEX).Another important bibliographic service is INSPEC (Information Services for the Physics and Engineering Communities), managed jointly by IEEE and the Institution of Electrical Engineers of Great Britain. Source databases are either numeric, full text, or a combination of the two. Examples are materials property databases[1] and component or product specifications. For example, an interactive database can be used to find information on a particular kind of pump. The Information Handling Services[2] (IHS) database called Product–Subject Index generates a locator code as well as the name of a manufacturer of the pump. The code could then be used to provide existing standards for the pump which are housed in the Industry and International Standards database. If the pump was intended for use by a government agency the appropriate data could be accessed from the Military and Federal Specifications and Standards database. If names of alternative manufacturers for the pump are required then the Vendor Information File could be accessed.

The vendors who provide these kinds of online computerized databases are called database utilities. They do not generate the data but they put it in machine-readable form and provide the necessary interfaces. Utilities charge a basic subscription fee for access to the service and then charge by the minute for using a specific database.

Material Properties Databases

Increasing recognition of the value of well-documented material property data for good engineering design has led to the formation of The National Materials Property Data Network.[3] MPD is a not-for-profit corporation with the mission

---

[1] J. S. Glazman and J. R. Rumble, "Computerization and Networking of Materials Data Bases," ASTM STP 1017, ASTM, Philadelphia, PA, 1989.

[2] Information Handling Services, P.O. Box 1154, Englewood, CO, 80150.

[3] J. G. Kaufman, *ASTM Standardization News*, March 1987, pp. 38–43. The address of MPD Network is P.O. Box 02224, Columbus, OH, 43202; phone (614)-421-3706.

of providing easy access to high quality numeric materials performance data for U.S. engineers and scientists. It is sponsored by the more than 50 corporations and organizations that provide financial support and contribute data. MPD will deal with all classes of materials and all types of materials data.

A materials property database is more than just a collection of numerical properties data. In addition to data, a good database must contain extensive *metadata*. Metadata is text or numerical information about the database and the primary data. It generally includes a detailed identification of the material, the source that supplied the data (including whether it was an accredited laboratory), and the test method and test conditions. There also must be a systematic means of collecting and updating data and a mechanism for evaluating and assigning quality indicators to the data. Generally we are dealing with a computer-based data storage system, although hardcopy may be provided. Finally, there must be a means to search and retrieve relevant data.

We should note that numerical data may be one dimensional, e.g., a value for elastic modulus or yield strength, multidimensional, e.g., a stress-strain curve or *S-N* curve, or image information, e.g., data from an ultrasonic scan. The data may be at different levels of statistical validity. Raw data consist of actual measured values from the laboratory. Lot data consist of averages on production lots of materials or parts. Statistically processed data have been subjected to standard statistical procedures. Literature data are values obtained from the published technical literature or from manufacturer's marketing publications. Specification data represent property limits from manufacturing or purchasing specifications.

Some of the reasons for the replacement of the traditional handbooks with computer-readable data sources are: 1) easier searches and faster retrieval, 2) easier data sharing and less redundant generation of data, 3) direct electronic transmission of data into CAD/CAM systems, 4) minimizing the loss of data through being buried in old reports, and 5) faster update of data. Industry gradually is realizing that the advantages of a shared database outweigh the traditional proprietary interests, especially when the high cost of data generation are fully considered. Progress has been hampered by lack of standards for data reporting, data structure, and data interchange, but these issues are being addressed by the recently formed ASTM Committee E-49 on Computerization of Material Property Data.

14-7 THE PATENT LITERATURE

The U.S. patent system is the largest body of information on technology in the world. At present there are over 4 million U.S. patents, and the number is increasing at the rate of 50,000 to 80,000 each year. Old patents can be very valuable for tracing the development of ideas in an engineering field, and new patents describe what is happening at the frontiers of many fields. Only about 20 percent of the technology that is contained in U.S. patents can be found elsewhere in the published literature.[1] Therefore, the engineer who ignores the patent literature is aware of only the tip of the iceberg.

The topic of patents was introduced in Sec. 2-11, where we considered the protection of intellectual property. In this section we consider patents chiefly from the viewpoint of their usefulness as engineering information.

The Patent Literature

The *Offical Gazette of the U.S. Patent Office* is issued each Tuesday with the weekly issuance of patents. It contains an abstract and selected figures from each patent that is issued that week. The Offical Gazette is very helpful for keeping current on the patent literature and for getting a quick overview of a patent. It is widely disseminated in the United States.

The "Annual Index" of patents is published each year in two volumes, one an alphabetical index of patentees and the other an index of inventions by subject matter. These can be used to obtain the patent number, and then the *Official Gazaette* can be used to learn about the patent. All U.S. patents may be examined in the public search room in Crystal City, VA (near Washington, D.C.). The main advantage of visiting the search room is that patents are grouped according to classes and subclasses. However, 58 libraries nationwide are designated as Patent Depository Libraries and contain most U.S. patents. In these libraries, the patents are arranged according to number.

Patents have been arranged into about 350 classes, and each class is subdivided into many subclasses.[2] The "Index to Classification," is a loose-leaf volume that lists the major subject headings into which patents have been divided. Once the Index identifies the appropriate class numbers, the searcher should go to the "Manual of Classification." This loose-leaf volume lists each class, with its subclasses, in numerical order. By searching the "Manual of Classification," you can identify the classes and subclasses that are likely to be of interest. It is important to realize that subjects will often be found in more than one patent class.

An online computerized classification system called CASSIS (Classification and Search Support Information System) is available at all of the Patent Depository Libraries around the nation. For a given patent number CASSIS will display all its locations in the Patent Classification System (PCS). For a given technology as described by a PCS classification, CASSIS will display the numbers of all patents assigned to that classification. Also, CASSIS will identify all classifications whose full titles contain designated key words. Finally, CASSIS will search the alphabetical list of subject headings in the index to the PCS to identify the classifications containing subject matter of interest.

---

[1] P. J. Terrango, *IEEE Trans. Prof. Comm.*, vol. PC-22, no. 2, pp. 101–104, 1979.

[2] K. J. Dodd, *IEEE Trans. Prof. Comm.*, vol. PC-22, no 2, pp. 95–100, 1971.

The only place where U.S. patents are arranged for inspection by class and subclass is the Public Patent Search Facilities of the Patent and Trademark Office, 2021 Jefferson Davis Highway (U.S. Route 1), Arlington, Virgina. The public is admitted to the search room and may request to see all patents included in any class and subclass. The patents are arranged chronologically in each subclass, together with cross-reference patents from other subclasses. Browsing through the patents is the best way to conduct a state-of-the-art-search. Copies of interesting patents can be ordered for $1.50 each, regardless of length. A state-of-the-art search preferably should be conducted by the engineer starting out on a particular R&D effort.

A *patentability search* is a search of the patent literature to determine whether an invention can be patented and what the scope of the patent protection would be. This search draws on the information assembled in the search, especially the prior art and background of the invention. By carefully studying the claims in the prior art, the patent attorney can construct claims that are neither too broad (so as to be precluded by prior patents) or too narrow (so as to limit the usefulness of the patent).

An *infringement search* is an exhaustive search of the patent literature to determine whether an idea is likely to infringe on patents held by others. Often it is undertaken when making a new product is contemplated.

If you find your patent or idea in possible infringement with an existing patent, you may fight back by challenging the validity of the patent in the courts. To do so requires a *validity search*. More than half of all patents challenged are ruled invalid by the courts. A patent is really not good until it is tried and held valid by the courts, although it is presumed valid when issued by the Patent Office.

Reading a Patent

Because a patent is a legal document, it is organized and written in a style much different from the style of the usual technical paper. Patents must stand on their own and contain sufficient disclosure to permit the public to practice the invention after the patent expires. Therefore, each patent is a complete exposition on the problem, the solution to the problem, and the applications for the invention in practical use.

Figure 14-2 reproduces a short patent for illustrative purposes. The first page of the patent (Fig. 14-2a) carries bibliographic information, information about the examination process, an abstract, and a drawing (if the patent is illustrated). At the very top we find the inventors, the patent number, and the date of issuance. Below the line on the left we find the title of the invention, the inventor(s) and address(es), the date the patent application was filed, and the application number. Next are listed the class and subclass for both the U.S. patent system and the international classification system and the U.S. classes in which the examiner searched for prior art. The references are the patents that the examiner cited as showing the most pertinent prior art at the time of the

United States Patent

Marcus et al.

[15] **3,656,458**

[45] **Apr. 18, 1972**

[54] **MOLLUSC CLIP**

[72] Inventors: **Douglas Larry Marcus; Clifford L. Sayre, Jr.,** both of 1415 Ladd Street, Silver Spring, Md. 20902

[22] Filed: **June 23, 1970**

[21] Appl. No.: **49,096**

[52] U.S. Cl. **119/4**
[51] Int. Cl. **A01k 61/00**
[58] Field of Search 119/4, 24/137, 137.5, 138

[56] **References Cited**

UNITED STATES PATENTS

| 2,825,952 | 3/1958 | Van Driel | 24/138 |
| 2,931,086 | 4/1960 | Rose | 24/137 |

| 3,203,061 | 8/1965 | Thomas | 24/137 X |

FOREIGN PATENTS OR APPLICATIONS

| 1,320,608 | 1/1963 | France |

Primary Examiner—Hugh R. Chamblee
Attorney—Abraham A. Saffitz

[57] **ABSTRACT**

A clip for securing clutch material for the intensive farming or raft culture of sessile molluscs such as the oyster. The clip provides an opening for receiving and holding the clutch material. The clip is also provided with a keyhole shaped opening for receiving a supporting line whereby the clip and clutch material is attached thereto.

10 Claims, 8 Drawing Figures

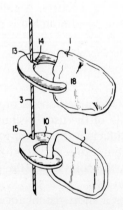

(a)

FIGURE 14-2
A United States patent.

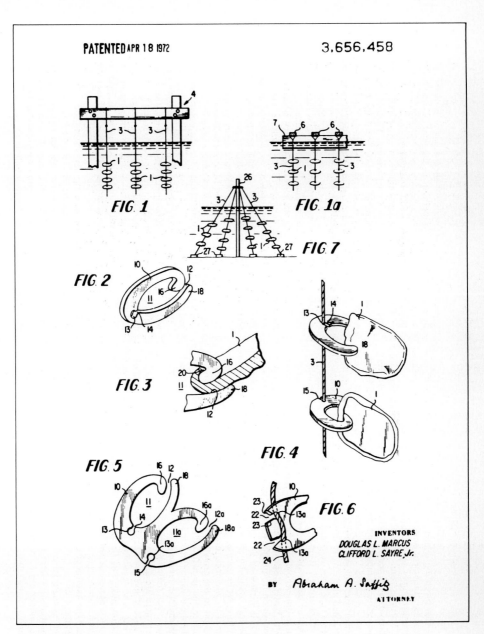

PATENTED APR 18 1972   3,656,458

FIG. 1

FIG. 1a

FIG. 7

FIG. 2

FIG. 3

FIG. 4

FIG. 5

FIG. 6

INVENTORS
DOUGLAS L. MARCUS
CLIFFORD L. SAYRE, Jr.

BY *Abraham A. Saffitz*
ATTORNEY

(*b*)

FIGURE 14-2
(*Continued.*)

3,656,458

1

MOLLUSC CLIP

This invention relates to the use and configuration of an improved attachment device for use in the raft culture or intensive farming of sessile molluscs such as oysters.

Various forms of shellfish culture have been practiced for thousands of years. A current practice which has been very successful in Japan and has been introduced into the United States is called "raft culture." The cultch material on which oyster spat will set, or which may already have juvenile oysters attached thereto, is drilled and strung onto rods, wires, or other type lines and suspended from rafts or frames where the oysters are left to mature. Raft culture has many advantages such as better use of available growing areas because of the three-dimensional use of space, improved feeding and growth rate, better disposal of fecal and waste products, improved protection from some forms of predators and disease, and simplified harvesting as compared to the tonging or dredging of natural oyster bottoms. The use of strings also permits easy shifting from seeding to growing areas, moving from feeding areas to wintering areas in regions where there is a dormant period in the growth cycle, and periodic sunning or airing to reduce the growth of fouling. Cultch materials are typically old shells which are cheap and readily available and for which the shellfish larvae appear to have a natural preference. Various other materials have been employed for cultch material, but these are usually less successful in terms of the density of the set of spat and they are invariably more expensive than old shells. The feasibility of raft culture as a technique is well-documented, but the commercial success (particularly in the United States) depends on developing methods and materials which can complete economically with harvesters who can reap the natural crop on public beds with only a minimum investment of capital. The handling and hand labor costs (as well as material costs) of the raft culture technique must be reduced to a bare minimum if farming or intensive growing techniques are to compete successfully against the traditional harvesting methods.

2

production by punching out of sheet stock as well as molding by any of the common methods in simple dies. Another merit is the fabrication from a single piece which simplifies production and eliminates the need for assembly.

With particular reference to the drawings, FIGS. 1 and 1a illustrate the present art of oyster culture. The old shells 1, or similar, carry cultch material, have drilled holes 2 therein by which the material may be strung onto lines, wires or rods 3. The strings of cultch material are suspended in appropriate water areas from frame work 4 anchored in the water bed, as illustrated in FIG. 1. FIG. 1a shows the strings of cultch material suspended from supporting means 6 fastened to an anchored float or raft or similar support means 7.

Referring to FIG. 2, the clip is a unitary structure, either stamped out from sheet material, or molded, and having a generally rectangular cross section, as illustrated in the figure. The clip comprises a body 10, having an opening 11 therein. Opening 11 is provided with an entrance portion 12. Remote from entrance 12 a hole or aperture 13 is provided in the body. A narrow slot formation 14 connects hole 13 with opening 11, whereby means 13 and 14 form a keyhole formation. If desired, slot 14 may connect hole 13 to the outer periphery of body 1, as illustrated.

Body 10 terminates at one end in a hook formation 16 which lies opposite wall portion 18 of the body, the hook and wall portion defining relatively narrow entrance 12.

The clip may also be fabricated by extruding the desired plastic in a body having a cross-sectional configuration of the clip illustrated in FIGS. 2, 5 and 6. After, or as, the body is extruded, it may be cross sectionally cut or sliced to form individual clips.

In use, the clip is strung on the supporting rod, wire or line means in strings by moving the clip into the line 3 through entrance 12 and forcing the line through slot 14 to bring the line into hole 13. The formation of means 13 and 14 within the body provides a spring-like action whereby slot 14 yields to permit passage of the line means into hole 13.

In lieu of slot **14** connecting hole **13** to opening **11**, a slot **15** may be used to connect hole **13** to the outer periphery of body **10**, as illustrated in FIGS. **4** and **5**. The outer location of slot **15** permits the attachment or detachment of the clip to line **3** without disturbing the cultch material.

FIG. **6** shows a further modification of the line securing means. Two adjacent holes **13a** are formed in the clip with outer peripheral slots **22**. Slots **22** have inclined sides **23** to accommodate lines of different diameters. The clip is secured to its line **24** by passing the line serially through both means **13a–22**, as illustrated in FIG. **6**. The double securing means **13a–22** makes the agreement between the hole size and line diameter less critical. Hence, one size of clip securing means could be used with a variety of line diameters.

By being able to fasten directly and securely to the line the use of drilled holes and hence the conventional drilling operation is eliminated. If the diameter of hole **13** is properly chosen, the clip will be held firmly to the supporting rod, wire, or line, and be maintained in such position against normal forces. The use of spacers to separate the shells and the operation of threading the spacers is eliminated. Spacing is an important factor to provide room between adjacent clusters of oysters to permit optimum growth. Some growers economize by eliminating spacers because the shells are cheaper and the assembly without spacers is simpler. However, in addition to crowding for food, unspaced clusters of juvenile oysters are liable to damage by abrasion and being knocked about by other clusters which are too close. The location of keyhole means **13–14** in the interior of the clip has the advantage of preventing the clip from falling off of the supporting line even if the clip works loose (although the spacing benefits would obviously be lost). The selection of some types of plastic lines for the support string will be a plastic which would swell due to water absorption, thus causing the hole in the clip to hold the line more securely after immersion.

The shell or similar cultch material is attached to the clip by inserting the shell within entrance **12** so that it is held in position by the clamping effect of hook **16** and wall portion **18**

Accordingly, the primary objectives of the present clip and its application are to reduce both the material and labor costs associated with the preparation, handling and harvesting of strings used in raft culture. The use of the clip avoids the necessity of using spacers (extra material and labor) and eliminates the drilling operation.

Other advantages and features will be apparent from the following description, and the figures of the attached drawing wherein:

FIGS. **1** and **1a** illustrate the general arrangement used in raft culture;

FIG. **2** is a perspective top view of the proposed clip;

FIG. **3** is a partial cross section through a shell and clip illustrating the unique attachment feature;

FIG. **4** illustrates the clip in use;

FIG. **5** is a top view of a clip having multiple openings for a number of cultch material pieces;

FIG. **6** is a plan view of a further modification of the clip's securing means to a line; and

FIG. **7** is a view of a modified form of a raft culture arrangement.

Although various materials might be suitable for the clip, common metals are not recommended. Metals may corrode. Certain metallic ions in solution may inhibit the set of spat and are known to be ingested, retained and concentrated by the shellfish to the detriment of good growth and taste of the product. The deflection of the clip due to the insertion of the oyster hinge produces strains which might lead to stress-corrosion cracking in some metals. However, metals which are resistant to corrosion, such as the stainless steels and similar metals may be utilized, especially in fresh waters or the like. Plastics such as polyethylene, polyvinylchloride or polystyrene are recommended as suitable materials. Another suitable material is asbestos impregnated with cement, such as Portland cement or epoxy resin. One of the merits of the design is its flat shape and uniform thickness which would permit

FIGURE 14-2
(Continued.)

(c)

thereon. The configuration of the body structure with opening 11 therein is such that it acts like a spring to exert a holding or clamping force on the portion of the cultch material between hook 16 and wall 18.

The particular shape of the engaging portion of the hook which is illustrated in FIG. 3 solves a very difficult problem. Many shells, oysters in particular, lack uniformity in size, shape and thickness. The design of the present clip takes advantage of a feature which all bivalve molluscs have in common — the hinge. At the hinge location 20 the shells have a recessed, curved section. The shape of the hook of the clip uses this feature to lock the clip to the recessed portion within the shell and to the outer side of the shell more securely than the simple pinching action of means 16 and 18 applied to the main body of the shell. The cross section of the lower portion of hook 16 of the clip is rectangular, preferably square. This feature permits the end of the hook to be flexible and capable of deflection either in the plane of the clip or sideways. This flexible construction is such that the clip may be applied by inserting the hinge recess directly, or by sliding the clip on from the side — whichever method is easier for each particular shell. Hence, the clip is adaptable to a wide variety of types of shells. The width of entrance 12, that is, the space between hook 16 and wall 18 is such that the spring force of the clip will not cause the shell to cock parallel to the clip and thus slide out. The opposing surfaces of hook 16 and wall 18 will always be in alignment and thereby prevent shell cocking forces.

In some cases the present practice of raft culture uses shells to which spat or juvenile oysters have already become attached. The use of clips in such cases would reduce the handling time out of the water as well as avoid the damage and mortality hazard of the drilling operation. The harvesting of raft-grown oysters would be improved by the use of clips. Mature oysters could be removed easily and the supporting lines returned to the water to permit undersized oysters to grow. With drilled cultch material selective harvesting is obviously more complicated. The plastic clips should be reusable since

figuration, that is, to form opening 11, slot 14, hole 13, hook 16 and wall portion 18. A wire clip having the disclosed configuration will have spring-like characteristics for holding the clip properly spaced along the line and maintain the shell in the proper position. The wire clip configuration should be fabricated in such a fashion as to avoid crevices where oxygen depletion or corrosion concentration cells occur.

FIG. 7 discloses an "umbrella" type arrangement of raft culture. A series of lines 3 radiate from a single pole-like support means 27. The ends of lines 3 may be secured to support means 27.

Many other variations and modifications of this invention will be apparent to those skilled in the art. Accordingly, the foregoing description is intended to be construed as illustrative only, rather than limiting. The invention is limited only by the scope of the following claims.

What we claim is new and desire to secure by Letters Patent of the United States is:

1. A clip for attaching a cultch material to a supporting line means comprising a body, said body having an opening for the reception of the cultch material, said opening being provided with an entrance formed by two facing portions of the body, one portion having a hook configuration and the other opposing portion being a substantially straight wall portion, said straight wall portion having an end terminating beyond said hook configuration, said portions receiving and holding the cultch material therebetween, and securing means for attaching the clip to the supporting line means, said securing means having a hole in the body and a slot means narrower than the hole connecting said hole with the opening, thereby permitting a supporting line to pass through the entrance, opening and slot into the hole.

2. A clip as set forth in claim 1 wherein the cultch material is an old shell of a mollusc and the hook portion is shaped to enter and grasp the natural depression formed in the shell at the hinge point.

3. A clip as set forth in claim 1 wherein the body of the clip is formed of a resilient plastic material.

they are essentially inert material and not damaged by ordinary use. Occasionally marine fouling may make some clips unfit for further service or breakage may occur during harvesting, but these are minor hazards. If polyethylene or polypropylene, or similar material, is used for supporting the strings, entire arrays of lines and clips might be reused. Experiments with these plastic lines have shown no loss of strength in oyster support strings over periods as long as 4 years — except for cases of abrasion against some adjacent structure. Fixing of the clip to the line also avoids parting of the support lines from abrasion due to wave or other motion which causes loose drilled shells to move up and down thus abrading the support line.

A clip with two or more sets of openings and hooks would be an obvious improvement for use with small shells or where denser populations of growing oysters were feasible. Such clip means is illustrated in FIG. 5, wherein body 10, in addition to means 11, 12, 13, 14, 16 and 18, is formed with a second set of like means 11a, 12a, 13a, 16a and 18a whereby the second set may accommodate a second shell. Means 13a and slot 15 are provided for the additional set, but it may be eliminated, if so desired, since only one means 13–14 is necessary for attachment of the multiple clip. Similarly, additional sets of means 11, 12, 16 and 18 may be provided in body 10 to hold additional cultch material.

A clip having the configuration of the disclosed clip may also be fabricated from suitable corrosion resistant wire or metal rodding. The wire needs to be bent to the disclosed con-

4. A clip as set forth in claim 1 wherein the clip is formed of a material comprising asbestos material impregnated with a cementitious material.

5. A clip as set forth in claim 1 wherein the body of the clip is formed of wire type material.

6. A clip as set forth in claim 1 wherein the plastic material is polyethylene.

7. A clip as set forth in claim 1 wherein the body is provided with additional formations, each formation providing an opening and opposing hook and wall portions forming an entrance thereto, to thereby provide additional reception openings for additional cultch material.

8. A clip as set forth in claim 1 wherein the slot means comprise sides inclined to each other with the wider spacing at the peripheral surface.

9. A clip as set forth in claim 8 wherein a duplicate hole and slot means are placed adjacent said hole and slot means and the line is serially passed through both holes.

10. A clip as set forth in claim 1 wherein a plurality of supporting lines are used for immersing a number of clips and their cultch material in water, said plurality of lines being supported at an end from a pole-like support and radiating therefrom in an umbrella-like fashion.

* * * * *

FIGURE 14-2
(*Continued.*)

639

invention. Then the names of the patent examiner and the patent attorney are given. This is followed by a brief abstract and a key figure. All of the figures used to illustrate the patent are on page 2 (Fig. 14-2*b*).

The body of the patent, on page 3 (Fig. 14-2*c*), starts with an identification of the field in which the invention lies. This is followed by more detailed background. This section closes with a paragraph that briefly states the objectives of the invention. Some U.S. patents are clearly divided with section headings into Field of Invention, Description of Prior Art, Summary of the Invention, Statement of the Objects of Invention, etc. However, it is more common for the reader to have to supply the section headings for himself as he reads the patent.

The next section is a brief description of the illustrations. The body of the patent up to the claims is a description of the *preferred embodiment* of the invention. This comprises a detailed description and explanation of the invention, often in legal terms and phrases that are strange-sounding to the engineer. The examples cited show as broadly as possible how to practice the invention, how to use the products, and how the invention is superior to prior art. Not all examples describe experiments that were actually run, but they do provide the inventor's teaching of how they should best be run. The last part of the patent (Fig. 14-2*d*) comprises the *claims* of the invention. These are the legal description of the rights of invention. The broadest claims are usually placed first, with more specific claims toward the end of the list. The strategy in writing a patent is to aim at getting the broadest possible claims. The broadest claims are often disallowed first, so it is necessary to write narrower and narrower claims so that not all claims are disallowed.

There is a very important difference between a patent disclosure and a technical paper. In writing a patent, the inventor and his attorney purposely broaden the scope to include all materials, conditions, and procedures that are believed to be equally likely to be operative as the conditions that were actually tested and observed. The purpose is to develop the broadest possible claims. This is a perfectly legitimate legal practice, but it has the risk that some of the ways of practicing the invention that are described in the embodiments might not actually work. If that happens, then the way is left open to declare the patent to be invalid.

Another major difference between patents and technical papers is that patents usually avoid any detailed discussion of theory or why the invention works. Those subjects are avoided to minimize any limitations to the claims of the patent that could arise through the argument that the discovery would have been obvious from an understanding of the theory.

14-8 STANDARDS AND SPECIFICATIONS

The dictionary defines a standard as something that is established by authority, custom or general consent as a model or example to be followed. We have standards of conduct as a society, and as a technological society we have evolved technical standards. The ASTM, the primary technical standard setting organization in the United States, defines a standard as "a rule for an orderly approach to a specific activity, formulated and applied for the benefit and with the cooperation of all concerned." Generally a *standard* is a document that establishes engineering and technical limitations and applications for items, materials, processes, methods, designs, and engineering practice. *Specifications* usually are more specific to a particular application than are standards.

Specifications and standards comprise one of the keystones of our industrialized system. Specifications are a part of nearly every buy-sell operation. The time saved through contract language documented in the form of a specification greatly decreases the cost of doing business. Specifications make possible the standardization on which our system of economical mass manufacture is based. There are approximately 85,000 governmental, public, and private standards in use in the United States.

We can identify a hierarchy of standards[1] that are used by society. *Value standards* are the highest level of standards in terms of their impact on society; standards dealing with the regulation of radioactivity or the need for clean air and water are examples. *Regulatory standards* frequently are derived from more basic value standards. There are three types of regulatory standards.

1. Industry regulations or codes that are produced (and paid for) by industry.
2. Consensus-type regulatory standards paid for by members of standards-writing bodies and/or the federal government.
3. Mandatory regulatory standards that are entirely the product of federal, state, or local government agencies.

The ASTM develops six different types of standards. A *standard test method* describes procedures for determining a property of a material or the performance of a product. A *standard specification* is a concise statement of the requirements to be satisfied by a product, material or process. *Standard practices* are procedures or guidelines which are often auxiliary to a test method or specification. Examples would be statistical procedures or sampling procedures. *Standard terminology* provides definitions and descriptions of terms, or explanations of symbols, abbreviations, and acronyms. A *standard guide* offers a series of options or instructions but does not recommend a specific course of action. *Standard classifications* define systematic arrangements or divisions of materials or products into groups based on similar characteristics.

A design code, such as the ASME Boiler Code or the local Building

---

[1] "Materials and Process Specifications and Standards," NMAB-330, National Research Council, Washington, D.C., 1977.

Code, is a set of rules of procedure and standards aimed at providing uniform methods for solving design problems. Codes evolve to protect the public interest and usually have some kind of legal standing.

There are more than 400 standards writing organizations in the United States. The ASTM recognizes four levels of standards based on the degree of consensus needed for their development and use. At the lowest level is the *company standard* which is used internally for design, production, purchasing, or quality control. The consensus here is among the employees of the company. The next higher level is the *industry standard* which is typically developed by a trade association or professional society. Here the consensus is among the members of the organization. A *government standard* reflects many degrees of consensus. Sometimes a government agency will adopt standards prepared by private organizations, but other times they may be written by a small group. A *full consensus standard* is the type of standard developed by the ASTM. It is developed by representatives of all sectors including producers, users, academia, government, and consumers.

Standards Organizations

The United States is the only industrialized country in which the national standards body is not a part of or supported by the national government. The American National Standards Institute (ANSI) is the coordinating organization for the voluntary standards system of the United States. It certifies the standards-making processes of other organizations, initiates new standards-making projects, represents the United States on the International Standards Committees of the International Organization for Standardization (ISO), and examines the standards prepared by other organizations to determine whether they meet the requirements for consensus so as to be included as an ANSI standard.

The American Society for Testing and Materials (ASTM) is the major organization that prepares standards in the field of materials and product systems. It is the source of more than half of the existing ANSI standards.

The Standards Development Services (SDSS) of the National Bureau of Standards manages the voluntary product standards program established by Part 10, Title 15 of the Code of Federal Regulations.

Trade associations produce or review voluntary standards. Those that have produced a substantial number of standards include:

- American Petroleum Institute
- Association of American Railroads
- Electronics Industries Association
- Manufacturing Chemists Association
- National Electrical Manufacturers Association

A number of professional and technical societies have made important contributions through standards activities. The most active are:

- American Association of State Highway and Transportation Officials
- American Concrete Institute
- American Society of Agricultural Engineers
- American Society of Mechanical Engineers
- Institute of Electrical and Electronics Engineers
- Society of Automotive Engineers

The ASME prepares the well-known Boiler and Pressure Vessel Code that is incorporated into the laws of most states. The ASME Codes and Standards Division also published performance test codes for turbines, combustion engines, and other large mechanical equipment.

Several other important standards-making organizations are:

- National Fire Protection Association (NFPA)
- Underwriters Laboratories, Inc. (UL)
- Factory Mutual Engineering Corp. (FMEC)

The Department of Defense (DOD) is the most active federal agency in developing specifications and standards. The General Services Agency (GSA) is charged with preparing standards for common items such as light bulbs and hand tools.

The following are key reference sources for information about specifications and standards.

- "Annual Catalog"
  American National Standards Institute
  10 East 40th Street
  New York, NY 10016
- "Annual Book of ASTM Standards"
  ASTM
  1916 Race Street
  Philadelphia, Pa. 19107

The Book of Standard is in 48 volumes and contains over 8500 standards

- "Index of Specification and Standards"
  U.S. Department of Defense
  Washington, D.C.
- "Index of Federal Specifications and Standards"
  U.S. General Services Administration
  GPO, Washington, D.C.

- "Index and Directory of U.S. Industry Standards," 2d ed., 1984
  Information Handling Services
  Englewood, CO 80150.

References to over 400 U.S. standards organizations and more than 26,000 individual standards

- "World Industrial Standards Speedy Finder," 1983
  ASTM
  Philadelphia, PA 19107

Can locate a specific standard by either number or product name for U.S., U.K., W. Germany, France, or Japan.

- R. B. Toth (ed.): "Standards Management," ANSI, New York, 1989.

The following specialized volumes dealing with metals are available.

- "Unified Numbering System for Metals and Alloys," 5th ed., ASTM, 1989.
- "Worldwide Guide to Equivalent Irons and Steels," ASM, 2d ed., 1987.
- "Worldwide Guide to Equivalent Nonferrous Metals and Alloys," ASM, 2d ed., 1987.

BIBLIOGRAPHY

Anthony, L. J.: "Information Sources in Engineering," Butterworth, Boston, 1985.
"Guide to Materials Engineering Data and Information," ASM International, Metals Park, Ohio, 1986.
Harvey, A. P. (ed.): European Sources of Scientific and Technical Information," 7th ed., Longman Group, U.K., 1986.
Mildren, K. W. (ed.): "Use of Engineering Literature," Butterworth, Boston, 1976.
Schenk, M. T., and J. K. Webster: "What Every Engineer Should Know about Engineering Information Resources," Marcel Dekker Inc., New York, 1984.
Subramanyam, K.: "Scientific and Technical Information Resources," Marcel Dekker Inc., New York, 1981.

CHAPTER
15

COMMUNICATING
THE DESIGN

15-1 THE NATURE OF COMMUNICATION

You've completed your design, and you've come up with some innovative concepts and a cost estimate that predicts a nice healthy profit. What next? Now you must be able to communicate your findings to the people who matter. There is an old adage that a tree falling in a forest doesn't make a sound unless there is someone to listen. Similarly, the best technical design in the world might never be implemented unless you can communicate it to the proper people in the right way. At another personal level, it is rare today for design to be done by a single individual. To be successful with a design project, you must be able to communicate with your peers, your subordinates, and your superiors.

Communication can be simply described as *the flow of intelligence from one mind to another*. Communication occurs through a common system of symbols, signs, and behavior that utilize one or more of the five human senses. We communicate by actual physical touch, as in a handshake or a pat on the back. We also communicate by visual movements of the body (body language), as with the wink of an eye or a smile. Sometimes we even communicate via taste and smell. In most technical communication, however, symbols or signs that are either heard with the ear or seen with the eyes are used.

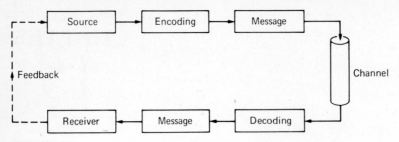

FIGURE 15-1
Basic elements of communications system.

The basic elements of all communication activities are shown in Fig. 15-1. The source of the communication arises somewhere in the organization: someone has information to communicate to someone else. An encoding process occurs when the information is translated into a systematic set of symbols (language) that expresses what the source wishes to transmit. The product of encoding is the message. The form of the message depends on the nature of the communication channel. The channel is the medium through which the message will be carried from source to receiver. The message can be a written report, a face-to-face communication, a telephone call, or a transmission via a computer network. It may or may not undergo a transformation in the decoder. When it reaches the receiver, it is interpreted in light of the person's previous experience or frame of reference. Feedback from the receiver to the source aids in determining how faithfully the message has been transmitted. Feedback gives the source an opportunity to determine whether the message has been received and whether it has produced the intended purpose.

Any communications system is less than perfect. Strong electric currents or atmospheric disturbances will produce some error in the transmission. These disturbances are called noise. In the same way, any communication between two persons will be subject to semantic noise and psychological noise. *Semantic noise* has to do with the meaning of words and ideas. In technical fields common words take on specialized meanings that can be completely unknown to the layman. *Psychological noise* arises from the particulars of the situation. If a person feels threatened by a new technology, he may have difficulty learning about it. Some people are intimidated by authority, and thus they have difficulty communicating with people above them in the organization. For communication to occur, the sender must have the authority of knowledge and a purpose for generating and transmitting the message. A communication that arises simply "to get something off your chest" usually will not succeed. In addition, the recipient must be willing to receive the message and be capable of understanding it.

Studies of communication in business organizations have shown that geographic proximity is a major factor in enhancing communication. There usually is more information than the channel has capacity to handle. Sometimes selective filtration of information occurs. At any rate, the managements of many organizations make decisions on the basis of inadequate information. Communication in an organization is a multidimensional process. Many managers attempt to communicate only with the people under them and forget to communicate laterally with those who are assisting them. Lateral communication between project teams is particularly important in engineering organizations.

15-2 PROPER RECORDING OF RESULTS

We digress briefly from discussing communication of results to consider how ideas and experimental results should be recorded during the progress of the design project. If the recording task is treated with care and thoroughness, communicating the results of a design project will be made easier.

The chief tool for recording results is the design notebook. It should be an $8\frac{1}{2}$-by11-in bound notebook (not spiral bound), preferably with a hard cover. It should be the repository for all of your planning (including plans that were not carried out), all analytical calculations, all records of experimental data, all references to sources of information, and all significant thinking about your project.

You should not use your notebook as a diary; but at the same time, you and your notebook should become an intimate communication system. Information should be entered directly into the notebook, not recopied from rough drafts. However, you should organize the information you enter in the notebook. Use main headings and subheadings; label key facts and ideas; liberally cross-reference your material; and keep an index at the front of the book to aid in your organization. About once a week, review what you have done and write a summary of your progress that emphasizes the high points. Whenever you do anything that may seem the least bit patentable, have your notebook read and witnessed by a knowledgeable colleague.

The following are good rules[1] for keeping a design notebook.

1. Keep an index at the front of the book.
2. Make your entries at the time you do the work. Include favorable and unfavorable results and things not fully understood at the time. If you make errors, just cross them out. Do not erase, and never tear a page out of the notebook.
3. All data must be in their original primary form (strip charts, oscilloscope pictures, photomicrographs, etc.), not after recalculation or transformation.

---

[1] Adapted from T. T. Woodson, "Engineering Design," app. F. McGraw-Hill Book Company, New York, 1966.

4. Rough graphs should be drawn directly in the notebook, but more carefully prepared plots on graph paper also should be made and entered in the book.

5. Give complete references to books, journals, reports, patents, and any other sources of information.

6. Entries should be made in ink and, of course, must be legible. Do not be obsessed with neatness at the expense of faithfully recording everything as it happens. Do not crowd your material on the pages. Paper is very much less expensive than engineering time.

A good engineering design notebook is one from which, several years after the project is completed, the project can be reconstructed. Critical decisions will be apparent, and the reasons for the actions taken will be backed up by facts. It should be possible to show where every figure, statement and conclusion of the published report of the project can be substantiated by original entries in the design note book.

15-3 WRITING THE TECHNICAL REPORT

In no other area of professional activity will you be judged so critically as your first technical report. The quality of a report generally provides an image in the reader's mind that, in large measure, determines the reader's impression of the quality of the work. Of course, an excellent job of report writing cannot disguise a sloppy investigation, but many excellent design studies have not received proper attention and credit because the work was reported in a careless manner. You should be aware that written reports carry a message farther than the spoken word and have greater permanence. Therefore, technical workers often are known more widely for their writings than for their talks.

Organization of Reports

Written communications take the form of letters, brief memorandum reports, formal technical reports, technical papers, and proposals. In terms of the communications model shown in Fig. 15-1, the source is the mind of the writer. The process of encoding consists of translating the idea from the mind to words on a paper. The channel is the pile of manuscript papers. Decoding the message depends on the reader's ability to understand the language and familiarity with the ideas presented in the message. The final receiver is the mind of the reader. Noise is present in the form of poor writing mechanics, incomplete diagrams, incorrect references, etc. Since there is no direct feedback, the writer must anticipate the needs of the receiver and attempt to minimize the noise.

The first principle of written communication is to know your audience so that you can anticipate and fulfill its needs. The purpose of engineering writing is to present information, not to entertain. Therefore, the information should be easy to find. Always when writing your report, keep in mind the busy reader who has only a limited amount of time for your report and may, in addition, not be familiar with your subject.

Memorandum reports. The memorandum report usually is written to a specific person or group of persons concerning a specific topic with which both the writer and recipient(s) are familiar. It is written in memorandum form.

- Date
- To:
- From:
- Subject:
- Introduction (Brief and indicating why the study was carried out)
- Discussion (Includes data and its analysis)
- Conclusions (Includes what was concluded from the study and recommendations made on the conclusions)

Memorandum reports are short (one to three pages). Sometimes a report that is more than one page has a summary section before the introduction so the reader does not have to read the entire memorandum to get its message. The purpose in writing a memorandum report is to get a concise report to interested parties as quickly as possible. The main emphasis is on results, discussion, and conclusions with a minimum of writing about experimental details unless, of course, those details are critical to the analysis of the data. Very often a more detailed report follows the memorandum report.

Formal technical reports. A formal technical report usually is written at the end of a project. Generally, it is a complete, stand-alone document aimed at persons having widely diverse backgrounds. Therefore, much more detail is required. The outline of a typical formal report[1] might be:

- Covering letter (letter of transmittal)
- Summary (containing conclusions)
- Introduction (containing background to the work to acquaint reader with the problem and the purpose for carrying on the work)

---

[1] The contribution of Professor Richard W. Heckel for much of the material in this section is acknowledged.

- Experimental Procedure
- Experimental Results
- Discussion (of results)
- Conclusions
- References
- Appendixes
- Tables
- Figures

The *covering letter* is provided so that persons who might receive the report without prior notification will have some introduction to it. The *summary* is provided early in the report to enable the reader to know if it is worth the effort to read the entire report or send it to someone else who may be interested. It should be concise and give the reader a description of the contents of the report. The *introduction* should contain the pertinent technical facts that might be unknown to the reader but will be used in the report. It should "set the stage" for which the work was undertaken. The *experimental procedure section* is usually included to indicate how the data were obtained and to describe any nonstandard types of apparatus or techniques that were employed. The *experimental results section* describes the results of the study. Data in the form of tables or figures are usually placed at the end of the report and are referred to by number in the results section. This section should also indicate any uncertainties in the data and possible errors in their sources. The *discussion section* is normally concerned with analyzing the data to make a specific point, develop the data into some more meaningful form, or relate the data to theory described in the introduction. All arguments based on the data are developed here. The section may also be used to discuss the effects of experimental error on the analysis of the data. The *conclusion section* states in as concise a form as possible the conclusions *that can be drawn from the study. No new information should be introduced here.* In general, this section is the culmination of the work and the report. The conclusions are directly related to the purpose in undertaking the study. *Appendixes* are used for mathematical developments, sample calculations, etc., that are not directly associated with the subject of the report and that, if placed in the main body of the report, would seriously impede the logical flow of thought. Final equations developed in the appendixes are then placed in the body of the report with reference to the particular appendix in which they were developed. The same procedure applies to mathematical calculations; the results are used in the report and the proper appendix is referenced. Tables and figures require special attention; see Sec. 15-6.

Technical papers usually have an outline similar to this:

- Abstract
- Introduction

- Experimental procedure ⎫ These may be combined
- Experimental results ⎭ in a single section.
- Discussion
- Summary and/or conclusions
- Acknowledgments
- Appendixes
- Tables
- Figures

The content of the above sections is the same as for a formal technical report. The *abstract section* is similar to the formal report summary except that it doesn't contain reference to orientation of the work to particular company problems. A technical paper often contains a *summary* at its end that allows the writer to "assemble" the paper prior to making *conclusions*. Some papers omit the summary section if it is not felt to be necessary and if it would be essentially the same as the abstract.

A proposal is a report written to a sponsor in solicitation of financial support. The object of a proposal is to convince the sponsor of the value of your idea and to convince him that your organization has the capability (manpower and facilities) to deliver the expected results. A typical proposal to a federal agency might be organized as follows:

- Introduction
- Purpose and objectives
- Technical background
- Program approach (your ideas and approach)
- Statement of work
- Program schedule
- Program organization
- Personnel qualifications
- Facilities and equipment
- Summary
- References
- Budget (often submitted as a separate document)
- Appendixes

The secret to writing winning proposals is to align your interests with those of the sponsor. Be sure the proposal carefully indicates the magnitude of the expected gain from doing the proposed work. It usually is helpful to know who will evaluate the proposal and what their standards for performance and excellence will be.

Steps in Writing a Report

The five operations involved in the writing of a high-quality report are best remembered with the acronym POWER.

P   Plan the writing
O   Outline the report
W   Write
E   Edit
R   Rewrite

The planning stage of a report is concerned with assembling the data, analyzing the data, drawing conclusions from the data analysis, and organizing the report into various logical sections. The planning of a report is usually carried out by considering the various facets of the work and providing a logical blend of the material. The initial planning of a report should begin *before* the work is carried out. In that way the planning of the work and planning of the report are woven together, which facilitates the actual writing operation.

Outlining the report consists of actually formulating a series of headings, subheadings, sub-subheadings, etc., which encompass the various sections of the report. The outline can then be used as a guide to the writing. A complete outline can be detailed to the point at which each line consists of a single thought or point to be made and will then represent one paragraph in the report. The main headings and subheadings of the outline are usually placed in the report to guide the reader.

The writing operation should be carried out in the form of a rough draft using the maximum technical and compositional skill at the command of the writer. However, do not worry about perfection at this stage. Once you get going, don't break stride to check out fine details of punctuation or sentence structure.

Editing is the process of reading the rough draft and employing self-criticism. It consists of strengthening the rough draft by analyzing paragraph and sentence structure, economizing on words, checking spelling and punctuation, checking the line of logical thought, and, in general, asking oneself the question "Why?" Editing can be the secret of good writing. It is better for the writer to ask himself embarrassing questions than to hear them from his technical readers, his supervisor, or his instructor. In connection with editing, it has often been said that the superior writer makes good use of both ends of the pencil.

It is generally good practice to allow at least a day to elapse after writing the rough draft before editing it. That allows the writer to forget the logical pattern used in writing the report and appear more in the role of an unbiased reader when editing. Many mistakes or weak lines of thought that would normally escape unnoticed are thereby uncovered. The rewriting operation consists of retyping or rewriting the edited rough draft to put it in a form suitable for the reader. An important tip for preparing a handwritten report draft is to use every other line on the paper. In that way you will be able to make corrections in the empty lines and use part of your rough draft without doing a complete rewrite. Of course, if you are able to do your rough draft on a word processor, the revision is much less painful.

Mechanics of Writing

The following suggestions are presented as a guide to writing and an aid in avodiing some of the most common mistakes. You also should avail yourself of one of the popular guides to English grammar and style.[1]

Title. The title should be a meaningful description of what you have written.

Basic ideas. State basic ideas early; give the reader an overview of your study at the beginning. Enough background must be given to allow the least-informed reader to understand why the design was undertaken. The reader must be able to fit whatever is being reported into a context that is meaningful.

Whole vs. part. Describe the whole before the part; be sure you present the reader with the whole picture before you lead into the details. You should describe the essence, the function, and the purpose of a device, process, or component before you go into details about the parts.

Important Information. Emphasize important information; beware of the common error of burying it under a mass of details. Put the important ideas early in your writing; use appropriate capitalization and underlining; and relegate information of secondary importance to the appendix.

Headings. Use headings liberally; headings and subheadings are signposts that help the reader understand the organization of your ideas. If you have prepared a good outline, the headings will be self-evident.

Fact vs. Opinion. Separate fact from opinion. It is important for the reader to know what your contributions are, what ideas you obtained from others (the references should indicate that), and which are opinions not substantiated by fact.

Paragraph structure. Each paragraph should begin with a topic sentence that provides an overall understanding of the paragraph. Since each paragraph

---

[1] W. Strunk and E. B. White, "The Elements of Style," 3d ed., Macmillan, Inc., New York, 1978; S. Baker, "Practical Stylist," 4th ed., Harper and Row, New York, 1977.

should have a single theme or conclusion, the topic sentence states that theme or conclusion. Any elaboration is deferred to subsequent sentences in the paragraph. Don't force the reader to wade through many sentences of disconnected verbiage to arrive at the conclusion in the last sentence of the paragraph. The reader should be able to get an understanding of the report by reading the first sentence of each paragraph.

Sentence length. Sentences should be kept as short as possible so that their structure is simple and readable. Generally a sentence should not exceed about 35 words. Long sentences require complex construction, provide an abundance of opportunity for grammatical errors, take considerable writing time, and slow the reader down. Long sentences are often the result of putting together two independent thoughts that could be stated better in separate sentences.

Pronouns. There is no room for any degree of ambiguity between a pronoun and the noun for which it is used. Novices commonly use "it," "this," "that," etc., where it would be better to use one of several nouns. It may be clear to the writer, but it is often ambiguous to the reader.

In general, personal pronouns (I, you, he, she, we, my, mine, our, us) are not used in technical reports. The only exception is when the writer *must* involve himself personally in the report, as when writing a report to be used as a basis for a patent. In that instance, the writer must state definitely that he was the inventor.

Tense. The choice of the tense of verbs is often confusing to student writers. The following simple rules are usually employed by experienced writers:

Past tense. Use to describe work done in the laboratory or in general, to past events. "Hardness readings *were* taken on all specimens."

Present tense. Use in reference to items and ideas in the report itself. "It *is* clear from the data in Figure 4 that strain energy *is* the driving force for recovery" or "The group recommends that the experiment be repeated" (present opinion).

Future tense. Use in making prediction from the data that will be applicable in the future. "The market data given in Table II indicate that the tonnage *will continue* to increase in the next ten years."

Spelling and punctuation. Errors in these basic elements of writing in the final draft of the report are inexcusable.

Appendixes, tables, and figures. Appendixes, tables, and figures are placed at the end of a report to speed the reading of the text and to allow separate preparation of tables and figures without having to bother with leaving the required space in the text. The following numbering conventions are often used:

Appendix A, B, C, D (capital letters)

Table I, II, III, IV (Roman numerals)

Figure 1, 2, 3, 4 (numbers)

Each appendix, table, or figure should have a title and should be self-explanatory. For example, the reader should not have to refer to the text for an understanding of the variables plotted on a graph. Graphs should be drawn on suitable graph paper (linear, semilog, log-log, etc.), and the axes should be properly labeled (including units). If data points are labeled by code numbers, the code should appear on the graph, not in the text. The independent and dependent variables are usually placed on the abscissa (x axis) and ordinate (y axis), respectively. Generally, only one table or figure is placed on a page. However, several small photographs may make up one figure.

Reference to data in tables and figures is often a cause of difficulty. The statement "It is obvious from the data that . . ." is used much too often when nothing is obvious before the data have been explained properly. The statement given above should be used *only* when an obvious conclusion can be made, not when it is the hope of the writer that the reader thinks something is obvious.

The *initial* reference to a set of data in a table or figure as: "The data are given in Table III" exerts an extreme hardship on the reader and in many instances forces him to grope unguided through the data. In such instances the reader must digest the data in a few minutes, whereas the digestion may have taken the writer days or weeks. (The writer also runs the risk of the reader's reaching an "improper" conclusion before being biased by the text). It is better form to first describe what a set of results shows *prior* to telling the reader where the data can be found. This saves the reader's time by indicating what to look for when exposed to the data.

References. References are usually placed at the end of the written text. Those to the technical literature (described as readily available on subscription and included in most library collections) are made by author and journal reference (with the title of article omitted) as shown by the following example. There is no single universally accepted format for references, but many journals follow the style given here.

Journal article
R. M. Horn and Robert O. Ritchie: *Metall. Trans. A,* 1978 vol. 9A, pp. 1039–1053.
Book
Thomas T. Woodson: "Introduction to Engineering Design," pp. 321–346, McGraw-Hill Book Company, New York, 1966.
A private communication
J. J. Doe, XYZ Company, Altoona, PA, unpublished research, 1981.

Internal reports
  J. J. Doe: Report No. 642, XYZ Company, Altoona, Pa., February 1980.

Be scrupulous about references as you acquire them. A few minutes spent in recording the full reference (including the inclusive page numbers) can prevent hours of time being spent in the library when you are finishing the final report.

15-4 CONDUCTING A MEETING

The business world is full of meetings, which are held to report on progress, inform and educate, stimulate, or solve problems (see Sec. 3-2). Many professional people feel that an inordinate amount of their time is spent in meetings. Perhaps, so, or perhaps they feel that way because too often a meeting is improperly planned and conducted. Therefore, the purpose of this section is to suggest some ways you can conduct informative and stimulating meetings.

The Business Meeting

Its purpose usually will dictate who is invited to the meeting. However, when considerable discretion is possible, consider such factors as organizational politics and representation of divergent views, including people who can implement your expected outcome or who are known for good ideas and judgment. You should limit the meeting to 12 to 15 people if your objective is discussion and an interchange of ideas.

  The first step in a successful meeting is the preparation of a meeting agenda, which each participant should receive from two to ten days prior to the meeting. The *agenda* should include: the date, time for starting and closing the meeting, the place of the meeting, the topic and subtopics to be discussed, and the participants. Any background reading material should be enclosed as an attachment to the agenda. The advantages of distributing an agenda are 1) by making an agenda, the chairman will be better organized and the meeting will go more smoothly, 2) the chairman will have better control over digressions by referring back to the agenda, 3) participants can prepare for the meeting, 4) participants will know who has been invited to the meeting so each can plan his strategy. Frequently the last item on the agenda is "other business." This builds some flexibility into the meeting.

  It is important to check out the conference room prior to the meeting to be sure that paper, visual aids, refreshments, etc. are available. The seating arrangement for the meeting can have important psychological consequences. For good discussion with a small group a round table is best. For a larger group a wide rectangular or U-shaped arrangement is effective. Avoid seating people face to face across a narrow table.

  The person who convenes the meeting usually is the leader. As leader it is important to sit where you can be easily seen by all participants. If the situation calls for a more authoritarian atmosphere, you should position yourself at the head of the table. For a more collegial atmosphere sit in the middle of the table. If you know you have a possible disruptive member seat him to your immediate right where you will not make eye contact.

As leader it is your responsibility to set the tone for the meeting. Start on time. To wait for the laggards is to set a bad example and encourage others to be late at the next meeting. Open the meeting with a brief statement of the meeting's purpose. Be positive and upbeat, and the meeting will start off on a spirited tone. In a large group you may want to use an overhead projector to reinforce the agenda and get your points across. For a smaller group handouts will be sufficient.

As leader you are responsible for keeping the meeting moving along. You must stimulate discussion when it lags and balance it when only a single viewpoint is put forth. You are the "traffic cop" who decides who speaks next, who steers the discussion back on track, and who breaks up hot controversies if they should develop. It is your responsibility to watch the time and finish on schedule. As the time approaches for ending the meeting you should allow enough time for "other business." In that way valuable contributions that were put off earlier as digressions are not lost. The leader should stay neutral, avoid lecturing, and patiently guide the group to a solution. Never monopolize the meeting, publicly rebuke a committee member, or permit unnecessary interruptions such as phone calls.

Before the meeting breaks up, the leader should summarize the key points and decisions that have been made. Any assignments that were made during the meeting are reviewed to be sure the responsibility for follow-up is clearly understood.

Most meetings worth holding should be made a matter of record. Depending on the nature of the meeting, the participants should receive a communication from the leader describing the conclusions reached and the future action to be taken. This may be a brief memorandum or more complete and formal minutes of the meeting describing what was said at the meeting. Be sure to give proper credit to the participants (by name) for their ideas and contributions to the meeting.

The Technical Meeting

Hopefully you will become active in one or more professional or technical societies. At some time you most likely will be called upon to chair a session at a local or national technical meeting. Ideas on presenting a technical talk are given in Sec. 15-5. Here we deal with the different but related situation of being the session chairman.

A technical meeting will usually deal with an audience of from 30 to 300 people. As session chairman you are responsible for introducing the speakers, conducting the discussion after each talk, and seeing to it that the meeting runs on schedule. Frequently you will have a cochairman to share those duties.

Usually you will be provided with biographical information on the speaker; but if not, you should obtain that information from all of the speakers before the meeting starts. Unless you are introducing the speaker for a prestigious major address, you should restrict your introduction to a few words about the speaker's affiliation and background. Some societies encourage the speakers and chairmen to get acquainted at an "author's breakfast" on the morning of the talk.

As session chairman you should check out the meeting room at least one-half hour prior to the start of the meeting to be sure that visual aids and operators are available. Pay particular attention to how the lights will be turned on and off to accommodate slides. Make sure there is an "electric pointer" for use by authors showing slides. The duties of the session chairman can be listed in the following order.

1. Start meeting on time.
2. Give brief statement of origin and purpose of the particular technical session.
3. Introduce each speaker.
4. Assist each speaker with microphone (if needed).
5. Keep track of time and warn speaker when his time is about to run out.
6. Conduct the discussion.
7. Arrange for break.
8. Thank speakers for contributions to the technical session.
9. Make announcements and adjourn the meeting.

15-5   ORAL PRESENTATIONS

Impressions and reputations (favorable or unfavorable) are made most quickly by audience reaction to an oral presentation. There are a number of situations in which you will be called upon to give a talk. Progress reports, whether to your boss in a one-on-one situation or in a more formal setting to your customer, are common situations in which oral communication is used. Selling an idea or a proposal to your management budget committee or a sponsor is another common situation. In the more technical arena, you may be asked to present a talk to a local technical society chapter or present a paper at a national technical meeting.

Oral communication has several special characteristics: quick feedback by questions and dialogue; impact of personal enthusiasm; impact of visual aids; and the important influence of tone, emphasis, and gesture. A skilled speaker in close contact with an audience can communicate far more effectively than the cold, distant, easily evaded written word. On the other hand, the organization and logic of presentation must be of a higher order for oral than for written communication. The listener to an oral communication has no opportunity to reread a page to clarify a point. Many opportunities for noise exist in oral communication. The preparation and delivery of the speaker, the environment of the meeting room, and the quality of the visual aids all contribute to the efficiency of the oral communication process.

The Business-Oriented Technical Talk

The purpose of your talk may be to present the results of the past 3 months of work by a 10-person design team, or it may be to present some new ideas on computer-aided design to an audience of upper management who are skeptical that their large investment in CAD will pay off. Whatever the reason, you should know the purpose of your talk and have a good idea of who will be attending your presentation. This information is vital if you are to prepare an effective talk.

The most appropriate type of delivery for most business-oriented talks is an *extemporaneous-prepared talk*. All the points in the talk are thought out and planned in detail. However, the delivery is based on a written outline, or alternatively, the text of the talk is completely written but the talk is delivered from an outline prepared from the text. This type of presentation establishes a more natural, closer contact with the audience that is much more believable than if the talk is read by the speaker.

Develop the material in your talk in terms of the interest of the audience. Organize it on a thought-by-thought rather than a word-by-word basis. Write your conclusions first. That will make it easier to sort through all the material you have and select only the pieces of information that support the conclusions. If your talk is aimed at selling an idea, list all of your idea's strengths and weaknesses. That will help you counter arguments against adopting your idea.

The opening few minutes of your talk are vital in establishing whether you will get the audience's attention. You need to "bring them up to speed" by explaining the reason for your presentation. Include background enough that they can follow the main body of your presentation, which should be carefully planned. Stay well within the time allotted for the talk so there is an opportunity for questions. Include humorous stories and jokes in your talk only if you are very good at telling them. If you are not, it is best to play it straight. Also, avoid specialized technical jargon in your talk. Before ending your presentation, summarize your main points and conclusions. The audience should have no confusion as to the message you wanted to deliver.

Visual aids are an important part of any technical presentation; good ones can increase the audience retention of your ideas by 50 percent. The type of visual aid to use depends upon the nature of the talk and the audience. For small informal meeting of up to 10 or 12 people, handouts of an outline, data, and charts usually are effective. For larger groups of up to 30 people a flip chart on an easel may be used. Transparencies used with an overhead projector are good for groups from 10 to 200 people. Slides are the preferred visual aids for large audiences. The important subject of visual aids is discussed at greater length in Sec. 15-6.

The usual reason a technical talk is poor is lack of preparation. It is a rare individual who is able to give an outstanding talk without practicing it. Once you have prepared the talk, the first stage is individual practice. Give the talk out loud in an empty room to fix the thoughts in your mind and check the timing. You may want to memorize the introductory and concluding remarks. If at all possible, videotape your individual practice. The dry run is a dress rehearsal before a small audience. If possible, hold the dry run in the same room where you will give the talk. Use the same visual aids that you will use in your talk. The purpose of the dry run is to help you work out any problems in delivery, organization, or timing. There should be a critique following the dry run, and the talk should be reworked and repeated as many times as are necessary to do it right.

When delivering the talk, if you are not formally introduced, you should give your name and the names of any other team members. You should speak loudly enough to be easily heard. For a large group that will require the use of a microphone. Work hard to project a calm, confident delivery, but don't come on in an overly aggressive style that will arouse adversary tendencies in your audience. Avoid annoying mannerisms like rattling the change in your pocket and pacing up and down the platform. Whenever possible, avoid talking in the dark. The audience might well go to sleep or, at worst, sneak out. Maintaining eye contact with the audience is an important part of the feedback in the communication loop.

The questions that follow a talk are an important part of the oral communication process; they show that the audience is interested and has been listening. If at all possible, do not allow interruptions to your talk for questions. If the "big boss" interrupts with a question, compliment him for his perceptiveness and explain that the point will be covered in a few moments. Never apologize for the inadequacy of your results. Let a questioner complete the questions before breaking in with an answer. Avoid being argumentative or letting the questioner see that you think the question is stupid. Do not prolong the question period unnecessarily. When the questions slack off, adjourn the meeting.

The Technical Society Talk

It is an honor to be asked to present the results of your work at a national meeting of a professional society, but you may be surprised to find that you have only 15 or 20 minutes in which to present your work. That calls for good planning and organization. Frequently an abstract of your talk will be available in the meeting announcement or printed book of abstracts that is available to registrants of the meeting. However, that does not absolve you from the responsibility for starting your talk with a good introduction, which should set forth the scope and objectives of the talk. It may deal with the history of events leading up to the work to be discussed. Any coauthors of the work described should be mentioned, and acknowledgments should be made to sponsors and those who may have given special help.

The main body of the technical talk covers the following items:

- Experimental or analytical procedure
- Results and observations
- Discussions and conclusions
- Ongoing research and/or future studies

The talk should end with a summary that repeats the main information given in the body of the talk. A good scheme for the organization of a technical talk is to:

1. Tell them what you are going to tell them.
2. Tell them.
3. Tell them what you told them!

15-6 VISUAL AIDS AND GRAPHICS

Except in a small meeting, visual aids make the difference between effective and ineffective oral communication. The selection of the visual aid medium will depend upon the size and importance of the audience and the number of times the talk will be given. Not only are good visual aids important in transmitting the message, so that the dual senses of hearing and seeing are employed, they also assist greatly in reducing the nervous tension (psychological noise) of inexperienced speakers.

1. Limit slides (or viewgraphs) to *no more than* one per minute.
2. Each slide should contain one idea.
3. Slides that present more than three curves on a single graph or 20 words or numbers on a slide are too complicated.
4. The first slide in the presentation should show the title of your talk and the names and affiliations of the authors.
5. The second slide should give a brief outline of your talk.
6. The last slide should summarize the message you delivered.
7. If you need to show a slide more than once, use a second copy so the projectionist will not have to back up, and in the process distract the audience.
8. Avoid leaving on the screen a slide that you have finished discussing while you go on to something else. Put in a blank slide, which will allow the audience to turn their attention to your words.

The chief purpose of a visual aid is to improve and simplify communication. There should be no intention to have a slide or viewgraph stand alone, unsupported by oral communication. The spoken word should complement the limited information that can be contained on the visual. On the other hand, it is a waste of time for the speaker to read the words on the slide to the audience.

Flip Charts

The flip chart can be an effective visual aid for oral presentations for small groups of up to 30 people. It employs a large pad of paper, typically 22 by 32 in. The pad is placed on an easel, and each page is flipped over to reveal the next message. Key words, bar graphs, or tables usually are hand-lettered with a felt-tip pen. Neatness and legibility are important. Use 2-in-high lettering for a room up to 30 ft deep and increase to 4-in letters for a room 60 ft deep.

Flip charts are the easiest to prepare and least expensive form of visual aid. Slides may be more impressive, but they are more expensive and require considerable lead time. A serious limitation of flip charts for technical presentations is that they cannot easily deal with illustrations of equipment or photographic data like fractographs or photomicrographs.

Slides

Slides prepared from 35-mm transparency film (2- by 2-in size) are a standard visual aid for technical presentations. The 2- by 2-in slide has largely replaced the $3\frac{1}{4}$ by 4-in glass lantern slide format.

Slides require professionally prepared drawings and artwork. A slide that is acceptable for a technical talk usually *cannot be made* by reproducing figures and tables from a report or book. The scale of dimensions and the density of information found on the printed page simply are wrong for producing a good slide for a talk. The dimensions of the transparency for a standard 2- by 2-in slide in the horizontal position are 24 by 36 mm. Therefore, the artwork to be photographed for a horizontal slide should be contained in a rectangle 6 in high by 9 in wide. To determine whether the artwork will make a slide that is adequately legible, view it from a distance of 6 ft. From that distance you will see the material as it will appear to a viewer at the rear of a large room. To check a 35-mm slide for legibility, view it unmagnified from a distance of 12 in.

Figure 15-2 shows an example of a poor illustration for a slide or a report. The use of fine-line graph paper gives a cluttered feeling, as does the inclusion of supplementary data in the body of the graph. Pay particular attention to the marginal comments around this figure.

Figure 15-3 shows the same data redrawn to make a good illustration. Note that reducing the number of grid lines results in a cleaner illustration. An even crisper tone is produced by eliminating the grid lines altogether and using ticks along the four axes of the graph. The letters in titles for a slide should be

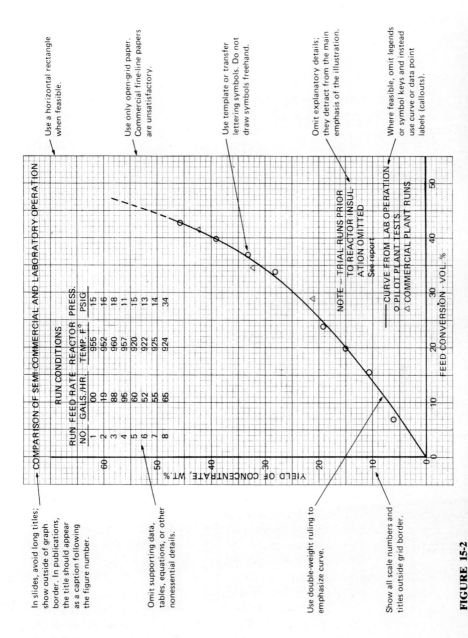

FIGURE 15-2

An example of poor technical illustration. (*ANSI Y15.1M–1979, "Illustrations for Publication and Preparation." Used by permission of American Society of Mechanical Engineers.*)

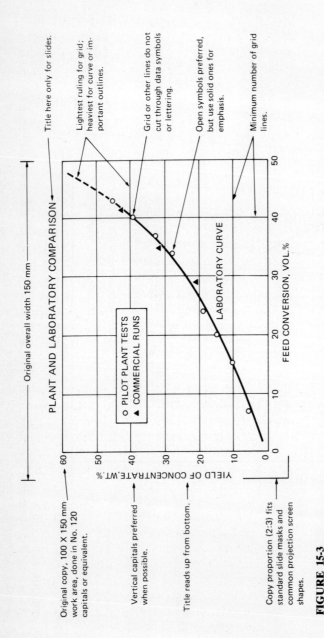

FIGURE 15-3

Figure 15-2 redrawn to be suitable for slide projection. (*ANSI Y15.1M–1979. Used by permission of the American Society of Mechanical Engineers.*)

at least $\frac{1}{4}$ in high. All other lettering should be at least $\frac{3}{16}$ in high. Data points should be indicated by open symbols. In drawing curves the lines should not run through the data points.

Here are a few tips concerning giving a talk while using slides as your visual aid. There is nothing more embarrassing than to be halfway through a slide talk and find that the next slide is projected upside down. To avoid that, put thumb tabs on your slides to aid the projectionist. If possible, load your slides early and run through the entire set to ensure they have been correctly oriented. Slides can be shown properly only in a darkened room. Sometimes there is no light for the speaker, which can be a problem if you were counting on using notes. The ideal arrangement is one in which a speaker can change the slides from the podium without having to call out to the projectionist.

Overhead Transparencies

Transparencies (viewgraphs) with an overhead projector provide a flexible visual aid system for groups up to about 200 people. Because transparencies can be viewed in a semilighted room, they permit the speaker to face the audience and maintain eye contact. They can be made fairly quickly by using a standard office copier, and they lend themselves to using overlays for revealing information. In small groups it is possible to pass out photocopies of transparencies as a permanent record of the talk. Transparencies are not as good as slides in reproducing photomicrographs and other photographs. Most of the tips for preparing material for slides apply to transparencies also.

Graphics for Reports

This section is concerned with graphics for technical reports and papers. The old adage that "one picture is worth 1000 words" is very pertinent with respect to technical publications. Combining graphics with words reinforces the communication by providing redundancy and reduction in noise. The common forms of graphic material used in technical publications are tables, charts, block diagrams, schematic diagrams, graphs or curves, drawings, and photographs.

The chief attention in preparing a table should be given to the headings. Try to present the facts you have with the smallest number of headings. For example, if nearly all the values under a heading are the same, that information can be included in a subtitle or footnote and the heading can be omitted. Be sure that the heading adequately describes the variables you are representing. Be sure to include proper units in the heading. The headings describe the columnar entries (variables) in the table. The arrangement of data for the rows of the table is called the stub. Some typical ways of arranging stub entries are 1) in the order the data were gathered, 2) by materials or general classes, 3) by observed phenomena, 4) by source of data.

If your report contains extensive tables of data, you must provide interpretation for the reader. This can take the form of summary graphs or summary tables that "boil down" the extensive data. Often detailed tables of data are included in an appendix and only the summary graphs and tables are included in the body of the report.

Tables consisting of computer-printout data are not acceptable for quality reports. Generally the computer printout must be reduced to fit into the report, so that the legibility is poor. If you know the computer printout must be used in a report, special care can be used in planning the format of the printout.

Graphs that will be used for slides or in a technical publication should be drawn two or three times the final publication size to make it possible to use consistent line weights and crisp lettering. Graphs that will be used as an overhead transparency should be prepared in actual size. The standard column width on an $8\frac{1}{2}$ by 11 journal page is 75 mm. Therefore, the original artwork for technical publications or 35-mm slides should be drawn to a size either 150 by 100 mm or 225 by 150 mm. Lettering should be done with a mechanical lettering set, transfer letters, or a phototypesetter. An ordinary typewriter is not suitable for lettering an illustration.

Photographs for reproduction should be glossy prints; those printed on matte paper will lose crispness in reproduction. It is a good idea to have two copies of each photograph. Dry-mount or tape one copy of a 200- by 250-mm print to an illustration board and keep the other copy in reserve. Indicate the vertical and horizontal crop marks on the illustration board, not on the photograph. Indicate the top of the photograph. The reproduction size should be given in millimeters.

BIBLIOGRAPHY

Andrews, D. C., and M. D. Blickle: "Technical Writing: Principles and Forms," Mecmillan Publishing Co., New York, 1978.

Eisenberg, A.: "Effective Technical Communication," McGraw-Hill Book Co., New York, 1982.

Ewing, D. W.: "Writing for Results," 2d ed., John Wiley & Sons, Inc., New York, 1979.

IEEE Transactions on Professional Communications, published quarterly.

Michaelson, H. B.: "How to Write and Publish Engineering Papers and Reports," 2d ed., ISI Press, Philadelphia, 1986.

Mills, G. H., and J. A. Walter: "Technical Writing," 4th ed., Holt, Rinehart and Winston, New York, 1978.

Olsen, L. A., and T. N. Huckin: "Principles of Communication for Science and Technology," McGraw-Hill Book Co., New York, 1983.

Rosenstein, A. B., R. R. Rathbone, and W. F. Schneerer: "Engineering Communications," Prentice-Hall, Inc., Englewood Cliffs, N.J., 1964.

Scott, B.: "Communication for Professional Engineers," Thomas Telford Ltd., London, 1984.

APPENDIX A

PROBLEMS

CHAPTER 1

1-1 There is a need in underdeveloped countries for building materials. One approach is to make building blocks (4 by 6 by 12 in) from highly compacted soil. Your assignment is to design a block-making machine with the capacity for producing 600 blocks per day at a capital cost of less than $150. Develop a needs analysis, a definitive problem statement, and a plan for the information that will be needed to complete the design.

1-2 The need for material conservation and decreased cost has increased the desirability of corrosion-resistant coatings on steel. Develop several design concepts for producing 12-in-wide low-carbon-steel sheet that is coated on one side with a thin layer, e.g., 0.001 in, of nickel.

1-3 The support of thin steel strip on a cushion of air introduces exciting prospects for the processing and handling of coated steel strip. Develop a feasibility analysis for the concept.

1-4 The steel wheel for a freight car has three basic functions: 1) to act as a brake drum, 2) to support the weight of the car and its cargo, and 3) to guide the freight car on the rails. Freight car wheels are produced by either casting or rotary forging. They are subject to complex conditions of dynamic thermal and mechanical stresses. Safety is of great importance, since derailment can cause loss of life and property. Develop a broad systems approach to the design of an improved cast-steel car wheel.

1-5 Review the design example given in Sec. 1-7. By doing additional reading on the subject, expand the needs analysis. Discuss the major steps in the design process that must be undertaken to produce a ceramic stator for a gas turbine engine.

1-6 Consider the design of aluminum bicycle frames. A prototype model failed in fatigue after 1600 km of riding, whereas most steel frames can be ridden for over 60,000 km. Describe a design program that will solve this problem.

CHAPTER 2

2-1 Discuss the spectrum of engineering job functions with regard to such factors as *a*) need for advanced education, *b*) intellectual challenge and satisfaction, *c*) financial reward, *d*) opportunity for career advancement, and *e*) people vs. "thing" orientation.

2-2 Strong performance in your engineering discipline ordinarily is one necessary condition for becoming a successful engineering manager. What other conditions are there?

2-3 Discuss the pro's and con's of continuing your education for an MS in an engineering discipline or an MBA on your projected career progression.

2-4 Discuss in some detail the relative roles of the project manager and the functional manager in the matrix type of organization.

2-5 Discuss why engineering qualifies as a profession. Compare the differences between and similarities of the engineering and medical professions.

2-6 Discuss the importance to the engineering profession of having a strong coordinating (umbrella) organization like AAES. How has the engineering profession suffered by not having such an organization in its past?

2-7 Discuss the ethics of the following situation. You are a design engineer for the Ajax Manufacturing Co., a large multiplant producer of plastic parts. As part of your employment, you were required to sign a secrecy agreement that prohibits divulging information that the company considers proprietary.

Ajax has modified a standard piece of equipment that greatly increases the efficiency in cooling viscous plastic slurries. The company decides not to patent the development but instead to keep it as a trade secret. As part of your regular job assignment, you work with this proprietary equipment and become thoroughly familiar with its enhanced capabilities.

Five years later you leave Ajax and go to work for a candy manufacturer as chief of production. Your new employer is not in any way in competition with Ajax. You quickly realize that Ajax's trade secret can be applied with great profit to completely different machine used for cooling fudge. You order the change to be made. Discuss the ethics.

2-8 Discuss the ethics in the following situation. You have been on the job for nine months as an assistant research engineer working with a world-famous authority on heat transfer. It is an ideal job, because you are learning a great deal under his sympathetic tutelage while you pursue an advanced degree part-time.

You are asked to evaluate two new flame-retardant paints A and B. Because of late delivery of some constituents of paint A, the test has been delayed and your boss has been forced to make a tentative recommendation of paint A to the design group. You are asked to make the after-the-fact tests "for the record." Much to your surprise, the tests show that your boss was wrong and that formulation B shows better flame resistance. However, a large quantity of paint A already has been purchased. Your boss asks you to "fudge the data" in favor of his original decision, and since there is reasonable possibility that your data were in error, you reluctantly change them to favor his decision. Discuss the ethics.

2-9 List the factors that are important in developing a new technologically oriented product.

2-10 List the key steps in the technology transfer (diffusion) process. What are some of the factors that make technology transfer difficult? What are the forms in which information can be transferred?

2-11 (*a*) Discuss the societal impact of a major national program to develop synthetic fuel (liquid and gaseous) from coal. (It has been estimated that to reach the level of supply equal to the imports from OPEC countries would require over 50 installations, each costing several billion dollars.)

(*b*) Do you feel there is a basic difference in the perception by society of the impact of a synthetic fuel program compared with the impact of nuclear energy? Why?

(*c*) The reason synthetic fuel from coal has not yet become a developed technology is that the cost still exceeds that of comparable natural fuel. What are some of the alternatives to synthetic fuel that may solve our nation's long-term energy problem?

2-12 Most likely you will conclude that the basic problem in developing synthetic fuel on a massive scale is economic. Propose a scheme by which the federal government could stimulate the development of synthetic fuels but which would keep government and industry in their historical roles.

2-13 Suppose you are the inventor of a new device called the helicopter. By describing the functional characteristics of the machine, list some of the societal needs that it is expected to satisfy.

CHAPTER 3

3-1 Select two pages at random from a large mail-order catalog. Select one item from each page and try to combine the two items into a useful innovation.

3-2 A technique for removing a blockage in the creative process is to apply transformation rules to an existing unsatisfactory solution. Some common transformation operators are 1) put to others uses, 2) modify, 3) magnify, 4) diminish, 5) substitute, 6) rearrange, 7) reverse, 8) combine. A related technique is to use Roget's *Thesaurus* to suggest leads for alternative solutions. A key word from the existing solution is looked up in the thesaurus, and it provides a number of related and opposite words that stimulate new approaches.

Apply these techniques to the following problem. As a city engineer you are asked to suggest ways to eliminate puddles from pedestrian walkways after a rainstorm. Start with the obviously inadequate solution of waiting for the puddles to evaporate.

3-3 As central station power plant operators consider reconverting from oil or gas to coal as the energy source, they sometimes find that there is not a suitable large land area near the plant that can be used for on-the-ground coal storage. Conduct a brainstorming session to propose alternative solutions to a conventional coal pile.

3-4 What are the questions that need to be asked and answered in order to prepare a problem statement? Develop a problem statement for the situation described in Prob. 3-3. Include the following elements in the problem statement: 1) need statement, 2) goals, 3) constraints and trade-offs, and 4) criteria for evaluating the design.

3-5 Assume that sometime in the future you are in a key decision-making position and you must decide whether electric or internal-combustion vehicles will be used exclusively in the United States. Develop the decision-making process by using the ideas in Chap. 3.

3-6 Develop a morphological box for internal-combustion engines.

3-7 The following factors may be useful in deciding which brand of automobile to purchase: interior trim, exterior design, workmanship, initial cost, fuel economy, handling and steering, braking, ride, and comfort. To assist in developing the weighting factor for each of those attributes, group the attributes into four categories of body, cost, reliability, and performance and use a relevance tree to establish the individual weighting factors.

3-8 The introduction of a new line of turboencabulators will result in a $10 million increase in profits if a competitor does not introduce a similar product. If the new product is introduced and the competitor introduces a similar product, we could lose $7 million. The prior estimate of being able to introduce the product without competition is 0.4.

   (*a*) What is the expected value from this scenario?

   (*b*) Further market study has indicated that a $10 million profit can be expected with 80 percent probability if competition does not come into the market and 50 percent probability if it does. Should we enter the market based on this new information?

3-9 A design decision concerning the transmission for an automatic washer must be made. The following attributes are considered important: cost, reliability (mean time between failure), torque, serviceability, weight, noise. Develop a decision matrix for this design.

3-10 Modify the decision matrix of Prob. 3-9 to include the certainty with which the technology can be implemented.

3-11 Construct a simple personal decision tree over whether to take an umbrella when you go to work on a cloudy day.

3-12 This decision concerns whether to develop a microprocessor-controlled machine tool. The high-technology microprocessor-equipped machine costs $4 million to develop, and the low-technology machine costs $1.5 million to develop. The low-technology machine is less likely to receive wide customer acclaim ($P = 0.3$) vs. $P = 0.8$ for the microprocessor-equipped machine. The expected payoffs (present worth of all future profits) are as follows:

	Strong market acceptance	Minor market acceptance
High technology	$P = 0.8$ PW = $16 M	$P = 0.2$ PW = $10 M
Low technology	$P = 0.3$ PW = $12 M	$P = 0.7$ PW = 0

If the low-technology machine does not meet with strong market acceptance (there is a chance its low cost will be more attractive than its capability), it can be upgraded with microprocessor control at a cost of $3.2 million. It will then have an 80 percent chance of strong market acceptance and will bring in a total return of $10 million. The non-upgraded machine will have a net return of $3 million. Draw the decision tree and decide what you would do on the basis of various decision criteria.

CHAPTER 4

4-1 The following is a generally accepted list of the category of models used in engineering practice:

- Proof-of-concept model
- Scale model
- Experimental model
- Prototype model

Going down in this hierarchy, the model increases in cost, complexity, and completeness. Define each category of model in some detail.

4-2 Classify each of the following models as iconic, analogic, or symbolic. Give your reasons.

(a) The front view and left side view of a new fuel-efficient automobile when the scale is 1 in = 1 ft.

(b) The relation between the flow rate Q through a packed bed in terms of area A, pressure drop Δp, and height of the bed is given by

$$Q = \frac{K_D A \Delta p}{L}$$

(c) A strip chart recording showing the temperature-time profile for a carburizing cycle.

(d) A flow chart showing movement of a cylinder block through a machine shop.

(e) A free body diagram like Fig. 4-1a.

(f) A set of N outcomes from a random experiment, represented by $A = a_1, a_2, \ldots, a_n$.

4-3 With the slowdown in the growth in the use of electric power, many utilities have been forced to take older coal-fired power plants off base-load operation (steady state) and use them only to supply daily peak demands. That means the plants will be used in a cycling mode of operation, being fired to full power only at the daily period of peak demand. By using the model for a coal-fired power plant in Fig. 4-5, identify a number of materials-related problems that would be expected to result from a cycling type of operation.

4-4 A novel idea for absorbing energy in an automotive crash is a "mechanical fuse" (see M. C. Shaw, *Mechanical Engineering*, pp. 22–29, April 1972). The idea is to use the energy absorbed by metal cutting to dissipate the kinetic energy of the moving vehicle in a bumper energy absorber. In the concept, a round steel bar is pushed through a circular cutting tool, thereby creating a chip that absorbs the energy of the impact. In effect, the bar is "skinned" by the cutting tool.

Develop a mathematical model for the mechanical fuse. It is appropriate to assume that the metal cutting force is independent of velocity and displacement. Assume that the specific cutting energy (300,000 lbf/in^3 for steel) is not affected by the circular geometry of the tool or by the impact load.

4-5 Use dimensional analysis to determine the dimensionless groups that describe the forced-convection heat transfer of a fluid in a tube. Experience shows that the heat transfer coefficient h is given by:

$$h = f(\bar{V}, \rho, k, \eta, C_p, D)$$

where $\bar{V}$ = mean velocity
ρ = mean density
k = thermal conductivity
η = viscosity
C_p = specific heat at constant pressure
D = diameter

4-6 Use dimensional analysis to develop Griffith's equation for the fracture strength of a brittle material. This will be an equation similiar to Eq. (7-2).

4-7 The following table gives the number of defects that are found to occur in parts of type A and type B.

Defects	No. of times occurring in part A	No. of times occurring in part B
0	5	2
1	5	3
2	15	5
3	30	10
4	20	20
5	10	40
6	5	10
7	5	5
8	3	3
9	2	2
	100	100

Find with the help of a Monte Carlo simulation the expected numbers of defects in the final assembly C, which is made up of parts A and B. The first three random numbers generated in the simulation are 14 15; 58 20; 82 14.

CHAPTER 5

5-1 We want to design a hot-water pipeline to carry a large quantity of hot water from the heater to the place where it will be used. The total cost is the sum of four items: 1) the cost of pumping the water, 2) the cost of heating the water, 3) the cost of the pipe, 4) the cost of insulating the pipe.

(*a*) By, using basic engineering principles, show that the system cost is

$$C = K_p \frac{1}{D^5} + K_h \frac{1}{\ln[(D+x)/D]} + K_m D + K_i x$$

where x is the thickness of insulation on a pipe of ID $= D$.

(*b*) If $K_p = 10.0$, $K_h = 2.0$, $K_m = 3.0$, $K_i = 1.0$, and $D_0 = x_0 = 1.0$, find the values of D and x that will minimize the system cost.

5-2 Determine the value of insulation thickness for a furnace wall that results in minimum cost, given the following data:

Wall temperature 500°F; outside air temperature 70°F
Air film coefficient $= 4$ Btu/(h)(ft^2)(°F)
Thermal conductivity of insulation $= 0.03$ Btu/(h)(ft)(°F)
Insulation cost per inch of thickness $= \$0.75$ per ft^2
Value of heat saved $= \$0.60$ per million Btu
Fixed charges $= 30$ percent per year
Hours of operation $= 87000$ h/year

Base your calculation on 100 ft^2 of furnace wall.

5-3 Find the maximum value of $y = 12x - x^2$ with the golden mean search method for an original interval of uncertainty of $0 \le x \le 10$. Carry out the search until the difference between the two largest calculated values of y is 0.01 or less.

5-4 A manufacturer of steel cans wants to minimize costs. As a first approximation, the cost of making a can consists of the cost of the metal plus the cost of welding the longitudinal seam and the top and bottom. The can may have any dimension D and L to give a volume V_0. The thickness is δ. By using the the method of Lagrange multipliers, find the dimensions of the can that will minimize cost.

5-5 By using the method of steepest ascent, find the minimum of the objective function $U = x^2 + 2y^2 + xy$. Start the search at location $x = 2.0$, $y = 2.0$.

5-6 A machine component requires a screw machine operation followed by welding and assembly into a larger subassembly. Two versions of the component are produced: one for ordinary service and one for heavy-duty operation. A single unit of the ordinary design requires 3 min of screw machine time, 2 min of spot welding, and 5 min for assembly. The profit for each unit is $1.20. Each heavy-duty unit requires 3 min of screw machine time, 7 min for welding and hard-facing, and 6 min for assembly. The profit for each unit is $1.50. The total weekly capacity of the machine shop is 1200 min; that of the welding department is 800 min; that of assembly is 1600 min. What is the optimum mix between ordinary service and heavy-duty components to maximize the total profit?

5-7 The time to machine a shaft on a lathe is given by

$$t = 1.2v^{-1}f^{-1} + 0.25 \times 10^{-5}v^{3.17}f^{0.88} + 1.05 \text{ min/shaft}$$

The maximum feed f is limited to 0.02 in/rev by available power considerations, and the velocity v is limited to 470 fpm by the maximum spindle speed. Find the optimum machining time.

5-8 The LP equations for Prob. 5–6 are:

$$U = 1.20x_1 + 1.50x_2$$
$$3x_1 + 3x_2 \le 1200$$
$$2x_1 + 7x_2 \le 800$$
$$5x_1 + 6x_2 \le 1600$$

where x_1 is the number of ordinary service components produced in a week and x_2 is the number of heavy-duty components. Solve this problem with the Simplex algorithm.

5-9 The economic model for machining costs (see Sec. 9-13) shows that the total unit cost C_u is the sum of the machining cost C_{mc}, the tooling cost C_t, and a constant cost (which includes material cost) C_m.

$$C_u = C_{mc} + C_t + C_m$$

A basic parameter is the machining time for one piece t_0, where $t_0 = \pi DL/12vf$

for straight lathe turning. Tool life T, feed f, and velocity v, are related through the Taylor tool life equation $vT^n f^m = K$.

These equations can be combined to give

$$C'_u = K_{01} v^{-1} f^{-1} + K_{02} v^{(1/n)-1} f^{(m/n)-1}$$

$$C_u = C'_u + C_0 t_h \qquad \text{subject to } f < f_{max}$$

where

$$K_{01} = \frac{C_0 \pi D L}{12}$$

$$K_{02} = \frac{\pi D L (C_0 t_c + C_t)}{2K^{1/n}}$$

$C_0 = \text{cost of operating time, \$/min}$

$t_{change} = \text{tool changing time, min/edge}$

$C_{tool} = \text{tool cost, \$/edge}$

Use the technique of geometric programming to find the minimum cost and optimum values of v and f, given the following values. (From D. S. Ermer, *Trans. ASME, J. Eng. Ind.*, vol. 93, pp. 1067–1072, 1971).

$$C_0 = \$0.10 \text{ per min} \qquad n = 0.25$$
$$C_{tool} = \$0.50 \text{ per edge} \qquad m = 0.29$$
$$t_{change} = 0.5 \text{ min} \qquad K = 140$$
$$t_{hand} = 2.0 \text{ min} \qquad f_{max} = 0.005 \text{ in/rev}$$
$$D = 6 \text{ in} \qquad L = 8 \text{ in}$$

CHAPTER 6

6-1 Think about why books are printed on paper. Suggest a number of alternative materials that could be used. Under what conditions (costs, availability, etc.) would the alternative materials be most attractive?

6-2 Consider a soft drink can as a materials system. List all the components in the system and consider alternative materials for each component.

6-3 Select a tool material for thread-rolling mild-steel bolts. In your analysis of the problem you should consider the following points: 1) functional requirements of a good tool material, 2) critical properties of a good tool material, 3) screening process for candidate materials and 4) selection process.

6-4 Consider the redesign of a household self-cleaning oven. Suggest some radical material changes emphasizing nonmetallic materials. See *Metals Progress*, January 1968, for a discussion of design criteria and currently used materials.

6-5 Rank-order the following materials for use as an automobile radiator: copper, stainless steel, brass, aluminum, ABS, galvanized steel.

6-6 Discuss the use of aluminum vs. steel for electric power transmission line towers.

6-7 Classify the common stainless steel alloys into a "family of alloys" and develop a procedure for selecting among them.

6-8 Chromium is a highly strategic metal whose supply could be shut off completely by a war in southern Africa. Discuss the reasons for the strategic nature of chromium, and develop available corporate strategies to deal with the problem. What potential substitutes for chromium are available?

6-9 In an aerospace application, total cost of a component can be expressed by

$$C_t = C_f + C_m W + PW$$

where C_f = cost of fabrication
C_m = material cost, \$/lb
W = weight of the component, lb
P = penalty factor by which performance is jeopardized, \$/lb

(a) Discuss the various strategies available to minimize C_t.

(b) Determine whether material B can be economically substituted for material A if 1) there is no weight penalty factor and 2) the weight penalty is \$10 per lb.

Material	C_m, \$/lb	W, lb
A	1	90
B	100	2

6-10 Titanium alloys are being considered for automotive engine valves. What advantages would the titanium provide? Suggest alloys for both the inlet and outlet valves.

6-11 The wider use of many materials in automobiles, see Table 6-8, has led to problems in recycling because of the difficulty in separating such a wide range of materials. For example, a midsize car may contain up to 250 lbs of as many as 20 varieties of plastics. Propose a solution for this problem.

6-12 The weight of material required in a structural application can be expressed by

$$W = \frac{Q}{S^\alpha/\rho}$$

were Q is a performance factor dependent on the loading and the dimensions of the member, S is an appropriate material property, ρ is the density of the material, and α is a constant. Evaluate the expression for the case of a beam of width a, depth b, and length L that is loaded at the center in three-point bending. The material is a brittle cast iron for which the rupture strength is given by $\sigma = 3PL/2ab^2$.

6-13 Two materials are being considered for an application in which electrical conductivity is important

Material	Working strength MN/m²	Electrical conductance%
A	500	50
B	1000	40

The weighting factor on strength is 3 and 10 for conductance. Which material is preferred based on the weighted property index?

6-14 An aircraft windshield is rated according to the following material characteristics. The weighting factors are shown in parentheses.

Resistance to shattering (10)   The candidate materials are:
Fabricability (2)
Weight (8)   *A* plate glass
Scratch resistance (9)   *B* PMMA
Thermal expansion (5)   *C* tempered glass
                                  D a special polymer laminate

The properties are evaluated by a panel of technical experts, and they are expressed as percentages of maximum achievable values.

	Candidate material			
Property	*A*	*B*	*C*	*D*
Resistance to shattering	0	100	90	90
Fabricability	50	100	10	30
Weight	45	100	45	90
Scratch resistance	100	5	100	90
Thermal expansion	100	10	100	30

Use the weighted property index to select the best material.

6-15 A cantilever beam is loaded with a force P at its free end to produce a deflection $\delta = PL^3/3EI$. If the beam has a circular cross section, $I = \pi r^4/4$. Develop a figure of merit for selecting a material that minimizes the weight of a beam for a given stiffness (P/δ). By using the material properties given below, select the best material a) on the basis of performance and b) on the basis of cost and performance.

	E			
Material	GNm⁻²	ksi	ρ, Mgm⁻³	Approx. cost, $/ton (1980)
Steel	200	29×10^3	7.8	450
Wood	9–16	1.7×10^3	0.4–0.8	450
Concrete	50	7.3×10^3	2.4–2.8	300
Aluminum	69	10×10^3	2.7	2,000
Carbon-fiber-reinforced plastic (CFRP)	70–200	15×10^3	1.5–1.6	200,000

6-16 Select the most economical steel plate to construct a spherical pressure vessel in which to store gaseous nitrogen at a design pressure of 100 psi at ambient weather conditions down to a minimum of −20°F. The pressure vessel has a radius of 138 in. Your selection should be based on the steels listed below and expressed in terms of cost per square foot of material. Use a value of 489 lb/ft^3 for steel.

ASTM spec.	Grade	Allowable stress, psi	Pricing, ¢/lb (est. 1981 prices)						
			Base	Spec. grade	Qual. extra	Width thick	Test	Heat-teat.	Total
A-36		12,650	24.1	0.30	—	2.0	—	—	26.4
A-285	C	13,750	24.1	3.00	—	2.0	—	—	29.0
A-442	60	15,000	24.1	—	3.60	3.0	0.50	—	31.2
A-533	B	20,000	34.4	12.60	2.10	6.2	0.50	15.2	71.0
A-157	B	28,750	34.4	10.70	2.10	8.2	2.00	15.2	72.6

6-17 Using the example of the hydraulic valve housing given in Sec. 6-11, put it into the framework of the solution steps for value analysis shown in Fig. 6-10.

CHAPTER 7

7-1 One evolution in the design of automobile engines has been the change from in-line long-stroke engines to compact four-and six-cylinder engines. As a result, the crankshaft material has been changed from quenched and tempered steel forgings to cast crankshafts made from pearlitic malleable cast iron or nodular iron. Discuss this change in materials and processing in terms of the service performance and the properties of the materials.

7-2 A small hardware fitting is made from free-machining brass. For simplicity consider that the production cost is the sum of three terms: 1) material cost, 2) labor costs and 3) overhead costs. Assume that the fitting is made in production lots of 500, 50,000, and 5×10^6 pieces by using, respectively, an engine lathe, a tracer lathe, and an automatic screw machine. By using the cost per part as an indicator, schematically plot the relative distribution of the cost due to materials, labor, and overhead for each of the production quantities.

7-3 Difficult-to-work materials mean that complex shapes often can be produced only by machining operations that convert a large part of the workpiece into chips. Let α be the fraction of the workpiece that is converted into chips to make the part. These chips will be cleaned, reprocessed, and melted and rolled to produce useful material and sold back to the manufacturer. If C is the cost of the material, this will go through the same cycle over and over to produce a real cost of material $C(1 + \alpha + \alpha^2 + \alpha^3 + \cdots) = C/(1 - \alpha)$. If the chips are sold at a ratio of β to the price of the material, determine a relation for the true cost of the material under a condition in which scrap (chips) is involved. If titanium alloys cost $10,000 per ton and $\alpha = 0.90$ and $\beta = 0.1$, what is the real cost of the workpiece material!

7-4 Explain why alloys designed for casting generally are not used for forgings, and vice versa.

7-5 You are concerned with the cast nodular-iron crankshafts. What design factors determine the manufacturing cost? Which of the costs are determined by the foundry and which by the purchaser?

7-6 Describe the manufacturing steps to produce an automobile rear-axle housing from a tubular blank.

7-7 Titanium alloys are more difficult to form into sheet-metal structural members than aluminum alloys in aircraft construction. Discuss the reasons for this difference and also discuss manufacturing methods that have been developed to overcome the difficulties.

7-8 A machine shaft has a diameter of 1.75 in and is 3.5 ft long. It must withstand a maximum bending stress of 80,000 psi. Since other parts of the same machine are made from 4140H steel, we would like to use that material for the shaft if it has sufficient hardenability. Use data available in the ASM Metals Handbook to determine whether 4140H steel is acceptable for the application.

7-9 A wheel spindle must be capable of developing RC 50 minimum when quenched to 90 percent martensite at the critical section. The critical section is 2.0 in in the final machined condition, but 0.20 in must be allowed for removal by machining from the forged and heat-treated surface. A steel with a 0.40 percent carbon content has been specified. The Jominy equivalent cooling rate at 0.20 in below the surface is found to be J9.5.

(*a*) What is the ideal critical diameter needed for this application?

(*b*) Suggest and prove out a more economical approach than using a medium-carbon alloy steel and an oil quench.

7-10 Compare steels A and B for the construction of a pressure vessel 30 in inside diameter and 12 ft long. The pressure vessel must withstand an internal pressure of 5000 psi. Design against a radial internal flaw of length $2a$. Use a factor of safety of 2. For each steel, determine a) critical flaw size and b) flaw size for a leak-before-break condition.

Steel	Yield strength, psi	K_{Ic}, ksi $\sqrt{in}$
A	260	80
B	110	170

7-11 A high-strength steel has a yield strength of 100 ksi and a fracture toughness $K_{Ic} = 150$ ksi $\sqrt{in}$. By use of a certain nondestructive evaluation technique, the smallest size flaw that can be detected routinely is 0.3 in. Assume that the most dangerous crack geometry in the structure is a single-edge notch so that $K_{Ic} = 1.12\sigma \sqrt{\pi a}$. The structure is subjected to cyclic fatigue loading in which $\sigma_{max} = 45$ ksi and $\sigma_{min} = 25$ ksi. The crack growth rate for the steel is given by $da/dN = 0.66 \times 10^{-8}(\Delta K)^{2.25}$. Estimate the number of cycles of fatigue stress that the structure can withstand.

7-12 Stress-corrosion failures occur in the 304 and 316 stainless-steel recirculation piping of boiling-water nuclear reactors (BWR). As of April 1981, a total of 254 incidents of cracking had been reported. What are the three conditions necessary for stress-corrosion cracking? Suggest remedies for the cracking.

7-13 What are the chief advantages and disadvantages of plastic gears? Discuss how material structure and processing are utilized to improve performance.

CHAPTER 8

8-1 (a) Calculate the amount realized at the end of 7 years through annual deposits of $1000 at 10 percent compound interest.

(b) What would the amount be if interest were compounded semiannually?

8-2 A young woman purchases a new car. After down payment and allowances, the amount to be paid is $4800. If money is available at 10 percent, what is the monthly payment to pay off the loan in 4 years?

8-3 A new machine tool costs $15,000 and has a $5000 salvage value at the end of 5 years. The interest rate is 10 percent. The annual cost of capital recovery is the annual depreciation charge (use straight-line depreciation) plus the equivalent annual interest charge. Work this out on a year-by-year basis and show that it equals the number obtained quickly by using the capital recovery factor.

8-4 A father desires to establish a fund for his new child's college education. He estimates that the current cost of a year of college education is $8000 and that the cost will escalate at an annual rate of 8 percent.

(a) What amount is needed on the child's eighteenth, nineteenth, twentieth, and twenty-first birthdays to provide for a four-year college education?

(b) If a rich aunt gives $5000 on the day the child is born, how much must be set aside at 10 percent on each of the first through seventeenth birthdays to build up the college fund?

8-5 Machine A costs $8500 and has annual operating costs of $4500. Machine B costs $7000 and has an annual operating cost of $4800. Each machine has an economic life of 10 years. If the minimum required rate of return is 10 percent, compare the advantages of machine A by a) present worth method, b) annual cost method, and c) rate of return on investment.

8-6 Make a cost comparison between two conveyor systems for transporting raw materials.

	System A	System B
Installed cost	$25,000	$15,000
Annual operating cost	6,000	11,000

The service life of each system is 5 years and the write-off period is 5 years. Use straight-line depreciation and assume no salvage value for either system. At what rate of return *after taxes* would B be more attractive than A?

8-7 A resurfaced floor costs $5000 and will last 2 years. If money is worth 10 percent after taxes, how long must a new floor costing $19,000 last to be economically justified? The tax rate is 52 percent. For tax purposes a new floor can be written off in 1 year. Use sum-of-the-year-digits depreciation. Use the capitalized cost method for your analysis.

8-8 You are concerned with the purchase of a heat-teating furnace for the gas carburizing of steel parts. Furnace A will cost $325,000 and will last 10 years; furnace B will cost $400,000 and will also last 10 years. However, furnace B will provide closer control on case depth, which means that the heat treater can shoot for the low side of the specification range on case depth. That will mean that the production rate for furnace B will be 2740 lb/h compared with 2300 lb/h for furnace A. Total yearly production is required to be 15,400,000 lb. The cycle time for furnace A is 16.5 h, and that for furnace B is 13.8 h. The hourly operating cost is $64.50 per h.

Justify the purchase of furnace B on the basis of *a*) payout time and *b*) discounted cash flow rate of return after taxes. Assume money is worth 10 percent and the tax rate is 50 percent.

8-9 The cost of capital has a strong influence on the willingness of management to invest in long term projects. If the cost of capital in America is 10% and in Japan 4%, what must the return be after 2 years on a two year investment of $1M for each of the situations to provide an acceptable return on the investment? Repeat the analysis for a 20 year period.

8-10 If the tax credit for capital investments is t_c, where t_c is a fraction from 0 to 1.0, develop an equation for the net effective capital investment where a tax credit applies.

8-11 In order to justify investment in a new plating facility, it is necessary to determine the present worth of the costs.
   Calculate the present worth given the following information:

Cost of equipment	$350,000
Planning period	5years
Fixed charges	20 percent of investment each year
Variable charges	40,000 first year, escalating at 6 percent each year with inflation starting at $t = 0$
Rate of return	$i = 10\%$

8-12 Determine the net present worth of the costs for a major construction project under the following set of conditions:
   (*a*) Estimated cost $300 million over 3 years (baseline case).
   (*b*) Project is delayed by 3 years with rate of inflation 10 percent and interest cost 16 percent.
   (*c*) Project is delayed 6 years with rate of inflation 10 percent and interest costs 16 percent.

8-13 Whether a maintenance operation is classified as a repair (expense charged against revenues in current year) or improvement (capitalized expense) can have a big influence on taxes. Determine the net savings for a $10,000 operation using the two different approaches if *a*) you are in a business that is in the 50 percent tax bracket and *b*) you are in a small business in the 20 percent tax bracket. Use a 10 percent interest rate and a 10 percent investment tax credit.

8-14 At what annual mileage is it cheaper to provide your field representatives with cars than to pay them $0.22 per mile for the use of their own cars? The costs of furnishing a car are as follows:

Purchase price	$9000
Life	4 years
Salvage	$1500
Storage	$150 per year
Maintenance	$0.08 per mile

(*a*) Assume $i = 10$ percent.
(*b*) Assume $i = 16$ percent.

8-15 To *levelize expenditures* means to create a uniform end-of-year payment that will have the same present worth as a series of irregular end-of-year payments. To illustrate, consider the estimated 5-year maintenance budget for a pilot plant. Develop a levelized cost assuming that $i = 0.10$ and the annual inflation escalation will be 8 percent.

Year	Maintenance budget estimate
1	25,000
2	150,000
3	60,000
4	70,000
5	300,000

CHAPTER 9

9-1 A company has received an order for four sophisticated space widgets. The buyer will take delivery of one unit at the end of the first year and one unit at the end of each of the succeeding 3 years. He will pay for a unit immediately upon receipt and not before. However, the manufacturer can make the units ahead of time and store them at no cost for further delivery.

The chief component of cost of the space widget is labor at $15 per h. All units made in the same year can take advantage of an 80 percent learning curve. The first unit requires 100,000 h of labor. Learning occurs only in one year and is not carried over from year to year. If money is worth 16 percent after a 52 percent tax rate, decide whether it would be more economical to build four units the first year and store them or build one unit in each of the four years.

9-2 The following data are presented for the cost of vacuum-melting specialty steel by the vacuum arc refining process (VAR) and the electroslag remelting process (ESR). Note that all costs are in 1968 dollars. Determine the cost of melting a pound of steel (in 1968 dollars) for the two processes.

Cost component	VAR	ESR
Direct labor, one operator and helper	$48,000	$48,000
Manufacturing overhead, 140% of direct labor	67,000	67,000
Melting power	0.3 kwh/lb	0.5 kwh/lb
	1000 lb/h	1250 lb/h
	1.2¢/kwh	1.2¢/kwh
Cooling water	3,500	4,100
Slag		28,000

The capital cost of a VAR system is $480,000, and that of an ESR system is $380,000. Each system has a 10-year useful life. Each uses 1000 ft^2 of factory space, which costs $25 per ft^2. Assume both furnaces operate for 15 eight-hour shifts per week for 50 weeks in the year.

9-3 The engineering department has developed a new design for a machine component in which plastic is substituted for metal. The part will be made from polycarbonate ($1.40 per lb) by a thermoplastic molding process. Each part uses $\frac{1}{8}$ lb of plastic resin. The 10-cavity mold that is used makes 1400 parts per hour at an operating cost of $20 per h. For an order of 10,000 parts, determine the selling price based on the example in Sec. 9-11.

9-4 Petroleum coke is used for anode electrodes in the production of aluminum. The coke must contain less than 1 percent sulfur, which places restrictions and costs on the process design. Determine the return on investment (ROI) for the following conditions:
Sales of 200,000 tons/years at a price of $27.50 per ton.

Operating costs	$18 per ton
Selling expense	$0.5 per ton
G&A	$0.5 per ton
Taxes and insurance	$0.5 per ton
Working capital	$300,00
Plant investment	$3,700,000

9-5 Determine which system is most economical on a lifecycle costing basis

	System A	System B
Initial cost	$300,000	$240,000
Installation	23,000	20,000
Useful life	12 years	12 years
Operators needed	1	2
Operating hours	2100	2100
Operating wage rate	$8 per h	$8 per h
Parts & supplies cost (% of initial cost)	1%	2%
Power	8 kw at 4 ¢/kWh	9 kw at 4 ¢/kWh
Escalation of operating costs	6%	6%
Mean time between failures	600 h	450 h
Mean time to repair	35 h	45 h
Maintenance wage rate	$10 per h	$10 per h
Maintenance escalation rate	6%	6%
Desired rate of return	10%	10%
Tax rate	45%	45%

9-6 Discuss the automobile safety standards and air pollution standards in terms of the concept of life cycle costs.

CHAPTER 10

10-1 The following restrictions exist in a scheduling network. Determine whether the network is correct; and if it is not, draw the correct network.

(a) A precedes C
   B precedes E
   C precedes D and E

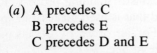

(b) A precedes d and E
   B precdes E and F
   C precedes F

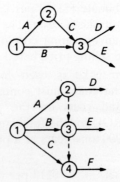

10-2 The development of an electronic widget is expected to follow the following steps.

Activity	Description	Time est., weeks	Preceded by
A	Define customer needs	4	
B	Evaluate competitor's product	3	
C	Define the market	3	
D	Prepare product specs	2	B
E	Produce sales forecast	2	B
F	Survey competitor's marketing methods	1	B
G	Evaluate product vs. customer needs	3	A, D
H	Design and test the product	5	A, B, D
I	Plan marketing activity	4	C, F
J	Gather information on competitor's pricing	2	B, E, G
K	Conduct advertising campaign	2	I
L	Send sales literature to distributors	4	E, G
M	Establish product pricing	3	H, J

Establish the arrow network diagram for this project and determine the critical path by using the CPM technique.

CHAPTER 11

11-1 The characteristics of a population may be described by either a *discrete variable,* which can assume only certain isolated values, or a *continuous variable,* which can assume any value between two limits. Identify each of the following variables as discrete or continuous, and write down the values that each variable can assume.

(*a*) Time required to repair a broken machine.
(*b*) Number of defective parts in a production lot of 500.
(*c*) Weight of a broken connecting rod.
(*d*) Diameter of a shaft finished by cylindrical grinding to $2.500^{+0.002}_{-0.010}$
(*e*) Number of ball bearings in a quality-control sample.

11-2 Extensive analysis of failure reports on an automotive component show that there is a 7 in 10 chance that the failure is caused by fatigue and a 4 in 10 chance that is due to corrosion. The probability that the cause of failure is both fatigue and corrosion is 0.2. What is the probability of failure being due to fatigue or corrosion?

11-3 Twelve experimental alloys are checked for the contaminants Fe and Bi. Eight had Fe and five had Bi, but the heats got mixed up and must be rechecked.

(*a*) Determine the probability of a heat having been thought to have Fe, given it has Bi on recheck?

(*b*) What is the probability of a heat having been thought to have Bi, given it has Fe on recheck?

11-4 Precision bearings are supplied from three sources. The average acceptability of the bearings from each source are given below.

Source	Relative probability of supply	Average acceptability
1	0.60	0.98
2	0.30	0.90
3	0.10	0.75

(a) What is the probability that any precision bearing received by the assembly plant will perform acceptably?

(b) What is the probability that a particular bearing came from supplier 2 when it is known to have performed acceptably?

11-5 The following data* were obtained for the transverse reduction of area (RAT) of a steel gun tube:

RAT, percent class interval	Midpoint of class interval, x_i	Frequency f_i	$x_i f_i$
29–31	30	2	60
31–33	32	3	96
33–35	34	6	204
35–37	36	4	144
37–39	38	10	380
39–41	40	21	840
41–43	42	27	1,134
43–45	44	46	2,024
45–47	46	61	2,806
47–49	48	62	2,976
49–51	50	32	1,600
51–53	52	2	104
53–55	54	1	54
		$277 = \Sigma f_i$	$12{,}422 = \Sigma f_i x_i$

(a) Plot a frequency histogram from these data.

(b) Plot the cumulative frequency distribution on regular coordinate paper.

(c) Determine the mean, median, and mode.

(d) Determine the range and standard deviation.

(e) Plot the cumulative frequency distribution on normal-probability paper, and determine the mean and standard deviation.

11-6 (a) For the data given in Prob. 11-5, what are the 95 percent confidence limits on the mean of the population?

(b) What are the odds that the mean of the population will be outside the region 44.84 ± 0.76?

11-7 Eight fatigue specimens were tested at the same stress. The number of

---

* E. G. Olds and Cyril Wells, *Trans. ASM,* vol. 42, p. 851, 1950.

cycles to failure, expressed as log N, were as follows:

i $\log N_i$
1   4.8388
2   4.9243
3   4.9445
4   4.9542
5   4.9731
6   4.9777
7   5.0334
8   5.0828

What are the mean fatigue life and its standard deviation?

11-8 What is the probability of getting exactly three heads in nine tosses of an unbiased coin?

11-9 If 10 percent of the parts produced by a screw machine are defective, what is the probability that, out of five parts selected at random, none of the parts will be defective?

11-10 Thirty parts are produced by a screw machine that has an historical defective rate of 10 percent. What is the chance of observing zero defects in a sample of five parts?

11-11 The surface defects in each 10-ft length of cold-drawn bar stock were found to follow in number the data given below. What is the probability of being at or beyond the specification limit of three defects per 10-ft length?

Defects per 10 ft	No. of 10-ft lengths
0	40
1	9
2	4
3	2
4	1
5	1
6	1
	—
	58

11-12 A new grinding machine for converting bar into precision shafting produces diameters that are normally distributed around the mean with $\sigma = 0.005$ in.

    (a) What symmetrical tolerance limits are necessary to have no more than 2 percent of the bars outside tolerance?

    (b) If the tolerances on a ground bars are $3.000^{+0.005}_{-0.010}$ in, what percentage of the bars will be acceptable?

11-13 In an experiment to determine whether there is an effect of specimen diameter on RAT the following data were obtained:

0.505-in diam	0.252-in diam
$\bar{x} = 40.156$	$\bar{x} = 40.544$
$s^2 = 5.7850$	$s^2 = 3.2765$
$n = 16$	$n = 16$

Is the difference in the means for the two tests statistically significant? (Note that the variances are not significantly different.)

11-14 Fatigue tests from two presumably identical bars of steel gave the following results for tests at 60,000 psi:

Bar 1	Bar 2
$\overline{\log N} = 5.0631$	$\overline{\log N} = 4.8306$
$s = 0.00615$	$s = 0.00315$
$n = 10$	$n = 12$

The F test shows that the variances are significantly different. Establish whether or not the two samples are from the same population.

11-15 A batch of H11 tool steel bearings from the same heat was divided into two equal groups and given two different austenitizing treatments before being quenched and tempered in an attempt to minimize retained austenite. Did the difference in "hardening temperature" make a significant difference as measured by the following Rockwell hardness readings?

Rockwell C Hardness

Treatment 1	Treatment 2
50	48
51	50
51	52
48	49
50	49
47	47
52	50
49	48
47	51
50	48

11-16 Using the hypothesis $H_0: \mu = 10$ and $H_1: \mu = 12$ with $\sigma = 2$, determine a) β if $\alpha = 0.05$ and $n = 20$ and b) α if $\beta = 0.05$ and $n = 20$.

11-17 If $\alpha = 0.05$ in Prob. 11-13, what is the value of β? To what level would the variability have to be reduced for β to equal 0.05?

11-18 The yield of a new pilot plant process has a mean of 92.3 percent and a standard deviation of 1.2 percent based on 10 trial runs.

    (a) What are the 95 percent confidence limits on this mean?

(*b*) If the mean and standard deviation are the result of many trials ($n > 100$), what are the 95 percent confidence limits?

(*c*) For the original 10 trial runs, what is the tolerance interval that contains at least 95 percent of the population with a 95 percent confidence?

(*d*) What is the prediction interval to contain all 10 future observations with 95 percent probability?

11-19 A statistically designed experiment was conducted to determine the influence of laboratory environment (vacuum or air) and stress level (two levels) on the fretting fatigue life of 7075 T6 aluminum alloy. (Poon and Hoeppner, *Trans. ASME,* vol. 103, *J. Eng. Materials Tech.,* pp. 218–222, July 1981). Ten specimens were tested at each of the two treatment levels for the factors. The ANOVA table is as follows:

Source of variation	Sum of squares
Environment (E)	7.11
Stress level (S)	11.89
E&S interaction	1.77
Error	0.27
Total	21.59

Do these data show a significant effect of environment, stress level, or their interaction on the fretting fatigue life at the 5 percent level of significance?

11-20 A statistical study was made on the influence of weather conditions (temperature *T*, wind velocity *V*) and rail size *S* on the strength of welded steel rail joints (S. M. Wu, *Welding J. Res. Suppl.,* April 1964). A 2^3 factorial design was conducted at the test conditions given below. Duplicate specimens were tested for weld strength at each treatment condition.

Std order	T	V	S	1	2	Avg.
1	0	0	0.364	84.0	91.0	87.5
a	70	0	0.364	95.5	84.0	87.3
b	0	20	0.364	69.6	86.0	77.8
ab	70	20	0.364	76.0	98.0	87.0
c	0	0	1.000	77.7	80.5	79.1
ac	70	0	1.000	99.7	95.4	97.6
bc	0	20	1.000	82.7	74.5	78.6
abc	70	20	1.000	93.7	81.7	87.7

Columns under "Treatment conditions": T, V, S. Columns under "Weld strength, ksi": 1, 2, Avg.

(*a*) Calculate the main effects and the two factor interaction effects.

(*b*) Determine the standard error.

(*c*) Construct an analysis of variance (ANOVA) table and determine which factors are significant.

11-21 The data given below show the thickness of an oxide film, in angstroms, as a function of the time, in minutes, that the metal specimen was at temperature

Time x	20	30	40	60	70	90	100	120	150	180
Film thickness y	3.5	7.4	7.1	15.6	11.1	14.9	23.9	27.1	22.1	32.9

(a) Establish the regression equation.

(b) Is this a good fit to the data?

11-22 Inertia welding is a special welding process that converts inertial energy into frictional heat that is used to create a solid-state bond at the weld interface. Basically, the process works as follows: A rotating part consisting of a flywheel and the workpiece is brought up to an initial speed of N rpm. During welding, the nonrotating part of the workpiece is pushed axially against the rotating part under a constant pressure P at the same time that the rotating part is disengaged and is quickly decelerated to a standstill by frictional forces at the interface between the two halves of the workpiece. The frictional heat and pressure combine to form a hot pressure weld.

The chief parameters in the process that control the quality of the weld are unit axial pressure A, initial linear rubbing velocity at the interface B, and the total moment of inertia C. It is decided to use a 2^3 factorial design and the method of steepest ascent to optimize these process parameters. The measured response is the fracture strength of 1020 steel cylinder cylinders produced by the inertia welding process. The average fracture strength of the 1020 steel tested without a weld is $\bar{x} = 85.2$ ksi and $s = 0.6$ ksi.

For the first set of experiments, a full factorial was used based on the following:

Factor	Code	Design center	Unit of variation
Pressure, psi	A	9.7	1.1
Velocity, fpm	B	283	25
Inertia, lb-ft^2	C	0.80	0.18

The values of the response, in standard order, for the factorial design conducted around the design center were 47.2, 96.4, 92.5, 90.6, 98.0, 104.6, 102.5, 102.1 ksi. These are the average of two replicates for each treatment. The first-order equation describing the response surface was:

$$y = 91.7 + 6.7x_1 + 5.2x_2 + 10.1x_3 \tag{I}$$

(a) Carry out an analysis to determine the coefficients in Eq. (I) and discuss what the various coefficients mean.

(b) Calculate two new treatments (trials 9 and 10) along the steepest ascent vector if we decide to set pressure at 10.4 and 11.0, respectively, for these new trials.

11-23 A manufacturer produces a special alloy steel with an average yield strength of 85,800 psi. A modification in composition is expected to increase the strength. The standard deviation of the yield strength is known to be 2000 psi, and the change in composition is not expected to change that value. If the change in composition does not increase the average yield strength, the manufacturer wants to know that with 95 percent confidence. If the average yield strength is increased by as much as 2000 psi, the manufacturer is willing to take only a 10 percent risk of not detecting it. How many specimens should be tested to meet all of the above requirements?

CHAPTER 12

12-1 Assume you are part of a federal commission established in 1910 to consider the risk to society of the expected widespread use of the motor car powered with highly flammable gasoline. Without the benefit of hindsight, what potential dangers can you contemplate? Use a worst-case scenario. Now, taking advantage of hindsight, what lesson can you draw about evaluating the hazards of future technologies?

12-2 Comment on the following statistics in terms of the societal perception of risk.

According to the Consumer Product Safety Commission, in 1974 there were about 40 serious injuries from fires started in wood stoves or fireplaces. By 1979 the number had jumped to 400.

In 1980 the National Fire Protection Association counted 26 fires started by wood stoves or fireplaces in which 3 or more people died.

12-3 A steel tensile link has a mean yield strength of $\bar{S}_y = 27,000$ psi and a standard deviation on strength of $S_y = 4000$ psi. The variable applied stress has a mean value of $\bar{\sigma} = 13,000$ psi and a standard deviation $s = 300$ psi.

(a) What is the probability of failure taking place? Show the situation with carefully drawn frequency distributions.

(b) The factor of safety is the ratio of the mean material strength divided by the mean applied stress. What factor of safety is required if the allowable failure rate is 5 percent?

(c) If absolutely no failures can be tolerated, what is the lowest value of the factor of safety?

12-4 A machine component has average life of 120 h. Assuming an exponential failure distribution, what is the probability of the component operating for at least 200 h before failing?

12-5 A nonreplacement test was carried out on 100 electronic components with a known constant failure rate. The history of failures was as follows:

1st failure after   93 h
2nd failure after   1,010 h
3rd failure after   5,000 h

4th failure after   28,000 h
5th failure after   63,000 h

The testing was discontinued after the fifth failure. If we can assume that the test gives an accurate estimate of the failure rate, determine the probability that one of the components would last for a) 10^5 h and b) 10^6 h.

12-6 The failure of a group of mechanical components follows a Weibull distribution, where $\theta = 10^5$ h, $m = 4$, and $t_0 = 0$. What is the probability that one of these components will have a life of 2×10^4 h?

12-7 A complex system consists of 550 components in a series configuration. Tests on a sample of 100 components showed that 2 failures occurred after 1000 h. If the failure rate can be assumed to be constant, what is the reliability of the system to operate for 1000 h? If an overall system reliability of 0.98 in 1000 h is required, what would the failure rate of each component have to be?

12-8 A system has a unit with MTBF = 30,000 h and a standby unit (MTBF = 20,000 h). If the system must operate for 10,000 h, what would be the MTBF of a single unit (constant failure rate) that, without standby, would have the same reliability as the standby system?

12-9 Make a preliminary hazard analysis for an oil-fired process heater in which the flame is accidentally extinguished.

12-10 Construct a qualitative fault tree diagram for a coal miner who is injured by a falling mine roof.

12-11 List a number of reasons why the determination of product life is important in engineering design.

12-12 There has been growing societal concern about oil spills from tankers. What major changes in design would you propose to significantly change the risk of a major environmental accident?

CHAPTER 13

13-1 By reading the literature, classify the macroappearance of fatigue failures with respect to type of loading (bending, torsion, axial) and presence of stress concentrations.

13-2 Using the principles of mechanical metallurgy, what would a torsion failure look like in a ductile material and a brittle material?

13-3 What steps of failure analysis would you undertake to determine what caused the crack in the Liberty Bell?

13-4 Read one of the following detailed accounts of a failure analysis.

(a) C. O. Smith, "Failure of a Twistdrill," *Trans. ASME, J. Eng. Materials. Tech.,* vol. 96, pp. 88–90, April 1974.

(b) C. O. Smith, "Failure of a Welded Blower Fan Assembly," ibid., vol. 99, pp. 83–85, January 1977.

(c) R. F. Wagner and D. R. McIntyre, "Brittle Fracture of a Steel Heat Exchanger Shell," ibid., vol. 102, pp. 384–387, October 1980.

13-5 The following case study deals with forensic engineering, the combination of technical analysis and legal discovery.

Background

A manufacturer of modular housing units provided units to be used by the U.S. Navy as modular air traffic control towers. The original Navy specifications called for painted steel siding, but the supplier induced the Navy to substitute panels made from aluminum bonded on one side to plywood. These panels were used for both interior and exterior sheathing.

The supplier began to manufacture panels in February 1976, and delivery to the Navy was completed in November 1976. However, in October 1976 the Navy notified the supplier that the panels were deteriorating. Deterioration was in the form of blistering of the surface paint, delamination of the aluminium from the plywood, and bubbling and cracks in the aluminum skin. These phenomena occurred in all towers constructed, in both interior and exterior panels.

The Dialog

The Navy demanded the panels be repaired, charging they were defective in manufacture. The supplier maintained that, because the panels had been stored in an open field and then improperly assembled by the Navy, moisture in the form of rainwater had been allowed to enter and cause the observed damage.

The Failure Analysis

Macro- and microexamination of the delaminated and cracked panels showed that the aluminum was failing because of corrosion at the plywood-aluminum interface.

Problem Analysis

(*a*) Give three possible sources of the corrosion.

(*b*) List six detailed questions about plywood manufacture and bonding to plywood that must be answered. The answers to the questions point the finger toward a fire retardant in the plywood as the only possible source of corrosion.

(*c*) What simple tests would you perform to check out the retardant hypothesis? The tests showed an unusually high concentration of phosphorus and strongly suggested that a fire retardant was involved. Checking the records of the supplier revealed that the panels had been painted on the plywood face and edges with a fire retardant paint containing phosphorus. In addition, x-ray fluorescence of the powdery white corrosion product with an electron beam microprobe showed the same elements as in the 3006 aluminum alloy, with the one exception that chlorine was present. Moreover, it was established from the manufacturer that the fire retardant paint had no history of causing corrosion when applied to bare aluminum.

(*d*) What are your tentative conclusions?

A wood technologist expert in adhesives and plywood manufacture was employed to assist the metallurgist in this investigation. He quickly noticed the previously overlooked fact that the corrosion was worse in the center of the panels than at the edges. He also focused on the source of water to act as a leachant for chloride. He pointed out that plywood has a high initial water content (8 to 12 percent), which can be increased by the water-base glue used in its manufacture. His expertise caused the emphasis to be focused on the manufacture of the panels.

(*e*) Based on this new technical input, what is your analysis of the cause of failure? How could this be backed up by tests? For the trial litigation and the final outcome of this case, read the discussion of this problem by W. G. Dobson, N. J. Dilloff, and H. B. Gatslick, *Metal Progress,* pp. 61–68, August 1980. Do not read the article until you have tried to answer the preceding questions.

13-6 Use the concept of statistical hypothesis testing to identify and classify the errors that can occur in quality-control inspection.

13-7 Contrast the NDE and quality assurance requirements for a structural component made from a reinforced composite material and one made from an aluminum alloy forging.

13-8 Discuss the concept of quality-control circles. What is involved in implementing a quality-circle program?

13-9 Discuss the process of using consumer complaints to establish that a product is hazardous and should be recalled.

13-10 Construct a model for the cost of poor quality production. Consider a single level process in which q sets of materials each costing M are introduced. They all get processed and tested at a cost of P each. A certain number w fail the quality test and must be reworked, at a cost R each. The probability of success in the process is y, the fractional yield of the process. Assume that binomial statistics applies.

CHAPTER 14

14-1 Prepare in writing a personal plan for combating technological obsolescence. Be specific about the things you intend to do and read.

14-2 Select a technical topic of interest to you.

(*a*) Compare the information that is available on this subject in a general encyclopedia and a technical encyclopedia.

(*b* Look for more specific information on the topic in a handbook.

(*c*) Find five current texts or monographs on the subject.

14-3 Use the indexing and abstracting services to obtain at least 20 current references on a technical topic of interest to you. If possible, repeat the search with a computerized search. Use appropriate indexes to find 10 government reports related to your topic.

14-4 Where would you find the following information?
   (*a*) The services of a taxidermist.
   (*b*) A consultant on carbon-fiber-reinforced composite materials.
   (*c*) The price of an X3427 semiconductor chip.
   (*d*) The melting point of osmium.
   (*e*) The proper hardening treatment for A1S1 4320 steel.

14-5 Discuss how priority is established in a patent litigation.

14-6 How would you obtain information about a U.S. patent given the following conditions:
   (*a*) you know the patent number?
   (*b*) you know only the patentee's name?
   (*c*) you know the patent number and want to know prior development in the field?
   (*d*) no patent numbers or names are given?

CHAPTER 15

15-1 See whether your library has copies of the notebooks of Leonardo da Vinci and Michael Faraday. Use the notebooks to discuss the proper recording of experimental ideas.

15-2 Carefully read a technical paper from a journal in your field and comment on how it conforms with the outline discussed in Sec. 15-3.

15-3 Write a memorandum to your supervisor justifying your project being 3 weeks late and asking for an extension.

15-4 Prepare a technical talk based on the paper you studied in Prob. 15-2.

APPENDIX

B

INTEREST
TABLES

TABLE B-1
4% compound interest rate

Discrete compound interest = 4%

	Single payment		Uniform annual series			Uniform gradient series	Depreciation series	
	Compound-interest factor $(1+i)^n$	Present-worth factor $\dfrac{1}{(1+i)^n}$	Unacost present-worth factor $\dfrac{(1+i)^n-1}{i(1+i)^n}$	Capital-recovery factor $\dfrac{i(1+i)^n}{(1+i)^n-1}$	Capitalized-cost factor $\dfrac{(1+i)^n}{(1+i)^n-1}$	Present-worth factor $\dfrac{F_{RP}-nF_{SP}}{i}$	Sum-of-digits present-worth $\dfrac{n-F_{RP}}{0.5n(n+1)i}$	Straight-line present-worth factor $\dfrac{1}{niF_{PR}}$
	P to S	S to P	R to P	P to R	P to K	G to P	SD to P	SL to P
n	F_{PS}	F_{SP}	F_{RP}	F_{PR}	F_{PK}	F_{GP}	F_{SDP}	F_{SLP}
1	1.0400E 00	9.6154E-01	9.6154E-01	1.0400E 00	2.6000E 01	1.0000E 00	9.6154E-01	9.6154E-01
2	1.0816E 00	9.2456E-01	1.8861E 00	5.3020E-01	1.3255E 01	9.2455E-01	9.4921E-01	9.4305E-01
3	1.1249E 00	8.8900E-01	2.7751E 00	3.6035E-01	9.0087E 00	2.7025E 00	9.3712E-01	9.2503E-01
4	1.1699E 00	8.5480E-01	3.6299E 00	2.7549E-01	6.8873E 00	5.2670E 00	9.2526E-01	9.0747E-01
5	1.2167E 00	8.2193E-01	4.4518E 00	2.2463E-01	5.6157E 00	8.5547E 00	9.1363E-01	8.9036E-01
6	1.2653E 00	7.9031E-01	5.2421E 00	1.9076E-01	4.7690E 00	1.2506E 01	9.0222E-01	8.7369E-01
7	1.3159E 00	7.5992E-01	6.0021E 00	1.6661E-01	4.1652E 00	1.7066E 01	8.9102E-01	8.5744E-01
8	1.3686E 00	7.3069E-01	6.7327E 00	1.4853E-01	3.7132E 00	2.2181E 01	8.8004E-01	8.4159E-01
9	1.4233E 00	7.0259E-01	7.4353E 00	1.3449E-01	3.3623E 00	2.7801E 01	8.6926E-01	8.2615E-01
10	1.4802E 00	6.7556E-01	8.1109E 00	1.2329E-01	3.0823E 00	3.3881E 01	8.5868E-01	8.1109E-01
11	1.5395E 00	6.4958E-01	8.7605E 00	1.1415E-01	2.8537E 00	4.0377E 01	8.4830E-01	7.9641E-01
12	1.6010E 00	6.2460E-01	9.3851E 00	1.0655E-01	2.6638E 00	4.7248E 01	8.3812E-01	7.8209E-01
13	1.6651E 00	6.0057E-01	9.9856E 00	1.0014E-01	2.5036E 00	5.4455E 01	8.2812E-01	7.6813E-01
14	1.7317E 00	5.7748E-01	1.0563E 01	9.4669E-02	2.3667E 00	6.1962E 01	8.1830E-01	7.5451E-01
15	1.8009E 00	5.5526E-01	1.1118E 01	8.9941E-02	2.2485E 00	6.9735E 01	8.0867E-01	7.4123E-01
16	1.8730E 00	5.3391E-01	1.1652E 01	8.5820E-02	2.1455E 00	7.7744E 01	7.9921E-01	7.2827E-01
18	2.0258E 00	4.9363E-01	1.2659E 01	7.8993E-02	1.9748E 00	9.4350E 01	7.8080E-01	7.0329E-01
20	2.1911E 00	4.5639E-01	1.3590E 01	7.3582E-02	1.8395E 00	1.1156E 02	7.6306E-01	6.7952E-01
25	2.6658E 00	3.7512E-01	1.5622E 01	6.4012E-02	1.6003E 00	1.5610E 02	7.2138E-01	6.2488E-01
30	3.2434E 00	3.0832E-01	1.7292E 01	5.7830E-02	1.4458E 00	2.0106E 02	6.8322E-01	5.7640E-01
35	3.9461E 00	2.5342E-01	1.8665E 01	5.3577E-02	1.3394E 00	2.4488E 02	6.4823E-01	5.3327E-01
40	4.8010E 00	2.0829E-01	1.9793E 01	5.0523E-02	1.2631E 00	2.8653E 02	6.1607E-01	4.9482E-01
45	5.8412E 00	1.7120E-01	2.0720E 01	4.8262E-02	1.2066E 00	3.2540E 02	5.8647E-01	4.6045E-01
50	7.1067E 00	1.4071E-01	2.1482E 01	4.6550E-02	1.1638E 00	3.6116E 02	5.5917E-01	4.2964E-01

Reproduced with permission from F. C. Jelen (ed.), "Cost and Optimization Engineering," McGraw-Hill Book Company, New York, 1970.

TABLE B-2
10% compound interest rate

Discrete compound interest = 10%

	Single payment		Uniform annual series			Uniform gradient series	Depreciation series	
	Compound-interest factor $(1+i)^n$	Present-worth factor $\dfrac{1}{(1+i)^n}$	Unacost present-worth factor $\dfrac{(1+i)^n-1}{i(1+i)^n}$	Capital-recovery factor $\dfrac{i(1+i)^n}{(1+i)^n-1}$	Capitalized-cost factor $\dfrac{(1+i)^n}{(1+i)^n-1}$	Present-worth factor $\dfrac{F_{RP}-nF_{SP}}{i}$	Sum-of-digits present-worth factor $\dfrac{n-F_{RP}}{0.5n(n+1)i}$	Straight-line present-worth factor $\dfrac{1}{niF_{PK}}$
	P to S	S to P	R to P	P to R	P to K	G to P	SD to P	SL to P
n	F_{PS}	F_{SP}	F_{RP}	F_{PR}	F_{PK}	F_{GP}	F_{SDP}	F_{SLP}
1	1.1000E 00	9.0909E-01	9.0909E-01	1.1000E 00	1.1000E 01	1.0000E 00	9.0909E-01	9.0909E-01
2	1.2100E 00	8.2645E-01	1.7355E 00	5.7619E-01	5.7619E 00	8.2645E-01	8.8154E-01	8.6777E-01
3	1.3310E 00	7.5131E-01	2.4869E 00	4.0211E-01	4.0211E 00	2.3291E 00	8.5525E-01	8.2895E-01
4	1.4641E 00	6.8301E-01	3.1699E 00	3.1547E-01	3.1547E 00	4.3781E 00	8.3013E-01	7.9247E-01
5	1.6105E 00	6.2092E-01	3.7908E 00	2.6380E-01	2.6380E 00	6.8618E 00	8.0614E-01	7.5816E-01
6	1.7716E 00	5.6447E-01	4.3553E 00	2.2961E-01	2.2961E 00	9.6842E 00	7.8321E-01	7.2588E-01
7	1.9487E 00	5.1316E-01	4.8684E 00	2.0541E-01	2.0541E 00	1.2763E 01	7.6128E-01	6.9549E-01
8	2.1436E 00	4.6651E-01	5.3349E 00	1.8744E-01	1.8744E 00	1.6029E 01	7.4030E-01	6.6687E-01
9	2.3579E 00	4.2410E-01	5.7590E 00	1.7364E-01	1.7364E 00	1.9421E 01	7.2022E-01	6.3989E-01
10	2.5937E 00	3.8554E-01	6.1446E 00	1.6275E-01	1.6275E 00	2.2891E 01	7.0099E-01	6.1446E-01
11	2.8531E 00	3.5049E-01	6.4951E 00	1.5396E-01	1.5396E 00	2.6396E 01	6.8257E-01	5.9046E-01
12	3.1384E 00	3.1863E-01	6.8137E 00	1.4676E-01	1.4676E 00	2.9901E 01	6.6491E-01	5.6781E-01
13	3.4523E 00	2.8966E-01	7.1034E 00	1.4078E-01	1.4078E 00	3.3377E 01	6.4798E-01	5.4641E-01
14	3.7975E 00	2.6333E-01	7.3667E 00	1.3575E-01	1.3575E 00	3.6800E 01	6.3174E-01	5.2619E-01
15	4.1772E 00	2.3939E-01	7.6061E 00	1.3147E-01	1.3147E 00	4.0152E 01	6.1616E-01	5.0707E-01
16	4.5950E 00	2.1763E-01	7.8237E 00	1.2782E-01	1.2782E 00	4.3416E 01	6.0120E-01	4.8898E-01
18	5.5599E 00	1.7986E-01	8.2014E 00	1.2193E-01	1.2193E 00	4.9640E 01	5.7302E-01	4.5563E-01
20	6.7275E 00	1.4864E-01	8.5136E 00	1.1746E-01	1.1746E 00	5.5407E 01	5.4697E-01	4.2568E-01
25	1.0835E 01	9.2296E-02	9.0770E 00	1.1017E-01	1.1017E 00	6.7696E 01	4.8994E-01	3.6308E-01
30	1.7449E 01	5.7309E-02	9.4269E 00	1.0608E-01	1.0608E 00	7.7077E 01	4.4244E-01	3.1423E-01
35	2.8102E 01	3.5584E-02	9.6442E 00	1.0369E-01	1.0369E 00	8.3987E 01	4.0247E-01	2.7555E-01
40	4.5259E 01	2.2095E-02	9.7791E 00	1.0226E-01	1.0226E 00	8.8953E 01	3.6855E-01	2.4448E-01
45	7.2890E 01	1.3719E-02	9.8628E 00	1.0139E-01	1.0139E 00	9.2454E 01	3.3949E-01	2.1917E-01
50	1.1739E 02	8.5186E-03	9.9148E 00	1.0086E-01	1.0086E 00	9.4889E 01	3.1439E-01	1.9830E-01

TABLE B-3
16% compound interest rate

Discrete compound interest = 16%

n	Single payment — Compound-interest factor $(1+i)^n$ — P to S — F_{PS}	Single payment — Present-worth factor $\frac{1}{(1+i)^n}$ — S to P — F_{SP}	Uniform annual series — Unacost present-worth factor $\frac{(1+i)^n-1}{i(1+i)^n}$ — R to P — F_{RP}	Uniform annual series — Capital-recovery factor $\frac{i(1+i)^n}{(1+i)^n-1}$ — P to R — F_{PR}	Capitalized-cost factor $\frac{(1+i)^n}{(1+i)^n-1}$ — P to K — F_{PK}	Uniform gradient series — Present-worth factor $\frac{F_{RP}-nF_{SP}}{i}$ — G to P — F_{GP}	Depreciation series — Sum-of-digits present-worth factor $\frac{n-F_{RP}}{0.5n(n+1)i}$ — SD to P — F_{SDP}	Depreciation series — Straight-line present-worth factor $\frac{1}{niF_{PR}}$ — SL to P — F_{SLP}
1	1.1600E 00	8.6207E-01	8.6207E-01	1.1600E 00	7.2500E 00	1.0000E 00	8.6207E-01	8.6207E-01
2	1.3456E 00	7.4316E-01	1.6052E 00	6.2296E-01	3.8935E 00	7.4316E-01	8.2243E-01	8.0262E-01
3	1.5609E 00	6.4066E-01	2.2459E 00	4.4526E-01	2.7829E 00	2.0245E 00	7.8553E-01	7.4863E-01
4	1.8106E 00	5.5229E-01	2.7982E 00	3.5738E-01	2.2336E 00	3.6814E 00	7.5114E-01	6.9955E-01
5	2.1003E 00	4.7611E-01	3.2743E 00	3.0541E-01	1.9088E 00	5.5858E 00	7.1904E-01	6.5486E-01
6	2.4364E 00	4.1044E-01	3.6847E 00	2.7139E-01	1.6962E 00	7.6380E 00	6.8907E-01	6.1412E-01
7	2.8262E 00	3.5383E-01	4.0386E 00	2.4761E-01	1.5476E 00	9.7610E 00	6.6103E-01	5.7694E-01
8	3.2784E 00	3.0503E-01	4.3436E 00	2.3022E-01	1.4389E 00	1.1896E 01	6.3479E-01	5.4295E-01
9	3.8030E 00	2.6295E-01	4.6065E 00	2.1708E-01	1.3568E 00	1.4000E 01	6.1020E-01	5.1184E-01
10	4.4114E 00	2.2668E-01	4.8332E 00	2.0690E-01	1.2931E 00	1.6040E 01	5.8713E-01	4.8332E-01
11	5.1173E 00	1.9542E-01	5.0286E 00	1.9886E-01	1.2429E 00	1.7994E 01	5.6547E-01	4.5715E-01
12	5.9360E 00	1.6846E-01	5.1971E 00	1.9241E-01	1.2026E 00	1.9847E 01	5.4510E-01	4.3309E-01
13	6.8858E 00	1.4523E-01	5.3423E 00	1.8718E-01	1.1699E 00	2.1590E 01	5.2594E-01	4.1095E-01
14	7.9875E 00	1.2520E-01	5.4675E 00	1.8290E-01	1.1431E 00	2.3217E 01	5.0789E-01	3.9054E-01
15	9.2655E 00	1.0793E-01	5.5755E 00	1.7936E-01	1.1210E 00	2.4728E 01	4.9086E-01	3.7170E-01
16	1.0748E 01	9.3041E-02	5.6685E 00	1.7641E-01	1.1026E 00	2.6124E 01	4.7479E-01	3.5428E-01
18	1.4463E 01	6.9144E-02	5.8178E 00	1.7188E-01	1.0743E 00	2.8593E 01	4.4525E-01	3.2321E-01
20	1.9461E 01	5.1385E-02	5.9288E 00	1.6867E-01	1.0542E 00	3.0632E 01	4.1878E-01	2.9644E-01
25	4.0874E 01	2.4465E-02	6.0971E 00	1.6401E-01	1.0251E 00	3.4264E 01	3.6352E-01	2.4388E-01
30	8.5850E 01	1.1648E-02	6.1772E 00	1.6189E-01	1.0118E 00	3.6423E 01	3.2020E-01	2.0591E-01
35	1.8031E 02	5.5459E-03	6.2153E 00	1.6089E-01	1.0056E 00	3.7633E 01	2.8556E-01	1.7758E-01
40	3.7872E 02	2.6405E-03	6.2335E 00	1.6042E-01	1.0026E 00	3.8299E 01	2.5737E-01	1.5584E-01
45	7.9544E 02	1.2572E-03	6.2421E 00	1.6020E-01	1.0013E 00	3.8660E 01	2.3405E-01	1.3871E-01
50	1.6707E 03	5.9855E-04	6.2463E 00	1.6010E-01	1.0006E 00	3.8852E 01	2.1448E-01	1.2493E-01

APPENDIX
C

STATISTICAL
TABLES

TABLE C-1
Cumulative distribution function for the standard normal distribution (SND)

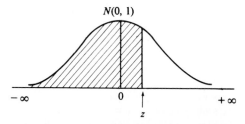

For the SND, the table gives the cdf value at z, that is,

$$\int_{-\infty}^{z} N(0, 1) \, dz$$

z	0.00	0.01	0.02	0.03	0.04	0.05	0.06	0.07	0.08	0.09
−3.5	0.00023	0.00022	0.00022	0.00021	0.00020	0.00019	0.00019	0.00018	0.00017	0.00017
−3.4	0.00034	0.00033	0.00031	0.00030	0.00029	0.00028	0.00027	0.00026	0.00025	0.00024
−3.3	0.00048	0.00047	0.00045	0.00043	0.00042	0.00040	0.00039	0.00038	0.00036	0.00035
−3.2	0.00069	0.00066	0.00064	0.00062	0.00060	0.00058	0.00056	0.00054	0.00052	0.00050
−3.1	0.00097	0.00094	0.00090	0.00087	0.00085	0.00082	0.00079	0.00076	0.00074	0.00071
−3.0	0.00135	0.00131	0.00126	0.00122	0.00118	0.00114	0.00111	0.00107	0.00104	0.00100
−2.9	0.0019	0.0018	0.0017	0.0017	0.0016	0.0016	0.0015	0.0015	0.0014	0.0014
−2.8	0.0026	0.0025	0.0024	0.0023	0.0023	0.0022	0.0021	0.0021	0.0020	0.0019
−2.7	0.0035	0.0034	0.0033	0.0032	0.0031	0.0030	0.0029	0.0028	0.0027	0.0026
−2.6	0.0047	0.0045	0.0044	0.0043	0.0041	0.0040	0.0039	0.0038	0.0037	0.0036
−2.5	0.0062	0.0060	0.0059	0.0057	0.0055	0.0054	0.0052	0.0051	0.0049	0.0048
−2.4	0.0082	0.0080	0.0078	0.0075	0.0073	0.0071	0.0069	0.0068	0.0066	0.0064
−2.3	0.0107	0.0104	0.0102	0.0099	0.0096	0.0094	0.0091	0.0089	0.0087	0.0084
−2.2	0.0139	0.0136	0.0132	0.0129	0.0125	0.0122	0.0119	0.0116	0.0113	0.0110
−2.1	0.0179	0.0174	0.0170	0.0166	0.0162	0.0158	0.0154	0.0150	0.0146	0.0143
−2.0	0.0228	0.0222	0.0217	0.0212	0.0207	0.0202	0.0197	0.0192	0.0188	0.0183
−1.9	0.0287	0.0281	0.0274	0.0268	0.0262	0.0256	0.0250	0.0244	0.0239	0.0233
−1.8	0.0359	0.0351	0.0344	0.0336	0.0329	0.0322	0.0314	0.0307	0.0301	0.0294
−1.7	0.0446	0.0436	0.0427	0.0418	0.0409	0.0401	0.0392	0.0384	0.0375	0.0367
−1.6	0.0548	0.0537	0.0526	0.0516	0.0505	0.0495	0.0485	0.0475	0.0465	0.0455
−1.5	0.0668	0.0655	0.0643	0.0630	0.0618	0.0606	0.0594	0.0582	0.0571	0.0559
−1.4	0.0808	0.0793	0.0778	0.0764	0.0749	0.0735	0.0721	0.0708	0.0694	0.0581
−1.3	0.0968	0.0951	0.0934	0.0918	0.0901	0.0885	0.0869	0.0853	0.0838	0.0823
−1.2	0.1151	0.1131	0.1112	0.1093	0.1075	0.1057	0.1038	0.1020	0.1003	0.0985
−1.1	0.1357	0.1335	0.1314	0.1292	0.1271	0.1251	0.1230	0.1210	0.1190	0.1170
−1.0	0.1587	0.1562	0.1539	0.1515	0.1492	0.1469	0.1446	0.1423	0.1401	0.1379
−0.9	0.1841	0.1814	0.1788	0.1762	0.1736	0.1711	0.1685	0.1660	0.1635	0.1611
−0.8	0.2119	0.2090	0.2061	0.2033	0.2005	0.1977	0.1949	0.1922	0.1894	0.1867
−0.7	0.2420	0.2389	0.2358	0.2327	0.2297	0.2266	0.2236	0.2207	0.2177	0.2148
−0.6	0.2743	0.2709	0.2676	0.2643	0.2611	0.2578	0.2546	0.2514	0.2483	0.2451
−0.5	0.3085	0.3050	0.3015	0.2981	0.2946	0.2912	0.2877	0.2843	0.2810	0.2776
−0.4	0.3446	0.3409	0.3372	0.3336	0.3300	0.3264	0.3228	0.3192	0.3156	0.3121
−0.3	0.3821	0.3783	0.3745	0.3707	0.3669	0.3632	0.3594	0.3557	0.3520	0.3483
−0.2	0.4207	0.4168	0.4129	0.4090	0.4052	0.4013	0.3974	0.3936	0.3897	0.3859
−0.1	0.4602	0.4562	0.4522	0.4483	0.4443	0.4404	0.4364	0.4325	0.4286	0.4247
−0.0	0.5000	0.4960	0.4920	0.4880	0.4840	0.4801	0.4761	0.4721	0.4681	0.4641

Reprinted, with permission, from E. L. Grant and R. S. Leavenworth, "Statistical Quality Control," McGraw-Hill Book Company, New York, 1972.

Table C-1

(*Continued*)

z	0.00	0.01	0.02	0.03	0.04	0.05	0.06	0.07	0.08	0.09
+0.0	0.5000	0.5040	0.5080	0.5120	0.5160	0.5199	0.5239	0.5279	0.5319	0.5359
+0.1	0.5398	0.5438	0.5478	0.5517	0.5557	0.5596	0.5636	0.5675	0.5714	0.5753
+0.2	0.5793	0.5832	0.5871	0.5910	0.5948	0.5987	0.6026	0.6064	0.6103	0.6141
+0.3	0.6179	0.6217	0.6255	0.6293	0.6331	0.6368	0.6406	0.6443	0.6480	0.6517
+0.4	0.6554	0.6591	0.6628	0.6664	0.6700	0.6736	0.6772	0.6808	0.6844	0.6870
+0.5	0.6915	0.6950	0.6985	0.7019	0.7054	0.7088	0.7123	0.7157	0.7190	0.7224
+0.6	0.7257	0.7291	0.7324	0.7357	0.7389	0.7422	0.7454	0.7486	0.7517	0.7549
+0.7	0.7580	0.7611	0.7642	0.7673	0.7704	0.7734	0.7764	0.7794	0.7823	0.7852
+0.8	0.7881	0.7910	0.7939	0.7967	0.7995	0.8023	0.8051	0.8079	0.8106	0.8133
+0.9	0.8159	0.8186	0.8212	0.8238	0.8264	0.8289	0.8315	0.8340	0.8365	0.8389
+1.0	0.8413	0.8438	0.8461	0.8485	0.8508	0.8531	0.8554	0.8577	0.8599	0.8621
+1.1	0.8643	0.8665	0.8686	0.8708	0.8729	0.8749	0.8770	0.8790	0.8810	0.8830
+1.2	0.8849	0.8869	0.8888	0.8907	0.8925	0.8944	0.8962	0.8980	0.8997	0.9015
+1.3	0.9032	0.9049	0.9066	0.9082	0.9099	0.9115	0.9131	0.9147	0.9162	0.9177
+1.4	0.9192	0.9207	0.9222	0.9236	0.9251	0.9265	0.9279	0.9292	0.9306	0.9319
+1.5	0.9332	0.9345	0.9357	0.9370	0.9382	0.9394	0.9406	0.9418	0.9429	0.9441
+1.6	0.9452	0.9463	0.9474	0.9484	0.9495	0.9505	0.9515	0.9525	0.9535	0.9545
+1.7	0.9554	0.9564	0.9573	0.9582	0.9591	0.9599	0.9608	0.9616	0.9625	0.9633
+1.8	0.9641	0.9649	0.9656	0.9664	0.9671	0.9678	0.9686	0.9693	0.9699	0.9706
+1.9	0.9713	0.9719	0.9726	0.9732	0.9738	0.9744	0.9750	0.9756	0.9761	0.9767
+2.0	0.9773	0.9778	0.9783	0.9788	0.9793	0.9798	0.9803	0.9808	0.9812	0.9817
+2.1	0.9821	0.9826	0.9830	0.9834	0.9838	0.9842	0.9846	0.9850	0.9854	0.9857
+2.2	0.9861	0.9864	0.9868	0.9871	0.9875	0.9878	0.9881	0.9884	0.9887	0.9890
+2.3	0.9893	0.9896	0.9898	0.9901	0.9904	0.9906	0.9909	0.9911	0.9913	0.9916
+2.4	0.9918	0.9920	0.9922	0.9925	0.9927	0.9929	0.9931	0.9932	0.9934	0.9936
+2.5	0.9938	0.9940	0.9941	0.9943	0.9945	0.9946	0.9948	0.9949	0.9951	0.9952
+2.6	0.9953	0.9955	0.9956	0.9957	0.9959	0.9960	0.9961	0.9962	0.9963	0.9964
+2.7	0.9965	0.9966	0.9967	0.9968	0.9969	0.9970	0.9971	0.9972	0.9973	0.9974
+2.8	0.9974	0.9975	0.9976	0.9977	0.9977	0.9978	0.9979	0.9979	0.9980	0.9981
+2.9	0.9981	0.9982	0.9983	0.9983	0.9984	0.9984	0.9985	0.9985	0.9986	0.9986
+3.0	0.99865	0.99869	0.99874	0.99878	0.99882	0.99886	0.99889	0.99893	0.99896	0.99900
+3.1	0.99903	0.99906	0.99910	0.99913	0.99915	0.99918	0.99921	0.99924	0.99926	0.99929
+3.2	0.99931	0.99934	0.99936	0.99938	0.99940	0.99942	0.99944	0.99946	0.99948	0.99950
+3.3	0.99952	0.99953	0.99955	0.99957	0.99958	0.99960	0.99961	0.99962	0.99964	0.99965
+3.4	0.99966	0.99967	0.99969	0.99970	0.99971	0.99972	0.99973	0.99974	0.99975	0.99976
+3.5	0.99977	0.99978	0.99978	0.99979	0.99980	0.99981	0.99981	0.99982	0.99983	0.99983

TABLE C-2
The t distribution

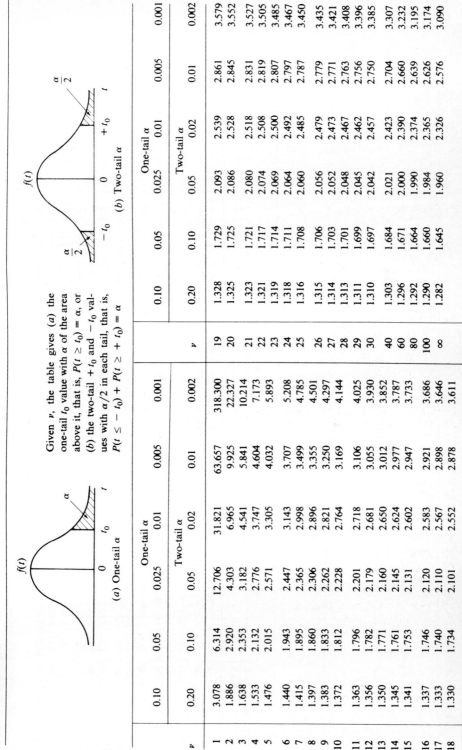

Given ν, the table gives (a) the one-tail t_0 value with α of the area above it, that is, $P(t \geq t_0) = \alpha$, or (b) the two-tail $+t_0$ and $-t_0$ values with $\alpha/2$ in each tail, that is, $P(t \leq -t_0) + P(t \geq +t_0) = \alpha$

	One-tail α					
	0.10	0.05	0.025	0.01	0.005	0.001
	Two-tail α					
ν	0.20	0.10	0.05	0.02	0.01	0.002
1	3.078	6.314	12.706	31.821	63.657	318.300
2	1.886	2.920	4.303	6.965	9.925	22.327
3	1.638	2.353	3.182	4.541	5.841	10.214
4	1.533	2.132	2.776	3.747	4.604	7.173
5	1.476	2.015	2.571	3.305	4.032	5.893
6	1.440	1.943	2.447	3.143	3.707	5.208
7	1.415	1.895	2.365	2.998	3.499	4.785
8	1.397	1.860	2.306	2.896	3.355	4.501
9	1.383	1.833	2.262	2.821	3.250	4.297
10	1.372	1.812	2.228	2.764	3.169	4.144
11	1.363	1.796	2.201	2.718	3.106	4.025
12	1.356	1.782	2.179	2.681	3.055	3.930
13	1.350	1.771	2.160	2.650	3.012	3.852
14	1.345	1.761	2.145	2.624	2.977	3.787
15	1.341	1.753	2.131	2.602	2.947	3.733
16	1.337	1.746	2.120	2.583	2.921	3.686
17	1.333	1.740	2.110	2.567	2.898	3.646
18	1.330	1.734	2.101	2.552	2.878	3.611
19	1.328	1.729	2.093	2.539	2.861	3.579
20	1.325	1.725	2.086	2.528	2.845	3.552
21	1.323	1.721	2.080	2.518	2.831	3.527
22	1.321	1.717	2.074	2.508	2.819	3.505
23	1.319	1.714	2.069	2.500	2.807	3.485
24	1.318	1.711	2.064	2.492	2.797	3.467
25	1.316	1.708	2.060	2.485	2.787	3.450
26	1.315	1.706	2.056	2.479	2.779	3.435
27	1.314	1.703	2.052	2.473	2.771	3.421
28	1.313	1.701	2.048	2.467	2.763	3.408
29	1.311	1.699	2.045	2.462	2.756	3.396
30	1.310	1.697	2.042	2.457	2.750	3.385
40	1.303	1.684	2.021	2.423	2.704	3.307
60	1.296	1.671	2.000	2.390	2.660	3.232
80	1.292	1.664	1.990	2.374	2.639	3.195
100	1.290	1.660	1.984	2.365	2.626	3.174
∞	1.282	1.645	1.960	2.326	2.576	3.090

Reprinted, with permission, from L. Blank, "Statistical Procedures for Engineering, Management and Science," McGraw-Hill Book Company, New York, 1980.

TABLE C-3
The F distribution (α = 0.10, 0.05, and 0.01)

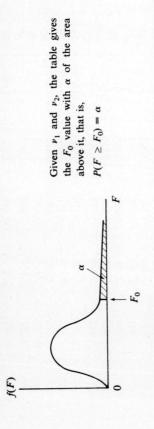

Given v_1 and v_2, the table gives the F_0 value with α of the area above it, that is,

$$P(F \geq F_0) = \alpha$$

v_1 (numerator)

v_2	α	1	2	3	4	5	6	7	8	9	10	11	12	14	15	19	20	24	30	50	100	500	∞
1	.10	39.9	49.5	53.6	55.8	57.2	58.2	58.9	59.4	59.9	60.2	60.5	60.7	61.1	61.2	61.6	61.7	62.0	62.3	62.7	63.0	63.3	63.3
	.05	161	200	216	225	230	234	237	239	241	242	243	244	245	246	248	248	249	250	252	253	254	254
2	.10	8.53	9.00	9.16	9.24	9.29	9.33	9.35	9.37	9.38	9.39	9.40	9.41	9.42	9.42	9.44	9.44	9.45	9.46	9.47	9.48	9.49	9.49
	.05	18.5	19.0	19.2	19.2	19.3	19.3	19.4	19.4	19.4	19.4	19.4	19.4	19.4	19.4	19.4	19.4	19.5	19.5	19.5	19.5	19.5	19.5
	.01	98.5	99.0	99.2	99.2	99.3	99.3	99.4	99.4	99.4	99.4	99.4	99.4	99.4	99.4	99.4	99.4	99.5	99.5	99.5	99.5	99.5	99.5
3	.10	5.54	5.46	5.39	5.34	5.31	5.28	5.27	5.25	5.24	5.23	5.22	5.22	5.20	5.20	5.18	5.18	5.18	5.17	5.15	5.14	5.14	5.13
	.05	10.1	9.55	9.28	9.12	9.10	8.94	8.89	8.85	8.81	8.79	8.76	8.74	8.71	8.70	8.67	8.66	8.64	8.62	8.58	8.55	8.53	8.53
	.01	34.1	30.8	29.5	28.7	28.2	27.9	27.7	27.5	27.3	27.2	27.1	27.1	26.9	26.9	26.7	26.7	26.6	26.5	26.4	26.2	26.1	26.1
4	.10	4.54	4.32	4.19	4.11	4.05	4.01	3.98	3.95	3.94	3.92	3.91	3.90	3.88	3.87	3.84	3.84	3.83	3.82	3.80	3.78	3.76	3.76
	.05	7.71	6.94	6.59	6.39	6.26	6.16	6.09	6.04	6.00	5.96	5.94	5.91	5.87	5.86	5.81	5.80	5.77	5.75	5.70	5.66	5.64	5.63
	.01	21.2	18.0	16.7	16.0	15.5	15.2	15.0	14.8	14.7	14.5	14.4	14.4	14.2	14.2	14.0	14.0	13.9	13.8	13.7	13.6	13.5	13.5
5	.10	4.06	3.78	3.62	3.52	3.45	3.40	3.37	3.34	3.32	3.30	3.28	3.27	3.25	3.24	3.21	3.21	3.19	3.17	3.15	3.13	3.11	3.10
	.05	6.61	5.79	5.41	5.19	5.05	4.95	4.88	4.82	4.77	4.74	4.74	4.68	4.64	4.62	4.57	4.56	4.53	4.50	4.44	4.41	4.37	4.36
	.01	16.26	13.27	12.06	11.39	10.97	10.67	10.46	10.29	10.16	10.05	9.96	9.89	9.77	9.72	9.58	9.55	9.47	9.38	9.24	9.13	9.04	9.02

The degrees of freedom are v_1 for the numerator and v_2 for the denominator.

ν_1 (numerator)

ν_2	α	1	2	3	4	5	6	7	8	9	10	11	12	14	15	19	20	24	30	50	100	500	∞
6	.10	3.78	3.46	3.29	3.18	3.11	3.05	3.01	2.98	2.96	2.94	2.92	2.90	2.88	2.87	2.84	2.84	2.82	2.80	2.77	2.75	2.73	2.72
	.05	5.99	5.14	4.76	4.53	4.39	4.28	4.21	4.15	4.10	4.06	4.03	4.00	3.96	3.94	3.88	3.87	3.84	3.81	3.75	3.71	3.68	3.67
	.01	13.74	10.92	9.78	9.15	8.75	8.47	8.26	8.10	7.98	7.87	7.79	7.72	7.60	7.56	7.42	7.40	7.31	7.23	7.09	6.99	6.90	6.88
7	.10	3.59	3.26	3.07	2.96	2.88	2.83	2.78	2.75	2.72	2.70	2.68	2.67	2.64	2.63	2.60	2.59	2.58	2.56	2.52	2.50	2.48	2.47
	.05	5.59	4.74	4.35	4.12	3.97	3.87	3.79	3.73	3.68	3.64	3.60	3.57	3.53	3.51	3.46	3.44	3.41	3.38	3.32	3.27	3.24	3.23
	.01	12.25	9.55	8.45	7.85	7.46	7.19	6.99	6.84	6.72	6.62	6.54	6.47	6.36	6.31	6.18	6.16	6.07	5.99	5.86	5.75	5.67	5.65
8	.10	3.46	3.11	2.92	2.81	2.73	2.67	2.62	2.59	2.56	2.54	2.52	2.50	2.47	2.46	2.43	2.42	2.40	2.38	2.35	2.32	2.30	2.29
	.05	5.32	4.46	4.07	3.84	3.69	3.58	3.50	3.44	3.39	3.35	3.31	3.28	3.24	3.22	3.16	3.15	3.12	3.08	3.02	2.97	2.94	2.93
	.01	11.26	8.65	7.59	7.01	6.63	6.37	6.18	6.03	5.91	5.81	5.73	5.67	5.56	5.52	5.38	5.36	5.28	5.20	5.07	4.96	4.88	4.86
9	.10	3.36	3.01	2.81	2.69	2.61	2.55	2.51	2.47	2.44	2.42	2.40	2.38	2.35	2.34	2.31	2.30	2.28	2.25	2.22	2.19	2.17	2.16
	.05	5.12	4.26	3.86	3.63	3.48	3.37	3.29	3.23	3.18	3.14	3.10	3.07	3.03	3.01	2.95	2.94	2.90	2.86	2.80	2.76	2.72	2.71
	.01	10.56	8.02	6.99	6.42	6.06	5.80	5.61	5.47	5.35	5.26	5.18	5.11	5.00	4.96	4.83	4.81	4.73	4.65	4.52	4.42	4.33	4.31
10	.10	3.28	2.92	2.73	2.61	2.52	2.46	2.41	2.38	2.35	2.32	2.30	2.28	2.25	2.24	2.21	2.20	2.18	2.16	2.12	2.09	2.06	2.06
	.05	4.96	4.10	3.71	3.48	3.33	3.22	3.14	3.07	3.02	2.98	2.94	2.91	2.86	2.85	2.78	2.77	2.74	2.70	2.64	2.59	2.55	2.54
	.01	10.04	7.56	6.55	5.99	5.64	5.39	5.20	5.06	4.94	4.85	4.77	4.71	4.60	4.56	4.43	4.41	4.33	4.25	4.12	4.01	3.93	3.91
11	.10	3.23	2.86	2.66	2.54	2.45	2.39	2.34	2.30	2.27	2.25	2.23	2.21	2.18	2.17	2.13	2.12	2.10	2.08	2.04	2.00	1.98	1.97
	.05	4.84	3.98	3.59	3.36	3.20	3.09	3.01	2.95	2.90	2.85	2.82	2.79	2.74	2.72	2.66	2.65	2.61	2.57	2.51	2.46	2.42	2.40
	.01	9.65	7.21	6.22	5.67	5.32	5.07	4.89	4.74	4.63	4.54	4.46	4.40	4.29	4.25	4.12	4.10	4.02	3.94	3.81	3.71	3.62	3.60
12	.10	3.18	2.81	2.61	2.48	2.39	2.33	2.28	2.24	2.21	2.19	2.17	2.15	2.11	2.10	2.07	2.06	2.04	2.01	1.97	1.94	1.91	1.90
	.05	4.75	3.89	3.49	3.26	3.11	3.00	2.91	2.85	2.80	2.75	2.72	2.69	2.64	2.62	2.56	2.54	2.51	2.47	2.40	2.35	2.31	2.30
	.01	9.33	6.93	5.95	5.41	5.06	4.82	4.64	4.50	4.39	4.30	4.22	4.16	4.05	4.01	3.88	3.86	3.78	3.70	3.57	3.47	3.38	3.36
14	.10	3.10	2.73	2.52	2.39	2.31	2.24	2.19	2.15	2.12	2.10	2.08	2.05	2.02	2.01	1.97	1.96	1.94	1.91	1.87	1.83	1.80	1.80
	.05	4.60	3.74	3.34	3.11	2.96	2.85	2.76	2.70	2.65	2.60	2.57	2.53	2.48	2.46	2.40	2.39	2.35	2.31	2.24	2.19	2.14	2.13
	.01	8.86	6.51	5.56	5.04	4.69	4.46	4.28	4.14	4.03	3.94	3.86	3.80	3.70	3.66	3.53	3.51	3.43	3.35	3.22	3.11	3.03	3.00
15	.10	3.07	2.70	2.49	2.36	2.27	2.21	2.16	2.12	2.09	2.06	2.04	2.02	1.98	1.97	1.93	1.92	1.90	1.87	1.83	1.79	1.76	1.76
	0.5	4.54	3.68	3.29	3.06	2.90	2.79	2.71	2.64	2.59	2.54	2.51	2.48	2.42	2.40	2.34	2.33	2.29	2.25	2.18	2.12	2.08	2.07
	.01	8.68	6.36	5.42	4.89	4.56	4.32	4.14	4.00	3.89	3.80	3.73	3.67	3.56	3.52	3.40	3.37	3.29	3.21	3.08	2.98	2.89	2.87
16	.10	3.05	2.67	2.46	2.33	2.24	2.18	2.13	2.09	2.06	2.03	2.01	1.99	1.95	1.94	1.90	1.89	1.87	1.84	1.79	1.76	1.73	1.72
	.05	4.49	3.63	3.24	3.01	2.85	2.74	2.66	2.59	2.54	2.49	2.46	2.42	2.37	2.35	2.29	2.28	2.24	2.19	2.12	2.07	2.02	2.01
	.01	8.53	6.23	5.29	4.77	4.44	4.20	4.03	3.89	3.78	3.69	3.62	3.55	3.45	3.41	3.28	3.26	3.18	3.10	2.97	2.86	2.78	2.75

(continued)

TABLE C-3
(Continued)

ν_2	α	ν_1 (numerator)																					
		1	2	3	4	5	6	7	8	9	10	11	12	14	15	19	20	24	30	50	100	500	∞
18	.10	3.01	2.62	2.42	2.29	2.20	2.13	2.08	2.04	2.00	1.98	1.96	1.93	1.90	1.89	1.85	1.84	1.81	1.78	1.74	1.70	1.67	1.66
	.05	4.41	3.55	3.16	2.93	2.77	2.66	2.58	2.51	2.46	2.41	2.37	2.34	2.29	2.27	2.20	2.19	2.15	2.11	2.04	1.98	1.93	1.92
	.01	8.29	6.01	5.09	4.58	4.25	4.01	3.84	3.71	3.60	3.51	3.43	3.37	3.27	3.23	3.10	3.08	3.00	2.92	2.78	2.68	2.59	2.57
19	.10	2.99	2.61	2.40	2.27	2.18	2.11	2.06	2.02	1.98	1.96	1.94	1.91	1.87	1.86	1.82	1.81	1.79	1.76	1.71	1.67	1.64	1.63
	.05	4.38	3.52	3.13	2.90	2.74	2.63	2.54	2.48	2.42	2.38	2.34	2.31	2.26	2.23	2.17	2.16	2.11	2.07	2.00	1.94	1.89	1.88
	.01	8.18	5.93	5.01	4.50	4.17	3.94	3.77	3.63	3.52	3.43	3.36	3.30	3.19	3.15	3.03	3.00	2.92	2.84	2.71	2.60	2.51	2.49
20	.10	2.97	2.59	2.38	2.25	2.16	2.09	2.04	2.00	1.96	1.94	1.92	1.89	1.85	1.84	1.80	1.79	1.77	1.74	1.69	1.65	1.62	1.61
	.05	4.35	3.49	3.10	2.87	2.71	2.60	2.51	2.45	2.39	2.35	2.31	2.28	2.22	2.20	2.14	2.12	2.08	2.04	1.97	1.91	1.86	1.84
	.01	8.10	5.85	4.94	4.43	4.10	3.87	3.70	3.56	3.46	3.37	3.29	3.23	3.13	3.09	2.96	2.94	2.86	2.78	2.64	2.54	2.44	2.42
24	.10	2.93	2.54	2.33	2.19	2.10	2.04	1.98	1.94	1.91	1.88	1.85	1.83	1.79	1.78	1.74	1.73	1.70	1.67	1.62	1.58	1.54	1.53
	.05	4.26	3.40	3.01	2.78	2.62	2.51	2.42	2.36	2.30	2.25	2.21	2.18	2.13	2.11	2.04	2.03	1.98	1.94	1.86	1.80	1.75	1.73
	.01	7.82	5.61	4.72	4.22	3.90	3.67	3.50	3.36	3.26	3.17	3.09	3.03	2.93	2.89	2.76	2.74	2.66	2.58	2.44	2.33	2.24	2.21
30	.10	2.88	2.49	2.28	2.14	2.05	1.98	1.93	1.88	1.85	1.82	1.79	1.77	1.73	1.72	1.68	1.67	1.64	1.61	1.55	1.51	1.47	1.46
	.05	4.17	3.32	2.92	2.69	2.53	2.42	2.33	2.27	2.21	2.16	2.13	2.09	2.04	2.01	1.95	1.93	1.89	1.84	1.76	1.70	1.64	1.62
	.01	7.56	5.39	4.51	4.02	3.70	3.47	3.30	3.17	3.07	2.98	2.91	2.84	2.74	2.70	2.57	2.55	2.47	2.39	2.25	2.13	2.03	2.01
50	.10	2.81	2.41	2.20	2.06	1.97	1.90	1.84	1.80	1.76	1.73	1.70	1.68	1.64	1.63	1.58	1.57	1.54	1.50	1.44	1.39	1.34	1.33
	.05	4.03	3.18	2.79	2.56	2.40	2.29	2.20	2.13	2.07	2.03	1.99	1.95	1.89	1.87	1.80	1.78	1.74	1.69	1.60	1.52	1.46	1.44
	.01	7.17	5.06	4.20	3.72	3.41	3.19	3.02	2.89	2.79	2.70	2.63	2.56	2.46	2.42	2.29	2.27	2.18	2.10	1.95	1.82	1.71	1.68
100	.10	2.76	2.36	2.14	2.00	1.91	1.83	1.78	1.73	1.70	1.66	1.63	1.61	1.57	1.56	1.50	1.49	1.46	1.42	1.35	1.29	1.23	1.21
	.05	3.94	3.09	2.70	2.46	2.31	2.19	2.10	2.03	1.97	1.93	1.89	1.85	1.79	1.77	1.69	1.68	1.63	1.57	1.48	1.39	1.31	1.28
	.01	6.90	4.82	3.98	3.51	3.21	2.99	2.82	2.69	2.59	2.50	2.43	2.37	2.26	2.22	2.09	2.07	1.98	1.89	1.73	1.60	1.47	1.43
500	.10	2.72	2.31	2.10	1.96	1.86	1.79	1.73	1.68	1.64	1.61	1.58	1.56	1.52	1.50	1.45	1.44	1.41	1.36	1.28	1.21	1.12	1.09
	.05	3.86	3.01	2.62	2.39	2.23	2.12	2.03	1.96	1.90	1.85	1.81	1.77	1.71	1.69	1.61	1.59	1.54	1.48	1.38	1.28	1.16	1.11
	.01	6.69	4.65	3.82	3.36	3.05	2.84	2.68	2.55	2.44	2.36	2.28	2.22	2.12	2.07	1.94	1.92	1.83	1.74	1.56	1.41	1.23	1.16
∞	.10	2.71	2.30	2.08	1.94	1.85	1.77	1.72	1.67	1.63	1.60	1.57	1.55	1.51	1.49	1.43	1.42	1.38	1.34	1.26	1.18	1.08	1.00
	.05	3.84	3.00	2.60	2.37	2.21	2.10	2.01	1.94	1.88	1.83	1.79	1.75	1.69	1.67	1.59	1.57	1.52	1.46	1.35	1.24	1.11	1.00
	.01	6.63	4.61	3.78	3.32	3.02	2.80	2.64	2.51	2.41	2.32	2.25	2.18	2.08	2.04	1.90	1.88	1.79	1.70	1.52	1.36	1.15	1.00

Reprinted with permission, from L. Blank, "Statistical Procedures for Engineering, Management and Science," McGraw-Hill Book Company, New York, 1980.

USEFUL
OPERATING
CHARACTERISTIC
CURVES

The abscissa scale on the operating characteristic curves is interpreted differently depending on the form of H_1:

Two-sided,	$H_1: \mu \neq \mu_0$	$d = \lvert \mu - \mu_0 \rvert / \sigma$
One-sided,	$H_1: \mu < \mu_0$	$d = \mu_0 - \mu / \sigma$
One-sided,	$H_1: \mu > \mu_0$	$d = \mu - \mu_0 / \sigma$
Two-sided, paired test	$H_1: \mu_1 - \mu_2 \neq \delta$	$d = \dfrac{\mu_1 - \mu_2}{(\sigma_1^2 + \sigma_2^2)^{1/2}}$

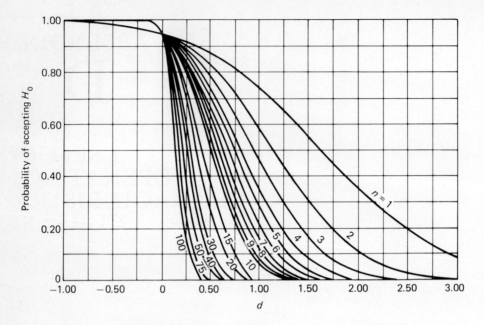

Operating characteristic curves for *one-sided normal* test with $\alpha = 0.05$. (A. H. Bowker and G. J. Lieberman, "Engineering Statistics," 2d ed., 1972, p. 190. Reproduced by permission of Prentice-Hall, Inc., Englewood Cliffs, N.J.)

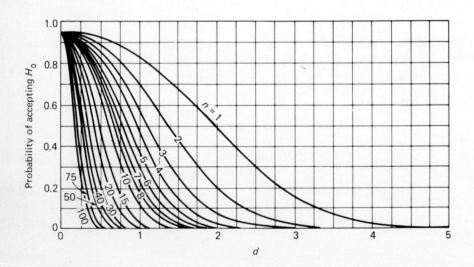

Operating characteristic curves for a *two-sided normal* test with $\alpha = 0.05$. (Reproduced with permission from "OC of the Common Statistical Tests of Significance." C. D. Ferris, F. E. Grubbs, and C. L. Weaver, *The Annals of Mathematical Statistics,* vol. 17, no. 2, June 1946.

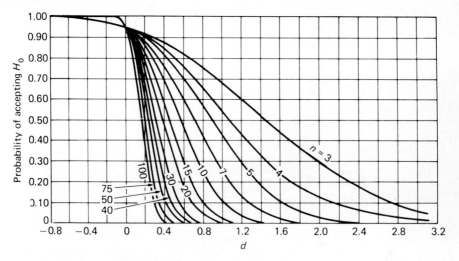

Operating characteristic curves for *one-sided normal* test with $\alpha = 0.05$. (A. H. Bowker and G. J. Lieberman, "Engineering Statistics," 2d ed., 1972, p. 190. Reproduced by permission of Prentice-Hall, Inc., Englewood Cliffs, N.J.)

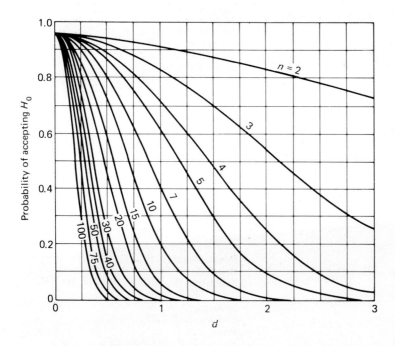

Operating characteristic curves for a *two-sided normal* test with $\alpha = 0.05$. (Reproduced with permission from "OC of the Common Statistical Tests of Significance." C. D. Ferris, F. E. Grubbs, and C. L. Weaver, *The Annals of Mathematical Statistics*, vol. 17, no. 2, June 1946.

AUTHOR INDEX

SUBJECT INDEX